Sustainable Agricultural Practices

Plant and Soil Microbiome

Sustainable Agricultural Practices

Edited by

Ajay Kumar

Amity Institute of Biotechnology, Amity University, Noida, India

James F. White

Department of Plant Biology, Rutgers University, New Brunswick, NJ, United States

Joginder Singh

Department of Botany, Nagaland University, Nagaland, India

Series Editors

Ajay Kumar

James F. White

Joginder Singh

ELSEVIER

ACADEMIC PRESS

An imprint of Elsevier

Academic Press is an imprint of Elsevier
125 London Wall, London EC2Y 5AS, United Kingdom
525 B Street, Suite 1650, San Diego, CA 92101, United States
50 Hampshire Street, 5th Floor, Cambridge, MA 02139, United States

ISBN: 978-0-443-19150-3

For Information on all Academic Press publications
visit our website at https://www.elsevier.com/books-and-journals

Publisher: Nikki Levy
Acquisitions Editor: Nancy Maragioglio
Editorial Project Manager: Anthony Marvullo
Production Project Manager: Neena S. Maheen
Cover Designer: Vicky Pearson Esser

Typeset by MPS Limited, Chennai, India

Contents

List of contributors

Ahmet Akköprü
Department of Plant Protection, Faculty of Agriculture, Van Yuzuncu Yil University, Van, Turkey

Angelika A. Astaykina
Soil Science Faculty, Lomonosov Moscow State University, Moscow, Russia

Neha Basera
Department of Biotechnology, School of Applied and Life Sciences, Uttaranchal University, Dehradun, Uttarakhand, India

Andrey A. Belov
Soil Science Faculty, Lomonosov Moscow State University, Moscow, Russia

Celia Beltran-García
Department of Biotechnology and Environment, Universidad Autónoma de Guadalajara, Zapopan, Jalisco, Mexico

Miguel J. Beltrán-García
Chemistry Department, Universidad Autónoma de Guadalajara, Zapopan, Jalisco, Mexico; Department of Biotechnology and Environment, Universidad Autónoma de Guadalajara, Zapopan, Jalisco, Mexico

Riddhi Bhatt
Department of Life Science, University School of Sciences, Gujarat University, Ahmedabad, Gujarat, India

Joorie Bhattacharya
Genetic Gains, International Crops Research Institute for the Semi-Arid Tropics, Hyderabad, Telangana, India

Francisco J. Blancas-Benitez
Division of Research and Postgraduate Studies, National Technological Institute of Mexico/Technological Institute of Tepic, Tepic, Nayarit, Mexico

Gökhan Boyno
Department of Plant Protection, Faculty of Agriculture, Van Yuzuncu Yil University, Van, Turkey

Paloma P. Casas-Junco
Food Technology Unit, Research and Postgraduate Secretariat, Autonomous University of Nayarit, Tepic, Nayarit, Mexico

Xiaoqian Chang
Department of Plant Biology, Rutgers University, New Brunswick, NJ, United States

Karan Chavda
Department of Biochemistry & Forensic Science, University School of Sciences, Gujarat University, Ahmedabad, Gujarat, India

Vladimir S. Cheptsov
Soil Science Faculty, Lomonosov Moscow State University, Moscow, Russia

Semra Demir
Department of Plant Protection, Faculty of Agriculture, Van Yuzuncu Yil University, Van, Turkey

Emre Demirer Durak
Department of Plant Protection, Faculty of Agriculture, Van Yuzuncu Yil University, Van, Turkey

Paolo Di Mascio
Department of Biochemistry, Institute of Chemistry, University of São Paulo, Paolo Di Mascio, São Paulo, Brazil

Rihab Djebaili
Department of Life, Health and Environmental Sciences, University of L'Aquila, L'Aquila, Italy

Z. Dube
Department of Agriculture, Faculty of Agriculture and Natural Science, University of Mpumalanga, Mbombela, South Africa

Beatrice Farda
Department of Life, Health and Environmental Sciences, University of L'Aquila, L'Aquila, Italy

Ana Y. Flores-Ramírez
Division of Research and Postgraduate Studies, National Technological Institute of Mexico/Technological Institute of Tepic, Tepic, Nayarit, Mexico

Vikas Ghattargi
Department of Serology and Microbiome, Mylab Discovery Solutions Pvt. Ltd., Pune, Maharashtra, India

Sougata Ghosh
Department of Physics, Faculty of Science, Kasetsart University, Bangkok, Thailand; Department of Microbiology, School of Science, RK University, Rajkot, Gujarat, India

Ramsés R. González-Estrada
Division of Research and Postgraduate Studies, National Technological Institute of Mexico/Technological Institute of Tepic, Tepic, Nayarit, Mexico

Dweipayan Goswami
Department of Microbiology & Biotechnology, University School of Sciences, Gujarat University, Ahmedabad, Gujarat, India

Luis G. Hernández-Montiel
Agriculture Program in Arid Zones, Northwest Biological Research Center, La Paz, Baja California Sur, Mexico

Jalaluddin
Laboratory of Algal Research, Centre of Advanced Study in Botany, Institute of Science, Banaras Hindu University, Varanasi, Uttar Pradesh, India

Komal E. Kaware
Department of Microbiology, Modern College of Arts, Science and Commerce, Ganeshkhind, Pune, Maharashtra, India

Kathryn Kingsley
Department of Plant Biology, Rutgers University, New Brunswick, NJ, United States

Shivani P. Kulkarni
Department of Microbiology, Modern College of Arts, Science and Commerce, Ganeshkhind, Pune, Maharashtra, India

Ajay Kumar
Amity Institute of Biotechnology, Amity University, Noida, Uttar Pradesh, India

Sunil Kumar
Department of Botany, Centre of Advanced Study, Institute of Science, Banaras Hindu University, Varanasi, Uttar Pradesh, India; Central Ayurveda Research Institute—Central Council for Research in Ayurvedic Sciences, Bhubaneswar, Odisha, India

Ana Maria Queijeiro López
Institute of Chemistry and Biotechnology (IQB) , Biochemistry Section, Federal University of Alagoas, Maceió-AL, Brazil

Ulises M. López-García
Division of Research and Postgraduate Studies, National Technological Institute of Mexico/Technological Institute of Tepic, Tepic, Nayarit, Mexico

C. Madakadze
Department of Plant and Soil Sciences, Faculty of Natural and Agricultural Sciences, University of Pretoria, Hartfield, South Africa

P. Maponya
Department of Crop Science, Agricultural Research Council – Vegetable, Industrial and Medicinal Plants, Pretoria, South Africa

América Martínez-Rodríguez
Department of Biotechnology and Environment, Universidad Autónoma de Guadalajara, Zapopan, Jalisco, Mexico; Department of Advanced Materials, Engineering Institute, Universidad Autónoma de Baja California, Mexicali, Baja California, Mexico

April Micci
Department of Plant Biology, Rutgers University, New Brunswick, NJ, United States

Jorge Vicente Miranda-Rivera
Department of Biotechnology and Environment, Universidad Autónoma de Guadalajara, Zapopan, Jalisco, Mexico

Sudhanshu Mishra
School of Biosciences, Apeejay Stya University, Sohna, Haryana, India

Beatriz Montaño-Leyva
Department of Research and Postgraduate in Food, University of Sonora, Hermosillo, Sonora, Mexico

Cristina Moreno-Hernández
Division of Research and Postgraduate Studies, National Technological Institute of Mexico/Technological Institute of Tepic, Tepic, Nayarit, Mexico

Solmaz Najafi
Department of Field Crops, Faculty of Agriculture, Van Yuzuncu Yil University, Van, Turkey

Rahul Nitnavare
Division of Plant and Crop Sciences, School of Biosciences, University of Nottingham, Nottingham, Leicestershire, United Kingdom; Department of Plant Sciences, Rothamsted Research, Harpenden, Hertfordshire, United Kingdom

Thabo Nkuna
Department of Plant and Soil Sciences, Faculty of Natural and Agricultural Sciences, University of Pretoria, Hartfield, South Africa; Department of Crop Science, Agricultural Research Council – Vegetable, Industrial and Medicinal Plants, Pretoria, South Africa

Akanksha Pandey
Department of Botany and Microbiology, Gurukula Kangri (Deemed to be University), Haridwar, Uttarakhand, India

Kapil Deo Pandey
Laboratory of Algal Research, Centre of Advanced Study in Botany, Institute of Science, Banaras Hindu University, Varanasi, Uttar Pradesh, India

Marika Pellegrini
Department of Life, Health and Environmental Sciences, University of L'Aquila, L'Aquila, Italy

Jignesh Prajapati
Department of Biochemistry & Forensic Science, University School of Sciences, Gujarat University, Ahmedabad, Gujarat, India

Minakshi Rajput
Department of Biotechnology, School of Applied and Life Sciences, Uttaranchal University, Dehradun, Uttarakhand, India

Vibhuti Rana
Department of Biotechnology, School of Applied and Life Sciences, Uttaranchal University, Dehradun, Uttarakhand, India

Rakesh Rawal
Department of Biochemistry & Forensic Science, University School of Sciences, Gujarat University, Ahmedabad, Gujarat, India; Department of Life Science, University School of Sciences, Gujarat University, Ahmedabad, Gujarat, India

Jesús E. Reyna-Ochoa
Division of Research and Postgraduate Studies, National Technological Institute of Mexico/Technological Institute of Tepic, Tepic, Nayarit, Mexico

Younes Rezaee Danesh
Department of Plant Protection, Faculty of Agriculture, Urmia University, Urmia, Iran; Department of Plant Protection, Faculty of Agriculture, Van Yuzuncu Yil University, Van, Turkey

Tomás Rivas-García
National Council of Humanities, Sciences, and Technologies (CONAHCYT), Chapingo Autonomous University, Texcoco, Estado de Mexico, Mexico

Jonathan M. Sanchez-Silva
Postgraduate Study and Research Center, Faculty of Chemical Sciences, Autonomous University of San Luis Potosí, San Luis Potosí, Mexico

Amanda Lys dos Santos Silva
Institute of Biological and Health Sciences (ICBS), Microbiology Section, Federal University of Alagoas, Maceió-AL, Brazil

Riya Sheth
Department of Life Science, University School of Sciences, Gujarat University, Ahmedabad, Gujarat, India

Vaishali Shukla
Department of Botany, Centre of Advanced Study, Institute of Science, Banaras Hindu University, Varanasi, Uttar Pradesh, India

Monika Singh
Department of Biotechnology, School of Applied and Life Sciences, Uttaranchal University, Dehradun, Uttarakhand, India

Rahul Prasad Singh
Department of Botany, Centre of Advanced Study, Institute of Science, Banaras Hindu University, Varanasi, Uttar Pradesh, India; Department of Botany, Amity Institute of Biotechnology, Amity University, Noida, Uttar Pradesh, India

Savita Singh
Department of Botany, Babu Shivnath Agrawal College, Mathura, Uttar Pradesh, India

Shivani Singh
Department of Botany, Babu Shivnath Agrawal College, Mathura, Uttar Pradesh, India

Zalak Solanki
Department of Life Science, University School of Sciences, Gujarat University, Ahmedabad, Gujarat, India

Rostislav A. Streletskii
Soil Science Faculty, Lomonosov Moscow State University, Moscow, Russia

Syeda Summiya
Institute of Plant Science, University of Sindh, Jamshoro, Pakistan

Sirikanjana Thongmee
Department of Physics, Faculty of Science, Kasetsart University, Bangkok, Thailand

Yashoda Nandan Tripathi
Department of Botany, Centre of Advanced Study, Institute of Science, Banaras Hindu University, Varanasi, Uttar Pradesh, India

Ram Sanmukh Upadhyay
Department of Botany, Centre of Advanced Study, Institute of Science, Banaras Hindu University, Varanasi, Uttar Pradesh, India

Benjamín Valdez-Salas
Department of Advanced Materials, Engineering Institute, Universidad Autónoma de Baja California, Mexicali, Baja California, Mexico

Fernando Velazquez
Department of Plant Biology, Rutgers University, New Brunswick, NJ, United States

Anna A. Vetrova
Federal Research Center "Pushchino Scientific Center for Biological Research of the Russian Academy of Sciences," G.K. Skryabin Institute of Biochemistry and Physiology of Microorganisms, Pushchino, Russia

James F. White
Department of Plant Biology, Rutgers University, New Brunswick, NJ, United States

CHAPTER

1

Multifactorial stress combination, plant microbiome recruitment, and reactive oxygen species/antioxidant feedbacks for plant stress alleviation

Miguel J. Beltrán-García[1,2], América Martínez-Rodríguez[2,3], Celia Beltran-García[2], Jorge Vicente Miranda-Rivera[2], Benjamín Valdez-Salas[3], Paolo Di Mascio[4] and James F. White[5]

[1]*Chemistry Department, Universidad Autónoma de Guadalajara, Zapopan, Jalisco, Mexico*

[2]*Department of Biotechnology and Environment, Universidad Autónoma de Guadalajara, Zapopan, Jalisco, Mexico*

[3]*Department of Advanced Materials, Engineering Institute, Universidad Autónoma de Baja California, Mexicali, Baja California, Mexico*

[4]*Department of Biochemistry, Institute of Chemistry, University of São Paulo, Paolo Di Mascio, São Paulo, Brazil*

[5]*Department of Plant Biology, Rutgers University, New Brunswick, NJ, United States*

1.1 Introduction

Crops are constantly subjected to stressful conditions that reduce plant yields. The plant possesses an array of protective mechanisms acquired during evolution to combat adverse environmental situations. The plant systemic response against one type (biotic or abiotic) or stress combinations revealed a complex mechanism of perception and acclimatization that include the accumulation of different transcripts and metabolites, as well as various physiological, cellular, and molecular responses to maintain homeostasis that enables adaptation and/or survival to changing environmental insults (Zandalinas et al., 2021a, 2022; Zandalinas and Mittler, 2022).

Plants are "holobionts," a biological concept proposed by Lynn Margulis that explains the symbiotic association with a complex network of microorganisms subjected to adaptation processes and adjustments that have allowed them to cushion environmental changes quickly and accurately (Baedke et al., 2020).

Sustainable Agricultural Practices. DOI: https://doi.org/10.1016/B978-0-443-19150-3.00007-2

Microorganisms form symbiotic associations at the roots, stem, or leaves. Plant microbiomes located at the phyllosphere and rhizosphere are heterogeneous and transient, cataloged in the short and long term. However, the endophyte microbiome is highly specific, and its diversity is linked to the plant immunity that allows its establishment in the endosphere and, at the same time, contributes to plant stress resistance (tolerating or avoiding). The intrinsic metabolic and genetic capabilities of plant microbiomes improve plants' ability to tolerate multifactorial stresses (Dixit et al., 2019; Phour and Sindhu, 2022; Salvi et al., 2022).

Plants may be subjected to multiple stresses; their responses include a phytohormone crosstalk signaling led by abscisic acid (ABA), jasmonic acid (JA), and the homeostasis of the reactive oxygen species (ROS) by which plants mitigate and recover their growth. In this chapter, we will discuss how plants respond to stress and highlight how the plant microbiome, as a second line of defense, protects plants from stress through phytohormone regulation and ROS control. However, we will emphasize on "cry for help" strategy, where environmental stress and phytohormones signaling can induce the plant recruitment of beneficial microorganisms. Many questions have been formulated about the functionality of these microbes embedded in the microbiome. However, many are considered useful in the SymComs formulation to counteract stress directly or indirectly.

1.2 Plant responses to multifactorial stress combination

In the environment, plants are exposed to different stress conditions that negatively affect their productivity. The concept of stress at the physiological level can be described as any condition that is lethal for the reproduction and survival of plants. Plant stress has been typically categorized as abiotic and associated with climate change (heat waves, cold snaps, flooding, and prolonged drought) and biotic (insect attack and microbial pathogen pest). Recently, pollutants have been added to the list of plant stressor linked to anthropogenic activities, such as heavy metals, microplastics, pesticides, and antibiotics. Moreover, the increasing UV radiation, ozone (O_3), NO_2, and CO_2 impacts crop productivity (Zandalinas et al., 2021a; Zandalinas and Mittler, 2022; Sinha, 2021). Furthermore, combining different stressors, including nitrogen fertilization, reduces the biodiversity of the plant microbiome (dysbiosis), which helps support plant growth and health (Beltran-Garcia et al., 2021a; Yang et al., 2021a).

The research groups of Sara Zandalinas and Ron Mittler have integrated all stress factors in a concept called "multifactorial stress combination," highlighting that each condition of each stress has effects on plant growth. However, the simultaneous addition of various abiotic stresses provokes devastating effects (Zandalinas et al., 2021a). A large majority of studies showed that stress combination could become synergistic for the reduction of photosynthetic carbon metabolism (e.g., salinity, drought, and high temperatures) compared to the application of stress individually in various plants (Anderson et al., 2021). Nevertheless, stress combinations can be plant-beneficial (antagonistic) (Pascual et al., 2022; Zandalinas et al., 2021b; Rivero et al., 2022).

RNA sequencing and meta-analysis studies of plants under stress combinations reveal that single stresses or combinations elicit a common set of specific transcripts, which differentially regulate transcription factor families such as Heat Shock Factors, MYB, WRKY, and ERF (Ethylene responsive factor) (Zandalinas et al., 2021a; Yang et al., 2021a). In 2015 Prasch and Sonnewald (2013) proposed that genetic responses to universal stress conditions are a part of several signaling crosstalk pathways that activate common transcription factors evolved to mitigate the combined stress effects. Notably, Zandalinas et al. (2021a) discovered that individually exposing plants to different abiotic stressors, including cadmium and paraquat (an herbicide that alters cellular REDOX status), had minimal effects on plant growth and survival, but applying stressor combinations was detrimental to the plants. In the same study, the molecular analysis showed that the stress combination elicited a unique response in *Arabidopsis* and influenced specific pathways and processes important for plant multifactorial stress acclimation. This elegant study revealed the complexity of the multifactorial stress combination because the number of genes that respond to each stress decreases while the number of unique genes increases.

These unique genes were related to heat shock factors, redox control, stress response and ABA signaling, 2Fe-2S cluster binding, and hydrolases and glutaredoxin activities. Recently, Mittler et al. (2022) proposed that understanding ROS metabolism opens an opportunity to protect plants from combined multifactorial stress. This proposal is reinforced by the role played by enzymes such as ascorbate peroxidase 1 (APX1) and NADPH oxidase (RBOHD), as well as proteins participating in the mitogen-activated protein kinases cascade and the glutathione metabolism, which are integrated into the cold stress acclimatization and pathogens responses (Dorion et al., 2021; Xu and Fu, 2022). Thus integrated environmental signals and the activation of ROS stress response networks contribute to plant resilience.

In addition to activating both molecular and physiological responses to stress, plants must reprogram their metabolism to mitigate stress and adapt to energy demands (Pascual et al., 2022; Balfagón et al., 2022; Anzano et al., 2022; Liu et al., 2022a; Nakashima et al., 2014). For example, plants typically produce two types of metabolites: primary metabolites that are involved in cellular processes (e.g., survival, growth, and reproduction) derived from central metabolism (i.e., sugars, amino acids, polyamines, phospholipids) and secondary metabolites that are involved in regulating plant–environment interactions such as phenolic compounds, flavonoids, terpenes, and nitrogen molecules that are produced and also regulates plant development and stress response.

It was recently proposed that some secondary metabolites are multifunctional and could function as powerful plant growth regulators and defend against abiotic and biotic stress or multifactorial stress combinations (Jayaraman et al., 2021; Meraj et al., 2020; Yu et al., 2022a). Within the wide range of secondary metabolites synthesized under combined stress, flavonoid biosynthesis has especially attracted attention in response to drought, where their synthesis involved the expression of transcription factor MYB, EGL3 (enhancer of GLABRA3), and ethylene-insensitive 3 (EIN3), as well as the activity of chalcone synthase and chalcone isomerase (Dixit et al., 2019; Jayaraman et al., 2021; Hou et al., 2022; Liu et al., 2019a; Nakabayashi et al., 2014; Waititu et al., 2021). Moreover, it has been hypothesized that the high production of flavonoids in response to drought stress serves to reshape the microbiome, attracting beneficial bacteria that help to increase the resistance of plants dehydration as was recently observed in Arabidopsis where the attraction by flavonoids of *Aeromonas* H1strain impulses the stomatic closure (He et al., 2022).

As we will discuss later, the responses to combined stress and the synthesis and exudation of metabolites may correlate with the recruitment of beneficial microbes that reshape plant microbiomes.

1.3 Plant stress and abscisic acid crosstalk

Previous studies showed that drought and other abiotic factors negatively affect crop productivity and plant holobiont (De Vries et al., 2020; Shree et al., 2022). Drought is considered a multidimensional stress affecting different and unrelated processes in plants. A wide range of metabolic and molecular reactions are activated, often occurring in parallel. According to Prasch and Sonnewald (2013), it is assumed that plants sense stress on a multidimensional level integrating diverse and complex mechanisms under specific stress conditions that accompany molecular, cellular, and physiological changes in their tissues. In addition to primary and secondary metabolites, phytohormones, including auxin/indole acetic acid (AUX), gibberellic acid (GA), abscisic acid (ABA), cytokinins (CTKs), salicylic acid (SA), brassinosteroids (BRs), strigolactone (SL) and jasmonic acid (JA), and the gaseous hormones ethylene (ET) and nitric oxide (NO) are also produced (Fig. 1.1).

Phytohormones are important for many dynamic and highly regulated molecular responses in the plant life cycle. In addition, phytohormones participate in response to various biotic and abiotic stresses and their combinations, including drought, heat, salinity, and pathogens (Zandalinas et al., 2021a; Jha et al., 2022; Breen et al., 2022; Wang et al., 2022; Yang et al., 2022; Kohli et al., 2013). Recently, it was published that phytohormones interact and integrate various input signals, plant growth readjustment, and stress tolerance acquisition (Jha et al., 2022). Plant response to different abiotic stress can be explained by phytohormones signaling crosstalk, led by ABA (Jha et al., 2022; Breen et al., 2022; Wang et al., 2022; Yang et al., 2022; Kohli et al., 2013; Li et al., 2019;

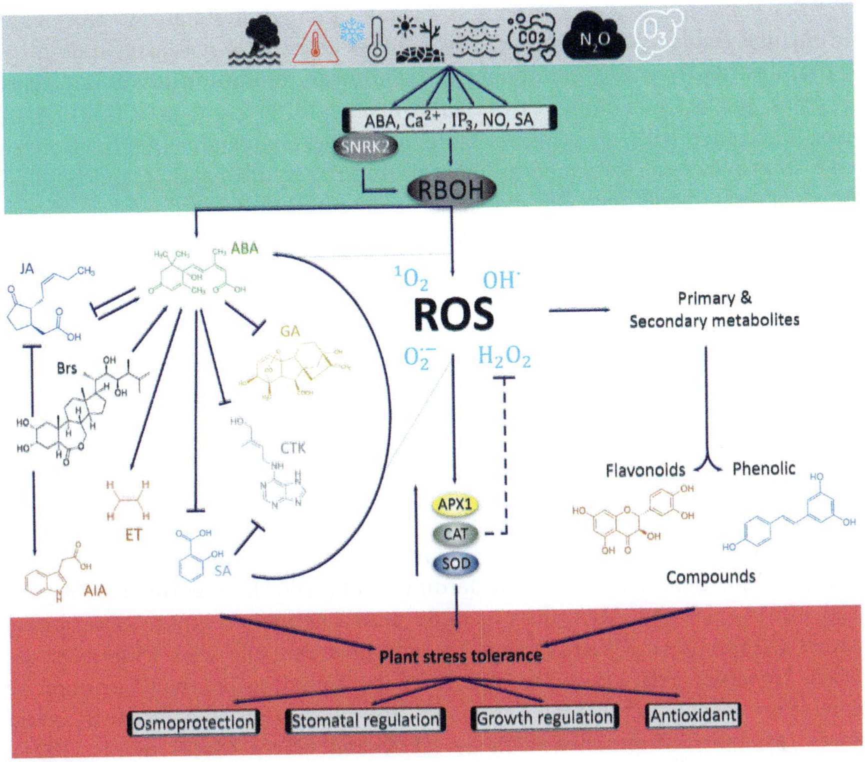

FIGURE 1.1

Phytohormone crosstalk and feedback loops ROS-antioxidants as systemic signaling signatures during multiple stress combination improves plant stress tolerance. Plant stress is perceived by membrane receptors activating second messengers such as calcium, inositol phosphate, and reactive oxygen species (ROS). ABA activates RBOH, inducing ROS and primary and secondary metabolites. Metabolic and phytohormone pathway integration allows plants to adjust growth and represents an antistress strategy.

Yang et al., 2021b; Salvi et al., 2021). ABA is widely involved in responses to plant defense against biotic, abiotic, or understress combinations and is the central regulator of guard cell physiology (Suzuki et al., 2016; Yuan et al., 2021; Waadt et al., 2022; Zhang et al., 2021; Ahres et al., 2021; Niharika et al., 2021; Xu et al., 2022; Ogran et al., 2021). For example, combining salinity and heat upregulates ABA gene expression and accumulation in *Arabidopsis*, Citrus, and Poplar. In tomatoes, ABA induces stomatic closure in response to the combined exposure to salinity and drought. Moreover, this phytohormone induces acclimation protein synthesis in *Arabidopsis* in response to a drought and heat combination (Zandalinas et al., 2021a; Pascual et al., 2022).

ABA is chemically classified as a sesquiterpene stimulated by stress conditions that reduce water accessibility to plant cells. It is synthesized de novo in the plastids from C_{40} carotenoid zeaxanthin to xanthonin involving many enzymes and ends in the cytosol (Salvi et al., 2021; Li et al., 2022). ABA also functioned like a priming molecule. The exogenous application of ABA improves heat tolerance by maintaining the hydro balance and regulating stomatic conductance. Moreover, drought priming, which involves physiological and molecular responses to the ABA, has been shown protection of plants from salinity and cold stress (Liu et al., 2019a; Singha et al., 2022; Guo et al., 2022; Liu et al., 2022a). However, other plant hormones play roles in stomatal modulation under stress. For example, JA and BRs can trigger stomatal closure and repress their opening during drought stress, but it has been observed under multiple stresses that ABA crosstalk with JA, SA, NO, and BRs is as crucial to regulate the stomatal closure (Singhal et al., 2021a; Yu et al., 2022b; Parwez et al., 2022; Iqbal et al., 2021; Muhammad et al., 2022).

These phytohormones in plants under stress have in common that they are synthesized faster and, in parallel, accumulate ROS (Bendou et al., 2022; Liu et al., 2022b; Suhita et al., 2004; Jardim et al., 1863). According to Agurla et al. (2018), JA and ABA share genetic networks and common transcription factors (MYB, NAC, WRKY), but JA acts upstream of ABA (Suhita et al., 2004). However, ABA is in the middle of the signaling pathways to improve plant stress response; this may explain the irregularity in ABA levels to antagonistic effects of hormones such as AUX and CTKs (Salvi et al., 2021). Drought stress upregulates JA, BR, and AUX synthesis and inhibits GA and SA. It has also been reported that ABA inhibits SA transduction upstream and downstream via the systemic acquired resistance signaling pathway (Parwez et al., 2022).

The synthesis of GA and its signaling pathway inhibition causes a reduction of vertical growth for plants conserving energy. Additionally, SA inhibition affects CTK synthesis, and attenuated concentrations of this phytohormone induce ABA synthesis and signaling, augmenting ABA effects. Moreover, BR signaling inhibits JA production and promotes ABA and AUX synthesis. ABA can also inhibit CTK signaling. Furthermore, salt stress increases JA and ABA concentrations but decreases AUX, GA, and SA levels (Fig. 1.2) (Salvi et al., 2021; Bari and Jones, 2009).

Understanding the crosstalk pathways of ABA, JA, and SA during stress signaling and the activation of molecular networks opens up new opportunities for future agricultural applications. Current knowledge on this topic has been reviewed by Waadt et al. (2022) and Gupta et al. (2020), and the collection of articles published in *Plant Cell Reports* Vol. 40 (2021) called "Plant Growth regulators and signaling molecules: Crosstalk in abiotic and biotic responses."

1.4 Abscisic acid signaling and reactive oxygen species connection to cope with plant stress

Plants subject to drought, salinity, and high light perceive the stimulus by membrane receptors, which in turn produce extracellular signals that are transformed into intracellular signals involving second messengers such as calcium ions (Ca^{2+}), inositol phosphate, ROS, and NO (Postiglione and Muday, 2020; Verma et al., 2016; Tang et al., 2021). Those second messengers activate signaling pathways mediated by mitogen-activated protein kinases and calcium-dependent protein kinases and phosphatases, which activate or deactivate transcription factors, such as DREB, MYB/MYC, NAC, ABRE, and WRKY, creating a regulatory network of genes against environmental stresses (Fig. 1.1).

In response to osmotic stress, ABA binds to the nucleocytoplasmic RCAR receptor (PYR/PYL/RCAR), causing a conformational change that promotes interaction with 2C phosphatases (PP2Cs) such as HAB1, HAB2, AHG1/3, HONSU, ABI1, and AB12 (Fidler et al., 2022). The interaction of ABA/RCAR-PP2Cs releases SnRK2s-related protein kinase 2 (SnRK2.6/OST1, SnRK2.3, and SnRK2.2). Moreover, SnRK2.6/OST1 phosphorylates NADPH oxidase resulting in ROS (e.g., superoxide anion anion [$O_2.^-$] and hydrogen peroxide [H_2O_2]) generation that upregulates ABA signaling, promoting stomatic closure, increasing cytosolic calcium, and inactivating ABI1 and ABI2 phosphatases (PPC2s) (Postiglione and Muday, 2020; Prasad et al., 2021). SnRK2-mediated phosphorylation of the SLAC1 ion channel and the KAT1 potassium channel causes plasma membrane depolarization through increased K^+ efflux and a loss of turgor in guard cells, leading to stomatic closure. Interestingly, ABA changes the expression and posttranscriptional modification of aquaporin (AQP) genes and maintains cellular turgidity during stoma closure that even protects against the attack of pathogens (Ding et al., 2022; Rodrigues et al., 2017; Xu et al., 2020; Sharipova et al., 2021).

ROS are a group of oxygen-derived molecules such as H_2O_2, superoxide anion (O_2^-), hydroxyl radical (OH), and singlet oxygen ($^1O_2$), as well as some organic and inorganic peroxides. The chemical nature of O_2 (i.e., bi-radical) allows it to accept electrons or high energy that completely modifies its reactivity against most cellular components, causing the oxidation of membrane lipids, proteins, RNA, DNA, and other small molecules (Sies and Jones, 2021). In plants, ROS-induced oxidative damage depends largely on where they formed, the target molecules' functionality, and the antioxidant response.

ROS have been classified as signaling molecules or "second messengers" that control molecular processes in response to biotic and abiotic stresses, promoting plant tolerance (Mittler et al., 2022; Jin et al., 2022). Genetic and biochemical approaches have demonstrated that ROS can act at the site where they are generated or at great distances from cell compartments or tissues because stable forms like H_2O_2 can be transported by AQPs (Rodrigues et al., 2017; Xu et al., 2020; Sharipova et al., 2021). ROS accumulation in cells under stress conditions alters the REDOX state of different proteins (e.g., membrane receptors and transcription factors) that facilitate the integration and transduction of multiresponse signals in stress pathways (Mittler et al., 2022; Peláez-Vico et al., 2022; Fichman and Mittler, 2021; Fichman et al., 2020). The altered genetic expression in response to ROS increases plant resilience (Fig. 1.1).

Beyond the cellular damage caused by ROS, we must not forget that ROS participates in fundamental cellular processes (e.g., proliferation, differentiation, and cell death), and their production also occurs during photosynthesis, mitochondrial respiration, or in both "homeostasis and recruitment of the plant microbiome" as has been published recently (Berrios and Rentsch, 2022; Song et al., 2021; Ali et al., 2022).

Recent advances have shown the existence of plant-sensing proteins under stress conditions such as temperature, light, osmotic stress, salinity, and drought induces rapid mobilization of calcium from the extracellular space to the cytosol after perception through calcium channels and then activating via phosphorylation of the respiratory burst oxidase homolog "RBOH" triggering ROS production (Zhang et al., 2022a; Kabała et al., 2022). RBOHs are regulated transcriptionally and protein level by posttranscriptional modifications, including both calcium binding and phosphatidic acid or via phosphorylation or nitrosylation (reviewed in Martin et al., 2022).

It has been reported that ROS signaling pathways activation is accompanied by the rapid increase of phytohormones, such as ABA, JA, ET, and SA and their crosstalk. The ABA signaling pathway in plants explicitly connects ROS and stress responsiveness, and it has been considered the rate-limiting step in guard cell physiology under stress (Postiglione and Muday, 2020; Zhao, 2022; Li, 2022; Mohanta et al., 2018). The induced generation of H_2O_2 (the most stable and widespread form of ROS) plays an important role in the ABA-dependent changes in the activity of ion channels bringing about stomatal closure, hydraulic conductivity, and AQP abundance (Rodrigues et al., 2017; Xu et al., 2020; Sharipova et al., 2021). Recently, further evidence of the crosstalk was reported, in which paraquat upregulated the expression of a small protein that increased drought tolerance through paraquat-induced ROS generation (H_2O_2) and ABA signaling (Sies and Jones, 2021; Faragó et al., 2022).

Numerous mechanisms are in place to maintain the ROS homeostasis required for efficient signaling and ensure that ROS levels do not reach damaging levels. Plants produce enzymes and specialized metabolites that function as antioxidants to prevent/restore ROS accumulation/homeostasis. Examples of ROS scavengers include enzyme-based antioxidants such as catalase (CAT), superoxide dismutase (SOD), ascorbate (APX) and guaiacol (GPX) peroxidases, monodehydroascorbate reductase (MDHAR), glutathione reductase (GR); and non-enzyme-based antioxidant molecules such as glutathione, carotenoids, proline, flavonoids, and phenolic compounds (Zandi and Schung, 2022; Sarkar et al., 2022). However, as we will discuss later, in addition to the plant antioxidant response, we must consider the plant microbiome functions that regulate ROS homeostasis under stressful conditions.

1.5 Plant microbiome: the invisible life that helps plants against stress

The plant microbiome is a dynamic and extraordinarily complex association of numerous microbes that help plants in various beneficial ways (De Faria et al., 2021; Semchenko et al., 2022). Plants have selected these microbial communities over millions of years of coevolution, contributing multiple functions to the holobiont. The plant microbiome is considered a second genome because it is important for regulating plant metabolism and critical for the plant immune system (Vannier et al., 2019). These microbial communities, including bacteria, fungi, viruses, and protists, expand their hosts' genomic and metabolic capacities against stress. However, a few numbers of those microbes have highly specific functions, and they are considered "keystones" that work in a complex network with other microorganisms of the same species as well as various species, genera, classes, and domains (e.g., Archaea and Eukarya) (Beltran-Garcia et al., 2021b; Tyagi et al., 2022).

An increasing number of publications show how those microbial components of plant microbiomes mitigate individual stresses, including drought, heat, cold, and salinity, but information about their roles in protecting plants under stressor combinations is lacking (Choi et al., 2021). These symbiotic microorganisms from the phyllosphere, rhizosphere, and endosphere have been considered the second line of defense against abiotic and biotic stress (Fan et al., 2020; Hanaka et al., 2021; Verma et al., 2021; Zheng et al., 2021; Munir et al., 2022).

Plant-colonizing bacteria support growth and abiotic stress response through direct phytohormone synthesis or by regulating the antagonism-synergism among phytohormones signaling pathways and their sensitivity. Indeed, endophytic bacteria directly synthesize AUX, NO, CTKs, ABA, BRs, SA, and GA following host colonization (Park et al., 2017; Khan et al., 2011; Mishra and Baek, 2021; Costa et al., 2022; Ma et al., 2022; Tsotetsi et al., 2022).

The *Pseudomonas fluorescens* G20−18 strain increases tomato growth and enhances drought tolerance. The influential responses for stress tolerance include chlorophyll synthesis in the leaves and ABA content. In addition, *Pseudomonas* activates antioxidant enzymes and the synthesis of phenolic compounds, anthocyanins, and flavonoids. Those genes differentially regulated were seven times higher under drought conditions in response to inoculation. *P. fluorescens* G20−18 produce CTKs (Mekureyaw et al., 2022). On the other hand, *Bacillus amyloliquefaciens* from rice seeds produce GA that not only boosts the physiological response of the host but also stimulates SA and repressed ABA levels (Shahzad et al., 2016).

Also, *Shewanella putrefaciens* and *Cronobacter dublinensis* endophytes enhance drought tolerance of pearl millet by modulating the expression levels of genes involved in ABA and AUX biosynthesis, and coding drought-responsive transcription factors were significantly higher under severe drought stress in endophyte-inoculated plants (Manjunatha et al., 2022). AUX-producing bacteria mediate drought and salt tolerance by increasing water permeability, water uptake and ROS detoxification, improving root architecture, and inducing the expression of stress-resistant genes (Khan et al., 2019; El Houda et al., 2022; Kour and Yadav, 2022).

Also, it has been reported that some species of endophytic fungi display anti-stress abiotic roles. For example, the endophytic fungus *Penicillium minioluteum* protects salt-injured soybean by reducing ABA levels, and *Penicillium glabrum* acts as a heat reliever in soybean and sunflowers by significantly decreasing the synthesis of ROS and lipid peroxidation and ABA and proline concentrations (Khan et al., 2011; Ismail et al., 2021). Recently, Lubna et al. (2022) found that by inoculating soybean with the endophyte fungus *Bypolaris* sp. under salinity stress, the endogenous ABA levels were decreased, whereas SA was increased in the fungal-inoculated plants.

Plant microbiomes can protect the hosts against biotic and abiotic stresses and improve the acquisition and absorption of nutrients (biological nitrogen fixation, phosphate solubilization, organic nitrogen transfer, and siderophore synthesis), can inhibit ethylene-induced senescence, and obtain ammonium through a synthesis of the ACC deaminase enzyme. Plant microbes synthesize various molecules as osmolites, such as betaines, polyamines, quaternary ammonium compounds, trehalose, glycerol, proline, amino acids, and melatonin to adjust plant metabolic water stress. These compounds also protect against the oxidation of membrane proteins and phospholipids functioning as ROS scavengers (Bashir et al., 2022; Saleem et al., 2019; Zhang et al., 2022b; Noman et al., 2021; Nanjani et al., 2022; Chen et al., 2022; Trivedi et al., 2020; Byregowda et al., 2022; Phour and Sindhu, 2022).

Interestingly, endophytes residing within the plant tissues modulated ROS concentration and protected the plant from the detrimental effects of ROS during stress. ROS levels are reduced by endophyte synthesizing SOD, APX, CAT, ascorbate glutathione, and tocopherols involved in ROS homeostasis directly or through the activation of plant antioxidant mechanisms, in which endophyte bacteria also induce secondary metabolites synthesis such as phenolic compounds, flavonoids, tannins, carotenoids, among others (Verma et al., 2021).

On the other hand, endophytes induce changes in the stressed plants by overexpressing stress-tolerant genes, including those involved in ABA, AUX GA and JA signaling genes, higher expression of ion channels KAT1 and KAT2 resulting in decreased Na^+/K^+ (Rani et al., 2022). This phenomenon is known as induced systemic tolerance (Chen et al., 2022; Mengistu, 2020;

Bacon and White, 2016) and refers to a microorganism-mediated response to the abiotic stress responses of plants.

1.6 Endophyte recruitment (cry for help strategy) under stress: from transitory altered state to plant resilience

The plant microbiome reduces the impact of environmental stresses. However, it is not static. Under stress conditions, changes in the microbiome assembly through microbial recruitment or the activation of the endophytic microbiome have been observed, which consequently mitigates stress.

Microorganisms, especially bacteria, are recruited from the soil, rhizosphere, and phyllosphere and then enter the endosphere. For example, the bacteria deposited on the rhizoplane is the potential source of about 90% of the root endosphere, which contributes more than 60% of the microbial composition of the leaf endosphere. However, their heterogeneity depends on plant–microbe and microbe–microbe interactions, which rely on creating specialized networks and microbial hubs (Santoyo, 2022; Lareen et al., 2016; Wei et al., 2021).

Microbes that reside in the endosphere are called endophytes (Rani et al., 2022). They consist of several microorganisms that spend their full or partial life cycles colonizing internal tissues. Bacterial endophytes to enter into the root endosphere require the secretion of enzymes that degrade cell walls (cellulases, xylanases, endoglucanases, and pectinases), gaps in the roots where lateral roots, wounds. However, it has also been reported that plants may internalize bacteria into root meriostematic cells where plant cell walls are not yet hardened as a nutrition strategy and utilize ROS to degrade bacterial cells in the rhizophagy cycle (Verma et al., 2021; Paungfoo-Lonhienne et al., 2010; Beltran-Garcia et al., 2014). Bacterial cells secrete exopolysaccharides that enhance the attachment to the root surface and protect bacterial cells from oxidative damage during the early colonization phase. Most endophytes spread systematically through the xylem (Afzal et al., 2019; Kandel et al., 2017).

Also, endophytic bacteria have been considered key for plant functionality at all stages and under stress conditions, being the biological material for developing bioinoculant formulations. At the phylogenetic level, endophytic bacteria include phyla such as Proteobacteria, Bacteroidetes, Firmicutes, and Actinobacteria. Conversely, epiphytic bacteria residing in the phyllosphere have a more dynamic environment than the endosphere and rhizosphere. At the phyla level, Proteobacteria appeared to be the dominant colonizers in the phyllosphere, followed by Bacteroidetes and Actinobacteria (Lindow and Brandl, 2003).

However, as we mentioned, the plant microbiome is not static. Their assembly can be altered throughout the plant's life and is influenced by multifaceted climate–plant, plant–microbe, and microbe–microbe interactions (Chaudry et al., 2021; Xun et al., 2021). Moreover, microbial community members strongly influence each other and display antagonistic, mutualistic, and competitive interactions. Typically, the endosphere has lower microbial diversity than the rhizosphere but is more abundant than in leaves. Plant colonization depends on the activation of mechanisms of attraction, recruitment, and exclusion (filtration) in the different niches of compartments of the plants (Kandel et al., 2017). Undoubtedly through genetic and metabolic filters, the host plant conducts a healthy microbiota that influences the life of the holobiont (Bordenstein and Theis, 2015).

One of the major questions in contemporary plant science involves determining the functional mechanisms that plants use to shape their microbiome. In the early stages of plant colonization, plants secreted exudates (primary and secondary metabolites), which serves as precommunication signals and energy source for the selective recruitment of microbial communities in the phyllosphere and rhizosphere (Chaudry et al., 2021; Xun et al., 2021). Indeed, many publications have unequivocally shown that secondary plant metabolites affect microbiome composition and function (Jacoby et al., 2021, 2020; Pantigoso et al., 2022). Those plant-derived metabolites shape microbial communities by allowing bacteria to metabolize and establish in the rhizosphere. However, biotic and abiotic stress or stress combination induces changes in the genetic and metabolic expression of plants, modifying the type of exudates in the roots and leaves and causing changes in the microbiome assembly by exerting selective pressure on microbes to survive under such conditions but having greater abundance with lower diversity (Rizaludin et al., 2021; Matilla and Roca, 2022). This microbial filtration is considered essential for the "cry for help" strategy, in which plants recruit useful microbes in a selective manner that allows the recovery of plants at the right time and the intensity of selective pressure.

Since stress conditions reduce the abundance and diversity and consequently their composition, we must consider compositional modifications as a signature of the stress. Rocca et al. (2019), through an analysis called "Microbiome Stress Project," showed that soil undergoes drastic changes in the composition of the microbiome in response to diverse types of stress, including heat, drought, and heavy metals. The authors conclude that disturbances, in general, have stochastic (random colonization to a new site and random shift in abundance) and rather than deterministic effects on the microbiome composition, leading to greater dispersion, an ecological phenomenon known as the "Anna Karenina principle," in which the microbiota under stress varies more among itself than the healthy microbiota. This ecological concept was recently transferred to the plant microbiota, conceptualizing the Anna Karenina principle as a "transient loss of host capacity to regulate its microbiota with a loss of function that leads to a reduction of the host fitness" (Arnault et al., 2022). As shown in Fig. 1.2, the transient loss of the microorganism/function (related to the term dysbiosis) increases plant symptoms. However, to compensate for this loss, the plant activates mechanisms known as the "Cry for help strategy" that attract different commensal microbes.

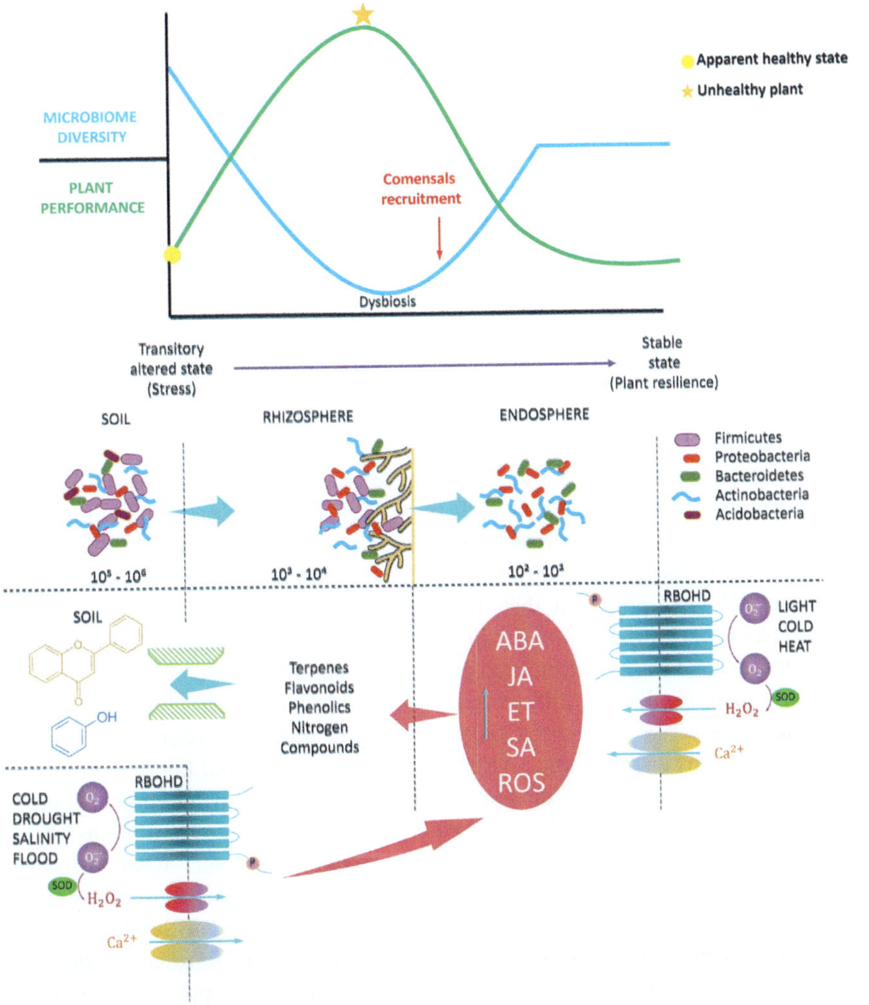

FIGURE 1.2

Microbial recruitment induced by stress response pathways such as phytohormones and ROS-derived from RBOHD activity. Microbiome diversity during plant stress can be altered due to a "transient loss of microorganisms" that impairs plant health. In compensation, plant roots activate the commensal recruitment from the soil to the endosphere, called the "Cry for help" strategy to release secondary metabolites. Those commensals that replace "the original" endophytes were selected across ROS gradients and then restores gradually plant health status.

Recently, an interesting paper by Liu et al. (2021) showed that roots of wheat-infected plants with *Fusarium pseudograminearum* experienced a decrease of the phylotypes such as Acidobacteria, Gemmatimonadetes, and Archaea, while they were enriched with Proteobacteria, Actinobacteria, and, to a lesser extent, Bacteroidetes. On the other hand, upon fungal infection, rhizosphere and endosphere were enriched with *Stenotrophomonas rhizophila*. While this bacterium does not inhibit the fungus under *in vitro* conditions, the plant signaling pathways regulated by JA and SA were significantly activated only in the presence of the pathogen. Moreover, an endophytic strain of *Bacillus megaterium* mitigates the effects of rice spikelet rot disease on wheat by improving the rhizosphere bacterial community and inducing salicylic acid-dependent resistance (Cheng et al., 2020). This phenomenon, linked to SA and JA signaling pathways, was also observed by inoculating heat-killed cells of the endophytic *Bacillus aryabhattai* in *Nicotiana* and Arabidopsis; however, no changes in the assembly of the microbiome were measured (Portieles et al., 2021).

One more example of this favorable recruitment of microorganisms was reported in winter wheat infected with the yellow mosaic virus. Following infection, the α-diversity of root-associated microbiota increases, whereas the microbial community becomes more homogeneous from the bulk soil to the root endosphere. It was proposed that after viral infection, specific genera such as *Streptomyces*, *Stenotrophomonas*, *Bradyrhizobium*, *Spingomonas*, and *Bacillus* were enriched and may contribute to pathogen suppression (Wu et al., 2021). This result revealed significant differences in microbial α-diversity associated with plant disease severity of bacteria from bulk soil to root endosphere, thus indicating that wheat has a strong selective effect on the belowground microbial community. Gu et al. (2022) prepared homogenized bulk-soil with identical microbial species to grow tomatoes, which later can show different configurations of the rhizospheric microbiome during vegetative growth. However, when *Ralstonia solanacearum* was added to the soil, new variations in the microbiome composition were observed with direct implications in the suppression of the disease. Five OTUs were enriched (*Lysinibacillus*, *Pseudoarthrobacter*, *Bordetella*, *Bacillus*, and *Chryseobacterium*). Culturing these species and their application with the pathogen reduces the severity of the disease by up to 100%.

The microbial loss is characterized by changes in the α-diversity of microbial communities, which explain drastic changes in the parameters of the niche, where only a few microbes can survive (α-diversity means microbial species diversity at the local scale). There are several examples that during the stress response phase of the plant, the decrease of certain microbiome components is observed, especially the Firmicutes and Acidobacteria, among others. Root exudates can regulate this reduction, decreasing competitive pressure, and leading to stochastic processes in microbial interactions. This decrease of those bacterial species belonging to Firmicutes or Actinobacteria in the microecosystem permit the growth of other bacterial species (e.g., Proteobacteria) that are metabolically competent and better able to adapt to the stress that affects the plant.

For example, in biotic stress conditions caused by fungi or bacteria at the phylogenetic level, a drastic decrease of Firmicutes and Actinobacteria has been observed with enrichment by Proteobacteria (Arnault et al., 2022; Lee et al., 2021; Runge et al., 2022; Erlacher et al., 2015). Sometimes, this situation becomes confusing because it is thought that members of the Proteobacteria phylotype may cause the disease symptoms. For example, in the *Agave tequilana* with early symptoms of bacterial soft rot associated to *Pectobacterium carotovorum* (formerly *Erwinia carotovorum*), Firmicutes content decreased up to 40% but under this condition leaf tissues are enrichment with Proteobacteria (Cristobal Fonseca, personal communication). In bananas, for example, the presence of *Fusarium* wilt disease increases the content of Proteobacteria and Bacteroidetes, with a decrease of the Firmicutes. Interestingly, *Chryseobacterium* belonging to the phylum Bacteroidetes, under severe conditions of *Fusarium* wilt, replaces strains belonging to Proteobacteria (*Pseudomonas, Stenotro phomonas, Enterobacter*). The abundance of *Chryseobacterium* and Proteo bacteria strains is related to antifungal protection and phosphate solubilization in various crops. In summary, the banana microbiome changes, indicating the plants' necessity for direct protection against fungal pathogens and nutrients (N and P) for energy synthesis and ethylene and AUX modulation (Beltran-Garcia et al., 2021b). Changes in the content of Proteobacteria, Bacteroidetes, and sometimes Actinobacteria also have been reported in influential publications demonstrating the critical role of bacterial taxa in the host plant's "cry for help" strategy (Mendes et al., 2011; Edwards et al., 2015; Chapelle et al., 2016; Berendsen et al., 2018; Carrión et al., 2019).

At the molecular level, the microbe-associated molecular patterns (MAMPS) conserved in these microbes can trigger MAMP-mediated plant defense. Plant immunity activation via this pathway allows plants to rapidly respond when health status changes, managing microbial colonization (diversity and abundance), whereby SA, JA, and ET play a key role in shaping the functionality and structure of the plant microbiome (Trivedi et al., 2020). However, at this stage, activation of MAMP-triggered immunity (MTI) limits pathogen proliferation and curtails plant growth, a phenomenon known as the growth-defense trade-off (Ma et al., 2021). A rheostat model has been proposed, in which there is a balance between commensals and MTI response modulation during the α-diversity reduction to buffer plant resistance to stress and defense-associated growth reduction.

On the other hand, abiotic stress restructures plant microbiomes to drive fitness benefits (Agler et al., 2016). The belowground microbiomes suffer alterations in the composition under abiotic stresses favoring Actinobacteria phylum and other gram positive over Proteobacteria communities. Significant changes in relative abundance were observed in the endosphere and rhizosphere of nitrogen-tolerant/stress-sensitive sorghum genotypes grown in nitrogen-deficient fields. *Streptomyces* belonging to the phylum Actinobacteria were enriched in the tolerant genotype to low soil nitrogen (Lopes et al., 2021). The "cry for help" also applies during drought and heat stress (Wipf et al., 2021). For example, several endosphere Actinobacteria were enriched during drought in the roots of rice, *Populus*, and grasses (Naylor et al., 2017; Fitzpatrick et al., 2018; Timm et al., 2018; Santos-Medellín et al., 2021). Recently, Omae and Tsuda (2022) proposed that Actinobacteria recruitment under drought is conserved between plant species.

In switch grass under drought conditions, Liu et al. (2021) showed that the bulk soil, rhizosheath soil, rhizoplane, and root endosphere harbored distinct and overlapping microbial communities. Rhizoplane and root endosphere displayed abundant Proteobacteria and Firmicutes compared to bulk soil where Cyanobacteria and Bacteroidetes were selectively enriched. Moreover, Proteobacteria and Firmicutes were selectively depleted in rhizosheath soil. After rewatering, *Citrobacter* and *Acinetobacter* were further enriched in rhizosheath soil (Cadell et al., 2020).

In addition, chemical communication in drought-stressed plants shapes resilient microbiomes. Drought-induced pipecolic acid and glycerol-3-phosphate synthesis influences the enrichment of Actinobacteria in the rhizosphere (Naylor, 2017; Caddell et al., 2019). On the other hand, iron homeostasis is impacted by the drought, the decrease or loss of iron transporter phytosiderophores impacting the enrichment of Actinobacteria, which naturally flourish in iron-deficient environments (Xu et al., 2021). Other microbial taxa were enriched plants during abiotic stress. For example, genetic regulation in rice under saline stress includes the microbial recruitment of species of *Dyella*, *Rhizobium*, and *Thiomonas*. Li et al. (2021) recently stated that the presence of a consortium of rhizospheric bacteria provides adaptation to saline stress through a synergistic regulation that includes bacteria at the level of Bacteriodetes, Firmicutes, and Proteobacteria (Li et al., 2021).

It has been proposed that microbial recruitment under stress can be bidirectional, activating a microbiome-root-shoot circuit and prioritizing plant growth over protection against pathogens in crosstalk to abiotic stress (Fig. 1.2). Hou et al. (2021) showed that low photosynthetic radiation perceived through phytochromes induces host-dependent and long-distance modulation of bacterial assemblages via JA signaling (i.e., growth-defense trade-off), indicating that aboveground stress responses in the plant can be modulated by signals from microbial root commensals (Hou et al., 2021).

Likewise, plants can obtain phosphate that favors microbial recruitment using microbial inoculants. For example, the inoculation of the phosphate-solubilizing *Pantoea agglomerans* in semiarid soil improves the availability of $(PO_3)^{-3}$ up to 69% through reduced pH due to the secretion of gluconic acid. In addition, the metagenomic analysis showed that *P. agglomerans* inoculation modified the structure and composition of the soil bacterial community, especially Firmicutes, while the abundance of Actinobacteria with Acidobacteria and Chloroflexi phyla decreased (Saadouli et al., 2021). Notably, there are several examples that microbial inoculants can modulate indigenous microbiomes enhancing important functions related to growth, performance, and protection (Berg et al., 2021; Noman et al., 2021).

Our knowledge of the microbiome assembly can support the management of plant resistance to stress, and we could potentially incorporate "microbial commensals," which will later be part of the endosphere. However, this is difficult because it depends on the seed-microbiome composition and the subsequent interaction with the soil microbiome after planting. If soil microbes detect "foreign microbes," they will attempt to eliminate their competition (microbe–microbe interaction).

Approaches at the molecular level of plant–microbe and microbe–microbe interactions will be critical next steps to improve plant productivity and environmental sustainability based on developing synthetic communities (SynComs) or host-mediated microbiome engineering, providing an innovative tool for engineering long-term beneficial effects.

1.7 Reactive oxygen species/antioxidants feedback for microbiome modulation, symbiosis, plant stress mitigation, and more: many questions that need further studies

In all aerobic organisms, the synthesis of ROS generation is an inherent fact of the cell life. Augmented ROS concentrations can have deleterious effects on cells, but at low concentrations, they function as signaling molecules for fundamental cellular processes (e.g., growth, differentiation, and death). ROS are important for plant growth, pathogen defense, and stress response and, along with other signaling molecules such as calcium, NO, and H_2S, serve as signal signatures for both intracellular signaling and cell-to-cell communications during both biotic/abiotic stresses independently or in combination with multifactorial stress (Marcec et al., 2019; Wang et al., 2021; Singhal et al., 2021b; Peláez-Vico, 2022). On the other hand, to maintain ROS homeostasis, plant cells activate specific and stringent scavenging systems, including enzymatic and nonenzymatic antioxidant systems, to promote plant development and stress management strategies that create a feedback loop (Fig. 1.3). Among these antioxidant molecules, flavonoids have attracted attention for their antioxidant and bacterial chemoattractant capacities under stress conditions (He et al., 2022).

The plant exudates contain flavonoids, coumarins, and other phenolic compounds that attract microbes but implicitly add exogenous ROS scavengers to modulate plant stress (Xing et al., 2014). The microbial ability to act as ROS scavengers is characteristic of beneficial microorganisms. Indeed, endophytes can activate signaling molecules such as H_2O_2, SA, and JA and upregulate the expression of antioxidant genes in plants (Peláez-Vico, 2022). However, under stress, if the endophytic microorganisms have a lower ability to counteract ROS, they will possibly be eliminated, causing transient dysbiosis in the endosphere (Song et al., 2021). To our knowledge, in the endosphere, there are bacteria (such as *Enterobacter*, *Stenotrophomonas*, and *Micrococcus*) without catalase activity but with outstanding growth promotion capabilities, including plant hormone synthesis (e.g., NO and auxins) but also can inhibit the ethylene synthesis (Beltran-Garcia et al., 2021a,b; Bacon and White, 2016; Beltran-Garcia et al., 2014). Therefore plants must recruit those microbes with abilities to overcome oxidative stress and replace native settlers, recovering from dysbiosis and improving the plant's tolerance to stress (Fig.1.2).

ROS are essential for modulating the plant microbiome under stressful conditions and during plant development (Berrios and Rentsch, 2022). ROS can be produced by plants or endophytic microbes, activating plant immunity and coordinating the abundance and diversity of microbial communities (Verma et al., 2021; Oukala et al., 2021). For ROS production, RBOHD (respiratory burst homolog) facilitates ROS production along with other signaling pathways, including pattern-triggered immunity, effector-triggered immunity, and vesicle trafficking pathways, which could explain how microbial homeostasis reverts dysbiosis (Chen et al., 2020). In addition, RBOHD is a convergence point for microbial recognition, stress signaling, and the plant response to biotic and abiotic stress (Pfeilmeier et al., 2021; Liu et al., 2022c).

The ROS production stimulated from the membrane RBOH and the signaling that activates the antioxidant response creates a feedback loop associated with symbiotic interactions between plants and microorganism. ROS has been shown to participate in many stages of the legume–rhizobia interaction, from rhizobia infection to nodule senescence. ROS occurs since the root perception of Nod factors (such as bacterial lipochitooligosaccharides); this fact has been measured through the induction of antioxidant genes in response to compatibility (Minguillón et al., 2022; Hawkins and Oresnik, 2022).

In *Phaseolus* beans, after 3 minutes of the NOD factor sense, the root hairs increase ROS release (Arthikala et al., 2017). The production of O_2^- and H_2O_2 are essential for effective infection and releasing rhizobia into the nodule cells. Interestingly, H_2O_2 has been specifically detected in the intercellular spaces and/or plant cell walls in indeterminate and determinate nodules. This ROS production depends on the activity of RBOHA/RBOHB isoforms, CuZnSOD, and other oxidase enzymes, which influences nodule formation and metabolism. In common beans, RBOHB plays a central role in infection thread formation and nodule organogenesis through crosstalk with flavonoids, carbon metabolisms, cell cycle regulation, and the AUX and CTKs at the beginning of nodulation (Fonseca et al., 2021). H_2O_2 has also been detected surrounding the bacteroids in nodules. Thus it is likely that nodule senescence is linked to an oxidative stress event involving a Fenton-type reaction with metals such as iron and copper (Liu et al., 2019b).

On the other hand, ROS exchange and reactive nitrogen species production are key for the communication between nodules and bacteroids, where antioxidants are critical to modulate their functions, preventing oxidative damage, and enhancing nitrogen fixation (Arthikala et al., 2014; Alqueres et al., 2010). The synthesis of SOD and CAT isoforms, especially, is essential for the activity of the bacteroid, and the deficiency in these antioxidants impacts infection and nodulation (Kunert and Foyer, 2022). Another example of the antioxidant enzymes was observed during the interaction between host and plant growth-promoting bacteria in *Arabidopsis*, rice, and tomato roots inoculated with *Azospirillum brasiliense* (Lade et al., 2018). In contrast, downregulated *sod*, *cat*, and *apx* expression was detected in maize roots inoculated with *A. brasiliense* (Mandon et al., 2021; Alberton et al., 2020).

In the same feedback context, ROS and antioxidants have been observed in tripartite symbiosis in legumes. The simultaneous symbiotic interaction of *Bradyrhizobium* and *Phomopsis liquidambaris*, an endophytic fungus, depends on the ROS production and the fungal biosynthesis of flavonoids that influence nodulation, increasing rhizobia responses to root exudates and early-stage peanut nodulation (Zhang et al., 2022c). Comparative transcriptomic and metabolic analyses showed that flavonoid production is a strategy for alleviating hyphal oxidative stress during rhizobia–fungal endophyte interactions (Nath et al., 2016). Therefore plants and their microbial partners have coopted the stress tolerance responses for signal exchange to establish a functional symbiosis.

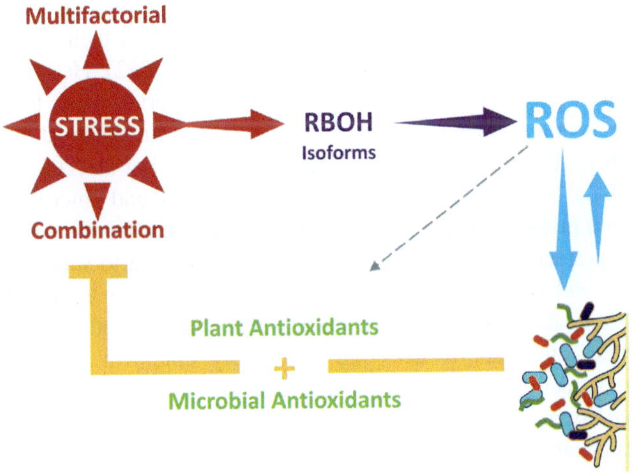

FIGURE 1.3

Plant–microbiome crosstalk for microbial modulation linked to ROS-antioxidant feedback to improve plant resilience. Plant stress activates ROS production and directly stimulates a plant antioxidant response or activates microbes that attenuate ROS. Under a feedback loop, plants allow colonization, establish symbiosis, and withstand stress.

Additionally, a tripartite interaction of plants with symbiotic associates was observed between arbuscular mycorrhizal fungi and the root symbiotic fungus *Piriformospora indica*, improving plant growth by relieving stress impacts and, consequently, plant fitness. Fungal colonization mainly improves the ROS kinetics signature (ROS metabolism and ROS homeostasis) in the cellular processes of plants and plant growth under biotic and abiotic stresses (Hamilton, 2012). The supporting idea of plant–microbial communication via ROS-antioxidants, and consequently the stress response, has also been observed during the symbiotic interaction of *Epichloë festucae* with cool season grasses (Rahnama et al., 2021). Under colonization, this asymptomatic fungus can produce antioxidants in response to biotic and abiotic stresses improving plant fitness (Kayano et al., 2018). However, *Epichloë* reduces ROS production to establish colonization by decreasing RBOH activity via two homologous Rho GTPases: Cdc42 and RacA that mediate the fungal infection (Qi et al., 2017).

However, as shown in Fig. 1.3, it will be proposed that modulation of plant microbiomes, their functionality, and total stress mitigation are linked to ROS/antioxidant feedback (Zandi and Schung, 2022). As we have mentioned throughout the chapter, the stress conditions influence the direct/indirect activation of different RBOH isoforms in the apoplast, increasing ROS production, especially H_2O_2 (Qi et al., 2017). The H_2O_2 is a stable molecule that crosses membranes using AQP channels. The increase of H_2O_2 concentration in the cytosol directly activates molecular stress responses linked to the production of antioxidants that will decrease the ROS signal and thus mitigate stress. However, how ROS works for microbiome modulation? It is an interesting issue that we need to study. We know that a few endophyte bacteria are recruited from the soil and then filtered based on their tolerance to ROS gradients to reach the endosphere (Berrios and Rentsch, 2022). However, we do not know the exact molecular mechanisms endophytic bacteria utilize to perceive and activate the antioxidant system to elicit an efficient response to the ROS on their way to the endosphere. Recently, Chang et al. (2021) and Micci et al. (2022), in studies of bacteria involved in the rhizophagy cycle in root hairs and microbes in leaf trichomes, showed histochemical evidence that microbes may secrete antioxidant forms of nitrogen to protect themselves from oxidative damage due to superoxide produced in plant tissues and cells. While this microbe secreted nitrogen may supply nitrogen to the host plant, there also are likely other antioxidants produced by endophytic microbes. These other antioxidants are completely unknown presently. We only know that the attraction and entry into the endosphere of these bacteria improve the plants' global stress response. It is worth mentioning that *Escherichia coli*, *Salmonella*, and other human pathogens increase their resistance to oxidative stress and antibiotics if they are primed with low concentrations of H_2O_2. In these bacteria, once activated, REDOX sensors such as SoxR, OxyR, and RsrA detoxify undesirable levels of ROS and protect cells with antioxidant strategies (Chang et al., 2021).

Once these bacteria colonize the endosphere, they produce ROS at low levels around bacteria to maintain symbiotic relationships and colonization. However, how bacteria in root cells can be adapted to support ROS production has yet to be fully discovered. Endophyte bacteria produce their antioxidants in response to ROS, but this is not permanent. It is also very interesting that in the rhizophagy cycle some of these "new endophytes" absorbed from soils at the root tip mersitems are afterward ejected from tips of growing root hairs to return to the soil where that may acquire additional nutrients before being attracted back to root tips where exudates are released to attract the microbes; while other bacteria may exit root cells into the intercellular spaces to reside permanently in the endosphere (White et al., 2021, 2019). Our knowledge of multifaceted plant–microbiome interactions and the consequences of plant fitness and productivity has emerged. However, future research should examine how these interactions change over time and space under multiple stress scenarios caused by climate change and anthropogenic activities, as has been interestingly proposed by Trivedi et al. (2022). What is needed now is to begin to evaluate what genes are expressed by bacterial endophytes while they are within plant tissues and cells—as they interact with host-produced ROS and other plant molecules. The majority of genomic analyses focus on host response to endophytic microbes—rather than the endophyte response to host cells and tissues.

Another symbiotic service involving ROS-Aox feedback offered by endophytic bacteria and cyanobacteria is related to nitrogen fixation and water use efficiency. For example, nitrogenase activity is extremely sensitive to inhibition by oxygen and ROS but is dependent on a high metabolic activity to produce the necessary ATPs via oxidative phosphorylation to sustain the biological fixation of nitrogen; so, in this paradoxical event requires upregulation of ROS detoxifying genes (Song et al., 2021; Alqueres et al., 2010; Raghavan et al., 2015; Kumar et al., 2018). In this sense, it is thought that the increase of phenolic compounds, flavonoids, saponins, and other molecules with prooxidant/antioxidant characteristics synthesized after plant inoculation with their endophytes allows the plant to fortify its own cells against ROS and reduces oxygen around bacteria, control oxygen concentrations, permitting nitrogen fixation by diazotrophic bacteria within plant cells (Chang et al., 2021; Micci et al., 2022; White et al., 2021).

1.8 Conclusions

Interestingly, the discrimination between signaling, oxidative stress, and plant–microbe symbiosis under ROS-antioxidant dynamics has increased our understanding of how microbes alleviate plant stress. However, they also bring us research opportunities that require decoding the language that connects plants with microbes, especially for microbiome modulation that facilitates biotechnological approaches for plant production exposed to multifactorial stress conditions using microbial inoculants. What is clear at this point is that we are only beginning to understand how endophytic microbes in plants result in oxidative stress resistance. A possible way forward in this research is to begin to dissect expression of genes within microbes located within plant cells and tissues, while simultaneously examining gene expression within plant cells and tissues that house endophytic microbes. This kind of analysis is rarely done but very needed to elucidate the specific interactions between plants and microbes. More research is needed to better understand the contributions of the plant microbiome to stress tolerance in plants.

Acknowledgments

The authors are grateful to Consejo Nacional de Ciencia y Tecnología (CONACYT) and Consejo Estatal de Ciencia y Tecnologia de Jalisco for support in the acquisition of the MALDI-TOF mass spectrometer (projects 2013-205520 and 2016-269607) and FODECIJAL 8022-2019, respectively. América Martínez-Rodríguez thanks CONACYT PhD fellowship (720754). M.J. Beltrán-García thanks Sistema Nacional de Investigadores (fellowship 32193) and Universidad Autónoma de Guadalajara for fellowship corresponding to "Programa de Carrera" UAG. We also thank the FODECIJAL program from COECYT-JAL for the economic support through the 10602-2023 project. Benjamín Valdez-Salas thanks Universidad Autónoma de Baja California for the research funding support. We thank Fundação de Amparo à Pesquisa do Estado de São Paulo (FAPESP) Grants 2012/13663-1 and CEPID Redoxoma 2013/07937-8, Conselho Nacional de Desenvolvimiento Cientifico e Tecnologico (CNPq) Grant 302120/2018-1, Coordenação de Aperfeiçoamento de Pessoal de Nivel Superior (CAPES) NAP Redoxoma (PRPUSP 2011.1.9352.1.8), and John Simon Guggenheim Memorial Foundation (PDM Fellowship) for financial support. Finally, we thank the staff of City Market Guadalajara for the facilities and attentions during the writing of this chapter.

References

Agler, M., et al., 2016. Microbial hub taxa link host and abiotic factors to plant microbiome variation. PLoS Biol. 14, e1002352.

Agurla, S., et al., 2018. Mechanism of stomatal closure in plants exposed to drought and cold stress. Adv. Exp. Med. Biol. 1081, 215.

Ahres, M., et al., 2021. The impact of far-red light supplementation on hormonal responses to cold acclimation in Barley. Biomolecules 11, 450.

Afzal, I., et al., 2019. Plant beneficial endophytic bacteria: mechanism, diversity, host range and genetic determinants. Microbiol. Res. 221, 36.

Alberton, D., et al., 2020. What did learn from plant-growth promoting rhizobacteria (PGPR)-grass association studies through proteomic and metabolomic approaches? Front. Sustain. Food Syst. 4, 607343.

Ali, B., et al., 2022. Role of endophytic bacteria in salinity stress amelioration by physiological and molecular mechanisms of defense: a comprehensive review. South. Afric. J. Bot. 151, 33.

Alqueres, S.M., et al., 2010. Antioxidant pathways are up-regulated during biological nitrogen fixation to prevent-ROS induced nitrogenase inhibition in Gluconobacter diazotrophicus. Arch. Microbiol. 192, 835.

Anderson, C.M., et al., 2021. High light and temperature reduce photosynthetic efficiency through different mechanisms in the C4 *Setaria viridis*. Commun. Biol. 4, 1092.

Anzano, A., et al., 2022. Plant metabolomics in biotic and abiotic stress: a critical overview. Phytochem. Rev. 21, 1.

Arnault, G., Mony, C., Vandenkoornhuyse, P., 2022. Plant microbiota dysbiosis and the Anna Karenina Principle. Trends Plant. Sci. S1360, 00218.

Arthikala, M., et al., 2017. Respiratory burst oxidase homolog gene a is crucial for *Rhizobium* infection and nodule maturation and function in common bean. Front. Plant. Sci. 8, 2003.

Arthikala, M.K., et al., 2014. RbohB, a *Phaseolus vulgaris* NADPH oxidase gene, enhances symbiosome number, bacteroid size, and nitrogen fixation in nodules and impairs mycorrhizal colonization. N. Phytol. 202, 886.

Bacon, C.W., White, J.F., 2016. Functions, mechanisms and regulation of endophytic and epiphytic microbial communities of plants. Symbiosis 68, 87.

Baedke, J., Fabregas-Tejeda, A., Nieves, A., 2020. The holobiont concept before Margulis. J. Exp. Zool. 334, 149.

Balfagón, D., et al., 2022. γ-Aminobutyric acid plays a key role in plant acclimation to a combination of high light and heat stress. Plant. Physiol. 188 (2026).

Bari, R., Jones, J.D., 2009. Role of plant hormones in plant defence responses. Plant. Mol. Biol. 69, 473.

Bashir, I., et al., 2022. Phyllosphere microbiome: diversity and functions. Microbiol. Res. 254, 126888.

Beltran-Garcia, M.J., et al., 2021a. Nitrogen fertilization and stress factors drive shifts in microbial diversity and plants. Symbiosis 84, 79.

Beltran-Garcia, M., et al., 2021b. Probiotic endophytes for more sustainable banana production. Microorganisms 9, 1805.

Beltran-Garcia, M.J., et al., 2014. Nitrogen acquisition in *Agave tequilana* from degradation of endophytic bacteria. Sci. Rep. 46, 6938.

Bendou, O., et al., 2022. Physiological and antioxidant response to different water deficit regimes of flag leaves and ears of wheat grown under combined elevated CO_2 and high temperature. Plants 11, 2384.

Berendsen, R., et al., 2018. Disease-induced assemblage of a plant-beneficial bacterial consortium. ISME J. 12, 1496.

Berg, G., et al., 2021. Microbiome modulation-toward a better understanding of plant microbiome response to microbial inoculants. Front. Sci. 12, 650610.

Berrios, L., Rentsch, J.D., 2022. Linking reactive oxygen species (ROS) to abiotic and biotic feedbacks in plant microbiomes: the dose makes the poison. Int. J. Mol. Sci. 23, 4402.

Bordenstein, S.R., Theis, K.R., 2015. Host biology in light of the microbiome: ten principles of holobionts and hologenomes. PLoS Biol. 13, e1002226.

Breen, S., et al., 2022. Chloroplasts play a central role in facilitating MAMP-triggered immunity, pathogen suppression of immunity and crosstalk with abiotic stress. Plant. Cell Env. 45, 3001.

Byregowda, R., et al., 2022. Is endophytic colonization of host plants a method of alleviating drought stress? Conceptualizing the hidden world of endophytes. Int. J. Mol. Sci. 23, 9194.

Caddell, D.F., Deng, S., Coleman-Derr, D., 2019. Role of the plant root microbiome in abiotic stress tolerance. In: Verma, S., White, Jr, J., (Eds.), Seed Endophytes. Sham, Switzerland: Springer Verlag. p. 273–311.

Cadell, D., et al., 2020. Drought shifts sorghum root metabolite and microbiome profiles and enriches the stress response factor pipecolic acid. bioRxiv.

Carrión, V., et al., 2019. Pathogen-induced activation of disease-suppressive functions in the endophytic root microbiome. Science 366, 606.

Chang, X., Kingsley, K.L., White, J.F., 2021. chemical interactions at the interface of plant root hair cells and intracellular bacteria. Microorganisms 9, 1041.

Chapelle, E., et al., 2016. Fungal invasion of the rhizosphere microbiome. ISME J. 10, 265.

Chaudry, V., et al., 2021. Shaping the leaf microbiota: plant-microbe-microbe interactions. J. Exp. Bot. 72, 36.

Chen, J., et al., 2022. Wheat microbiome: structure, dynamics, and role in improving performance under stress environments. Front. Microbiol. 12, 821546.

Chen, T., et al., 2020. A plant genetic network for preventing dysbiosis in the phyllosphere. Nature 580, 653.

Cheng, T., et al., 2020. Endophytic *Bacillus megaterium* triggers salicylic acid-dependent resistance and improves the rhizosphere bacterial community to mitigate rice spikelet rot disease. Appl. Soil. Ecol. 156, 103710.

Choi, J., et al., 2021. Inoculation of ACC deaminase-producing *Brevibacterium linens* RS16 enhances tolerance against combined UV-B radiation and heat stresses in rice (*Oryza sativa L.*). Sustainability 13, 10013.

Costa, S., et al., 2022. *Pseudomonas putida* and its close relatives: mixing and mastering the perfect tune for plants. Appl. Microbiol. Biotechnol. 106, 3351.

De Faria, M.R., et al., 2021. The rhizosphere microbiome: functions, dynamics, and role in plant protection. Trop. Plant. Pathol. 46, 13.

De Vries, F.T., et al., 2020. Harnessing rhizosphere microbiomes for drought-resilient crop production. Science 368, 270.

Ding, L., et al., 2022. The plasma membrane aquaporin ZmPIP2;5 enhances the sensitivity of stomatal closure to water deficit. Plant. Cell Environ. 45, 1.

Dixit, A.K., et al., 2019. AtGBF3 confers tolerance to *Arabidopsis thaliana* against combined drought and *Pseudomonas syringae* stress. Environ. Exp. Bot. 168, 103881.

Dorion, S., Ouellet, J.C., Rivoal, J., 2021. Glutathione metabolism in plants under stress: beyond reactive oxygen species detoxification. Metabolites 11, 641.

Edwards, J., et al., 2015. Structure, variation, and assembly of the root-associated microbiomes of rice. Proc. Natl. Acad. Sci. U S A 112, 911.

El Houda, N., et al., 2022. Alleviation of salt stress via habitat-adapted symbiosis. Forests 13, 586.

Erlacher, A., et al., 2015. Biotic stress shifted structure and abundance of *Enterobacteriaceae* in the lettuce microbiome. PLoS One 10, e0118068.

Fan, D., Subramanian, S., Smith, D.L., 2020. Plant endophytes promote growth and alleviate salt stress in *Arabidopsis thaliana*. Sci. Rep. 10, 12740.

Faragó, D., et al., 2022. Small paraquat resistance proteins modulate paraquat and ABA responses and confer drought tolerance to overexpressing Arabidopsis plants. Plant. Cell Env. 45, 1985.

Fichman, Y., Mittler, R., 2021. Integration of electric, calcium, reactive oxygen species and hydraulic signals during rapid systemic signaling in plants. Plant. J. 107, 7.

Fichman, Y., et al., 2020. MYB30 orchestrates systemic reactive oxygen signaling and plant acclimation. Plant. Physiol. 184, 666.

Fidler, J., et al., 2022. PYR/PYL/RCAR receptors play a vital role in the abscisic-acid-dependent responses of plants to external or internal stimuli. Cells 11, 1352.

Fitzpatrick, C., et al., 2018. Assembly and ecological function of the root microbiome across angiosperm plant species. Proc. Natl. Acad. Sci. U S A 115, E1157.

Fonseca, C., et al., 2021. An NADPH oxidase regulates carbon metabolism and the cell cycle during root nodule symbiosis in common bean (*Phaseolus vulgaris*). BMC Plant. Biol. 21, 274.

Gu, Y., et al., 2022. Small changes in rhizosphere microbiome composition predict disease outcomes earlier than pathogen density variations. ISME J. 16, 2448.

Guo, J., et al., 2022. Parental drought priming enhances tolerance to low temperature in wheat (*Triticum aestivum*) offspring. Funct. Plant. Biol. 49, 946.

Gupta, A., et al., 2020. Phytohormones regulate convergent and divergent responses between individual and combined drought and pathogen infection. Crit. Rev. Biotechnol. 40, 320.

Hamilton, C.E., 2012. Endophytic mediation of reactive oxygen species and antioxidant activity in plants: a review. Fungal Divers. 54, 1.

Hanaka, A., et al., 2021. Plant tolerance to drought stress in the presence of supporting bacteria and fungi: an efficient strategy in horticulture. Horticulturae 7, 390.

Hawkins, J., Oresnik, I., 2022. The rhizobium-legume symbiosis: co-opting successful stress management. Front. Plant. Sci. 12, 796045.

He, D., et al., 2022. Flavonoid-attracted *Aeromonas* sp. from the Arabidopsis root microbiome enhances plant dehydration resistance. ISME J. 16, 2622.

Hou, Q., et al., 2022. Genome-wide characterization of chalcone synthase genes in sweet cherry and functional characterization of CpCHS1 under drought stress. Front. Plant. Sci. 13, 989959.

Hou, S., et al., 2021. A microbiota-root-shoot-circuit favours *Arabidopsis* growth over defence under suboptimal light. Nat. Plants 7, 1078.

Iqbal, N., et al., 2021. Crosstalk between abscisic acid and nitric oxide under heat stress: exploring new vantage points. Plant. Cell Rep. 40, 1429.

Ismail, H.M., et al., 2021. *Penicilium glabrum* acted as heat stress relieving endophyte in soybean and sunflower. Pol. J. Environm. Stud. 30, 3099.

Jacoby, R.P., et al., 2020. Recent advances in the role of plant metabolites in shaping the root microbiome. F1000 Res. 26, 9.

Jacoby, R.P., Kopirova, A., Kopriva, S., et al., 2021. Pinpointing secondary metabolites that shape the composition and function of the plant microbiome. J. Exp. Bot. 72, 57.

Jardim, D., Margis, M., Sachetto, G., 1863. Salicylic acid and adenine nucleotides regulate the electron transport system and ROS production in plant mitochondria. Biochim. Biophys. Acta Bioenerg. 148559, 2022.

Jayaraman, K., et al., 2021. Stress-inducible expression of chalcone isomerase2 gene improves accumulation of flavonoids and imparts enhanced abiotic stress tolerance to rice. Environ. Exp. Bot. 190, 104582.

Jha, U., Nayyar, H., Siddique, K.H.M., 2022. Role of phytohormones in regulating heat stress acclimation in agricultural crops. J. Plant. Growth. Regul. 41, 1.

Jin, M.K., et al., 2022. ROS as a key player in quinolone antibiotic stress on *Arabidopsis thaliana*: from the perspective of photosystem function, oxidative stress and phyllosphere microbiome. Sci. Total. Environ. 848, 157821.

Kabała, K., et al., 2022. Role of plasma membrane NADPH oxidase in response to salt stress in cucumber seedlings. Antioxidants 11, 1534.

Kandel, S.L., Joubert, P.M., Doty, S.L., 2017. Bacterial endophyte colonization and distribution within plants. Microorganisms 5, 77.

Kayano, Y., Tanaka, A., Takemoto, D., 2018. Two closely related Rho GTPases, Cdc42 and RacA of the endophytic fungus *Epichloë festucae* have contrasting roles for ROS production and symbiotic infection synchronized with the host plant. PLoS Pathog. 14, e1006840.

Khan, A., et al., 2011. Salinity stress resistance offered by endophytic fungal interaction between *Penicillium minioluteum* LHL09 and glycine max. L. J. Microbiol. Biotechnol. 21, 893.

Khan, M., et al., 2019. Alleviation of salt stress response in soybean plants with the endophytic bacterial isolate *Curtobacterium sp.* SAK1, Ann. Microbiol. 69, 797.

Kohli, A., et al., 2013. The phytohormone crosstalk paradigm takes center stage in understanding how plants respond to abiotic stresses. Plant. Cell Rep. 32, 945.

Kour, D., Yadav, A.N., 2022. Bacterial mitigation of drought stress in plants: current perspectives and future challenges. Curr. Microbiol. 79, 248.

Kumar, A., Kirti, A., Rajaram, H., 2018. Regulation of multiple abiotic stress tolerance by LexA in the cyanobacterium Anabaena sp. strain PCC7120. Biochim. Biophys. Acta Gene Regul. Mech. S1874–9399, 30185.

Kunert, K.J., Foyer, C., 2022. Redox metabolism in soybean and its significance in nitrogen-fixing nodules. Adv. Bot. Res. 102, 177.

Lade, S.B., et al., 2018. Host-specific proteomic and growth analysis of maize and tomato seedlings inoculated with *Azospirrillum brasiliense* Sp7. Plant. Physiol. Biochem. 129, 381.

Lareen, A., Burton, F., Schafer, P., 2016. Plant root-microbe communication in shaping root microbiomes. Plant. Mol. Biol. 90, 575.

Lee, S.M., et al., 2021. Disruption of *Firmicutes* and *Actinobacteria* abundance in tomato rhizosphere causes the incidence of bacterial wilt disease. ISME J. 15, 330.

Li, H., et al., 2021. Salt-induced recruitment of specific root-associated bacterial consortium capacle of enhancing plant adaptability to salt stress. ISME J. 15, 2865.

Li, H., et al., 2019. Identification and analysis of genes involved in auxin, abscisic acid, gibberellin, and brassinosteroid metabolisms under drought stress in tender shoots of tea plants. DNA Cell Biol. 38, 1292.

Li, J., et al., 2022. Abscisic acid is required for cold-induced accumulation of ginsenosides Rg1 and Re in Panax ginseng adventitious roots. Plant. Cell Tissue Organ. Cult. 149, 1.

Li, Y., 2022. SISnRK2.3 interacts with SISUI1 to modulate high temperature tolerance via Abscisic acid (ABA) controlling stomatal movement in tomato. Plant. Sci. 321, 111305.

Lindow, S., Brandl, M., 2003. Microbiology of the phyllosphere. Appl. Environ. Microbiol. 69, 1875.

Liu, H., et al., 2021. Evidence for the plant recruitment of beneficial microbes to suppress soil-borne pathogens. N. Phytol. 229, 2873.

Liu, Y., et al., 2019a. Physiological and proteomic responses of mulberry trees (*Morus alba. L.*) to combined salt and drought stress. Int. J. Mol. Sci. 20, 2486.

Liu, X., et al., 2019b. Bacterioferritin comigratory protein is important in hydrogen peroxide resistance, nodulation, and nitrogen fixation in *Azorhizobium caulinodans*. Arch. Microbiol. 201, 823.

Liu, L., et al., 2022a. Comprehensive analysis of the MYB transcription factor gene family in *Morus alba*. BMC Plant. Biol. 22, 281.

Liu, X., et al., 2022a. Priming effect of exogenous ABA on heat stress tolerance in rice seedlings is associated with the upregulation of antioxidative defense capability and heat shock-related genes. Plant. Growth Regul. 98, 23–38.

Liu, J., et al., 2022b. Salicylic acid, a multifaceted hormone, combats abiotic stresses in plants. Life 12, 886.

Liu, Z., et al., 2022c. FLS2-RBOHD-PIF4 module regulates plant response to drought and salt stress. Int. J. Mol. Sci. 23, 1080.

Lopes, L.D., et al., 2021. Sweet sorghum genotypes tolerant and sensitive to nitrogen stress select distinct root endosphere and rhizosphere bacterial communities. Microorganisms 9, 1329.

Lubna, et al., 2022. Endophytic fungus *Bipolaris sp.* CSL-1 induces salt tolerance in Glycine max. L via modulating its endogenous hormones, antioxidative system and gene expression. J. Plant. Interact. 17, 319.

Ma, C., et al., 2022. Inoculation with carbofuran-degrading rhizobacteria promotes maize growth through production of IAA and regulation of the release of plant-specialized metabolites. Chemosphere 307, 136027.

Ma, K., et al., 2021. Coordination of microbe-host homeostasis by crosstalk with plant innate immunity. Nat. Plants 7, 814.

Mandon, K., et al., 2021. Redox regulation in diazotrophic bacteria in interaction with plants. Antioxidants 10, 880.

Manjunatha, B., et al., 2022. Plant growth-promoting rhizobacteria *Shewanella putrefaciens* and *Cronobacter dublinensis* enhance drought tolerance of pearl millet by modulating hormones and stress-responsive genes. Physiol. Plant. 174, e13676.

Marcec, M., et al., 2019. Mutual interplay of Ca^{2+} and ROS signaling in plant immune response. Plant. Sci. 283, 343.

Martin, R.E., Postiglione, A.E., Muday, G.K., 2022. Reactive oxygen species function as signaling molecules in controlling plant development and hormonal responses. Curr. Opin. Plant. Biol. 69, 102293.

Matilla, M.A., Roca, A., 2022. Multi-host lifestyle in plant-beneficial bacteria: an evolutionary advantage for survival and dispersal? Environ. Microbiol. 24, 3307.

Mekureyaw, M., et al., 2022. The cytokinin-producing plant beneficial bacterium *Pseudomonas fluorescens* G20-18 primes tomato (*Solanum lycopersicum*) for enhanced drought stress responses. J. Plant. Physiol. 270, 153629.

Mendes, R., et al., 2011. Deciphering the rhizosphere microbiome for disease-suppressive bacteria. Science 332, 1097.

Mengistu, A.A., 2020. Endophytes: colonization, behaviour, and their role in defense mechanism. Int. J. Microbiol. 2020, 6927219.

Meraj, T.A., et al., 2020. Transcriptional factors regulate plant stress responses through mediating secondary metabolism. Genes 11, 346.

Micci, A., et al., 2022. Histochemical evidence for nitrogen-transfer endosymbiosis in non-photosynthetic cells of leaves and inflorescence bracts of angiosperms. Biology 11, 876.

Minguillón, S., et al., 2022. Signaling by reactive molecules and antioxidants in legume nodules. N. Phytol. 236, 815.

Mishra, A., Baek, K., 2021. Salicylic acid biosynthesis and metabolism: a divergent pathway for plants and bacteria. Biomolecules 11, 705.

Mittler, R., et al., 2022. Reactive oxygen species signalling in plant stress responses. Nat. Rev. Mol. Cell. Biol. 23, 663.

Mohanta, T., et al., 2018. Early events in plant abiotic stress signaling interplay between calcium, reactive oxygen species and phytohormones. J. Plant. Growth Regul. 37, 1033.

Muhammad, M., et al., 2022. Mechanisms of abscisic acid-mediated drought stress responses in plants. Int. J. Mol. Sci. 23, 1084.

Munir, N., et al., 2022. Mechanisms and strategies of plant microbiome interactions to mitigate abiotic stresses. Agronomy 12, 2069.

Nakabayashi, R., et al., 2014. Enhancement of oxidative and drought tolerance in Arabidopsis by overaccumulation of antioxidant flavonoids. Plant. J. 77, 367.

Nakashima, K., Yamaguchi, K., Shinozaki, K., 2014. The transcriptional regulatory network in the drought response and its crosstalk in abiotic stress responses including drought, cold, and heat. Front. Plant. Sci. 5, 170.

Nanjani, S., et al., 2022. Genome analysis uncovers the prolific antagonistic and plant growth-promoting potential of endophyte *Bacillus velezensis* K1. Gene 836, 146671.

Nath, M., et al., 2016. Reactive oxygen species generation-scavenging and signaling during plant-arbuscular mycorrhizal and *Piriformospora indica* interaction under stress condition. Front. Plant. Sci. 7, 1574.

Naylor, D., et al., 2017. Drought and host selection influence bacterial community dynamics in the grass root microbiome. ISME J. 11, 2691.

Niharika, K.M., et al., 2021. Mitigating strategies of gibberellins in various environmental cues and their crosstalk with other hormonal pathways in plants: a review. Plant. Mol. Biol. 39, 34.

Noman, M., et al., 2021. Plant-microbiome crosstalk: dawning from composition and assemble of microbial community improvement of disease resilience in plants. Int. J. Mol. Sci. 22, 6852.

Ogran, A., et al., 2021. Water deficiency and induced defense against a generalist insect herbivore in desert and mediterranean populations of *Eruca sativa*. J. Chem. Ecol. 47, 768.

Omae, N., Tsuda, K., 2022. Plant-microbiota interactions in abiotic stress environments. Mol. Plant. Microbe Interact. 35, 511.

Oukala, N., Aissat, K., Pastor, V., 2021. Bacterial endophytes: the hidden actor in plant immune responses against biotic stress. Plants 19, 1012.

Pantigoso, H., Newberger, D., Vivanco, J., 2022. The rhizosphere microbiome: plant-microbial interactions for resource acquisition. J. Appl. Microbiol. 133, 2864.

Park, Y., et al., 2017. *Bacillus aryabhattai* SRB02 tolerates oxidative and nitrosative stress and promotes the growth of soybean by modulating the production of phytohormones. PLoS One 12, e0173203.

Parwez, R., et al., 2022. Abscisic acid signaling and crosstalk with phytohormones in regulation of environmental stress responses. Environ. Exp. Bot. 199, 104885.

Pascual, L.S., et al., 2022. Climate change-associated multifactorial stress combination: a present challenge for our ecosystems. J. Plant. Physiol. 276, 153764.

Paungfoo-Lonhienne, C., et al., 2010. Turning the table: plants consume microbes as a source of nutrients. PLoS One 5, e11915.

Peláez-Vico, M.A., et al., 2022. ROS and redox regulation of cell-to-cell and systemic signaling in plants during stress: ROS and redox regulation of stress responses. Free. Rad. Biol. Med. 193, 354.

Peláez-Vico, M.A., 2022. ROS and redox regulation of cell-to-cell systemic signaling in plants during stress. Free Rad. Biol. Med. 193, 354.

Pfeilmeier, S., et al., 2021. The plant NADPH oxidase RBOHD is required for microbiota homeostasis in leaves. Nat. Microbiol. 6, 852.

Phour, M., Sindhu, S.S., 2022. Mitigating abiotic stress: microbiome engineering for improving agricultural production and environmental sustainability. Planta. 256, 85.

Portieles, R., et al., 2021. Heat-killed endophytic bacterium induces robust plant defense responses against important patogens. Sci. Rep. 11, 12182.

Postiglione, A.E., Muday, G.K., 2020. The role of ROS homeostasis in ABA-induced guard cell signaling. Front. Plant. Sci. 11, 968.

Prasad, A., Muthappa, S.K., Prasad, M., 2021. Complex molecular mechanisms determine fitness of plants to biotic and abiotic stresses. J. Plant. Biochem. Biotechnol. 30, 633.

Prasch, C.M., Sonnewald, U., 2013. Simultaneous application of heat, drought, and virus to Arabidopsis plants reveals significant shifts in signaling networks. Plant. Physiol. 162 (1849).

Qi, J., et al., 2017. Apoplastic ROS signaling in plant immunity. Curr. Opin. Plant. Biol. 38, 92.

Raghavan, P.S., Rajaram, H., Apte, S.K., 2015. Membrane targeting of MnSOD is essential for oxidative stress tolerance of nitrogen fixing cultures of Anabaena sp. strain PCC7120. Plant. Mol. Biol. 88, 503.

Rahnama, M., Fleetwood, D., Johnson, R., 2021. Histological methods to detect early-stage plant defense responses during artificial inoculation of *Lolium perenne* with *Epichloë festucae*. Bio Protoc. 11, e4013.

Rani, S., et al., 2022. Endophytism: a multidimensional approach to plant-prokaryotic microbe interaction. Front. Microbiol. 13, 861235.

Rivero, R.M., et al., 2022. Developing climate-resilient crop: improving plant-tolerance to stress combination. Plant. J. 109, 373.

Rizaludin, M.S., et al., 2021. The chemistry of stress: understanding the cry for help of plant roots. Metabolites 11, 357.

Rocca, J., et al., 2019. The microbiome stress project: toward a global meta-analysis of environmental stressors and their effects on microbial communities. Front. Microbiol. 9, 3272.

Rodrigues, O., et al., 2017. Aquaporins facilitate hydrogen peroxide entry into guard cells to mediate ABA- and pathogen-triggered stomatal closure. Proc. Natl. Acad. Sci. U S A 114, 9200.

Runge, P., et al., 2022. Distinct phyllosphere microbiome of wild tomato species in central Peru upon dysbiosis. Microb. Ecol.

Saadouli, I., et al., 2021. The impact of the inoculation of phosphate-solubilizing bacteria *Pantoea agglomerans* on phosphorous avalaibility and bacterial community dynamics of semi-arid soils. Microorganisms 9, 1661.

Saleem, M., Hu, J., Jousset, A., 2019. More than the sum of its parts: microbiome biodiversity as a driver of plant growth and soil health. Annu. Rev. Ecol. Evol. 50, 145.

Salvi, P., et al., 2021. Phytohormone signaling and crosstalk in regulating drought stress response in plants. Plant. Cell Rep. 40, 1305.

Salvi, P., et al., 2022. Advancement in the molecular perspective of plant-endophytic interaction to mitigate drought stress in plants. Front. Microbiol. 13, 981355.

Santos-Medellín, C., et al., 2021. Prolonged drought imparts lasting compositional changes to the rice root microbiome. Nat. Plants 7, 1065.

Santoyo, G., 2022. How plants recruit their microbiome? New insights into beneficial interactions. J. Adv. Res. 40, 45.

Sarkar, R., et al., 2022. Comprehensive characterization and molecular insights into the salt tolerance of a Cu,Zn-superoxide dismutase from an Indian mangrove, *Avicennia marina*. Sci. Rep. 12, 1745.

Semchenko, M., et al., 2022. Deciphering the role of specialist and generalist plant-microbial interactions as drivers of plant-soil feedback. Ney Phytol. 234, 1929.

Shahzad, R., et al., 2016. Seed-borne endophytic *Bacillus amyloliquefaciens* RWL-1 produces gibberellins and regulates endogenous phytohormones of *Oryza sativa*. Plant Physiol. Biochem 106, 236.

Sharipova, G., et al., 2021. Involvement of reactive oxygen species in ABA-induced increase in hydraulic conductivity and aquaporin abundance. Int. J. Mol. Sci. 22, 9144.

Shree, B., Jayakrishnan, U., Bhushan, S., 2022. Impact of key parameters involved with plant-microbe interaction in context to global climate change. Front. Microbiol. 13, 1008451.

Sies, H., Jones, D.P., 2021. Reactive oxygen species (ROS) as pleiotropic physiological signalling agents. Nat. Rev. Mol. Cell Biol. 21, 363.

Singha, A., et al., 2022. Drought priming alleviated salinity stress and improved water use efficiency of wheat plants. Plant. Growth Regul. 94, 357.

Singhal, R.K., et al., 2021a. Roles of nitric oxide in conferring multiple abiotic stress tolerance in plants and crosstalk with other plant growth regulators. J. Plant. Growth Regul. 40, 2303.

Singhal, R., et al., 2021b. Crucial cell signaling compounds crosstalk and integrative multi-omics techniques for salinity stress tolerance in plants. Front. Plant. Sci. 12, 670369.

Sinha, R., et al., 2021. The impact of stress combination on reproductive processes in crops. Plant. Sci. 311, 111007.

Song, Y., et al., 2021. FERONIA restricts *Pseudomonas* in the rhizosphere microbiome via regulation of reactive oxygen species. Nat. Plants. 7, 644.

Suhita, D., et al., 2004. Cytoplasmic alkalization precedes reactive oxygen species production during methyl jasmonate- and abscisic acid-induced stomatal closure. Plant. Physiol. 134, 1536.

Suzuki, N., et al., 2016. ABA is required for plant acclimation to a combination of salt and heat stress. PLoS One 11, e0147625.

Tang, H., et al., 2021. A PIP-mediated osmotic stress signaling cascade plays a positive role in the salt tolerance of sugarcane. BMC Plant. Biol. 21, 589.

Timm, C., et al., 2018. Abiotic stresses shift belowground populus-associated bacteria toward a core stress microbiome. mSystems 3, e00070.

Trivedi, P., et al., 2022. Plant-microbiome interactions under a changing world: responses, consequences and perspectives. N. Phytol. 234, 1951.

Trivedi, P., et al., 2020. Plant-microbiome interactions: from community assembly to plant health. Nat. Rev. Microbiol. 18, 607.

Tsotetsi, T., et al., 2022. *Bacillus* for plant growth promotion and stress resilience: what have we learned? Plants 11, 2482.

Tyagi, R., et al., 2022. Management of abiotic stresses by microbiome-based engineering of the rhizosphere. J. Appl. Microbiol. 133, 254.

Vannier, N., Agler, M., Hacquard, S., 2019. Microbiota-mediated disease resistance in plants. PLoS Pathog. 15, e1007740.

Verma, H., et al., 2021. The potential application of endophytes in management of stress from drought and salinity in crop plants. Microorganisms 9, 1729.

Verma, V., Ravindran, P., Kumar, P.P., 2016. Plant hormone-mediated regulation of stress responses. BMC Plant. Biol. 16, 86.

Waadt, R., et al., 2022. Plant hormone regulation of abiotic stress responses. Nat. Rev. Mol. Cell Biol. 23, 680.

Waititu, J.K., et al., 2021. Transcriptome analysis of tolerant and susceptible maize genotypes reveals novel insights about the molecular mechanisms underlying drought responses in leaves. Int. J. Mol. Sci. 22, 6980.

Wang, C., et al., 2021. Hydrogen sulfide in plants: crosstalk with other signal molecules in response to abiotic stresses. Int. J. Mol. Sci. 22, 12068.

Wang, J., et al., 2022. Effects of individual and combined polystyrene nanoplastics and phenanthrene on the enzymology, physiology, and transcriptome parameters of rice (*Oryza sativa L.)*. Chemosphere 304, 135341.

Wei, G., et al., 2021. Compartment niche shapes the assembly and network of *Cannabis sativa*-associated microbiome. Front. Microbiol. 12, 714993.

White, J.F., et al., 2021. Endophytic bacteria in grass crop growth promotion and biostimulation. Grass Res. 1, 5.

White, J.F., et al., 2019. Endophytic microbes and their potential applications in crop management. Pest. Manag. Sci. 75, 2558.

Wipf, H., Bùi, T., Coleman, D., 2021. Distinguishing between the impacts of heat and drought stress on the root microbiome of *Sorghum bicolor*. Phytobiomes J. 5, 166.

Wu, C., et al., 2021. Enrichment of microbial taxa after onset of wheat yellow mosaic disease. Agric. Ecosys. Environm. 322, 107651.

Xing, F.H., et al., 2014. Rhizosphere interactions: root exudates, microbes, and microbial communities. Botany 92, 267.

Xu, H., et al., 2022. Tissue specificity and responses to abiotic stresses and hormones of PIN genes in rice. Biologia 77, 1459.

Xu, L., et al., 2021. Genome-resolved metagenomics reveals role of iron metabolism in drought-induced rhizosphere microbiome dynamics. Nat. Commun. 12, 3209.

Xu, Y., Fu, X., 2022. Reprogramming of plant central metabolism in response to abiotic stresses: a metabolomics view. Int. J. Mol. Sci. 23, 5716.

Xu, Y., et al., 2020. A novel aquaporin gene MaSIP2-1 confers tolerance to drought and cold stresses in transgenic banana. Mol. Breed. 40, 62.

Xun, W., et al., 2021. Rhizosphere microbiome: functional compensatory assembly for plant fitness. Comput. Struct. Biotech. J. 19, 5487.

Yang, T., et al., 2021a. Anthropogenic drivers of soil microbial communities and impacts on soil biological functions in agroecosystems. Glob. Ecol. Conserv. 27, e01521.

Yang, Y., Miao, W., Chen, J., 2021b. Roles of jasmonates and brassinosteroids in rice responses to high temperature stress – a review. Crop. J. 9, 977.

Yang, X., et al., 2022. Primary root response to combined drought and heat stress is regulated via salicylic acid metabolism in maize. BMC Plant. Biol. 22, 417.

Yu, W., et al., 2022a. Partial root-zone simulated drought induces greater flavonoid accumulation than full root-zone simulated water deficiency in the leaves of *Ginkgo biloba*. Environ. Exp. Bot. 201, 104998.

Yu, Y., et al., 2022b. The role of cytokinins in plant under salt stress. J. Plant. Growth Regul. 41, 1.

Yuan, W., et al., 2021. Abscisic acid is required for root elongation associated with Ca^{2+} influx in response to water stress. Plant. Physiol. Biochem. 169, 127.

Zandi, P., Schung, E., 2022. Reactive oxygen species, antioxidant responses and implications from a microbial modulation perspective. Biology 11, 155.

Zandalinas, S.I., Mittler, R., 2022. Plant response to multifactorial stress combination. N. Phytol. 234, 1161.

Zandalinas, S.I., et al., 2022. Plant responses to climate change: metabolic changes under combined abiotic stresses. J. Exp. Bot. 73, 3339.

Zandalinas, S.I., et al., 2021a. The impact of multifactorial stress combination on plant growth and survival. N. Phytol. 230, 1034.

Zandalinas, S.I., Fritschi, F.B., Mittler, R., 2021b. Global warming, climate change, and environmental pollution: recipe for a multifactorial stress combination disaster. Trends Plant. Sci. 26, 588.

Zhang, A., et al., 2021. OsIAA20, an Aux/IAA protein, mediates abiotic stress tolerance in rice through an ABA pathway. Plant. Sci. 308, 110903.

Zhang, H., et al., 2022a. Abiotic stress responses in plants. Nat. Rev. Genet. 23, 104.

Zhang, X., et al., 2022b. Regulation of rhizospheric microbial network to enhance plant growth and resist pollutants: unignorable weak electric field. Sci. Total. Environ. 855, 158888.

Zhang, W., et al., 2022c. Priming of rhizobial nodulation signaling in the mycosphere accelerates nodulation of legume hosts. N. Phytol. 235, 1212.

Zhao, H., 2022. Expression of the sweet potato MYB transcription factor IbMYB48 confers salt and drought tolerance in Arabidopsis. Genes 13, 1883.

Zheng, Y., et al., 2021. Patterns in the microbial community of salt-tolerant plants and the functional genes associated with salt stress alleviation. Microbiol. Spectr. 9, e0076721.

CHAPTER

Impact of COVID-19 on global agriculture and food sector

2

Rahul Nitnavare[1,2], Joorie Bhattacharya[3], Sirikanjana Thongmee[4] and Sougata Ghosh[4,5]

[1]*Division of Plant and Crop Sciences, School of Biosciences, University of Nottingham, Nottingham, Leicestershire, United Kingdom*

[2]*Department of Plant Sciences, Rothamsted Research, Harpenden, Hertfordshire, United Kingdom*

[3]*Genetic Gains, International Crops Research Institute for the Semi-Arid Tropics, Hyderabad, Telangana, India*

[4]*Department of Physics, Faculty of Science, Kasetsart University, Bangkok, Thailand*

[5]*Department of Microbiology, School of Science, RK University, Rajkot, Gujarat, India*

2.1 Introduction

The world has witnessed the appalling impact of the 2019 novel coronavirus (severe acute respiratory syndrome coronavirus 2 [SARS-CoV-2]) that emerged in Wuhan, China and rapidly spread all over the world by April 2020 (Ghosh et al., 2020). The novel Corona virus disease 2019 (COVID-19) has posed major challenges in the economic sector globally and has also caused huge amounts of debt and loss. Past experiences with pandemics have reflected their grave impact on collective human health and economic progress (Webster and Ghosh, 2020). With the increase in infectivity of the virus, a huge effect was seen in the agricultural sector as well. A major chunk of the import and export activities were disrupted due to restrictions imposed on travel as well as interstate/country movement. Additionally, it was also observed that there were difficulties in the transportation of produce to different regions. A large quantity of labor shortage was also seen, which caused loss during the harvest period. According to reports by FAO (2020a), the highest impact felt on agriculture was on the supply and demand of food. The food supply primarily comprises the manufacturing, packaging, and distribution of food and food products. Due to the isolation regulations imposed during the pandemic, the generation of food initially saw a decline. However, with the progress of the disease, supply was the primary sector which needed to be revived. Therefore governments across the globe ensured that the food supply was kept constant and maintained. On the other hand, food demand includes the consumer end of the chain.

Sustainable Agricultural Practices. DOI: https://doi.org/10.1016/B978-0-443-19150-3.00010-2

Interestingly, during the pandemic, food demand saw a variation with an initial rise and subsequently deterioration due to job losses and associated uncertainties. Additionally, the basic food necessities did not see an upsurge in price. However, economists predicted that there might be an increase in price for high-valued products (Siche, 2020; FAO, 2020a). During the period ensuing COVID-19, a large section of the society was experiencing chronic hunger and malnutrition due to an imbalance in the supply and demand chain. This group includes those belonging to countries with high hunger and malnutrition index, low-income and small-holder farmers, and children belonging to low-income families (FAO, 2020b). Unfortunately, the pandemic has affected the vulnerable section of society the most. Food security refers to the availability of sufficient nutritious food in a consistent manner. The disruption in global trade has altered food production and distribution and hence has led to a state of food insecurity (Poudel et al., 2020).

Most of the global regions have taken several primary measures to curb the impending increase in infection, such as travel restrictions, border closure, and effective implementation of social distancing protocols. However, this has affected the agricultural sector, including livestock, poultry, and fishery as well. The travel bans have limited the delivery of stocks and impacted food distribution. Further, there has been a negative consequence in terms of shortage of agricultural accessories and inputs like fertilizers, seeds, and pesticides to farmers, thus reducing food production. Countries dependent on the import of products have been especially affected. The rise in infection and the poor infrastructure associated with healthcare has led to a high transmission rate among workers and labors, causing a colossal setback in the agricultural industry as well. The travel restrictions have subsequently also restricted the movement of seasonal and farm workers (Poudel et al., 2020).

Food security has been at its most vulnerable stage during the pandemic, with the supply chain being hit the highest. To maintain food security, FAO is currently working to balance out the supply and demand chain to feed the millions facing chronic hunger (FAO, 2020c).

2.2 Background

The emergence of COVID-19 has led to immense fragility and vulnerability in societal and economic organizations. The first reported case of COVID-19 was reported in Wuhan, China, after which the outbreak caused over 120 million cases and 2.6 million deaths globally (World Health Organization, 2020). The epidemic portrayed the unpreparedness of the countries for a health crisis along with the lack of equipment to deal with the imposed lockdown. Therefore a huge number of the population was affected tremendously. The section vastly hit were the urban dwellers, which led to understanding the significance of food availability and ecosystem. The production and distribution of staple food commodities form an integral part of the food web chain involving farmers, industries, markets, and the final consumers. Globalization allowed the free movement of such commodities across the world. However, the advent of the SARS-CoV-2 restricted the import, export, supply, and access of such goods. To ensure food security in their respective regions, countries have imposed limitations on food exports of staple commodities produced in their own country for maintenance of the food reserve.

Similarly, certain countries relaxed regulations for food imports for commodities not produced in their countries. Reports of FAO, however, imply that the fluctuations in imports and exports would not affect the food price index and global market rate due to demand contractions. The sudden imposition of restrictions of movement, social distancing rules, and the overall shutdown of cities had led to a general panic among the masses. This consequently caused the public to hoard goods and food items, especially food with a longer shelf-life. An overall surge in demand occurred during COVID-19 times as compared to a year prior. Further, an increase in the use of home delivery services was observed, which would most likely continue postpandemic as well. The pandemic would prompt the governments to rethink their production policies and ensure food security for individual countries. Interestingly, the self-production of commodities such as fruits and vegetables intensified during COVID-19 (Pulighe and Lupia, 2020).

According to FAO (2014), a total of 820 million people experience chronic hunger worldwide, and the numbers are being aggravated due to the limitations on food supply due to the pandemic. Agroecologists have constantly emphasized that this is due to improper food distribution across social scales and not due to insufficient food production. The decline in trucking has caused a reduction in transportation of bulk fresh food supplies as well as loss of employment of an enormous chunk of society. Surprisingly, there has also been observed tremendous wastage of food due to a decrease in demand from restaurants, hotels, and other similar organizations (Purdy, 2020). Food access has also been affected in regions dependent on migrant farm workers. The opportunity for social distancing is less as they reside in crowded households and use local commute for travel, placing them in a susceptible group. An important experience out of this pandemic would be the necessity of encouraging small producers and urban farmers to get involved in the majority of food production to ensure a consistent supply of fresh food materials at affordable prices. Effective change in public policies by policymakers would eliminate inequalities in the food distribution system in the global market (Altieri and Nicholls, 2020).

The economic repercussions of COVID-19 have been greater than the known global crises. The first sector to be impacted throughout this was the agricultural trade relations. While the governments have desperately tried to immunize the disruptions in food supply by keeping them functioning, the panic buying, restrictions in movement, unequal distribution, and constraints in labor availability have led to the development of friction in supply and demand as the virus spread.

Incidentally, agroecologists anticipate a collective future response to such crises. The first would be the requirement of international cooperation to ensure an improved economy as well as a reduction in dependence on other countries. While the two alternatives are paradoxical to each other, the responses would be situation-appropriate. The purpose of international cooperation is the prompt exchange of medical and food supplies during emergencies.

On the other hand, the COVID-19 crisis has also exemplified the dependence of food security on foreign agricultural supplies, and therefore revision of policies would ensure self-sufficiency and self-production (Kerr, 2020). While a great part of the impact of COVID-19 on agriculture was felt in the distribution system and the associated logistics, food production was also greatly impacted. For example, certain countries witnessed postharvest losses due to a lack of farm labor, while some saw dumping and wastage of products such as milk, vegetables, livestock, and poultry as well. Food production forms the foundation and root of the food system, and problems pertaining to them would cause serious sustainability issues (Pu and Zhong, 2020).

2.3 Control measures

Several precautionary measures were taken worldwide to ensure the containment of the SARS-CoV-2 to restrict COVID-19 from cross-border spreading. This section summarizes the global control measures.

2.3.1 Travel restriction

The travel restrictions imposed post-COVID-19 infection was the first obvious regulation to be implemented worldwide. Governments had imposed restrictions on nonessential travel for 14 days to avoid the transmittance of the virus. This ban has been followed up for both domestic and international travels. Certain countries completely banned the travel of people from regions with a high incidence of infection. Many countries have also issued mandatory quarantine protocols for travelers entering the country, which require them to stay at home or in temporary lodgings (Gostin and Wiley, 2020). The advisories published by various countries, as reported in Al Jazeera, are on similar lines, viz, banning commercial flights except for cargo during peak days of infection (Al Jazeera, 2020). Further, outbound flights were allowed by some countries to allow the movement of people back to their home countries.

These restrictions impacted the mobility of the domestic and farm workforce considerably. Countries that depended on seasonal labor for most of their agricultural produce saw grave effects on their agricultural system. This resulted in a delay in harvesting the produce and food losses, primarily of perishable goods. Parenthetically, the willingness of seasonal labor to work in countries with such restrictions is also ambiguous. Additionally, the constraints also disrupted general logistics and transportation, which disallowed the import of agricultural inputs such as seeds, fertilizers, and pesticides. In essence, the travel restrictions had a cascading effect on the agricultural sector, especially on supply chains. Immediately after countries realized the occurrence of loss, they started developing strategies to mitigate these losses.

The first strategy was to start employing domestic workers instead of seasonal labor. However, this also led to increased training costs. Some countries allowed the movement of seasonal labor with stringent health measures. As mentioned earlier, the willingness of labor to work is another aspect which can be overcome by assuring them with a safe working environment along with medical arrangements and a stricter employee health policy. While the various tactics employed by different countries are still at their preliminary stage, they can serve as a basis for future research in ensuring safe operational environments and smooth functioning of the food supply chain (Bochtis et al., 2020; Devi, 2020).

2.3.2 Border closure

Another mitigation strategy adopted by several countries postpandemic was closing intercountry borders, which were on similar lines to travel restrictions. Many countries banned the movement of people from neighboring countries and across continents, especially those regions with a high incidence of COVID-19 cases. The implications of these restrictions have disrupted the food system web and consequently aggravated the hunger index. In one such incidence, Kazakhstan, one of the major wheat exporters, has closed its boundaries to evade a decrease in staple food supplies. Several countries have taken similar measures. The application of such strategies for prolonged periods will lead to extended damage to the food supply chain (Jámbor et al., 2020).

2.3.3 Social distancing

The drastic implementation of social distancing measures by various countries was executed by shutting down educational organizations, shops, and an overall reduction in activities which would require people to work in close proximity. However, the reduction of labor has led to a negative impact on the economy. It was also estimated that social distancing for 6 weeks is capable of bringing down the gross domestic product (GDP) by an appalling 5.6% (Barrot et al., 2020). As a part of the social distancing regulations, curfews were also implemented until absolutely urgent. The unintended consequences of various forms of social distancing outweighed its purpose, and therefore it is vital that policymakers ensure that the trajectory of food systems is maintained and are immune to such global shocks (Dada and Ogunyiola, 2020).

2.4 Immediate impacts

Most nations are dependent on agriculture as their primary contributor to the economy. Due to the impending cases of COVID-19 infections, a great impact is felt on the agriculture supply chains (ASCs). Reports of FAO (2020a) have implied the effect of COVID-19 on two specific aspects of ASCs, which are supply and demand. Additionally, until a reliable vaccination system is developed, the various governmental bodies have resorted to other means of reducing infection, such as social distancing and travel restrictions. Apart from ASCs, COVID-19 also has shown its effects on the transport system, labor availability, farm systems, and the market. All these factors collectively impair the agricultural system and, therefore, have already caused severe damage to the economy (Sharma et al., 2020).

2.4.1 Food production

The obvious solution to the pandemic was announcing a lockdown which worked impeccably in the short term. However, with the prolonged period taken for the release of the vaccine, several effects were seen on the food production and supply end of agriculture, as illustrated in Fig. 2.1. Stark consequences were also observed in manufacturing operations, which led to effects on the global supply chains. Fresh vegetables and

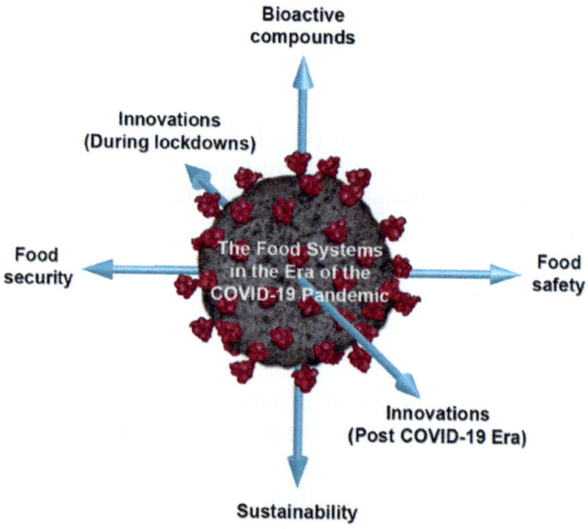

FIGURE 2.1

The food systems in the era of the COVID-19 pandemic and the prospect of implementing innovations during lockdowns and post-COVID-19 era.

Reprinted with permission from Galanakis, C.M., Rizou, M., Aldawoud, T.M.S., Ucak, I., Rowan, N.J., 2021. Innovations and technology disruptions in the food sector within the COVID-19 pandemic and post-lockdown era. Trends Food Sci. Technol. 110, 193–200. Copyright © 2021 Elsevier Ltd.

fruits, food grains, and similar perishable items have witnessed inevitable scarcity during the pandemic. In usual circumstances, the food deficit observed in regions is often overcome by supply from regions producing surplus food material (Singh et al., 2021). Further, the lack of farm labor and labor, in general, has hampered the harvesting, processing, and distribution systems (Henry, 2020).

To overcome these barriers and setbacks created during the pandemic, policymakers and economists have devised certain strategies, as depicted in Fig. 2.2

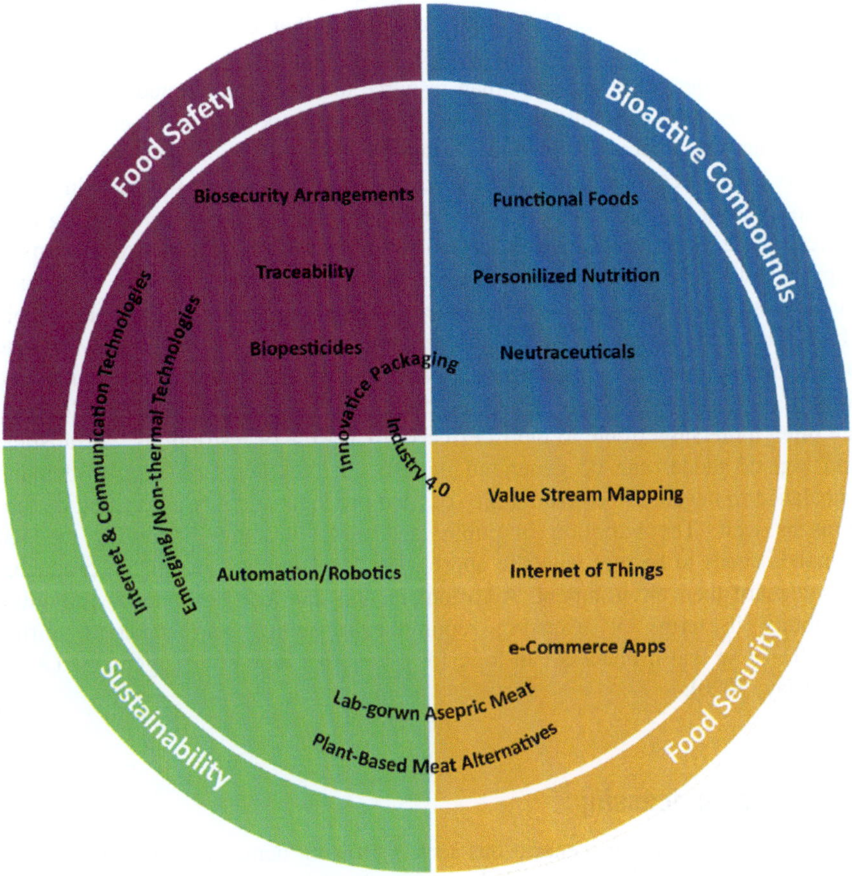

FIGURE 2.2

Foreseen innovations and disruptive technologies to tackle challenges of the four directions affected by the COVID-19 pandemic: food safety, bioactive compounds, food security, and sustainability.

Reprinted with permission from Galanakis, C.M., Rizou, M., Aldawoud, T.M.S., Ucak, I., Rowan, N.J., 2021. Innovations and technology disruptions in the food sector within the COVID-19 pandemic and post-lockdown era. Trends Food Sci. Technol. 110, 193–200. Copyright © 2021 Elsevier Ltd.

(Galanakis et al., 2021). The first and foremost strategy suggested is the development of self-sufficiency of countries in generating enough products to sustain themselves. This would essentially minimize the dependency on other countries. Additionally, it is understood that restrictions on labor have had a huge impact on food production. Therefore it is not a long-term solution if we want to sustain the agricultural economy. Hence, ensuring the basic protocols for hygiene requirements in the food production business, such as handwashing and applying masks and gloves, would suffice to curb the infection. Taking into review the industrial revolution, the automation of industries had incremental changes on production by pacing up production. This aspect can be further explored for minimizing human contact and, eventually, the state of infection.

As the challenges based on human resource management erupted significantly, organizations should partake in adopting modified work policies to ensure a safe environment and uphold a balance between food production and the safety of workers. The decentralization of production facilities, which would bring such facilities closer to the consumer, can also be developed to mitigate the long-term effects of the pandemic. Decentralization would also impart flexibility in terms of supply and demand and would provide fresh food articles to the consumer. Further, such facilities also minimize the cost of transport and shorten the supply chain. The performance of ASCs is dependent immensely on the current demand of the market. For example, during the pandemic the demand was mostly inclined toward sanitization products. However, food articles, especially perishable ones, are often more susceptible to the ill effects of COVID-19 restrictions. A statistical and logistic model can be generated initially for tackling such supply–demand imbalance to negate the unavailability of food products and any associated disruptions as well. The variation in public and private sectors pertaining to food demands needs to be met with the appropriate legal framework for the producers to safeguard their environment. Additionally, this can also be done by providing flexibility in terms of licensing requirements, selling and transport during COVID-19 so that producers can look for alternative market arrangements (Aday and Aday, 2020).

2.4.2 Food processing

The new-age agriculture scenario has seen a growing demand for processed food while that for staple food items such as grains and cereals has remained more or less constant. Over the years, even a dominantly agrarian country like India saw an unprecedented boom in large-scale integrated food processing units to deal with the increasing food demands and globalization (Thulasiraman et al., 2021). The impact of COVID-19 on the food processing sector has been quite extensive and significant as well. Several food processing facilities have been closed due to growing infection of the workers. As a consequence of this, large amounts of food have also been wasted due to the dumping of their produce.

Consequently, the availability of processed food also saw a decline despite the growing demand (Henry, 2020). With the arrival of the pandemic, consumer behavior has also changed drastically and thus, regulatory reforms have been made in food packaging by shifting product processing for retail instead of bulk processing for the service industry (Goddard, 2020). Numerous downstream processing units rely on small employment and, therefore, also suffer on a larger scale due to labor shortage. For example, perishable food items processing account for such operations; thus, disruptions of any sort affect the supply chain on a large scale. Such organizations, therefore, need to keep mitigation strategies in such crises (Hobbs, 2020). Additionally, nonperishable food items, for example, freeze-dried ready meals, saw an increase in consumers due to panic buying and their long shelf-life (Galanakis et al., 2021).

It is also crucial to understand that infection of manufacturers poses the issue of infection to a wide range of people as they are connected directly to producers, distributors, and sometimes consumers (Aday and Aday, 2020). The identified factors for COVID-19 risk in processing facilities are physical distance, hygiene, and effective social distancing protocols, as depicted in Fig. 2.3 (Nakat and Bou-Mitri 2021).

The advent of the pandemic also shed light on the drawbacks in the processing and production supply chain, calling for a more balanced approach to food production. The food processing industry, though large and effective, also brought

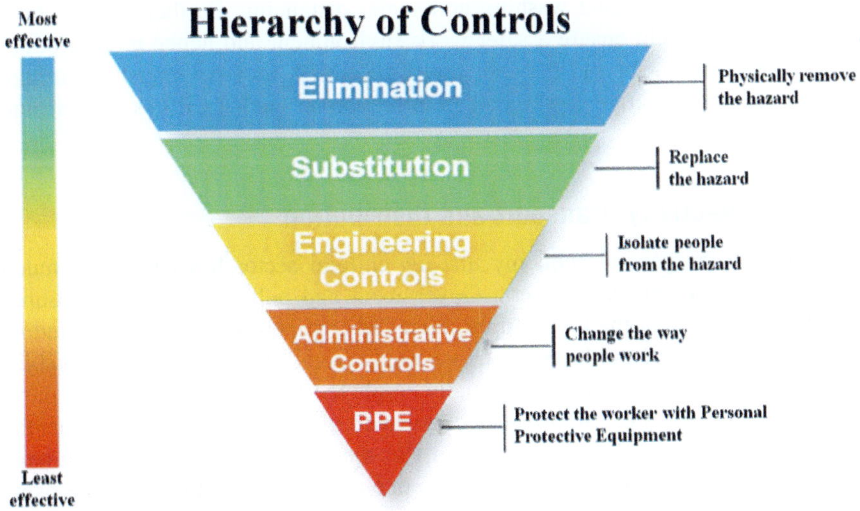

FIGURE 2.3

Hierarchy of controls as presented by the National Institute for Occupational Safety and Health of the United States (NIOSH).

Reprinted with permission from Nakat, Z., Bou-Mitri, C., 2021. COVID-19 and the food industry: readiness assessment. Food Control 121, 107661. Copyright © 2020 Elsevier Ltd. (Permission required).

along issues such as high energy consumption and greenhouse gas emission, authenticity concerns, and the development of food insecurity during pandemics. Therefore the novel virus enforces the need for a more sustainable approach (Thulasiraman et al., 2021).

To overcome such drawbacks, a few solutions can be taken up, such as a balance between supply chains and integrated food processing facilities. Reviving traditional food processing units and adopting a short food supply chain mechanism can bring in new possibilities for a more advanced process-supply system. This has also been described through the introduction of disruptive technologies. Disruptive technologies are essentially novel technologies which create a new market with enhanced value and cause disruptions in the existing market. For a post-COVID-19 scenario, disruptive technologies for better reach and connectivity are considered to be most effective. Disruptive innovations in the manufacturing and processing sector will generate newer opportunities as well as cater to the existing diversity in the market, as shown in Fig. 2.4. The development of fortified food instead of processed and packaged food, encouraging local food supply chains and the introduction of traditional technologies would eventually generate localized opportunities and lead to new start-ups, which would change the ASCs scenario for a more sustainable future (Galanakis et al., 2021; Thulasiraman et al., 2021).

The immediate response to the pandemic for such facilities was implemented with the introduction of current good manufacturing practices (cGMP). This requires a more frequent sanitization protocol in the food industry with the employment of basic hand washing and social distancing. Further, regular testing of workers would ensure a safer work environment and hence more productivity. These strategies and the fact that COVID-19 does not transmit through food deliver a collective lower risk of infection in such facilities (Nakat and Bou-Mitri, 2021).

2.4.3 Connectivity: transport and distribution

Compared to the general economy, the agricultural sector has performed much better as it is the primary concern. Government and policymakers have ensured smooth functioning of the food supply chain and continuous engagement of industry. Most agricultural products are transported via rail, trucks, and water bodies. Special refrigerated containers are usually employed for perishable items or moving long distances. For short-distance transportation of goods, trucking is the majorly preferred mode of commute. Most of these modes of transportation have been heavily affected during the pandemic due to travel restrictions and the overall closure of borders. Further, the pandemic also saw an exponential increase in door-to-door delivery services with consumer behavior changes. Large metric tonnes of food items such as grains and cereals are transported through railways. Nowadays, most rail movements have become less labor-intensive and more automated. Therefore the impact of social distancing protocols was not felt as intensely as it did in other sectors. This also holds true for ocean freight, where

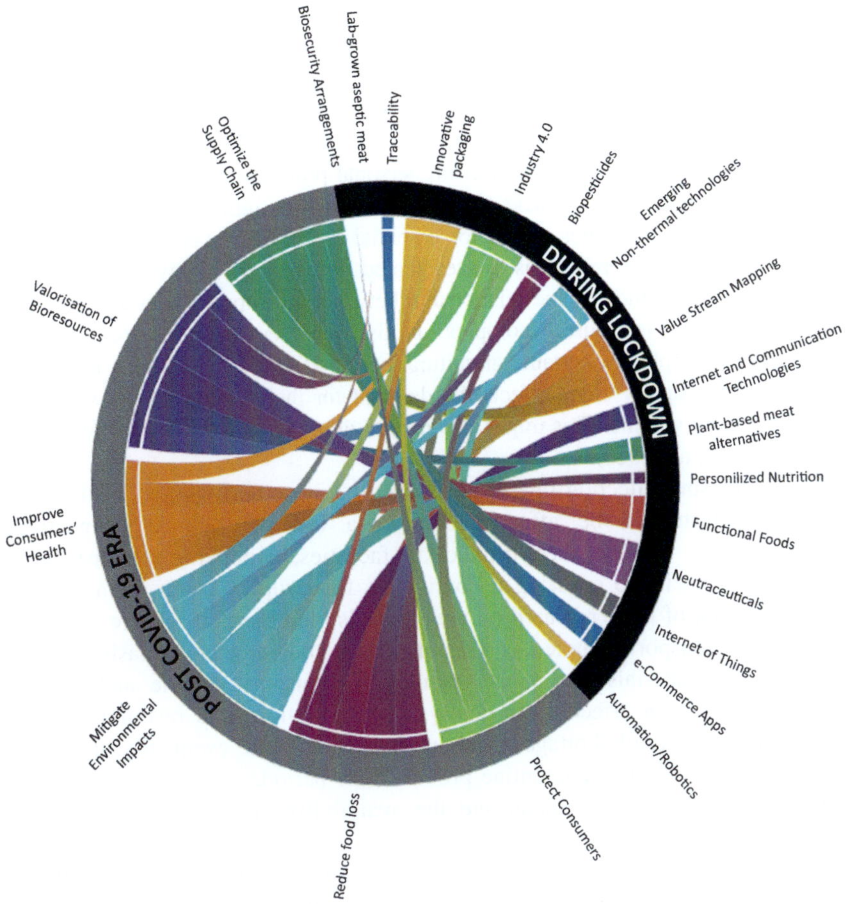

FIGURE 2.4

The prospect of implementing the foreseen innovations in the food sector within the era of the COVID-19 pandemic and respective applications during the lockdown and post-COVID-19 periods.

Reprinted with permission from Galanakis, C.M., Rizou, M., Aldawoud, T.M.S., Ucak, I., Rowan, N.J., 2021. Innovations and technology disruptions in the food sector within the COVID-19 pandemic and post-lockdown era. Trends Food Sci. Technol. 110, 193–200. Copyright © 2021 Elsevier Ltd.

smaller crews are present on the ships. Trucking as a means of movement of vital food commodities is practiced across various regions. During the pandemic, the transportation of such services was modified and regulated in favor of delivery of farm inputs, produce, and processed goods. However, in spite of this, the drivers who travel from distribution centers to outlets were the most vulnerable. Therefore social distancing protocols were executed to ensure minimum human-to-human contact (Gray, 2020).

An emergency relief response was developed at the initial stages of the pandemic directing transportation carrying essential food commodities to be exempted from travel restrictions. Transportation of agricultural goods as well as livestock was included in this declaration. These laws also emphasized adequate testing of drivers as well as complying with general traffic rules. It was also observed that several institutions provided a safe work environment, personal protective equipment, and regular cleaning and sanitization of trucks and other commutes right from the onset of the pandemic. Borders were not closed for essential items. However, during the initial stages of the pandemic, rest areas for designated drivers were highly restricted, making the travel unconducive. The relationship between the supply chain and the transportation sector is immensely complex. It also includes the influence on freight volume. For example, for certain agricultural commodities, the disruption in multiple junctions of the supply chain affects the demand for their transportation, which can affect the freight volume. This might also lead to dumping and wastage of food materials (Walters et al., 2020).

The increasing infection rate caused several drivers and workers in the transportation industry to be affected due to a lack of vehicle connectivity. Maintaining the link between production facilities, suppliers, and distribution facilities are crucial to ensuring a continuous food supply chain (Kumar et al., 2020). The lack of prior research on the virus also led to retail logistic imbalance pertaining to transport volume. As a result of panic buying and increasing demand for dry food materials during the pandemic, transport volume and capacity dynamics have been affected drastically. The restrictions and restraints in trade demand and an overall shortage of labor affect ASCs and, eventually, the freight volume. Along with this, quarantine protocols for periods as long as 14 days also impact the fulfilment of demand and the availability of transportation facilities (Loske, 2020).

While travel restrictions have exhibited a primary significance in minimizing the effects of COVID-19, it has affected the agricultural sector in various forms. Regulatory reforms to enhance the smooth functioning and movement of transportation services should be carried out to minimize the effect on the economy. Several proposed models aim to restore connectivity and a steady food supply. Fig. 2.5 shows the synchronization of trucks and drones to facilitate last-mile delivery in the highly infected regions known as hot zones (Singh et al., 2021). In this model, Drone-A (DA) and Drone-B (DB) are committed to delivering food items and medicines to customers C1, C2, C3, C4, C5, C6, and C7 on a scheduled path in a synchronized manner within a specific time.

2.4.4 Labor availability

As a result of the pandemic, safety regulations have been set up for workers to ensure the continuity of food production (Aday and Aday, 2020). These workers are involved in every step of the food supply chain, from production to distribution. The pandemic had a tremendous effect on workers of all sectors connected

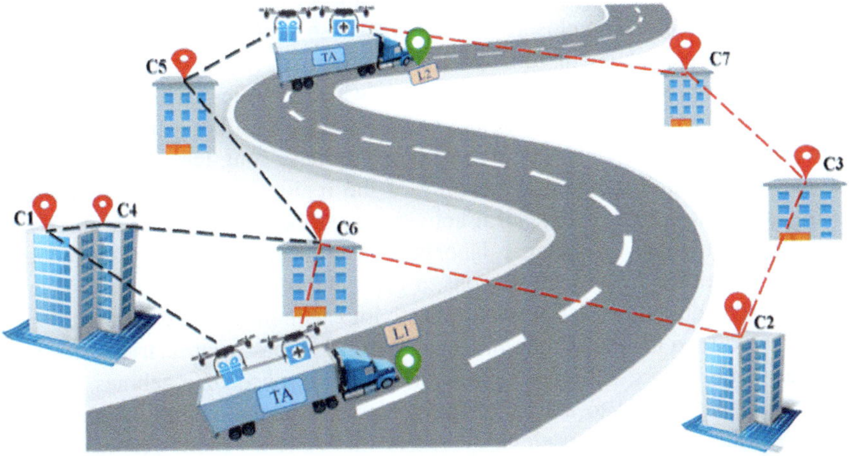

FIGURE 2.5

Coordination of drones and delivery trucks for last-mile delivery.

Reprinted with permission from Singh, S., Kumar, R., Panchal, R., Tiwari, M.K., 2020. Impact of COVID-19 on logistics systems and disruptions in food supply chain. Int. J. Prod. Res. 59 (7), 1993–2008. Copyright © 2020 Informa UK Limited, trading as Taylor & Francis Group.

to the food industry. According to the norms of basic human rights, every individual has the right to work in a just and favorable work environment and to protect against employment. Keeping this in context, the pandemic has questioned the very foundation of human rights as several workers were obligated to work in spite of the pandemic in unsafe work conditions. Also, the employees ignored necessary guidelines and response action to minimize the risk of COVID-19. Due to this, a large number of such workers have faced potential health threats as well as deaths. While workers in the medical sector have been given an ample amount of support, workers in agriculture have not received the same level of protection and safety standards (Ramos et al., 2020). The restrictions imposed on work environments with essential staff having to mandatorily work during the pandemic with government-advised regulation have a grave impact on the agricultural sector. With globalization, flexibility in labor activities also ensued with demand for skilled labor. This led to a large amount of labor immigration, with productivity being a key factor in ensuring the maximum benefit in the agricultural sector. Farm types for crop operations, such as harvesting, depend greatly on migrant labor and therefore faced huge setbacks during the pandemic (Cortignani et al., 2020). Other notable farms which depend on migrant labor for productivity are fruits, vegetables, and horticulture. The labor working in these sectors works in a gray and black economy and is the worst affected (Phillipson et al., 2020).

The restriction on movement affected labor availability, especially for the agriculture sector, as it is highly labor-intensive and depends largely on temporary labor. The shortage of labor observed had an impact on productivity and eventually on the economy. The closure of the border also prolonged the harvest season for various crops, which in turn led to compromising the quality of harvested crops. The effect was also extended to sectors which supplies farm essentials such as fertilizers and pesticides. Workers who derive their livelihood from daily wage were impacted severely and would most probably have long-term implications on their income as well as on the economy of the regions (Ayanlade and Radeny, 2020). In recent times, several farm activities have been automatized, and therefore the requirement of manual labor for such operations has also reduced. However, in certain developing nations, a great chunk of the agricultural population still relies on manual labor for their activities without farm equipment.

Further, due to lockdown, the interstate movement has been restricted for farm equipment. On an extended note, even with the presence of machines, shutting down of shops has led to a lack of mechanics and unavailability of spare parts rendering the machines useless. Workers in various food processing and manufacturing industries have been reduced to half, which has, in turn, affected the overall production. Transient workers have traveled home due to the situation, implying that various agriculture sectors have been impacted (Ray et al., 2020). In a nutshell, the reverse migration coupled with restrictions in transportation and travel has consequences on the overall agricultural economy.

Considering the long-term impact these restrictions might have on the economy, it is highly necessary to develop strategies to minimize loss. The primary step to be taken is to ensure a safe work environment. In the future as well, the workers related to the agricultural sector will continue to remain vulnerable to crisis and therefore enforcing health safety and well-being measures for the workforce on and off the farm to prevent susceptibility (Ramos et al., 2020). Agriculture product collection centers should be built at regular intervals and more frequently to enhance their accessibility to small-scale farmers. This would also reduce the need for mobility. These centers should also be a built-in high capacity to store larger quantities of food products to meet growing demand during a crisis. It is also imperative for the governments of respective countries to have incentives for small-scale farmers to deal with financial setbacks and continue production. Another crucial measure is encouraging local workers to take up agriculture as a profession and participate in farm activities to minimize the requirement for migrant labor. The increase in employment opportunities would also encourage workers to work in their native regions and not be forced to migrate for job opportunities. The pandemic has shown labor-replacing mechanization strategy as a long-term solution for coping with labor shortage (Aday and Aday, 2020).

Restructuring the work environment is essential so that workers can maintain safe distances. This can be continued for the long term as well. Additionally, considering the exponentially growing infectious cases and the already impacted economy, on-site screening services to identify active symptoms for farm workers should be built into a regular routine. For the longest time, the disparity has existed between the worker and other working classes. It is essential to develop an environment of equity and trust, as well as provide appropriate compensations to safeguard farm workers (Ramos et al., 2020). Appropriate recovery policies are extremely crucial at this point in time as it is predicted that several farm workers may switch crops, and some young farmers may also switch professions completely (Ayanlade and Radeny, 2020). Spreading awareness regarding the different regulations and also providing personal protective equipment to farmers and workers to prepare them for current crisis as well as impending crises is key (Ray et al., 2020). Fortunately, several governments had already introduced relief and emergency schemes to support the income of workers who faced loss during the pandemic (Solomon et al., 2020).

2.4.5 Impairment of farm systems

The effect of COVID-19 on the overall farm system was felt due to the unavailability of labor and farm inputs such as fertilizers and pesticides. Farm systems form a major chunk of the food supply chain. The shock of the pandemic has been felt in various structures of the farm systems, right from production to harvesting. Financial instability is one of the immediate consequences of the pandemic and would affect several farmers' families. Many farmers and farm workers have loans and interest payments which have been further burdened due to the loss in productivity (Deaton and Deaton, 2020).

Approximately 4 months into the pandemic, the UN Food and Agricultural Organization announced that the stocks for staple grain items such as wheat and rice would suffice to feed the people during the pandemic; depending on the length of the pandemic (FAO, 2020d). However, this did not hold true for more perishable food items. Such food items are dependent on storage availability as well as transportation and distribution. The restriction in movement also contributed to this limitation as less supply would mean more wastage of the perishable food items (Harris et al., 2020).

An astounding 25% of the global population is directly associated with farming (Timilsina et al., 2020). The impact on farm systems would be variable depending on the criteria for essential products, and the contracts for the products are subjective to moderate or long-term delivery (Phillipson et al., 2020). Most of the farm systems in numerous regions are family farms which rely on rural localities for labor and capital. Family farming accounts for 80% of the world's food production and occupies approximately 75% of agricultural land. Therefore they have a pivotal role to play in the global food supply. The closing down of local markets and the disruption of supply during COVID-19 have hampered family farming to a great extent. The necessity of developing family farms lies in uplifting rural agricultural standards and the overall poverty along with urban population food supply. Moreover, the numerical evidence implies its role in strategic food production, and therefore government and policymakers should take initiatives to invest in these farm systems (Cavalli et al., 2020). The import of fertilizers, agrochemicals, seeds, and farm machinery, which are essential for the functioning of farm systems, has been affected significantly due to movement restrictions. Further other logistical issues such as international and domestic shipping have led to the termination of dealerships. While farm systems nowadays have resorted to machine-based automatized farming, the disruptions in movement and transportation have likewise slowed the farm operations due to the inability to transport agriculture machinery.

The exact impact of COVID-19 on-farm systems is a complex network and can affect various factors directly or indirectly. For example, a shortage of farm labor will cause delays in farming operations which can further decline the farm productivity, eventually causing a reduction in income from farm systems. This can additionally cause liquidity and capital shortages along the food value chain. In another exemplary sequence of events, the restrictions imposed on transportation can cause import disruptions of farm materials and raw materials, which will lead to volatility in input prices. Fluctuations in price will correlatively increase the production estimates and eventually cause a decline in incomes from farm products. As evident, all the circumstances related to variations in farm systems due to COVID-19 regulations and restrictions would ultimately lead to capital constraints along the food value chain (Amjath-Babu et al., 2020). Family farms are, in general, more resilient to the crisis as compared to larger farms which greatly depend on external labor for their activities (Stephens et al., 2020). Farm resilience requires a structuralist approach wherein it is dependent on available resources, workload, and an overall societal structure. Considering the unusual dynamics farmers and farm workers have to face during a particular crisis, optimization of farm protocols to increase efficiency and minimizing production cost is the need of the hour. Radical changes implemented by policymakers which would create harmony and mutualism between the farmer and the agro-ecosystem will eventually also help build resilience (Darnhofer, 2020).

The pandemic also brought in the need for location-specific and nutrition-sensitive agriculture, which was termed farm-system-for-nutrition (FSN). This approach focuses on a more local-based resource procurement strategy and combines arable farming, horticulture, and animal husbandry for a more sustainable method of farming while conserving resources. FSN encourages the small-holder farmer to play a more proactive role in the ASCs and grow nutritionally enhanced food for a more resilient farm system against crisis like the COVID-19 pandemic (Bhavani and Gopinath, 2020). Automation and digitization of agriculture is a much-needed paradigm shift. Overall development of resilience would help to overcome the multidimensional impact felt on-farm systems by the current global pandemic.

2.4.6 Market and retailer

Market and retail are part of the end sections of the ASCs and are part of the demand–supply chain. During the pandemic, the demand for food produced saw immense variation, with the government regulations restricting movement and transportation. This caused an immediate effect on the import and export of food products. While a demand shock was felt globally with the reduction in food production, many countries limited the export of essential food materials to other countries to ensure ample reserve. A labor market issue was also felt globally with the restrictions implemented on travel and movement. Along with this, the horticulture market, which produces a substantial fresh food supply, also saw a decline. Major ports worldwide were congested with containers which were not able to be shipped due to the ongoing restrictions due to the pandemic and therefore had to be redirected to minor ports. This led to considerable monetary loss for the provider (Sharma et al., 2020).

While the global market had experienced shock earlier, the COVID-19 pandemic mediated crash in the market associated with a health crisis was witnessed for the first time. The negative shock in market returns and capital was linked directly to the confirmed number of infected cases of COVID-19 (Smales, 2021). In spite of the government categorizing food under essential commodity and hence relaxing the movement restrictions felt otherwise, the global food market was impacted significantly. Producers and manufacturers were compelled to direct the supplies from food service outlets to retail channels. The impact of COVID-19 was manifested across the supply spectrum of food value chains. All of these factors cumulatively affected the quantity of goods arriving at retail and food outlets as well as the prices of imports and exports. In the Indian market, the market is majorly dependent on cash transactions, and the disruptions due to COVID-19 extended throughout the value chain, including traders and retailers (Varshney et al., 2020).

Distance between the producers and retailers was found to be a major criterion for the availability of farm produce in the market. This can be owed to the restrictions on travel. On the demand side, panic-buying by the consumers caused a situation of stockout. Further, in both online and wholesale markets greater fall in products was observed due to interruptions in the supply chain. Interestingly, the availability of nonperishable items remained persistent throughout the lockdown, with the market increasing postlockdown. However, the availability of perishable items continued to remain low. This was due to the disruptions occurring in the supply chains of perishable items due to their limited shelf-life. It is evident that perishable goods supply chains are most susceptible to the consequences of the lockdown. While food unavailability is caused in the higher income category, it causes income loss for the farmers and working-class (Mahajan and Tomar, 2020). The economic recession which was generated due to COVID-19 has caused a rise in the price of high-end food commodities. Food commodities are usually rigid, and a rise in price has long-term impacts on the economy. Despite the disruptions in the food value chain and the consequential increase in food demand, and glooming food insecurity, global food consumption has remained unaffected owing to the inelastic demand for food commodities (Elleby et al., 2020).

Keeping these factors under consideration, agro-economists have suggested strategies to ensure that the global agricultural retail and market remains consistent. The first step would be to vigilantly keep track of the demand and market. It has been observed that maintaining the transparency of information would strengthen management and also discourage people from panic-buying along with ensuring profit for the producers. For this purpose, global agricultural organizations such as the Consultative Group on International Agricultural Research (CGIAR) and Food and Agricultural Organization (FAO) should closely monitor the fluctuations in the global as well as local market scenarios pertaining to supply and demand. Further, the demand for online deliveries of food and grocery items has opened an entirely new retail market. This also allows minimum human-to-human contact, thus mitigating the spread of infection. Keeping food trade primary is essential in times like these, and policies should be adopted with changing times to minimize negative effects and not cause further damage to the global agricultural economy (Fan et al., 2020).

2.4.7 Consumer

Pandemics have a direct effect on the primary reactions of the population due to the uncertainty of food security. Consumers, in general, avoided shopping at supermarkets and grocery stores which would generally remain crowded (Butu et al., 2020). Additionally, people have also reduced their consumption of food from restaurants and other food outlets. Online food delivery has seen a rise throughout the pandemic, with an increasing number of people opting for online delivery systems for grocery and other essential food items as well. The constraint related to this would be the feasibility of an online food delivery model for the long term compared to the current short-term scenario. Online delivery systems also require investments in the form of an established online web portal and sufficient delivery capacity (Hobbs, 2020). The food industry is shifting its product runs to focus on more basic food forms as well as baking ingredients, which implies the growing self-sufficiency of the general masses. Another form of the consumer which was observed was the stocking up of nonperishable and packaged food items. However, the con observed was the tremendous waste generated by perishable food items or food items containing an expiry date. Altered work timings and the new work-from-home culture have also given rise to modified food behavior. In addition to all this, experts predict that the disruptions in the agriculture market and food production would lead to higher food costs as well as price volatility in the long run. The duration and gravity of the pandemic would also define whether these adopted changes in online shopping and stocking up would persist or not (Cranfield, 2020).

The above-stated behavior is most likely on the short-term side, while long-term impacts are the decline in food demand as a result of a fall in consumer income. This may, in turn, influence the retailer behavior with small-food businesses facing challenges as compared to large-scale food businesses. Consumers have also started to opt for local markets and food supply chains. While consumers predominantly opt for local food outlets for groceries, localized food supply chains are often less cost-efficient than mainstream supply chains and also offer limited variety. Thus there is the possibility of consumers opting back to mainstream systems. The localized food system is fast adjusting to the demands and requirements of the consumers. However, mainstream systems also pose a competitive advantage and are getting accustomed to the initial demand shock. Therefore the near future and continuity of the pandemic would be able to define and shape consumer behavior (Hobbs, 2020).

It has been predicted that consumers would shift their preferences for functional food containing bioactive ingredients to build up immunity which would subsequently require the food industry to shift its focus to the extraction of nutraceutical compounds and production of more sustainable food (Nakat and Bou-Mitri, 2021). In the search for more nutrition-based food, consumers have looked for options meeting both nutrition and budget constraints. Therefore consumers are purchasing fruits, vegetables, legumes, and whole grains, which would suffice for the nutritional demands of the body, and have refrained from consuming unhealthy products. Food security is directly connected with consumer behavior, and it has been seen that consumers prioritize food safety and availability. In the times of COVID-19, food security is defined by the consumers' accessibility to food rather than the overall availability.

Consumers play a key role in the food supply chain, and hence the pandemic caused a disruption in the food value chain due to fluctuations in demand and panic buying. Consumer behavior has been observed in two major forms where one category was greatly concerned about food security and reduced food wastage, and the other category saw the dumping of perishable food items (Aday and Aday, 2020). As consumer behavior has modified and improvised, businesses have also modified. The mutualistic behavior of consumers and suppliers would ensure continuity in the food value chain. Companies have started to make their policies more flexible, and their infrastructure has been modified for a more resilient set-up to ensure productivity even in times of crisis. Additionally, the supply chains and logistics need to be adjusted taking into account the volatility of consumer demands. Public policy should be regulated to corroborate with consumer demands as they form the end yet, one of the essential factors of the food value chain (Sheth, 2020).

2.5 Conclusions and future perspectives

Since the COVID-19 cases are rising exponentially, the primary focus for the government and policymakers has more or less shifted toward the medical sector. However, due to this, there is a setback observed in various other economic sectors, including the agriculture sector. The GDP of several developed as well as developing nations is primarily dependent on agriculture. The pandemic has caused severe damage and vulnerabilities in the agricultural supply chain by limiting the workforce and availability of farm produce. The availability of food, especially in urban households, became a growing issue for administrators and policymakers while the resilience was continuously tested due to generation of unemployment. Several strategic measures have been suggested and also implemented in numerous regions by the government. To curb the growing demand for nutritionally rich foods, it is suggested to go local and procure food items from local markets and suppliers. Also, the government encourages the practice of home-grown fruits and vegetables for urban households for a more sustainable approach. This can also be correlated to consumer behavior changes where the pandemic triggered panic-buying and hoarding of packaged and nonperishable food items among the masses. A more localized purchase would ensure availability and prevent the loss of local suppliers and manufacturers. The restrictions on movement and travel had deleterious effects on the availability of food and made it difficult to maintain a balance between supply and demand. This can be overcome by relaxing the transportation of agricultural produce as well as other resources such as seeds, fertilizers, and pesticides across borders. Several governments have made reforms regarding the categorization of food products and also include farm machinery under essential commodities. This has smoothened out the functioning of the supply chain as appropriate resources are now reaching the producers. It is also extremely vital that the government takes initiatives in releasing policies to compensate for the loss faced by millions of farm workers during the pandemic due to lack of transport and a safe work environment. Also, due to the shutting down of several food industries as well as shops, workers had to migrate back to their native regions, which posed additional problems.

Many farm workers were also not able to rejoin work due to COVID restrictions. An unconducive work environment had also led to several workers being exposed to health threats which were a direct noncompliance of basic human rights. The progressing paradigm shift in agriculture requires the government to develop more diversified food supply systems which are localized. The pandemic has also revealed the need for ample stocking of food in case of crises like COVID-19. Decentralization of agriculture for a more resilient farm system is the utmost need of the hour. Reforms in the form of structure, automation, and digitization of agriculture are among the few key changes required. The pandemic has shed light on some of the more prominent drawbacks of the current global agricultural scenario, and therefore, a new system for maximum benefit across all strata is long overdue.

Acknowledgment

Dr. Sougata Ghosh acknowledges Kasetsart University, Bangkok, Thailand for a Post Doctoral Fellowship and funding under the Reinventing University Program (Ref. No. 6501.0207/9219 dated 14th September, 2022). Dr. Ghosh is also thankful to The Program Management Unit for Human Resources & Institutional Development and Innovation

(PMU-B) for funding the proposal entitled "Developing High Performance" workforce in Postdoc and Postmaster under Research Topic "Study and Development of Porous Nanofibers, Graphene Oxide Quantum Dot Nanofiber Composite, and Metal Doped Graphene Oxide Quantum Dot Nanofiber Composite for Carbon Dioxide Reduction and Capture" under the Program of National Postdoctoral and Postgraduate System approved by PMU-B Board Committees (Contract No. B13F660065).

References

Aday, S., Aday, M.S., 2020. Impact of COVID-19 on the food supply chain. Food Qual. Saf. 4 (4), 167–180.

Al Jazeera, 2020. Coronavirus: travel restrictions, border shutdowns by country | Coronavirus pandemic news. Available online: https://www.aljazeera.com/news/2020/03/coronavirus-travel-restrictions-border-shutdowns-country-200318091505922.html.

Altieri, M.A., Nicholls, C.I., 2020. Agroecology and the reconstruction of a post-COVID-19 agriculture. J. Peasant Stud. 47 (5), 881–898.

Amjath-Babu, T.S., Krupnik, T.J., Thilsted, S.H., McDonald, A.J., 2020. Key indicators for monitoring food system disruptions caused by the COVID-19 pandemic: insights from Bangladesh towards effective response. Food Secur. 12 (4), 761–768.

Ayanlade, A., Radeny, M., 2020. COVID-19 and food security in Sub-Saharan Africa: implications of lockdown during agricultural planting seasons. NPJ Sci. Food 4, 13.

Barrot, J.-N., Grassi, B., Sauvagnat, J., 2020. Sectoral effects of social distancing. (April 2, 2020). HEC Paris Research Paper No. FIN-2020-1371, Available at SSRN: https://ssrn.com/abstract = 3569446 or http://doi.org/10.2139/ssrn.3569446.

Bhavani, R.V., Gopinath, R., 2020. The COVID19 pandemic crisis and the relevance of a farm-system-for-nutrition approach. Food Secur. 12 (4), 881–884.

Bochtis, D., Benos, L., Lampridi, M., Marinoudi, V., Pearson, S., Sørensen, C.G., 2020. Agricultural workforce crisis in light of the COVID-19 pandemic. Sustainability 12 (19), 8212.

Butu, A., Brumă, I.S., Tanasă, L., Rodino, S., DinuVasiliu, C., Doboş, S., et al., 2020. The impact of COVID-19 crisis upon the consumer buying behavior of fresh vegetables directly from local producers. Case study: the quarantined area of Suceava County, Romania. Int. J. Environ. Res. Public Health 17 (15), 5485.

Cavalli, S.B., Soares, P., Martinelli, S.S., Schneider, S., 2020. Family farming in times of Covid-19. Rev. Nutr. 33, 1–8.

Cortignani, R., Carulli, G., Dono, G., 2020. COVID-19 and labour in agriculture: economic and productive impacts in an agricultural area of the Mediterranean. Ital. J. Agron. 15 (1653), 172–181.

Cranfield, J.A.L., 2020. Framing consumer food demand responses in a viral pandemic. Can. J. Agric. Econ. 68 (2), 151–156.

Dada, O., Ogunyiola, A., 2020. A preliminary impact assessment of social distancing on food systems and greenhouse gas emissions. J. Hum. Behav. Soc. Env. 31 (1–4), 254–261.

Darnhofer, I., 2020. Farm resilience in the face of the unexpected: lessons from the COVID-19 pandemic. Agric. Hum. Values 37, 605–606.

Deaton, B.J., Deaton, B.J., 2020. Food security and Canada's agricultural system challenged by COVID-19. Can. J. Agric. Econ. 68, 143–149.

Devi, S., 2020. Travel restrictions hampering COVID-19 response. Lancet 395 (10233), 1331–1332.

Elleby, C., Domínguez, I.P., Adenauer, M., Genovese, G., 2020. Impacts of the COVID-19 pandemic on the global agricultural markets. Environ. Resour. Econ. 76, 1067–1079.

Fan, S., Si, W., Zhang, Y., 2020. How to prevent a global food and nutrition security crisis under COVID-19? China Agric. Econ. Rev. 12 (3), 471–480.

FAO, 2014. Biodiversity and ecosystem services in agricultural production systems. Agroecology for Food Security and Nutrition. Available in: http://www.fao.org/publications/card/en/c/d1f541b5-39b8-4992-b764-7bdfffb5c63f/.

FAO – Food and Agriculture Organization, 2020a. Q&A:COVID-19 pandemic – impact on food and agriculture. Available in: http://www.fao.org/2019-ncov/q-and-a/en/.

FAO – Food and Agriculture Organization, 2020b. FAO alertasobre el impacto del COVID19 en la alimentación escolar de América Latina y el Caribe. Available in: http://www.fao.org/americas/noticias/ver/es/c/1267028/.

FAO, 2020d. Local food systems and COVID-19; a glimpse on India's responses. Available from: http://www.fao.org/in-action/food-forcities-programme/news/detail/en/c/1272232/.

FAO, 2020c. FAO Director-General urges G20 to ensurethat food value chains are not disrupted during COVID-19 pandemic. Available in: http://www.fao.org/news/story/en/item/1268254/icode/.

Galanakis, C.M., Rizou, M., Aldawoud, T.M.S., Ucak, I., Rowan, N.J., 2021. Innovations and technology disruptions in the food sector within the COVID-19 pandemic and post-lockdown era. Trends Food Sci. Technol. 110, 193–200.

Ghosh, S., Mostafavi, E., Vernet-Crua, A., Webster, T.J., 2020. Nanomaterial approaches for the prevention, diagnosis and treatment of COVID-19: a paradigm shift. J. Nanomater. Mol. Nanotechnol. 9, 4.

Goddard, E., 2020. The impact of COVID-19 on food retail and food service in Canada: preliminary assessment. Can. J. Agric. Econ. 1–5.

Gostin, L.O., Wiley, L.F., 2020. Governmental public health powers during the COVID-19 pandemic: stay-at-home orders, business closures, and travel restrictions. J. Am. Med. Assoc. 323 (21), 2137–2138.

Gray, R.S., 2020. Agriculture, transportation, and the COVID-19 crisis. Can. J. Agric. Econ. 68 (2), 239–243.

Harris, J., Depenbusch, L., Pal, A.A., Nair, R.M., Ramasamy, S., 2020. Food system disruption: initial livelihood and dietary effects of COVID-19 on vegetable producers in India. Food Secur. 12, 841–851.

Henry, R., 2020. Innovations in agriculture and food supply in response to the COVID-19 pandemic. Mol. Plant. 13 (8), 1095–1097.

Hobbs, J.E., 2020. Food supply chains during the COVID-19 pandemic. Can. J. Agric. Econ. 68 (2), 171–176.

Jámbor, A., Czine, P., Balogh, P., 2020. The impact of the coronavirus on agriculture: first evidence based on global newspapers. Sustainability 12 (4535), 1–10.

Kerr, W.A., 2020. The COVID-19 pandemic and agriculture: short- and long-run implications for international trade relations. Can. J. Agric. Econ. 68, 225–229.

Kumar, A., Luthra, S., Mangla, S.K., Kazançoğlu, Y., 2020. COVID-19 impact on sustainable production and operations management. Sustain. Oper. Comput. 1, 1–7.

Loske, D., 2020. The impact of COVID-19 on transport volume and freight capacity dynamics: an empirical analysis in German food retail logistics. Transp. Res. Interdiscip. Perspect. 6, 100165.

Mahajan, K., Tomar, S., 2020. COVID-19 and supply chain disruption: evidence from food markets in India. Am. J. Agr. Econ. 103 (1), 35–52.

Nakat, Z., Bou-Mitri, C., 2021. COVID-19 and the food industry: readiness assessment. Food Control. 121, 107661.

Phillipson, J., Gorton, M., Turner, R., Shucksmith, M., Aitken-McDermott, K., Areal, F., et al., 2020. The COVID-19 pandemic and its implications for rural economies. Sustainability 12 (10), 3973.

Poudel, P.B., Poudel, M.R., Gautam, A., Phuyal, S., Tiwari, C.K., Bashyal, N., et al., 2020. COVID-19 and its global impact on food and agriculture. J. Biol. Today's World 9 (5), 221.

Pu, M., Zhong, Y., 2020. Rising concerns over agricultural production as COVID-19 spreads: lessons from China. Glob. Food Sec 26, 100409.

Pulighe, G., Lupia, F., 2020. Food first: COVID-19 outbreak and cities lockdown a booster for a wider vision on urban agriculture. Sustainability 12, 5012.

Purdy, C., 2020. Covid-19 is about to reach us farms in a major test for food supply chains. Available in: https://qz.com/1829558/covid-19-is-about-to-reach-us-farms/.

Ramos, A.K., Lowe, A.E., Herstein, J.J., Schwedhelm, S., Dineen, K.K., Lowe, J.J., 2020. Invisible no more: the impact of COVID-19 on essential food production workers. J. Agromed. 25 (4), 378–382.

Ray, P., Acharya, K.K., Thakur, A., 2020. COVID-19 impact and implication to agriculture and food security. FSR 1 (7), 5–8.

Sharma, R., Shishodia, A., Kamble, S., Gunasekaran, A., Belhadi, A., 2020. Agriculture supply chain risks and COVID-19: mitigation strategies and implications for the practitioners. Int. J. Logist. Res. Appl. 1–27.

Sheth, J., 2020. Impact of Covid-19 on consumer behavior: will the old habits return or die? J. Bus. Res. 117, 280–283.

Siche, R., 2020. What is the impact of COVID-19 disease on agriculture? Sci. Agropecu. 11 (1), 3–6.

Singh, S., Kumar, R., Panchal, R., Tiwari, M.K., 2021. Impact of COVID-19 on logistics systems and disruptions in food supply chain. Int. J. Prod. Res. 59 (7), 1993–2008.

Smales, L.A., 2021. International review of financial analysis investor attention and global market returns during the COVID-19 crisis. Int. Rev. Financ. Anal. 73, 101616.

Solomon, S., Rao, G.P., Swapna, M., 2020. Impact of COVID-19 on Indian sugar industry. Sugar Tech. 22 (4), 547–551.

Stephens, E.C., Martin, G., van Wijk, M., Timsina, J., Snow, V., 2020. Impacts of COVID-19 on agricultural and food systems worldwide and on progress to the sustainable development goals. Agric. Syst. 183, 102873.

Thulasiraman, V., Nandagopal, M.S.G., Kothakota, A., 2021. Need for a balance between short food supply chains and integrated food processing sectors: COVID-19 takeaways from India. J. Food Sci. Technol. 1–9. Available from: https://doi.org/10.1007/s13197-020-04942-0.

Timilsina, B., Adhikari, N., Kafle, S., Paudel, S., Poudel, S., Gautam, D., 2020. Addressing impact of COVID-19 post pandemic on farming and agricultural deeds. AJARR 11 (4), 28–35.

Varshney, D., Roy, D., Meenakshi, J.V., 2020. The roles of commodity characteristics, disease caseload. Indian. Econ. Rev. 55, S83–S103. Available from: https://doi.org/10.1007/s41775-020-00095-1.

Walters, L., Wade, T., Suttles, S., 2020. Food and agricultural transportation challenges amid the COVID-19 pandemic. Choices Mag. 35 (3), 1–8. Available from: https://www.choicesmagazine.org/UserFiles/file/cmsarticle_752.pdf.

Webster, T.J., Ghosh, S., 2020. Potential promising nanomedicines to combat COVID-19. Am. Assoc. Pharm. Scientists Outsourcing Community Newsl. 1, 3–7.

World Health Organization, 2020. COVID-19 weekly epidemiological update 22. World Heal. Organ. 1–3. Available in: https://www.who.int/docs/default-source/coronaviruse/situation-reports/weekly_epidemiological_update_22.pdf.

CHAPTER

3

State of the soil health before maize intercropped with *Eucalyptus grandis* under agroforestry environment in Limpopo Province, South Africa

Thabo Nkuna[1,2], C. Madakadze[1], P. Maponya[2] and Z. Dube[3]

[1]Department of Plant and Soil Sciences, Faculty of Natural and Agricultural Sciences, University of Pretoria, Hartfield, South Africa

[2]Department of Crop Science, Agricultural Research Council – Vegetable, Industrial and Medicinal Plants, Pretoria, South Africa

[3]Department of Agriculture, Faculty of Agriculture and Natural Science, University of Mpumalanga, Mbombela, South Africa

3.1 Introduction

Diversification and intensification through inclusion of vegetable or cereal crops in agroforestry-based cropping systems represent a key technology in the drive toward the sustainable intensification of agriculture in South Africa. Prior to introducing vegetable crops, there is a need to evaluate soil health status as an indicator for successful crop production under agroforestry environment. Functional diversity can be measured to determine the biological status of soil microbial populations, since it relates to the actual or potential activities of organisms that contribute to ecosystem dynamics (Habig, 2019). Soil microbial activities, populations, and communities are governed by environmental variables and agricultural system (Melero et al., 2005). Microorganisms are typically in great abundance and high diversity in common soil and their integrated activity drives nutrient cycling on the ecosystem scale. One of the basic functions of soil microorganisms is the decomposition and transformation of organic materials, which are mostly derived from above- and below-ground plant residues. The soil microbiota maintain an essential role in biogeochemical cycling, as microbiota are responsible for degradation of organic compounds and govern plant productivity of terrestrial ecosystems. Eighty to ninety percent of the chemical processes in soil are reactions meditated by microbes (Nannipieri et al., 2003; Grundmann, 2004). Earlier researches have shown that

Sustainable Agricultural Practices. DOI: https://doi.org/10.1016/B978-0-443-19150-3.00004-7

organic applications increase nutrient status, microbial activity (Elliott and Lynch, 1994), and productive potential of soil while the use of chemical fertilizers alone in cropping system resulted in poor microbial activities and productive potentials of soils (Kang et al., 2005). Common soil amendments used in Nigeria include the use of organic and inorganic fertilizers for sustainable crop production. However, past research focus as regards organic and inorganic soil management practices have focused on increasing crop yields as well as monitoring soil physical and chemical properties while the cumulative effect on soil biological properties are often neglected. Soil biological properties such as the soil microbial biomass carbon are more susceptible to variations and changes in the environment and are quick to respond to these changes than the physical and chemical properties, including soil organic carbon and are seen as promising tools to indicate the quality and status of soils (Hole et al., 2005; Nogueira et al., 2006; Kaschuk et al., 2010).

A range of microbiological and molecular techniques is available to study soil microbial populations. Functional diversity can be measured to determine the biological status of soil microbial populations, since it relates to the actual or potential activities of organisms that contribute to ecosystem dynamics. The biogeochemical cycling of nutrients, such as carbon, nitrogen, and phosphorus is a fundamental soil function and, therefore, of great interest to assess the relative activity of soil microbial communities (Poore and Fries, 1985). In this context, microbial community level physiological profiles and enzymatic activity assays are often used to determine the functional diversity of soil microbial populations. In both types of analyses, the ability of the microbial population to utilize a specific substrate is measured. Substrate availability is the main factor that influences both size and activity of microbial communities (Chang et al., 2007). The catabolic diversity of bacterial communities can be determined based on sole-carbon substrate utilization, using the Biolog system. Communities of organisms will give a characteristic reaction pattern, called metabolic fingerprint and from these data, carbon source utilization profiles (CSUP) are generated. Contrary to CSUP analysis, microbial enzymatic activity assays consist of several culture independent methods, that is, determination of microbial activity without culturing microorganisms. Soil microbial enzymes fulfill crucial biochemical functions in organic matter decomposition in soil systems (Arancon et al., 2006). The activity of any enzyme assayed in a soil sample is the sum of active and potentially active enzymes from all the different sources. Enzymatic activities, that is, microbial activity, in relation to the cycling of carbon, nitrogen (ammonification, nitrification, denitrification), or release of inorganic phosphorus in soil have been used to evaluate the fertility of the soil or to describe the functioning of the ecosystem. β-Glucosidase has been reported to be a useful indicator of soil quality due to the important role it plays in catalyzing the hydrolysis and biodegradation of various β-glucosides present in plant debris decomposing in the ecosystem (Moges, 1998).

Phosphatases are believed to play critical roles in P cycles as evidence has shown that they are correlated to P stress and plant growth. Phosphatase activity has also been correlated with soil phosphorous state, with inorganic phosphorous having an inverse effect on phosphatase production. Phosphatase excreted by plant roots has been reported to be exclusively acid phosphatase which might be stimulated in cases of phosphorous deficiency, to increase availability of phosphate in the soil. Alkaline phosphatase, thus, would appear to be derived solely from soil microorganisms, allowing the distinction between acid phosphatase activity of roots and alkaline phosphatase activity of microorganisms. Urease activity plays a vital role in the regulation of N supply to plants, especially after urea fertilization. Due to the influence of pH, temperature, organic matter content, and soil moisture on microbial enzymatic activity, they are considered early indicators of ecosystem stress and can act as biological indicators of soil degradation, compared to classical and slowly changing soil properties such as organic matter. The objective of this study was to determine the soil microbial community possible effects impeding agroforestry in South Africa.

3.2 Materials and methods

Soil samples were collected during 2019 from various areas in Limpopo and Mpumalanga and kept cold until the time of analysis. The following soil samples were collected aseptically from the surface soil depths (0–20 cm) in each experimental site, whereby five composite samples were gotten in each sample field. Stones, plant, and root debris were removed and subsamples of the soils were air dried and grounded to pass through a 2 mm sieve for chemical analysis. Subsamples for microbiological analysis were collected in Ziploc bags and transported to the laboratory on ice packs, these were stored at 4°C before microbiological analysis.

1. Ratombo
2. Safcol Graskop
3. Safcol Block J4
4. Safcol Block J25
5. Serala Georges Valley
6. MTO Plantations Whiteriver.

3.2.1 Nitrogen determination

The soil samples were analyzed for nitrogen content in the form of ammonium (NH_4-N) and nitrate (NO_3-N) (mg/kg).

3.2.2 Determination of functional diversity

CSUP are assessed when carbon sources are utilized. Soil samples were diluted in sterile distilled water and inoculated into Biolog EcoPlates (Biolog Inc., Hayward, the United States) containing 31 carbon sources and a control well, in triplicate. The plates were incubated at 28°C. Respiration of carbon sources by microbial populations reduce the tetrazolium dye, causing a color change which was measured twice daily over a period of 5–10 days at 590 nm to determine average well color development (AWCD). The functional diversity of the soil microbial populations was determined using the amount and equitability of carbon substrates metabolized as indicators of richness and evenness, respectively.

3.2.3 Determination of soil microbial enzymatic activity

The ability of the soil microbial population to obtain carbon, phosphorus, and nitrogen was assayed by measuring the β-glucosidase, alkaline phosphatase, acid phosphatase, and urease activities in the soil. β-Glucosidase and phosphatase activities were calculated by determining the release of p-nitrophenyl after the incubation of soil with p-nitrophenyl glucoside and p-nitrophenyl phosphate, respectively. Urease activity was determined where released ammonia was measured after the incubation of soil samples with a urea solution, where results were then calculated with reference to the calibration curve.

3.2.4 Statistical analyses

Data could not be subjected to statistical analyses due to the absence of sample replications. CSUP and the overall microbial enzyme assays were statistically analyzed by cluster analyses (vertical hierarchical tree plots). Dendograms were 3,18 2,86 2,77 4,08 2,58 3,49 RATOMBO SAFCOL GRASKOP SAFCOL BLOCK J4 SAFCOL BLOCK J25 SERALA GEORGES VALLEY MTO PLANTATIONS WHITERIVER mg/kg NH_4-N 15,8 34,98 8,06 14,81 16,31 5,31 RATOMBO SAFCOL GRASKOP SAFCOL BLOCK J4 SAFCOL BLOCK J25 SERALA GEORGES VALLEY MTO PLANTATIONS WHITERIVER mg/kg NO_3-N constructed using Ward's clustering algorithm, and the Euclidean distance measure, that is, the geometric distance between variables in a multidimensional space. Homogenous grouping with Fisher least significant difference (LSD) could not be calculated for the received soil samples. Biodiversity was determined using the Shannon–Weaver diversity index and evenness index, indicating species richness and abundance, respectively.

3.3 Results and discussion

Nitrogen determination soil nitrogen content was determined in the form of ammonium (NH_4-N) and nitrate (NO_3-N) (mg/kg). Results are presented, respectively.

The ammonium–nitrogen content was the highest in the Safcol Block J25 and MTO Plantations samples, while the ammonium–nitrogen content in the Safcol

Graskop, Safcol Block J4, and the Serala Georges Valley samples was the lowest. The nitrate–nitrogen content was the highest in the Safcol Graskop sample, while the Safcol Block J4 and MTO Plantations demonstrated the lowest nitrate–nitrogen content. The Ratombo, Safcol Block J25, and Serala Georges Valley samples displayed nitrate–nitrogen contents that were higher than the content present in the Safcol Block J4 and MTO Plantations samples, but lower compared to the Safcol Graskop sample.

3.3.1 Carbon source utilization profiles

The mechanism of color development in Biolog EcoPlates is related to differences in CSU, that is, food source consumption, which, in turn, appears to relate to the diversity of viable microorganisms able to utilize the substrates ("food sources") within the wells of the EcoPlate as a sole-carbon source. The utilization of carbon source by active microbial communities between sampling sites are illustrated by means of principal component analysis (PCA) in Fig. 3.1.

From the PCA in Fig. 3.1, microbial communities in the Safcol Graskop site clustered separately from all the other sampling sites, while no clear clustering could be distinguished between the remaining sites. These results depict the CSU of the soil microbial populations present, clearly indicating differences in CSUP between the different sampling sites, also implying differences in microbial functioning between the different sampling sites. Since the ordination of the PCA might be ambiguous in demonstrating clear distinctions between groups, cluster analysis was performed as an alternative measure to enable a 2D visualization of the different groups illustrated in Fig. 3.1. Dendograms were constructed with the aid of cluster analysis to assign similar treatments into the same clusters as illustrated in Fig. 3.2.

The soil microbial profiles from the different sampling sites revealed that the CSUP at the Georges Valley, Safcol Block J4, and Ratombo sites (Blue Block) were similar, while the CSUP at the MTO Plantations and Safcol Block J25 sites (Green Block) differed from each other, as well as from the Georges Valley, Safcol Block J4, and Ratombo sites. The CSUP at the Safcol Graskop site (Red Block) clustered separately from all the other sites (Jimenez and Ladha, 1993).

Diversity Indices The Shannon–Weaver substrate and diversity index (H') is used to quantify the functional diversity of soil microbial communities based on the amount of different carbon sources utilized by soil microbial communities in Biolog EcoPlates, that is, comparable to species richness in the soil. Values of the index typically range between 1.5 and 3.5, but rarely increase above 4.5. Depending on the sampling sites, varying degrees of carbon sources were utilized, with values ranging from 0.47 (very low) to 2.59 (high) (Table 3.1). On the other hand, the evenness index is used as an indication of how abundant species are within a soil microbial community, that is, how close in abundance the different microbial species are in a soil microbial community. If the abundance of different species in a community is measured, it will invariably be found that some species

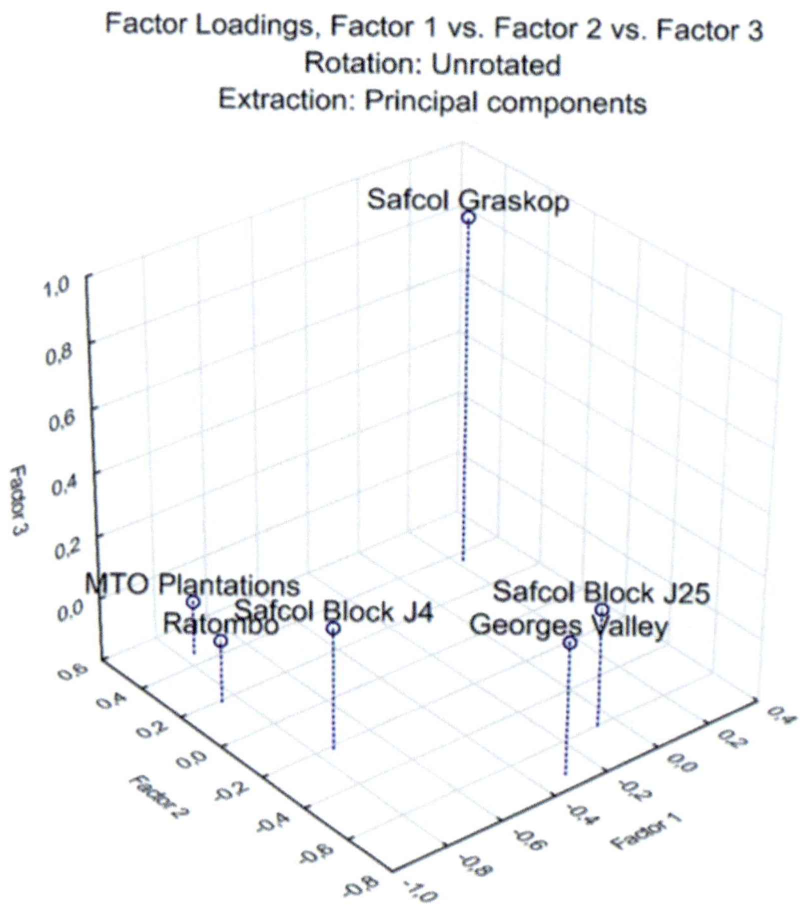

FIGURE 3.1

Principal component analysis (PCA) ordination plot illustrating the differences in carbon source utilization profiles of active microbial communities at the various collected sites.

are rare, whereas others are more abundant/dominant. Substrate evenness assumes a value between 0 and 1, with 1 representing a situation in which all species are equally abundant within a microbial population present in the samples. This means less variation in microbial populations between species, thus, less dominance and higher diversity. Substrate evenness indices obtained for the different sampling sites ranged between 0.30 (low) and 0.75 (high) (Table 3.1).

According to the diversity indices at the various sampling sites, species richness and abundance were the lowest at the Safcol Graskop and MTO Plantations sites, whereas the Safcol Block J4, Ratombo, and Georges Valley sites displayed the highest species richness and abundance. Fig. 3.3 provides a visual

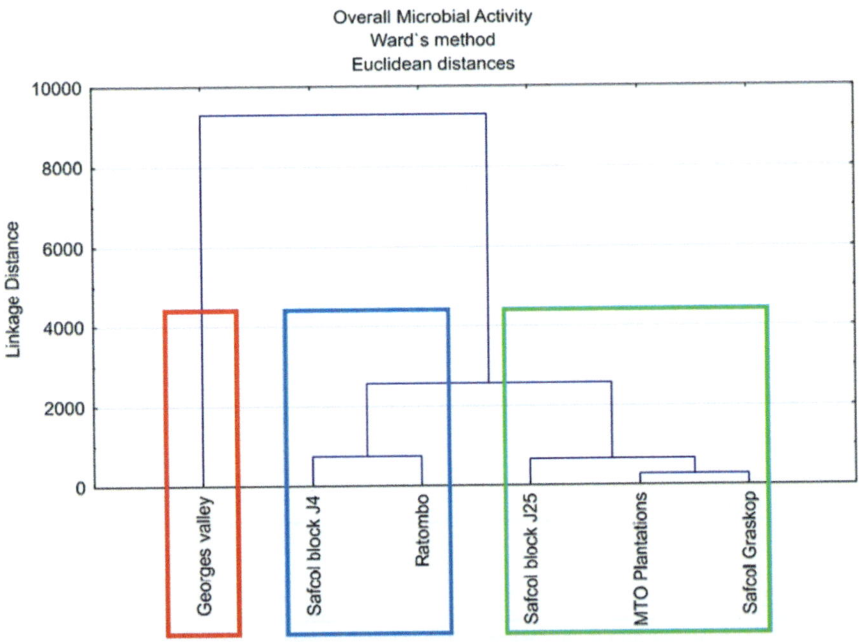

FIGURE 3.2

Dendogram illustrating the carbon source utilization profiles between the various sampling sites.

Table 3.1 Shannon–Weaver diversity index illustrating soil microbial species richness and the evenness index illustrating species abundance.

Sampling sites	Shannon–Weaver diversity index (H)	Evenness index (E)
Safcol Graskop	0.470	0.428
Ratombo	2.243	0.689
Safcol Block J4	2.160	0.699
Safcol Block J25	1.532	0.489
Georges valley	2.585	0.753
MTO White River	0.653	0.297

representation of the measured microbial diversity. The Safcol Block J4, Ratombo, and Georges Valley sites clustered together in the ideal quadrant and the Safcol Block J25 site in the transition quadrant, whereas the MTO Plantations and Safcol Graskop sites clustered together in the less ideal quadrant. Sampling sites in the less ideal quadrant are characterized by a low microbial species richness being dominated by some of the microbial species present. However, sampling sites clustering in the ideal quadrant are characterized by a high microbial

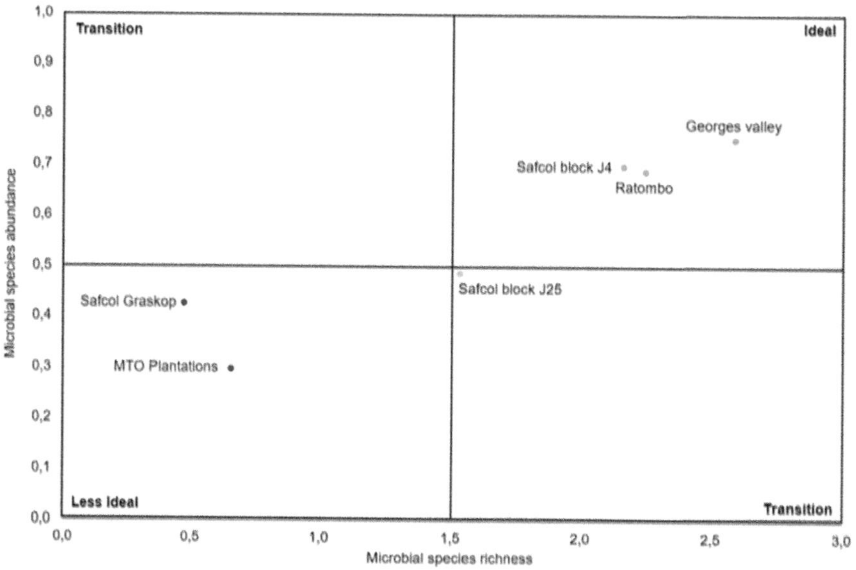

FIGURE 3.3

The microbial diversity profile representing the microbial species richness (Shannon–Weaver diversity index) and microbial abundance (evenness index) indices for the various sampling sites.

species richness with an equal abundance of the microbial species present. These results/observations could be attributed to the region, soil type, crops planted, or historical/current agricultural practices at the sampled sites. The lowest overall microbial diversity was demonstrated at the MTO Plantations site, while the Georges Valley site demonstrated the highest overall microbial diversity.

3.3.2 Soil microbial enzymatic activity, viz. microbial activity

The activities of four soil microbial enzymes were analyzed for this trial: β-glucosidase (Fig. 3.4), alkaline phosphatase (Fig. 3.5), acid phosphatase (Fig. 3.5), and urease (Fig. 3.6). Results are presented as a means of determining the potential of a soil to degrade or convert (mineralize) substrates from an organic form into plant-available nutrients (Table 3.2). By implication: the higher the microbial activity (i.e., mineralization rate), the faster the nutrients that are released from organic substrates, will be made available to the soil ecosystem and rhizosphere to be taken-up by plant roots (García-Gil et al., 2000).

After data collection and analysis, the results indicated from the PCA, those microbial communities in all the sampling sites were not showing any differences, while no clear clustering could be distinguished between the remaining sites. These results depict the CSU of the soil microbial populations present, clearly

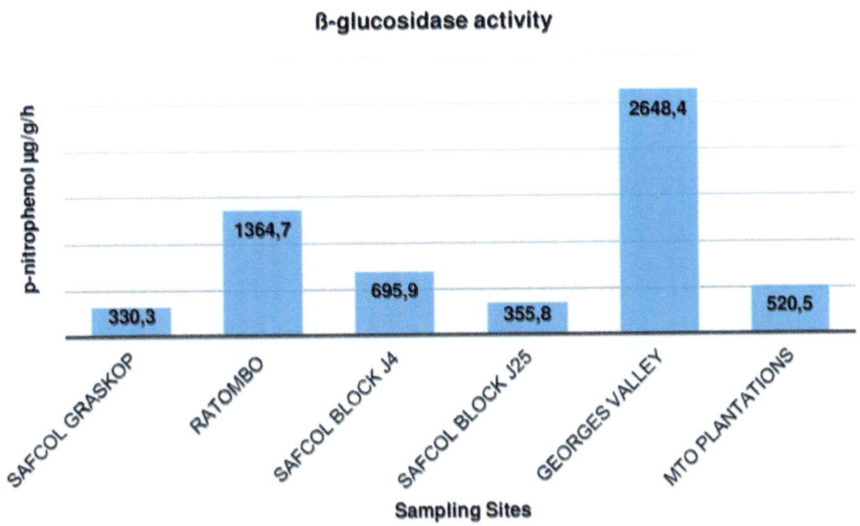

FIGURE 3.4

Bar graph illustrating soil microbial β-glucosidase activity in soils at the various sampling sites.

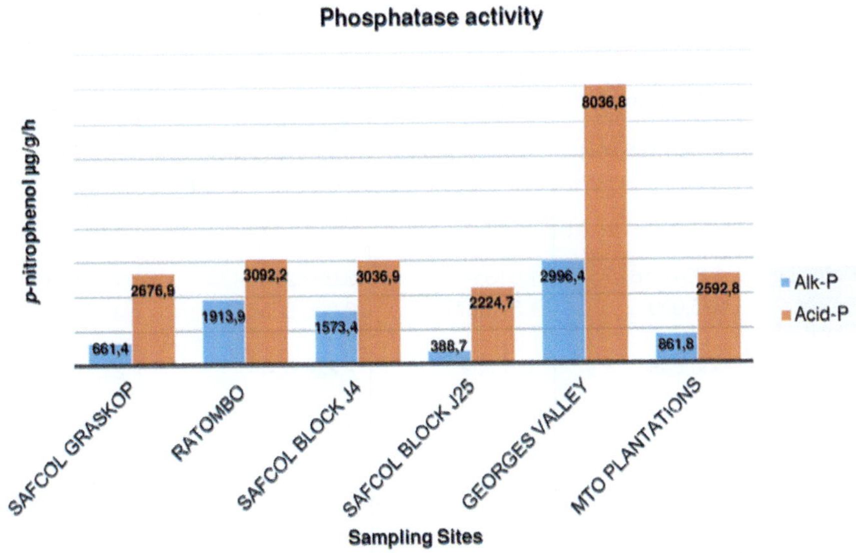

FIGURE 3.5

Bar graph illustrating soil microbial acid and alkaline phosphatase activities in soils at the various sampling sites.

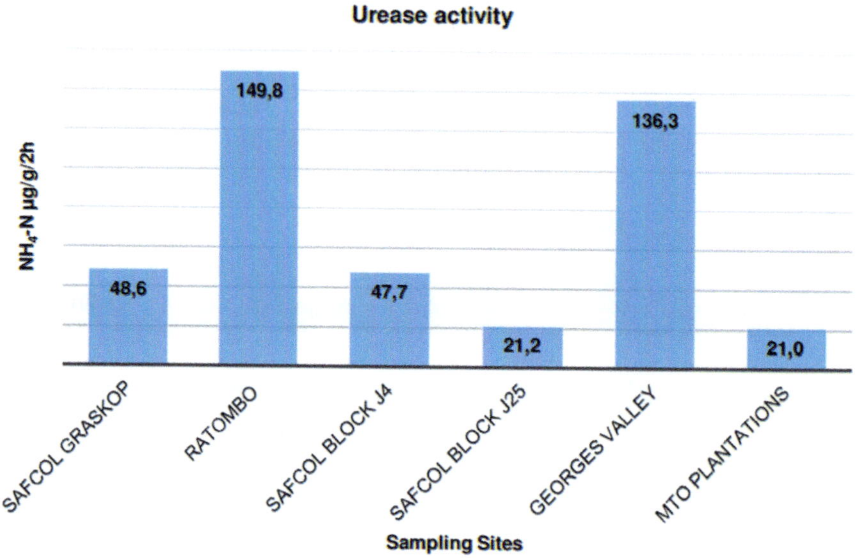

FIGURE 3.6

Bar graph illustrating soil microbial urease activity in soils at the various sampling sites.

Table 3.2 Soil microbial enzymatic activities between treatments in different sampling sites.

Sampling sites	β-Glucosidase activity (*p*-nitrophenol µg/g/h)	Alkaline phosphatase activity (*p*-nitrophenol µg/g/h)	Acid phos phatase activity (*p*-nitrophenol µg/g/h)	Urease activity (NH_4-N µg/g/h)
Safcol Graskop	330,322	661,440	2,676,949	48,584
Ratombo	1,364,678	1,913,923	3,092,214	149,806
Safcol Block J4	695,909	1,573,377	3,036,925	47,699
Salcol Block J25	355,839	388,709	2,224,654	21,194
Georges Valley	2,648,438	2,996,396	8,036,752	136,349
MTO White River	520,546	861,794	2,592,838	21,031

indicating differences in CSUP between the different sampling sites, also implying differences in microbial functioning between the different sampling sites. Various observation was confirmed using dendogram were the CSUP at the

Safcol Block J4 and Safcol Block J25 sites differed from each other. According to the diversity indices at the various sampling sites, species richness and abundance at Safcol Block J4 sites displayed the highest species richness and abundance whereas Safcol Block J25 site displayed the lowest overall microbial activity. Findings suggested that both soil microbial functional diversity and activity were sensitive compared to both sites. Consequently, the latter sites might hinder crop production directly through lower mineralization rates, or indirectly through the presence of a low microbial diversity.

Soil microbial communities associated with the various treatments differed in their ability/potential to mineralize/convert carbon (β-glucosidase), phosphorous (acid and alkaline phosphatase), and nitrogen (urease). Soil microbial communities at the Safcol Graskop and Safcol Block J25 sites exhibited the lowest β-glucosidase and alkaline phosphatase activities, whereas the Safcol Block J25 and MTO Plantations sites exhibited the lowest acid phosphatase and urease activities. On the other hand, the soil microbial communities at the Georges Valley and Ratombo sites exhibited the highest β-glucosidase, phosphatase, and urease activities. The soil microbial communities at all the sampling sites exhibited the potential to mineralize phosphorous at higher rates under low soil pH conditions than under high soil pH conditions. On average, the Safcol Block J25 site displayed the lowest overall microbial activity, while the overall microbial activity in the Georges Valley site was notably higher compared to the other sampling sites. Cluster analysis was performed to construct dendograms as an alternative measure to assign similar microbial activities at the various sampling sites into the same clusters (Kandeler and Gerber, 1988).

The cluster analyses revealed that the overall microbial activity at the Safcol Block J4 and Ratombo sites (Blue Block) were similar, while the microbial activity at the Safcol Graskop, MTO Plantations, and Safcol Block J25 sites (Green Block) were also similar to a lesser degree to the microbial activity at the Safcol Block J4 and Ratombo sites. The overall microbial activity at the Georges Valley site (Red Block) clustered separately from all the other sites. These observations are also supported by the results in Table 3.2, indicating that the overall microbial activity at the Georges Valley site was the highest compared to all the sampling sites, whereas the overall microbial activity in the remaining sampling sites were relatively similar. The separate clustering of the Safcol Block J25, MTO Plantations, and Safcol Graskop sampling sites from the Safcol Block J4 and Ratombo sampling sites, could be explained by the lower overall microbial activities in the former cluster than in the latter cluster.

Mofokeng et al. (2020) revealed that soil microbial enzymatic activities were assayed by measuring β-glucosidase, alkaline phosphatase, acid phosphatase, and urease activities in the soil. The bioslurry crust had the highest nutrient content in comparison to bioslurry and bioslurry sediment. PCA showed differences in CSUP between the different samples. Soil microbial populations present in the bioslurry crust and soil supplied with bioslurry were more similar, compared to those present in the untreated soil. The bioslurry crust sample demonstrated the highest overall enzyme activities. The study showed that the bioslurry and bioslurry crust could be used as organic soil amendments, with positive effects on soil health and fertility, contributing toward sustainability of the environment and ultimately to food security.

3.4 Conclusions

Both soil microbial functional diversity and activity were sensitive to, and differed among, the various sampling sites. Given the results obtained, the biological soil health seems to be more favorable at the Georges Valley and Ratombo sites due to the high microbial diversity and activity, compared to the soil health status of the Safcol Graskop, MTO Plantations, and even Safcol Block J25 sites. Consequently, the latter sites might hinder crop production directly through lower mineralization rates, or indirectly through the presence of a low microbial diversity. It is highly recommended that trends in CSUP and enzymatic activity be monitored over an extended period of time to attain a more complete reflection on the impact of different treatments/crops/agricultural practices might have on microbial diversity and activity as an indicator of soil fertility and health. Integration and comparison of available data with more soil quality indicators could contribute to more refined answers and possible predictions. In conclusion, it is highly recommended that trends in CSUP and enzymatic activity be monitored over an extended period of time to attain a more complete reflection on the impact of different treatments or crops practices might have on microbial diversity and activity as an indicator of soil fertility and health.

References

Arancon, N.Q., Edwards, C.A., Bierman, P., 2006. Influences of vermicompost on field strawberries: part 2. Effects on soil microbiological and chemical properties. Bioresour. Technol. 97, 831–840. Available from: https://doi.org/10.1016/j.biortech.2005.04.016.

Chang, E.H., Chung, R.S., Tsai, Y.H., 2007. Effect of different application rates of organic fertilizer on soil enzyme activity and microbial population. Soil. Sci. Plant. Nutr. 53, 132–140. Available from: https://doi.org/10.1111/j.1747-0765.2007.00122.x.

Elliott, L.F., Lynch, J.M., 1994. Biodiversity and New Delhi, Prentice Hall India (P) Limited. Soil resilience. In: Greenland, D.J., Szabolc, I. (Eds.), 23. Allen, S.E., H.M.

Grundmann, G.L., 2004. Spatial scales of soil bacterial diversity – the size of a clone. FEMS Microbiol. Ecol. 48, 119–127.

García-Gil, J.C., Plaza, C., Soler-Rovira, P., Polo, A., 2000. Long-term effects of municipal solid waste compost application on soil enzyme activities and microbial biomass. Soil. Biol. Biochem. 32, 1907–1913. Available from: https://doi.org/10.1016/S0038-0717(00)00165-6.

Habig, J., 2019. Soil enzyme analyses and community-level physiological profiling (biolog ecoplate) – agroforestry project. MicroLife Res. Cent. 1–12.

Hole, D.G., Perkins, A.J., Wilson, J.D., Alexander, I.H., Grice, P.V., Evans, A.D., 2005. Does organic farming benefit biodiversity? Biol. Conserv. 122, 113–130.

Jimenez, R.R., Ladha, J.K., 1993. Automated elemental analysis: a rapid and reliable but expensive measurement of total carbon and nitrogen in plant and soil samples. Commun. Soil Sci. Plant Anal. 24, 1897–1924. Available from: https://doi.org/10.1080/00103629309368926.

Kandeler, E., Gerber, H., 1988. Short-term assay of soil urease activity using colorimetric determination of ammonium. Biol. Fertil. Soils 6, 68–72. Available from: https://doi.org/10.1007/BF00257924.

Kang, G.S., Beri, V., Rupela, O.P., Sidhu, B.S., 2005. A new index to assess soil quality and sustainability of wheat based cropping systems. Biol. Fertil. Soils 41, 389–398.

Kaschuk, G., Alberton, O., Hungria, M., 2010. Three decades of soil microbial biomass studies in Brazilian ecosystems: lessons learned about soil quality and indications for improving sustainability. Soil Biol. Biochem. 42, 1–13.

Melero, S., Porras, J.C.R., Herencia, J.F., Madejon, E., 2005. Chemical and biochemical properties in a silty loam soil under conventional and organic management. Soil. Tillage Res. 90, 162–170.

Moges, Y., 1998. The role of exotic plantation forests in fostering the regeneration of native trees in an Afromontane Forest Area in Ethiopia. Master's Thesis, Wageningen Agricultural University, Wageningen.

Mofokeng, M.M., Habig, J., Amoo, S.O., Du Plooy, C.P., Mashela, P.W., Moeletsi, M.E., et al. 2020. Differences in soil microbial communities and enzyme activity due to the application of bioslurry under cultivation. https://doi.org/10.1080/02571862.2020.1761468.

Nannipieri, P., Ascher, J., Ceccherini, M.T., Landi, L., Pietramellara, G., Renella, G., 2003. Microbial diversity and soil functions. Eur. J. Soil Sci. 54, 655–670.

Nogueira, M.A., Albino, U.B., Brandão-Junior, O., Braun, G., Cruz, M.F., Dias, B.A., et al., 2006. Promising indicators for assessment of agro-ecosystems alteration among natural, reforested and agricultural land use in Southern Brazil. Agric. Ecosyst. Environ. 115, 237–247.

Poore, M.E.D., Fries, C., 1985. The ecological effects of Eucalyptus. FAO Forestry Paper Series 59, FAO, Rome.

CHAPTER

4 Arbuscular mycorrhizal fungi in sustainable agriculture

Semra Demir[1], Younes Rezaee Danesh[1,2], Emre Demirer Durak[1], Solmaz Najafi[3] and Gökhan Boyno[1]

[1]Department of Plant Protection, Faculty of Agriculture, Van Yuzuncu Yil University, Van, Turkey

[2]Department of Plant Protection, Faculty of Agriculture, Urmia University, Urmia, Iran

[3]Department of Field Crops, Faculty of Agriculture, Van Yuzuncu Yil University, Van, Turkey

4.1 Introduction

Nearly all ecosystems include mycorrhizal symbioses, which are essential for enhancing soil quality and plant health through important ecological processes. It has been demonstrated that mycorrhizal fungi perform a wide range of advantageous tasks for both their host plants and their surroundings, including better nutrient and water absorption, increased tolerance to biotic and abiotic stressors, and improved soil structure to prevent soil erosion. Arbuscular mycorrhizal fungi (AMF) and other fungal symbionts in soils can be severely impacted by intensive agricultural practices such the use of copious amounts of mineral fertilizers and pesticides, which can then have a detrimental influence on plant nutrition and growth. In fact, the loss of fungal variety impairs key ecosystem functions, including ecological variability and productivity, and it may also reduce plant biodiversity, which would result in the deterioration of the land. Microbial guilds are the primary drivers of terrestrial ecosystems. However, our inadequate knowledge of the variety and operation of such microbial communities significantly hinders us. Numerous varied and competitive fungal species, including soil decomposers, pathogens, endophytes, and obligate symbionts, grow on mineral particles and decomposing organic matter and reside near or inside plant roots (Frankland, 1998). Mycorrhizal fungi play a critical part in the biogeochemical cycles of the soil's enormous intertwining webs of hyphae, which are permeated by the soil and play a vital role in the sustainability of ecosystems. Since it has been established that these mycorrhizal species make up half of the fungal species in the majority of temperate, montane, and boreal forests, they are not only marginal oddities. The so-called mycorrhiza, which is the location of nutrition and carbon transfer between the two symbiotic partners, is a unique composite organ formed by the hyphae of mycorrhizal fungus and the roots of plants.

Sustainable Agricultural Practices. DOI: https://doi.org/10.1016/B978-0-443-19150-3.00017-5

Due to their mutualistic relationships, terrestrial plants may thrive even under less-than-ideal conditions (Read and Perez-Moreno, 2003). They contribute the majority of the host plant's nitrogen and phosphorus needs and receive 15%–25% of the net productivity of carbon, making them important drivers of ecosystem function. One of the most common relationships in all terrestrial ecosystems is without a doubt the symbiotic link between roots and these fungi (Smith and Read, 2008). Understanding the inputs and outputs in forest ecosystems under global change therefore requires an ever-increasing understanding of the processes for which these soil fungi is accountable and how (van der Heijden et al., 2014). The review will concentrate on what is currently known about the function of AMF in agricultural systems and how agricultural practices affect this interaction. We will go into greater depth about the value of AMF in sustainable agriculture, and how farmers may be able to maximize the advantages of the arbuscular mycorrhiza (AM) relationship in such systems.

4.2 Arbuscular mycorrhizal fungal symbiosis

Communities of AMF invade the roots of the majority of plant species and the soils of the majority of terrestrial ecosystems (Smith and Read, 2008). AM is an almost universal interaction of plants with fungal symbionts. Their primary function is to aid plants in absorbing low-available nutrients, particularly phosphorus (P), which helps shield plants from nutrient deprivation, a significant abiotic stress that plants must contend with (Delavaux et al., 2017). Mycorrhiza, however, protects plants from a variety of additional environmental challenges, including water scarcity, heavy metal toxicity, or pathogen attack (Augé, 2001). These advantages are somewhat connected to enhanced mineral nutrition, but they can also be caused by factors unrelated to diet, such as systemic resistance to infections. Mycorrhizal expenses, which are covered by carbon (C) from photosynthetically absorbed plant material transported to the obligately symbiotic AMF, can be insignificant or even made up for by greater photosynthetic rates under a variety of lighting conditions. In this way, mycorrhiza might be compared to a sophisticated insurance plan for changing circumstances. Appreciation of the evolutionary success of mycorrhizas under a variety of settings where the relationship does not look unambiguously mutualistic requires an understanding of the intricacy of mycorrhizal effects (Dickie et al., 2015; Veresoglou et al., 2022). In terms of agroecosystems, the complex stress protection is also practically significant since it may supplement our understanding of the function of mycorrhizas in plant growth (Verbruggen and Kiers, 2010). The AM symbiosis is a prehistoric symbiosis (Pirozynski and Dalpe, 1989). Aglaophyton fossils separated from the Rhynie chert have been found to include hyphae and arbuscules, and this proof shows that AM symbiosis existed in the early Devonian (Remy et al., 1994; Taylor et al., 1995). Additionally, scientific studies based on the divergence of the 18s rDNA nucleotide sequence imply that the Glomales evolved 350–460 million years ago and that the symbiosis was crucial to plants successfully colonizing the land (Remy et al., 1994; Taylor et al., 1995).

AMF (Glomeromycota) have been published under various taxonomic names at all levels, leading to inconsistent systematic systems and substantial confusion among scientists dealing with these significant plant symbionts. A group of biologists with more than a century of collective experience in Glomeromycota systematics has examined all available molecular-phylogenetic evidence within the framework of phylogenetic hypotheses that combine morphological characters when compatible. As a result of these studies, the current Glomeromycota classification has been widely accepted (Schüßler et al., 2001; Redecker et al., 2013). About 10% of plants have lost the AM symbiosis during the course of evolution, including whole angiosperm families. Of the majority of fungus found in agricultural soils, AM fungi make up 5%–36% of the soil's total biomass and 9%–55% of the biomass of soil microorganisms. Due to their antagonistic and competitive effects on pests and pathogens, these fungi are an essential part of agricultural systems because they can boost plant growth, plant reproductive capacity, plant water stress tolerance, plant health, and plant growth (Demir and Akköprü, 2007). In addition, this colonization makes the plant more resistant to biotic and abiotic stress factors (Newsham et al., 1994; Boyno et al., 2022). With the plant root system, AM fungi create a vast hyphal (extra-radical hyphae) network that significantly contributes to the enhancement of soil texture and water relations (Tisdall, 1991; Bethlenfalvay and Schuepp, 1994). The improved absorption of immobile soil nutrients, particularly phosphorus, is the primary advantage of the mycorrhizae symbiosis for the host plant (Jakobsen, 1999). Because the hyphae compete for the mineralized organic soil nitrogen, arbuscular mycorrhizal associations boost nitrogen buildup in plant tissues (Ibijbijen et al., 1996; Aysan and Demir, 2009). Additionally, these fungus collaborate with other soil creatures crucial to nutrient cycles. For instance, coinfection with AMF can improve Rhizobium's biological nitrogen fixation in legume hosts (Erman et al., 2011). These ecological functions play a crucial part in low-input farm management systems, which rely on natural nutrient cycles to supply the nutrients needed for plant growth. As a result, these fungi are a crucial and essential part of ecosystems and may be useful in developing sustainable agricultural systems (Schreiner and Bethlenfalvay, 1995).

4.3 Arbuscular mycorrhizal fungi and sustainable agriculture

The amount of agricultural production has increased significantly since the middle of last century due to the improvement of plant varieties, machinery, and chemical inputs worldwide (Craswell and Karjalainen, 1990). However, in recent years, due to the land degradation caused by the implementation of unsustainable agricultural practices, the yield of agricultural products has dropped sharply (Bindraban et al., 2012; Grassini et al., 2013; Gomiero, 2016). To meet the growing needs of the human population, it is necessary to increase food production with sustainable manners (Godfray et al., 2010). Unfortunately, to increase the production of agricultural products, the producers emphasize on increasing the production costs by using various agricultural operations in the field, including using of chemical inputs such as nitrogen and phosphorus fertilizers, as well as pesticides, which causes pollution and degradation of the environment (Clermont-Dauphin and Meynard, 1997). Currently, the consumption of nitrogen fertilizer has reached nine times and phosphorus fertilizer has reached more than four times in developing countries (Schultz et al., 1995). Therefore it is necessary to use sustainable agricultural methods at the global level to produce high-quality food sources by using minimal chemical inputs (Boiffin et al., 2001). Edaphic factors have numerous effects in sustainable agricultural systems, which include the control of pathogens in the rhizosphere, as well as increasing the growth of microbial agents in the soil and their effectiveness in antagonistic or parasitic reactions in root growing zone (Jawson et al., 1993; Knudsen et al., 1995). One of the sustainable methods in agriculture to overcome this problem is the use of beneficial microbial agents in the soil. Plant–microbe interactions are interesting events in agricultural sustainability. In this regard, symbiotic microorganisms such as AMF would be considered as the source of many beneficial effects, so, could be assumed as the key factor in sustainable agriculture (Godfray et al., 2010). Mycorrhizal symbiosis plays a vital role in sustainable agricultural systems, especially in the lack of food supplies. In these situations, the AMF extra- and intraradical mycelia transfer the nutrients in the usable form of the host plant. AM fungi have effects on plant growth, nutrient supply, and plant production even in phosphorus-rich soils, but the positive effects of these fungi are more obvious when the host plant is cultivated in low-fertile soils (Solaiman and Hirata, 1997; Balzergue et al., 2013). Attention to AM fungi and their importance in sustainable agriculture becomes more apparent when their role in improving soil quality and structure is also clarified. These fungal symbiosis efficiently improve the plant productivity and decrease the application of chemical inputs such as pesticides and fertilizers. These fungi have extensive dissemination and have significant contribution to soil microbes' biomass and soil nutrient cycling processes. They increase the absorption of essential elements, especially phosphorous, and micronutrients such as Zn and Cu; they also influence the plant production under different biotic and abiotic stresses (Sylvia and Williams, 1992; Davet, 1996; Smith and Read, 1997).

4.3.1 Functions of arbuscular mycorrhizal fungi in mineral nutrients acquisition

More than 90% of all plant species' roots have been documented to have mutualistic interactions with AM, a heterogeneous collection of various fungus taxa.

This symbiotic contact relies on a tightly controlled molecular discourse and demands a high level of synchronization between the two partners to be successful. The production of signaling molecules by fungi—myc factors that set off symbiotic reactions in the host plant occur where plant roots produce strigolactones that encourage fungal metabolism and branching. The essential aspect of this symbiotic connection, which takes place in host cortical cells that contain arbuscules, is the transfer of limited soil nutrients including phosphorus (P), nitrogen (N), and sulfur (S) in return for fixed carbon (Bhandari and Grag, 2017). The exchange of nutrients between the plant and the fungus is what drives AM interactions. Glomeromycotan fungi are obligatory symbionts that must have access to their plant hosts' carbon supply to complete their life cycle. In exchange, the fungus gives the plant nutritional advantages by supplying minerals, including the physiologically necessary elements phosphorus (P) and nitrogen (N). The bulk of this nutrition exchange is thought to take place inside root cortical cells that include arbuscules, heavily branching hyphal structures. Arbuscules never penetrate the host cytoplasm; instead, as they grow, they are encircled by freshly synthesized host membrane tissue (Marschner and Dell, 1994; Siddiqui and Pichtel, 2008; Barea et al., 2014; Jakobsen and Hammer, 2015).

AM symbiosis is a complex and very dynamic interaction that calls for a high level of coordination between the two partners and is based on a finely regulated molecular dialog (Barea et al., 2014; Mohanta and Bae, 2015). Three distinct development stages may be distinguished in the creation of AM symbiosis: (1) asymbiotic hyphal growth stage, when spores germinate and generate hyphae independently for a brief time; (2) presymbiotic hyphal development stage, where host signal perception stimulates hyphal growth; and (3) the symbiotic stage, where the fungus infiltrates the plant root and creates both intraradical mycelium (to exchange nutrients), as well as extra-radical hyphae (to recruit nutrients in the soil and form new spores) (Bhandari and Grag, 2017). The surrounding membranes and the molecular elements of the interface assist and control the processes of nutrition exchange. The interfacial apoplast is kept at a low pH by membrane-localized ATPase proteins that are encoded by plants and fungi. This electrochemical gradient fuels the activity of several membrane-localized proton:nutrient symporters. The components of arbuscular membranes have started to be defined using a mix of physiological, biochemical, and genetic research; the best known of these are the Pht1-type phosphate transporter proteins (Javot et al., 2007; Ruairidh et al., 2008). The first step in phosphate acquisition through the mycorrhizal route is the fungal extra-radical hyphae's absorption of free phosphate from the soil (Bucher, 2007). By extending outside the host root system, these fungal hyphae increase the amount of soil that may be used for phosphate absorption. The Pht1 family of fungal high-affinity phosphate transporters mediates uptake at the soil-hypha interface. Phosphorus is taken up by the fungus and then moved to the fungal vacuole, where it is polymerized to create polyphosphate chains and moved through the vacuolar compartment to the intraradical hyphae. Phosphorus is then delivered to the interfacial apoplast after the polyphosphate has been hydrolyzed. Plant mycorrhizal Pht1 transporters move phosphate across the peri-arbuscular membrane from the interfacial apoplast. Phosphate is transported throughout the plant via the vasculature after entering the cytoplasm of the plant (Harrison and van Buuren, 1995; Bucher, 2007).

According to studies, the host plant gives the AM fungus up to 20% of its photosynthetically fixed carbon, which is utilized to sustain and grow the fungus' hyphal network in the soil. However, the persistence of this cooperation has presented a conundrum for evolutionary theory since it is difficult to explain how self-centered people may take advantage of mutualisms and enjoy benefits without incurring any costs (Leigh, 2010). Recent investigations have shown that the flow of C to the fungus may be downregulated under sufficient nutritional regimes and that the fungus is also able to limit transfer of nutrients to less than beneficial hosts. Thus it has been proposed that biological market dynamics govern the C to nutrient exchange in mycorrhizal symbiosis and that reciprocal reward systems enable a "fair trade" between the two parties participating in AM symbiosis (Valentine et al., 2013; Fellbaum et al., 2014; Bücking and Kafle, 2015). Different macro- and micronutrients can be absorbed by AM fungi, but the likelihood of P uptake is higher than that of other nutrients. This could be attributed to the fungi's production of enzymes like phosphatase, which increases the solubility of insoluble P and subsequently the plant's ability to absorb it (Smith and Read, 2008). However, symbiotic effectiveness often declines when nutrient concentrations in the rhizosphere are higher due to the existence of nutrient receptors in plant cellular membranes that are negatively impacted . Thus the host's need on glomeromycotan fungus grows in low- and medium-nutrient environments (Valentine et al., 2013; Bhandari and Grag, 2017).

4.3.2 Functions and management of arbuscular mycorrhizal fungi against abiotic stresses

The abiotic stresses such as the presence of heavy metals, soil salinity, high temperature, extreme pH levels, transplant shock, and drought cause widespread losses to agricultural production (Modi et al., 2020; Patel et al., 2021; Singh et al., 2018). AM associations have been shown to benefit plants in resistance to abiotic and biotic factors by increasing mineral nutrients (primarily phosphorus) and water absorption (Smith and Read, 1997). Like the host plants, AM fungus vary in a variety of ways to adapt to abiotic stress. The development of arbuscules, spore germination, colonization, and sporulation are a few of the key ways that AM fungus adapts to stressful environments. While some studies have indicated that abiotic stress has an overall negative effect on mycorrhizal colonization (Zhang et al., 2019; Liu et al., 2020; Feng et al., 2020) and/or spore density (Krishnamoorthy et al., 2014; Zhang et al., 2016), others have stated that it promotes mycorrhizal colonization (Nakatani et al., 2011; Li et al., 2016; Mo et al., 2016) and/or spore density (Yang et al., 2015). Additionally, in AM fungal propagation systems, sporulation is typically encouraged by a period of drought stress before to harvest (Selvakumar et al., 2018). Arbuscules are regarded as the fundamental components of the AM symbiosis because they serve as the primary locations for nutrition exchange between two symbiotic partners (Harrison, 2012). It has been stated in many studies that abiotic stress greatly inhibits the formation of arbuscules (Feng et al., 2020; Liu et al., 2020).

Mycorrhizal fungi's hyphae are significant soil stabilizers. In many soils, the roots and hyphae, along with other organic elements, play a significant role in stabilizing soil aggregates, preventing erosion, and managing water (Piotrowski et al., 2004). Additionally, the AMF mycelial network is functionally related to the plants in their communities. It is thought that AMF stimulates growth by increasing the intake of nearly all essential nutrients while decreasing the uptake of Na and Cl (Evelin et al., 2012). Several studies have clarified how AMF works to absorb soil nutrients, particularly N and P, which can successfully encourage the growth of host plants. Many scientists have explained that AMF have the ability to absorb and transfer N to the nearby plants, which N is a premier growth limiting factor (Smith et al., 2011; Battini et al., 2017; Turrini et al., 2018). The creation of extensive hyphal networks and the excretion of glomalin, which in turn increase water and micro- and macronutrient intake, can be linked to the mycorrhizal fungi's ability to promote plant growth and improve soil quality (Gholamhoseini et al., 2013). Mycorrhizal plants are used especially against soil contamination mainly caused by metal toxicity. This effect, called phytoremediation, is caused by fungi and varies with host plant species, soil characteristics, and even AMF ecotypes (Heggo et al., 1990; Turnau et al., 2001; Ahmed et al., 2006; Yang et al., 2015). This protection is by improving the absorption of metal and metalloid ions and AMF can tolerate prolonged exposure to metals and metalloids to some extent (Oliveira et al., 2001). Saline soil is a common problem in cultivation, especially in regions with arid or semiarid climates. But, there are some salt-tolerant plants (halophytes) in nature that are highly colonized by AMF (Bothe, 2012). In some studies, it has been stated that AM can change plant physiology in a way that allows the plant to grow more efficiently in salty land (Al-Garni, 2006; Talaat and Shawky, 2011; Campanelli et al., 2013). By penetrating hyphae toward wet places, AMF aid in the management of water stress in a plant. It is common knowledge that plants colonized with AMF are better suited to grow in less fertile soil because of the increased absorption of mineral elements by hyphae (Bhantana et al., 2021). AM fungi can enhance plant tolerance to cold and heat stress. By improving secondary metabolites against cold stress and increasing protein stores, AMF strengthen the plant immune system and maintain the moisture balance in the host (Malhi et al., 2021). AMF-inoculated plants typically grow better under heat stress than noninoculated plants do (Gavito et al., 2005). For agriculture, a healthy AM community is crucial because symbiotic plants produce more biomass and crop yields, especially in nutrient-poor soils . With different fungus and host plant species, soil characteristics, and even AMF ecotypes, the symbiotic efficacy of AMF varies (Yang et al., 2015). For the optimum agricultural production, it is thought that AMF might potentially reduce the need of chemical fertilizers by up to 50%, but this estimate depends on the type of plant species and the prevailing stressful regimes (Begum et al., 2019). Throughout the last few decades due to changes in the global climate, excessive human activity, and environmental pollution, abiotic stresses are getting more and more severe. AMF's ability to improve crop yield and quality may aid in increased biotic and abiotic stress management (Singh et al., 2018).

4.3.3 Functions and management of arbuscular mycorrhizal fungi against biotic stresses

As a result of the negative effects of pesticide use against diseases and pests in agriculture on health and the environment, researchers have turned to beneficial organisms such as AMF. AMF are generally regarded as useful organisms in overcoming plant protection problems since they have been researched in various plant pathogen-host species combinations. The protection offered by AMF is particular to host plant species and microbial isolate (Pozo and Azcón-Aguilar, 2007). Different plants or genotypes of the same plant may respond to mycorrhizal life in different ways (Declerck et al., 1995; Sensoy et al., 2007). AMF interaction can increase the host plant's resistance to both above- and below-ground pests and diseases (Singh et al., 2020; Singh et al., 2017; Kumar et al., 2017). Numerous studies have shown that AMF may defend plants from a variety of plant pathogens, phytopathogenic insects, soil-borne nematodes, and parasitic plants (Whipps, 2004; Pozo and Azcón-Aguilar, 2007). The beneficial role of mycorrhizal fungi in promoting plant tolerance to biotic stress caused by soil-borne potential pathogens interacting with a wide variety of plant species has been supported by numerous researches (Pathak et al., 2022).

The effects of especially *Glomus* and *Gigaspora* species against various phytopathogenic fungi in different hosts were investigated in studies. Most studied pathogenic fungi: *Fusarium oxysporum* (Caron et al., 1985; Al-Momany and Al-Raddad, 1988; Sundaresan et al., 1993; Datnoff et al., 1995; Dugassa et al., 1996; Ozgonen and Erkilic, 2007; Bhagawati et al., 2000; Siddiqui and Singh, 2004; Akkopru and Demir, 2005; Hao et al., 2005; Kumari and Prabina, 2019a; Devi et al., 2022), other *Fusarium* genus (Siddiqui and Mahmood, 1995; Siddiqui et al., 1998; Elsayed Abdalla and Abdel-Fattah, 2000; Boby et al., 2008; Singh et al., 2010; Martínez-Medina et al., 2011), *Sclerotinia sclerotiorum* (Aysan and Demir, 2009; Mora-Romero et al., 2015), *Sclerotium rolfsii* (Kulkarni et al., 2011), *Macrophomina phaseolina* (Devi and Goswami, 1992; Akhtar and Siddiqui, 2006, 2007, 2008; Spagnoletti et al., 2020), *Rhizoctonia solani* (Guenoune et al., 2001; Abdel-Fattah and Shabana, 2002; Yao et al., 2002; Berta et al., 2005; Turhan and Demir, 2013; Demir et al., 2015; Demirer-Durak et al., 2018), *Verticillium dahliae* (Liu, 1995; Orak and Demir, 2011; Coşkun et al., 2021), *Pythium ultimum* (Rosendahl and Rosendahl, 1990; St-Arnaud et al., 1994), *Pythium aphanidermatum* (Kumari and

Srimeena, 2019b), *Aphanomyces euteiches* (Kjoller and Rosendahl, 1997; Bodker et al., 1998; Thygesen et al., 2004), and *Armillaria mellea* (Nogales et al., 2009). Less research has been done, and it appears to be less conclusive, on the effects of AM on diseases that are found above ground. Considering the few data on aerial fungal pathogens, it was determined that the studies mostly focused on *Alternaria alternata* (Nair et al., 2015), *Alternaria solani* (Fritz et al., 2006; De la Noval et al., 2007; Song et al., 2015; Boyno et al., 2022), *Botrytis cinerea* (Shaul et al., 1999; Møller et al., 2009; Pozo et al., 2010; Fiorilli et al., 2011), *Colletotrichum orbiculare* (Lee et al., 2005; Chandanie et al., 2006; Saldajeno and Hyakumachi, 2011), *Oidium* genus (Dugassa et al., 1996; Feldmann and Boyle, 1998); *Blumeria graminis* (Mustafa et al., 2016Mustafa *et al.*, 2017), *Erysiphe graminis* (Gernns et al., 2001), *Erysiphe pisi* (Liu et al., 2018), *Cladosporium fulvum* (Wang et al., 2017), and oomycota from the genus *Phytophthora* (Vestberg et al., 1994; Pozo et al., 2002; Ozgonen and Erkilic, 2007; de la Noval et al., 2007; Alejo-Iturvide et al., 2008; Gallou et al., 2011; Li et al., 2013).

Mostly *Glomus* and *Gigaspora* species have been studied against nematode damage in various plants. The most studied genus *Meloidogyne*, especially *Meloidogyne. incognita* (Pandey, 2005; Kantharaju et al., 2005; Siddiqui and Akhtar, 2006; Shreenivasa et al., 2007; Siddiqui and Akhtar, 2009; Vos et al., 2012), was followed by *Meloidogyne javanica* (Siddiqui and Mahmood, 1998; Calvet et al., 2001) and *Meloidogyne hapla* (Waceke et al., 2002; Masadeh et al., 2004), respectively. Other nematode genera examined for interaction were *Radopholus* (Umesh et al., 1988; Smith and Kaplan, 1988), *Heterodera* (Jain and Sethi, 1987; Tylka et al., 1991; Siddiqui et al., 2000), *Rotylenchulus* (Sitaramaiah and Sikora, 1982), and *Pratylenchus* (Hussey and Roncadori, 1978). In most of the studies, it was stated that there was a decrease in gall and egg numbers, nematode population, and damage in mycorrhizal plants. In some studies, it has been determined that AMF has no effect against nematode damage (O'Bannon and Nemec, 1979; Pinochet et al., 1997). Otherwise, the effect of AM fungi on viruses is more disease-promoting (Schönbeck, 1980; Dehne, 1982; Whipps, 2004). The viruses, which are obligate pathogens can benefit more from host infected plant and increase their concentrations since the mycorrhizal fungi promote plant growth and nutrition status of host plant (Linderman, 1994). Studies with Tobacco mosaic virus (Jabaji-Hare and Stobbs, 1984; Shaul et al., 1999), Potato leaf curl (Sipahioglu et al., 2009), Tomato spotted wilt virus (Miozzi et al., 2011), and Zucchini yellow mosaic virus (Coşkun et al., 2015) have reported increased disease severity and/or viral concentration. Between cucumber plants colonized by *Funneliformis mosseae* and control plants, Elsharkawy et al. (2012) found no significant differences in Cucumber mosaic virus infection. However, in another study, it was stated that *F. mosseae* had a positive effect against the disease caused by the Tomato yellow leaf curl Sardinia virus (Maffei et al., 2014). According to the results, tomato plants with mycorrhizal fungi showed fewer symptoms and contained less viral DNA. Jayaram and Kumar (1995) reported that yellow mosaic virus on mungbean decreased spore formation and mycorrhizal colonization. In another study, it was stated that the disease symptoms and viral concentration decreased in the presence of *Rhizophagus irregularis* and Tobacco mosaic virus, Cucumber green mottle mosaic virus in the plant compared to the control (Stolyarchuk et al., 2009). Thiem et al. (2014) stated that plants inoculated with *R. irregularis* and Potato virus Y observed slight symptoms and significant stimulation of shoot growth.

AMF has a protective effect against phytopathogenic bacteria, especially *Xanthomonas campestris*, by induction of systemic resistance (Liu et al., 2007; Mora-Romero et al., 2015; Cervantes-Gámez et al., 2016). It has been determined that mycorrhizae are also effective in suppressing diseases caused by phytoplasmas. Lingua et al. (2002) stated that the disease severity caused by the Stolbur group decreased in plants with *F. mosseae*. Likewise, *F. mosseae* decreased the amount of plants infected with chrysanthemum yellows phytoplasma (D'Amelio et al., 2007). The feeding mode and lifestyle of the pest are important in the relationship between mycorrhizal plants and insects (Koricheva et al., 2009). Generally, rhizophagous insects were negatively affected in mycorrhizal plant-insect experiments, both in controlled and greenhouse/field conditions, while the effect was weaker in shoot-fed ones. Furthermore, leaf-chewing insects are typically adversely impacted by the relationship, although aphids typically perform better on mycorrhizal plants. The detrimental impact on leaf-chewing insects is probably due to their susceptibility to jasmonate- or jasmonic acid-dependent defenses that are amplified in plants with mycorrhizal inoculation. In fact, phloem-feeding insects typically perform better on crops with AMF colonization than on non-AM crops (Hartley and Gange, 2009; Koricheva et al., 2009). A quick flux for aphid-repellent volatiles compounds can be elicited prior to the attack of the plant because aphid-infested crops can communicate signals via common mycelial networks to an uninfected plant. This can prevent the invasion from spreading or becoming more severe, which will reduce productivity (Babikova et al., 2014). The relationship of AMF with some parasitic plants that cause economic problems has also been investigated by some researchers. Some studies revealed that AMF adversely affected the germination of *Striga hermonthica* seeds and delayed the period of emergence for *Striga* (Othira, 2012; Isah et al., 2013; Manjunatha et al., 2018). The increased resilience of mycorrhizal plants to soil pathogens may be caused by many mechanisms acting simultaneously. Several mechanisms may be involved in mycorrhizal symbiosis against plant pathogens: (1) creating a physical barrier and preventing the attack and growth of pathogenic microorganisms; (2) making the cell wall thicker by creating a large amount of complex carbohydrates and depositing lignin, which together prevent root disease penetration; (3) causing host plants to produce biochemical substances (such as alkaloids) that increase the host's capacity for microbial resistance; (4) promoting flavonoid wall infusions; (5) increasing the concentration of orthodihydroxy phenols in roots, which discourages pathogenic invasion; (6) synthesizing antifungal and antibacterial antibiotics and toxins; (7) increased microbial activity of the rhizosphere and competition between them; and (8) changing the amount and structure of chemicals produced by the plant root.

When the results from the studies done so far are summarized: (1) AM interactions reduce plant pathogen damage, particularly those caused by fungi and nematodes; (2) although not equally efficient in all crops, AM symbiosis increases resistance or tolerance in roots; (3) protection is not effective against all pathogens; and (4) soil and other environmental factors influence disease protection. Understanding the mechanisms of plant disease resistance in mycorrhiza-infected plants is essential to develop an effective crop production system and sustainable agriculture (Kumar et al., 2021). To create innovative classes of biocides that can also reduce risk to both ecological health and human health, compositions with mycorrhizae are very necessary.

4.3.4 Functions of arbuscular mycorrhizal fungi in soil structure and quality

Degradation of the ecosystem affects the physical, chemical, and biological properties of the soil. The health and production of plants depends on the soil, and the quality of the soil is completely dependent on the survival and diversity of its biota (Doran and Linn, 1994). The importance of soil protection as a source of agricultural activities and as a dynamic and complex system for the stability and fertility of agricultural ecosystems is inevitable. The goal of sustainability in agriculture is maximum plant production with minimum soil damage. In sustainable agricultural systems, the role of mycorrhizal fungi is important as the interface between plants and soil (Miller and Jastrow, 1994). AM fungi strengthen the soil particles and help in soil aggregation and conservation processes (Dodd, 2000). These fungi contribute in soil aggregate formation by different mechanisms: (1) The growth of fungal hyphae in the soil matrix forms a skeleton-like structure that entangles the main soil particles and keeps them firmly together (Miller and Jastrow, 1994). (2) The roots and fungal hyphae together create a physical and chemical environment to produce amorphous and organic materials for the binding of soil particles (Tisdall, 1991). (3) The roots and fungal hyphae produce macroaggegate structures from microaggegates that increase the carbon and nutrient storage capacity, which provide a suitable habitat for soil microorganisms (Cambardella and Elliott, 1994). (4) AM fungal hyphae can produce the glomalin which accumulates in soil and helps in soil aggregation and soil stability (Wright and Upadhyaya, 1998). (5) These fungi release exudates in the soil which helps in soil aggregation stability as well as better microorganisms growth (Johnson et al., 2002). The effects of AM fungi in soil structure improvement has direct impact on soil microbial communities by improving soil aeration and moisture infiltration and indirect effect on host plant root growth stimulation. So, the mycorrhizal plants are useful for soil reclamation, especially sand dunes. Glomalin is a kind of glycoprotein which is produced by AM fungal hyphae and improves soil aggregation and stabilization (Wright and Upadhyaya, 1998). There is a strong correlation between glomalin as well as glomalin-related soil proteins with soil aggregation and stabilization (Wright and Anderson, 2000; Rillig, 2004). It has been demonstrated that soil ecosystem productivity, soil aeration, drainage, as well as microbial activities have been enhanced by glomalin production (Lovelock et al., 2004).

4.3.5 Functions of arbuscular mycorrhizal fungi in contaminated soils bioremediation

The heavy metals pollution has negative impacts on soil microorganisms' activities as microbial processes. All the microorganisms such as AM fungi can show resistance against heavy metals by two manners: "tolerance" when can be survived in high metal concentrations and "avoidance" when be able to restrict metal uptake. Using of plants to remove or reduce soil contamination by organic compounds or heavy metals (phytoremediation) has been considered as a useful, cost-effective, and eco-friendly strategy, and AM fungi potentially play a key role in such strategies (Leyval et al., 2002; Turnau et al., 2006; Khade and Adholeya, 2007; Sheoran et al., 2010). This AM fungal capacity commonly referred as bioremediation (Leyval et al., 2002; Göhre and Paszkowski, 2006) can be done by two ways: AM fungi can accumulate and sequester toxic ions and protect the host plant from them or they deliver the toxic ions to the host plant such as essential minerals (Cu and Zn) and accumulate the heavy metals in the host (Weissenhorn et al., 1995; Diaz et al., 1996; Gonzalez-Chavez et al., 2004). In the first technique, the host plant can be produced under polluted condition but with minimal contamination. In the second technique, the host plants should be harvested and destroyed to remove the pollution in the site (phytoextraction) (Burns et al., 1996; Khan et al., 2000). In both methods, the AM fungi should be heavy metal-tolerant strain. The phytoextraction technique requires host plants which tolerate high toxic metals concentration and at the same time have a large shoot biomass to accumulate heavy metals ions. There are several laboratory studies to explore the effect of AM fungi in bioremediation of polluted soils; however, only few field studies have been carried out in large-scale conditions (Burns et al., 1996; Adriano et al., 2004; Chibuike, 2013) and there are few companies that offer AM fungal products for bioremediation purposes worldwide. Some of the challenges include that different AM fungal strains have different responses to metal toxicity. It means that AM fungal strain which colonizes host plant and the abundance of its extra-radical mycelium determines the AM fungal potential for bioremediation (Diaz et al., 1996), which depend on fungal strain adaptation in polluted condition (Nelson and Safir, 1982). The most heavy metals accumulating host plants mainly are not AM fungal host plants (e.g., the crucifers). Also, in most cases, the AM fungal colonization is reduced in high heavy metals concentrations. *Glomus caledonium* seems an effective fungal strain for bioremediation of contaminated soils (Liao et al., 2003).

4.4 Commercial application of arbuscular mycorrhizal fungi

The development of AMF technology is needed to produce more effectively, apply combinations of minimum effective propagation to crops, identify species, and assesses mycorrhizal viability, all of which are priorities in mycorrhizal research for the sustainability of agriculture (Benami et al., 2020). The production and use of AMF inoculums has been the mycorrhizal market's primary emphasis outside of these categories due to the difficulty of these operations. Throughout the history of this sector, producing mycorrhizal inoculants that are affordable has been a challenging goal. Mycorrhizal inoculation in agricultural areas has, nevertheless, shown considerable yield improvements in a number of crop kinds, as recorded in several field studies (Hijri, 2016; Pellegrino et al., 2015; Benami et al., 2020). However, it might be claimed that the development of next-generation mycorrhizal technology should not just be focused on production and inoculum issues (Rillig et al., 2016).

Commercialization of AMF inoculums makes it possible to use new and more productive isolates of AMF to replace the less productive natural AMF isolates already present in the soil. It is thought that the effect of inoculated AMFs will decrease when they stay in the soil for a long time, but they can still be sporulated. Therefore commercialized AMF inoculums have a special place in sustainable agriculture (Faye et al., 2013). Only one application of these commercial inoculums to plants may be sufficient for many years. However, it can be said that the application of AMF inoculum to plants at optimum levels may also be beneficial (Sharma et al., 1996; Adholeya et al., 2005; Faye et al., 2013). The application of these commercialized inoculums can be divided into four main methods: broadcasting, in-furrow application, seed coating, and seedling inoculation (Muresu et al., 2003; Adholeya et al., 2005; Malusá et al., 2012; Basiru et al., 2021) (Fig. 4.1).

4.4.1 Broadcasting method

Inoculum and seeds are dispersed across the soil's surface and briefly mixed together (Koziol, 2017). This can be done manually, with fertilization equipment, or by utilizing a field seeder. When the seeds germinate, the propagules in the inoculum will grow and colonize the seedlings (Adholeya et al., 2005; Benami et al., 2020). In contrast, only propagules that are close to the seeds will colonize the seedlings, necessitating a significant inoculum to achieve colonization.

4.4.2 In-furrow application method

Other methods that are actively used and promoted globally include various types of in-furrow application (Bashan, 1998; Benami et al., 2020). This method

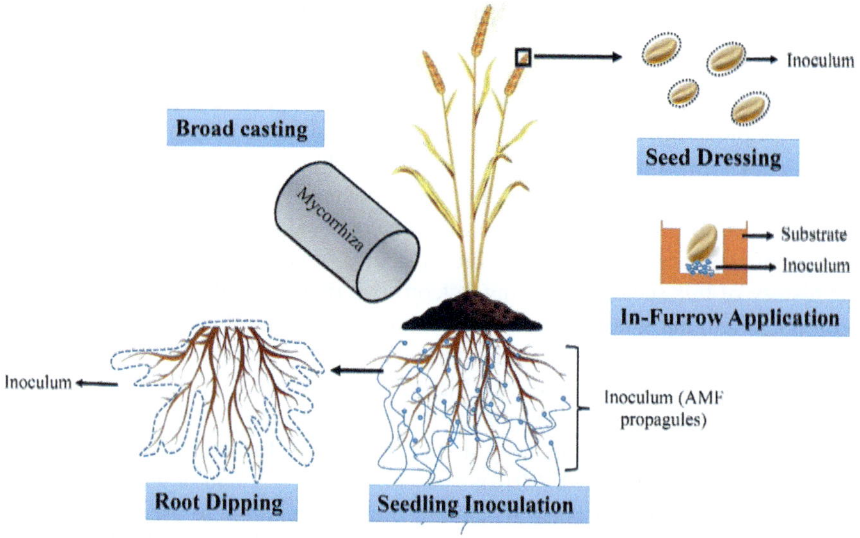

FIGURE 4.1

Schematic representation of application methods of commercialized arbuscular mycorrhizal fungi inoculums.

involves digging a shallow ditch slightly below the seed sowing depth and dropping the inoculums inside of it. On top of this inoculum, seeds are spread and then covered with dirt. The inoculum layer will colonize the developing roots when the seeds germination occurs. Though labor-intensive across a vast region, this technique achieves significant levels of mycorrhizal colonization (Adholeya et al., 2005).

4.4.3 Seed dressing method

An alternative inoculation method is this application, also referred to as seed coating. Here, the inoculum is created by combining it with a sticky carrier, such as acacia gum, to coat the seed with inoculum. After drying, the coated seed is sown in the fields. Because the inoculum is in direct touch with the seed in this manner, seedlings will be quickly colonized (Fernández et al., 2011). Additionally, it takes little work and inoculum (Sieverding, 1991).

4.4.4 Root dipping method

In general, this approach is advised for micro-propagation and plants cultivated in nurseries (Pellegrino et al., 2015; Hijri, 2016; Benami et al., 2020). These plants' roots are first thoroughly rinsed in water before being submerged for 5 minutes in an inoculum that has been combined with a carrier. After the inoculum sticks to the roots, the seedlings are transplanted using conventional procedures. By placing the inoculum in direct contact with the seedling's roots, colonization will take place (Constantino et al., 2008).

4.4.5 Seedling inoculation method

Producing AMF-inoculated seedlings for large-scale field planting is more labor-intensive (Adholeya et al., 2005). Colonized seedlings are raised in nurseries and are based on the idea that AMF has colonized the seedlings (Sulistiono, 2020). Typically, this method is used to inoculate young forest tree seedlings.

4.5 Future perspectives and challenges

The management of an appropriate nutrient supply constitutes a crucial requirement for having a sustainable agricultural system, and in this context, mycorrhizae participation cannot be neglected. AMF spores attack the roots of plants and improve soil quality, moisture retention, production, and management of nutrients for sustainability. AMF play a significant role in minimizing the negative effects of agricultural chemical inputs (such as pesticides and nutrients) since there is growing awareness and demand for the need to promote sustainability in agricultural growth. Viable AMF is an economical and harmless way to achieve higher yields that can lead to the development of a low-input farming system.

Although research on AMF has gained great momentum recently, bottlenecks in commercialization of these studies need to be carefully identified A major drawback is the difficulty to grow AMF in pure culture in particular. It can only be grown with plants by adding inoculum under specific circumstances, and it is difficult to produce in large quantities under laboratory conditions. Currently, a nonsterile medium, soil, and propagule (spores, hyphae, and colonized root pieces) are used to form the mycorrhizal inoculum in the majority of samples. However, in the coming days, it is expected that these bottlenecks can be overcome by using techniques based on organic processes. It is also obvious that current and future developments in AMF field research should focus on maximizing production, enhancing cost-effectiveness, improving its quality and profit return, conserving biodiversity, and protecting the environment given the constraints currently in place related to a large population, a need for food, and ecological changes. By integrating diverse disciplines in specific ways to create an interdisciplinary link and affiliation of all scholars in the subject, these constraints can be overcome. By carefully applying the principles in accordance with the climate, soil, and current markets, sustainable agricultural systems are likely to suit the needs of the general public in terms of production, efficiency, and management.

The favorable effect of AMF colonization is now most likely not nutritional because of their impact on soil aggregates, activity, and plant defenses. Therefore research in the near future must concentrate on enhancing AM effects on nutrient uptake to maintain productivity while also maximizing the sustainability of output. It is possible to make significant progress toward obtaining food supply in the near future in a more sustainable agricultural system by identifying and improving those characteristics in new cultivars that are linked with AMF accessibility, functionality, and climate resilience.

References

Abdel-Fattah, G.M., Shabana, Y.M., 2002. Efficacy of the arbuscular mycorrhizal fungus *Glomus clarum* in protection of cowpea plants against root-rot pathogen *Rhizoctonia solani*. J. Plant. Dis. Protec. 109, 207–215.

Adholeya, A., Tiwari, P., Singh, R., 2005. Large-scale inoculum production of arbuscular mycorrhizal fungi on root organs and inoculation strategies. In Vitro Culture of Mycorrhizas. Springer, Berlin, Heidelberg, pp. 315–338.

Adriano, D.C., Wenzel, W.W., Vangronsveld, J., Bolan, N.S., 2004. Role of assisted natural remediation in environmental cleanup. Geoderma. 122, 121–142.

Ahmed, F.R.S., Killham, K., Alexander, I., 2006. Influences of arbuscular mycorrhizal fungus Glomus mosseae on growth and nutrition of lentil irrigated with arsenic contaminated water. Plant. Soil. 283, 33–41.

Akhtar, M.S., Siddiqui, Z.A., 2006. Effects of phosphate solubilizing microorganisms on the growth and root-rot disease complex of chickpea. Mikol. Fitopatol. 40, 246–254.

Akhtar, M.S., Siddiqui, Z.A., 2007. Effects of *Glomus fasciculatum* and *Rhizobium* sp. on the growth and root-rot disease complex of chickpea. Arch. Phytopathol. Plant. Protec. 40, 37–43.

Akhtar, M.S., Siddiqui, Z.A., 2008. Biocontrol of a root-rot disease complex of chickpea by *Glomus intraradices*, *Rhizobium* sp. and *Pseudomonas straita*. Crop. Protec 27, 410–417.

Akkopru, A., Demir, S., 2005. Biocontrol of Fusarium wilt in tomato caused by *Fusarium oxysporum* f. sp. *lycopersici* by AMF *Glomus intraradices* and some rhizobacteria. J. Pyhtopathol 153, 544–550.

Alejo-Iturvide, F., Márquez-Lucio, M.A., Morales-Ramírez, I., Vázquez-Garcidueñas, M. S., Olalde-Portugal, V., 2008. Mycorrhizal protection of chili plants challenged by *Phytophthora capsici*. Eur. J. Plant. Pathol. 120, 13–20.

Al-Garni, S.M.S., 2006. Increasing NaCI-salt tolerance of a halophytic plant *Phragmites australis* by mycorrhizal symbiosis. Am.-Eurasian J. Agric. Environ. Sci. 1, 119–126.

Al-Momany, A., Al-Raddad, A., 1988. Effect of vesicular-arbuscular mycorrhizae on *Fusarium* wilt of tomato and pepper. Alexand. J. Agric. Res. 33, 249–261.

Augé, J.M., 2001. Water relations, drought and vesicular-arbuscular mycorrhizal symbiosis. Mycorrhiza. 11, 3–42.

Aysan, E., Demir, S., 2009. Using arbuscular mycorrhizal fungi and *Rhizobium leguminosarum*, biovar phaseoli against *Sclerotinia sclerotiorum* (Lib.) de bary in the common bean (*Phaseolus vulgaris* L.). Plant. Pathol. J. 8, 74–78.

Babikova, Z., Gilbert, L., Bruce, T., Dewhirst, S.Y., Pickett, J.A., Johnson, D., 2014. Arbuscular mycorrhizal fungi and aphids interact by changing host plant quality and volatile emission. Func. Ecol. 28, 375–385.

Balzergue, C., Chabaud, M., Barker, D.G., Becard, G., Rochange, S.F., 2013. High phosphate reduces host ability to develop arbuscular mycorrhizal symbiosis without affecting root calcium spiking responses to the fungus. Front. Plant. Sci. 4, 426.

Barea, J.M., Pozo, M.J., López-Ráez, J.A., Aroca, R., Ruíz-Lozano, J.M., Ferrol, N., et al., 2014. Arbuscular mycorrhizas and their significance in promoting soil-plant system sustainability against environmental stresses. Beneficial Plant-Microbial Interactions Ecology and Applications. CRC, Taylor & Francis, pp. 353–387.

Bashan, Y., 1998. Inoculants of plant growth-promoting bacteria for use in agriculture. Biotechnol. Adv. 16 (4), 729–770.

Basiru, S., Mwanza, H.P., Hijri, M., 2021. Analysis of arbuscular mycorrhizal fungal inoculant benchmarks. Microorganisms. 9 (1), 81.

Battini, F., Grønlund, M., Agnolucci, M., Giovannetti, M., Jakobsen, I., 2017. Facilitation of phosphorus uptake in maize plants by mycorrhizosphere bacteria. Sci. Rep. 7, 4686.

Begum, N., Qin, C., Ahanger, M.A., Raza, S., Khan, M.I., Ashraf, M., et al., 2019. Role of arbuscular mycorrhizal fungi in plant growth regulation: implications in abiotic stress tolerance. Front. Plant. Sci. 10, 1068.

Benami, M., Isack, Y., Grotsky, D., Levy, D., Kofman, Y., 2020. The economic potential of arbuscular mycorrhizal fungi in agriculture. Grand Challenges in Fungal Biotechnology. Springer, Cham, pp. 239–279.

Berta, G., Sampo, S., Gamalero, E., Musasa, N., Lemanceau, P., 2005. Suppression of Rhizoctonia root-rot of tomato by *Glomus mosseae* BEG 12 and *Pseudomonas fluorescens* A6RI is associated with their effect on the pathogen growth and on the root morphogenesis. Eur. J. Plant. Pathol. 111, 279–288.

Bethlenfalvay, G.J., Schuepp, H., 1994. Arbuscular mycorrhizae and agrosystem stability. Impact of arbuscular mycorrhizae on sustainable agriculture and natural ecosystems. Birkhause 117–131.

Bhagawati, B., Goswami, B.K., Singh, S., 2000. Management of disease complex of tomato caused by *Meloidogyne incognita* and *Fusarium oxysporum* f. sp. *lycopersici* through bioagent. Indian. J. Nematol. 30, 16–22.

Bhandari, P., Grag, N., 2017. Dynamics of arbuscular mycorrhizal symbiosis and its role in nutrient acquisition: an overview. Mycorrhiza – Nutrient Uptake, Biocontrol, Ecorestoration. Springer, pp. 21–44.

Bhantana, P., Rana, M.S., Sun, X.C., Moussa, M.G., Saleem, M.H., Syaifudin, M., et al., 2021. Arbuscular mycorrhizal fungi and its major role in plant growth, zinc nutrition, phosphorous regulation and phytoremediation. Symbiosis. 84 (1), 19–37.

Bindraban, P.S., van der Velde, M., Ye, L., van den Berg, M., Materechera, S., Kiba, D.I., et al., 2012. Assessing the impact of soil degradation on food production. Cur. Opin. Environ. Sustain. 4, 478–488.

Boby, V.U., Balakrishna, A.N., Bagyaraj, D.J., 2008. Interaction between *Glomus mosseae* and soil yeasts on growth and nutrition of cowpea. Microbiol. Res. 163, 693–700.

Bodker, L., Kjoller, R., Rosendahl, S., 1998. Effect of phosphate and arbuscular mycorrhizal fungus *Glomus intraradices* on disease severity of root rot of peas (*Pisum sativum*) caused by *Aphanomyces euteiches*. Mycorrhiza 8, 169–174.

Boiffin, J., Malézieux, E., Picard, D., 2001. Cropping systems for the future. In: Nosberger, J., Geiger, H.H., Struik, P.C. (Eds.), Crop Science: Progress and Prospects. CAB International, Oxford, UK, pp. 261–279.

Bothe, H., 2012. Arbuscular mycorrhiza and salt tolerance of plants. Symbiosis. 58, 7–16.

Boyno, G., Demir, S., Danesh, Y.R., 2022. Effects of biological control agents against *Alternaria solani* (Ell. and G. Martin) Sor. in tomato. Eur. J. Plant. Pathol. 162 (1), 19–29.

Bucher, M., 2007. Functional biology of plant phosphate uptake at root and mycorrhiza interfaces. N. Phytol. 173, 11–26.

Bücking, H., Kafle, A., 2015. Role of arbuscular mycorrhizal fungi in the nitrogen uptake of plants: current knowledge and research gaps. Agronomy. 5, 87–612.

Burns, R.G., Rogers, S.L., McGhee, I., 1996. Remediation of inorganics and organics in industrial and urban contaminated soils. In: Naidu, R., Kookana, R.S., Oliver, D.P., McLaughlin, M.J. (Eds.), Contaminants and the Soil Environment in the Australia Pacific region. Kluwer Academic Publishers, London, pp. 361–410.

Calvet, C., Pinochet, J., Hernandez-Dorrego, A., Estaun, V., Camprubi, A., 2001. Field microplot performance of the peach-almond GF-677 after inoculation with arbuscular mycorrhizal fungi in a replant soil infested with root-knot nematode. Mycorrhiza. 10, 295–300.

Cambardella, C.A., Elliott, E.T., 1994. Carbon and nitrogen dynamics of soil organic matter fractions from cultivated grassland soils. Soil. Sci. Soc. Am. J. 58, 123–130.

Campanelli, A., Ruta, C., Mastro, G.D., Morone-Fortunato, I., 2013. The role of arbuscular mycorrhizal fungi in alleviating salt stress in *Medicago sativa* L. var. icon. Symbiosis. 59, 65–76.

Caron, M., Fortin, J.A., Richard, C., 1985. Influence of substrate on the interaction of *Glomus intraradices* and *Fusarium oxysporum* f. sp. *radicis-lycopersici* on tomatoes. Plant. Soil. 87, 233–239.

Cervantes-Gámez, R.G., Bueno-Ibarra, M.A., Cruz-Mendívil, A., et al., 2016. Arbuscular mycorrhizal symbiosis-induced expression changes in *Solanum lycopersicum* leaves revealed by RNA-seq analysis. Plant. Mol. Biol. Rep. 34, 89–102.

Chandanie, W.A., Kubota, M., Hyakumachi, M., 2006. Interactions between plant growth promoting fungi and arbuscular mycorrhizal fungus *Glomus mosseae* and induction of systemic resistance to anthracnose disease in cucumber. Plant. Soil. 286, 209–217.

Chibuike, G.U., 2013. Use of mycorrhiza in soil remediation: a review. Sci. Res. Essays. 8, 1697. 1687.

Clermont-Dauphin, C., Meynard, J.M., 1997. L'emploi des pesticides et des engraisen Agriculture. In: Stengel, P., Gelin, S. (Eds.), Sols: Interfaces Fragiles. INRA, Paris, pp. 175–186.

Constantino, M., Gomez-Alvarez, R., Álvarez-Solís, J.D., Geissen, V., Huerta, E., Barba, E., 2008. Effect of inoculation with rhizobacteria and arbuscular mycorrhizal fungi on growth and yield of Capsicum chinense Jacquin. J. Agric. Rur. Develop. Tropics. Subtropics. 109 (2), 169–180.

Coşkun, A., Demir, S., Sipahioğlu, H.M., 2015. The effect of arbuscular mycorrhizal fungus (AMF) on Zucchini yellow mosaic virus (ZYMV) in summer squash cultivars. J. Turk. Phytopath. 44 (1–2-3), 11–21.

Coşkun, F., Demir, S., Alptekin, Y., 2021. The effectiveness of arbuscular mycorrhizal fungi and salicylic acid against *Verticillium dahliae* infecting pepper (*Capsicum annuum* L.). Appl. Ecol. Environment. Res. 19 (6), 5045–5057.

Craswell, E.T., Karjalainen, U., 1990. Recent research on fertilizer problems in Asian agriculture. Fert. Res. 26, 243–248.

D'Amelio, R., Massa, N., Gamalero, E., D'Agostino, G., Sampo, S., Berta, G., et al., 2007. Preliminary results on the evaluation of the effects of elicitors of plant resistance on chrysanthemum yellows phytoplasma infection. Bull. Insec. 60, 317.

Datnoff, L.E., Nemec, S., Pernezny, K., 1995. Biological control of Fusarium crown and root rot of tomato in Florida using *Trichoderma harzianum* and *Glomus intraradices*. Biol. Cont. 5, 427–431.

Davet, P., 1996. Vie microbienne du sol et production végétale. INRA, Paris.

De la Noval, B., Pérez, E., Martínez, B., León, O., Martínez-Gallardo, N., Délano-Frier, J., 2007. Exogenous systemin has a contrasting effect on disease resistance in mycorrhizal tomato (*Solanum lycopersicum*) plants infected with necrotrophic or hemibiotrophic pathogens. Mycorrhiza. 17, 449–460.

Declerck, S., Plenchette, C., Strullu, D.G., 1995. Mycorrhizal dependency of banana *(Musa acuminate*, AAA group) cultivar. Plant. Soil. 176, 183–187.

Dehne, H.W., 1982. Interactions between vesicular-arbuscular mycorrhizal fungi and plant pathogens. Phytopathology. 72, 1115–1119.

Delavaux, C.S., Smith-Ramesh, L.M., Kuebbing, S.E., 2017. Beyond nutrients: a meta-analysis of the diverse effects of arbuscular mycorrhizal fungi on plants and soils. Ecology. 98 (8), 2111–2119.

Demir, S., Akköprü, A., 2007. Using of arbuscular mycorrhizal fungi (AMF) for biocontrol of soil-borne fungal plant pathogens. Biological Control of Plant Diseases. Haworth Press, NY, USA, pp. 17–37.

Demir, S., Şensoy, S., Ocak, E., Tüfenkci, Ş., Durak, E.D., Erdinc, C., et al., 2015. Effects of arbuscular mycorrhizal fungus, humic acid, and whey on wilt disease caused by *Verticillium dahliae* Kleb. in three solanaceous crops. Turk. J. Agric. For. 39 (2), 300–309.

Demirer-Durak, E., Demir, S., Ocak, E., 2018. Evaluation of arbuscular mycorrhizal fungi (AMF) and whey on Rhizoctonia disease of potato caused by *Rhizoctonia solani* Kuhn. Fres. Environ. Bull. 27 (12A), 8773–8778.

Devi, T.P., Goswami, B.K., 1992. Effect of VA-mycorrhiza on the disease incidence due to *Macrophomina phaseolina* and *Meloidogyne incognita* on cowpea. Ann. Agric. Res. 13, 253–256.

Devi, N.O., Tombisana Devi, R.K., Debbarma, M., Hajong, M., Thokchom, S., 2022. Effect of endophytic Bacillus and arbuscular mycorrhiza fungi (AMF) against Fusarium wilt of tomato caused by *Fusarium oxysporum f. sp. lycopersici*. Egypt. J. Biol. Pest. Control. 32 (1), 1–14.

Diaz, G., AzconAguilar, C., Honrubia, M., 1996. Influence of arbuscular mycorrhizae on heavy metal (Zn and Pb) uptake and growth of *Lygeum spartum* and *Anthyllis cytisoides*. Plant. Soil. 180, 241–249.

Dickie, I.A., Alexander, I., Lennon, S., Opik, M., Selosse, M.A., van der Heijden, M.G.A., et al., 2015. Evolving insights to understanding mycorrhizas. N. Phytol. 205, 1369–1374.

Dodd, J.C., 2000. The role of arbuscular mycorrhizal fungi in natural ecosystems. Out. Agric. 29 (1), 55–62.

Doran, J.W., Linn, D.M., 1994. Microbial ecology of conservation management systems. In: Hatfield, J.L., Stewart, B.A. (Eds.), Soil Biology: Effects on Soil Quality. Lewis, Boca Raton, FL, pp. 1–57.

Dugassa, G.D., Von Alten, H., Schönbeck, F., 1996. Effects of arbuscular mycorrhiza (AM) on health of *Linum usitatissimum* L. infected by fungal pathogens. Plant. Soil. 185, 173–182.

Elsayed Abdalla, M., Abdel-Fattah, G.M., 2000. Influence of endomycorrhizal fungus *Glomus mosseae* on the development of peanut pod rot disease in Egypt. Mycorrhiza. 10, 29–35.

Elsharkawy, M.M., Shimizu, M., Takahashi, H., Hyakumachi, M., 2012. The plant growth-promoting fungus *Fusarium equiseti* and the arbuscular mycorrhizal fungus *Glomus mosseae* induce systemic resistance against *Cucumber mosaic virus* in cucumber plants. Plant. Soil. 361, 397–409.

Erman, M., Demir, S., Ocak, E., Tüfenkçi, Ş., Oğuz, F., Akköprü, A., 2011. Effects of rhizobium, arbuscular mycorrhiza and whey applications on some properties in chick pea (*Cicer arietinum* L.) under irrigated and rainfed conditions 1 - yield, yield components, nodulation and AMF colonization. Field Crop. Res. 122, 14–24.

Evelin, H., Giri, B., Kapoor, R., 2012. Contribution of *Glomus intraradices* inoculation to nutrient acquisition and mitigation of ionic imbalance in NaCl-stressed *Trigonella foenum-graecum*. Mycorrhiza. 22, 203–217.

Faye, A., Dalpé, Y., Ndung'u-Magiroi, K., Jefwa, J., Ndoye, I., Diouf, M., et al., 2013. Evaluation of commercial arbuscular mycorrhizal inoculants. Can. J. Plant. Sci. 93 (6), 1201–1208.

Feldmann, F., Boyle, C., 1998. Concurrent development of arbuscular mycorrhizal colonization and powdery mildew infection on three *Begonia hiemalis* cultivars. J. Plant. Dis. Protect. 105, 121–129.

Fellbaum, C.R., Mensah, J.A., Cloos, A.J., Strahan, G.E., Pfeffer, P.E., Kiers, E.T., et al., 2014. Fungal nutrient allocation in common mycorrhizal networks is regulated by the carbon source strength of individual host plants. N. Phytol. 203, 646–656.

Feng, Z., Liu, X., Zhu, H., Yao, Q., 2020. Responses of arbuscular mycorrhizal symbiosis to abiotic stress: a lipid-centric perspective. Front. Plant. Sci. 11, 578919.

Fernández, F., Dell'Amico, J.M., Angoa, M.V., de la Providencia, I.E., 2011. Use of a liquid inoculum of the arbuscular mycorrhizal fungi *Glomus hoi* in rice plants cultivated in a saline Gleysol: a new alternative to inoculate. J. Plant. Breed. Crop. Sci. 3 (2), 24–33.

Fiorilli, V., Catoni, M., Francia, D., et al., 2011. The arbuscular mycorrhizal symbiosis reduces disease severity in tomato plants infected by *Botrytis cinerea*. J. Plant. Pathol. 93, 237–242.

Frankland, J.C., 1998. Fungal succession – unravelling the unpredictable. Mycol. Res. 102, 1–15.

Fritz, M., Jakobsen, I., Lyngkjær, M.F., et al., 2006. Arbuscular mycorrhiza reduces susceptibility of tomato to *Alternaria solani*. Mycorrhiza. 16, 413–419.

Gallou, A., Mosquera, H.P.L., Cranenbrouck, S., et al., 2011. Mycorrhiza induced resistance in potato plantlets challenged by *Phytophthora infestans*. Physiol. Mol. Plant. Pathol. 76, 20–26.

Gavito, M.E., Olsson, P.A., Rouhier, H., Medinapeñafiel, A., Jakobsen, I., Bago, A., 2005. Temperature constraints on the growth and functioning of root organ cultures with arbuscular mycorrhizal fungi. N. Phytol. 168, 179–188.

Gernns, H., Alten, H., Poehling, H.M., 2001. Arbuscular mycorrhiza increased the activity of a biotrophic leaf pathogen – is a compensation possible? Mycorrhiza. 11, 237–243.

Gholamhoseini, M., Ghalavand, A., Dolatabadian, A., Jamshidi, E., Khodaei- Joghan, A., 2013. Effects of arbuscular mycorrhizal inoculation on growth, yield, nutrient uptake and irrigation water productivity of sunflowers grown under drought stress. Agric. Water Manag 117, 106–114.

Godfray, H.C.J., Beddington, J.R., Crute, I.R., Haddad, L., Lawrence, D., Muir, J.F., et al., 2010. Food security: the challenge of feeding 9 billion people. Science. 327, 812–818.

Göhre, V., Paszkowski, U., 2006. Contribution of the arbuscular mycorrhizal symbiosis to heavy metal phytoremediation. Planta. 223, 1115–1122.

Gomiero, T., 2016. Soil degradation, land scarcity and food security: reviewing a complex challenge. Sustainability. 8, 281.

Gonzalez-Chavez, M.C., Carrillo-Gonzalez, R., Wright, S.F., Nichols, K.A., 2004. The role of glomalin, a protein produced by arbuscular mycorrhizal fungi, in sequestering potentially toxic elements. Environ. Pollut. 130, 317–323.

Grassini, P., Eskridge, K.M., Cassman, K.G., 2013. Distinguishing between yield advances and yield plateaus in historical crop production trends. Nat. Commun 4, 2918.

Guenoune, D., Galili, S., Phillips, D.A., Volpin, H., Chet, I., Okon, Y., et al., 2001. The defense response elicited by the pathogen *Rhizoctonia solani* is suppressed by colonization of the AM fungus *Glomus intraradices*. Plant. Sci. 160, 925–932.

Hao, Z., Christie, P., Qin, L., Wang, C., Li, X., 2005. Control of Fusarium Wilt of cucumber seedlings by inoculation with an arbuscular mycorrhical fungus. J. Plant. Nutr. 28, 1961–1974.

Harrison, M.J., 2012. Cellular programs for arbuscular mycorrhizal symbiosis. Curr. Opin. Plant. Biol. 15, 691–698.

Harrison, M.J., van Buuren, M.L., 1995. A phosphate transporter from the mycorrhizal fungus *Glomus versiforme*. Nature. 378, 626–629.

Hartley, S.E., Gange, A.C., 2009. Impacts of plant symbiotic fungi on insect herbivores: mutualism in a multitrophic context. Ann. Rev. Entomol. 54, 323–342.

Heggo, A., Angle, J.S., Chaney, R.L., 1990. Effects of vesicular-arbuscular mycorrhizal fungi on heavy metal uptake by soybeans. Soil. Biol. Biochem. 22, 865–869.

Hijri, M., 2016. Analysis of a large dataset of mycorrhiza inoculation field trials on potato shows highly significant increases in yield. Mycorrhiza. 26 (3), 209–214.

Hussey, R.S., Roncadori, R.W., 1978. Interacton of *Pratylenchus brachyurus* and *Gigaspora margarita* on cotton. J. Nematol. 10, 16–20.

Ibijbijen, J., Urquaiga, S., Ismali, M., Alve, J.R., Boddey, R.M., 1996. Effect of arbuscular mycorrhizal fungi on growth, mineral nutrition, and nitrogen fixation of three varieties of common bean (*Phaseolus vulgaris*). N. Phytol. 134, 353–360.

Isah, K., Kumar, N., Lagoke, S.O., Atayese, M., 2013. Management of *Striga hermonthica* on sorghum (*Sorghum bicolor*) using arbuscular mycorrhizal fungi (*Glomus mosae*) and NPK fertilizer levels. Pak. J. Biol. Sci. 16, 1563–1568.

Jabaji-Hare, S.H., Stobbs, L.W., 1984. Electron microscopic examination of tomato roots coinfected with *Glomus sp*. and tobacco mosaic virus. Phytopathology. 74, 277–279.

Jain, R.K., Sethi, C.L., 1987. Pathogenicity of *Heterodera cajani* on cowpea as influenced by the presence of VAM fungi, *Glomus fasciculatum* or *G. epigaeus*. Ind. J. Nematol. 17, 165–170.

Jakobsen, I., 1999. Transport of phosphorus and carbon in arbuscular mycorrhizas. Mycorrhiza: Structure, Function, Molecular Biology. Springer, Heidelberg, pp. 535–542.

Jakobsen, I., Hammer, E.C., 2015. Nutrient dynamics in arbuscular mycorrhizal networks, Mycorrhizal Networks, Ecological Studies, 224. Springer, pp. 91–131.

Javot, H., Pumplin, N., Harrison, M.J., 2007. Phosphate in the arbuscular mycorrhizal symbiosis: transport properties and regulatory roles. Plant. Cell Env. 30, 310–322.

Jawson, M.D., Franzlubbers, A.J., Galusha, D.K., Aiken, R.M., 1993. Soil fumigation within monoculture and rotations—response of corn and mycorrhizae. Agron. J. 85, 1174–1180.

Jayaram, J., Kumar, D., 1995. Influence of mungbean yellow mosaic virus on mycorrhizal fungi associated with *Vigna radiata* var. PS 16. Ind. Phytopathol. 48, 108–110.

Johnson, D., Leake, J.R., Ostle, N., Ineson, P., Read, D.J., 2002. In situ $13CO_2$ pulse-labelling of upland grasslands demonstrates a rapid pathway of carbon flux from arbuscular mycorrhizal mycelia to the soil. N. Phytol. 153, 327–334.

Kantharaju, V., Krishnappa, K., Ravichardra, N.G., Karuna, K., 2005. Management of root-knot nematode, *Meloidogyne incognita* on tomato by using indigenous isolates of AM fungus, Glomus fasciculatum. Ind. J. Nematol. 35, 32–36.

Khade, S.W., Adholeya, A., 2007. Feasible bioremediation through arbuscular mycorrhizal fungi imparting heavy metal tolerance: a retrospective. Bioremed. J. 11, 33–43.

Khan, A.G., Kuek, C., Chaudhry, T.M., Khoo, C.S., Hayes, W.J., 2000. Role of plants, mycorrhizae and phytochelators in heavy metal contaminated land remediation. Chemosphere. 41, 197–207.

Kjoller, R., Rosendahl, S., 1997. The presence of arbuscular mycorrhizal fungus *Glomus intraradices* influences enzymatic activities of the root pathogen *Aphanomyces euteiches* in pea roots. Mycorrhiza. 6, 487–491.

Knudsen, I.M.B., Debosz, K., Hockenhull, J., Jensen, D.F., Elmholt, S., 1995. Suppressiveness of organically and conventionally managed soils towards brown foot rot of barley. Appl. Soil. Ecol 12, 61–72.

Koricheva, J., Gange, A.C., Jones, T., 2009. Effects of mycorrhizal fungi on insect herbivores: a meta-analysis. Ecology. 90, 2088–2097.

Koziol, L., 2017. User Manual: A Practical Guide to Inoculation With Arbuscular Mycorrhizal Fungi in Ecological Restoration. University of Kansas, Lawrence, United States.

Krishnamoorthy, R., Kim, K., Kim, C., Sa, T., 2014. Changes of arbuscular mycorrhizal traits and community structure with respect to soil salinity in a coastal reclamation land. Soil. Biol. Biochem. 72, 1–10.

Kulkarni, S.A., Kulkarni, S., Sreenivas, M.N., 2011. Interaction between vesicular-arbuscular (V-A) mycorrhizae and *Sclerotium rolfsii* Sacc. in groundnut. J. Farm. Sci. 10, 919–921.

Kumar, A., Verma, H., Singh, V.K., Singh, P.P., Singh, S.K., Ansari, W.A., et al., 2017. Role of *Pseudomonas* sp. in sustainable agriculture and disease management. Agriculturally Important Microbes for Sustainable Agriculture. Springer, Singapore, pp. 195–215.

Kumar, A., Singh, S.K., Kant, C., Verma, H., Kumar, D., Singh, P.P., et al., 2021. Microbial biosurfactant: a new frontier for sustainable agriculture and pharmaceutical industries. Antioxidants 10 (9), 1472.

Kumari, S.M.P., Prabina, B.J., 2019a. Protection of tomato, *Lycopersicon esculentum* from wilt pathogen, *Fusarium oxysporum f.sp. lycopersici* by arbuscular mycorrhizal fungi, *Glomus* sp. Int. J. Curr. Microbiol. Appl. Sci. 8, 1368–1378.

Kumari, S.M.P., Srimeena, N., 2019b. Arbuscular mycorrhizal fungi (AMF) induced defense factors against the damping-of disease pathogen, *Pythium aphanidermatum* in chilli (*Capsicum annum*). Int. J. Curr. Microbiol. Appl. Sci. 8, 2243–2248.

Lee, C.S., Lee, Y.J., Jeun, Y.C., 2005. Observations of infection structures on the leaves of cucumber plants pre-treated with arbuscular mycorrhiza *Glomus intraradices* after challenge inoculation with *Colletotrichum orbiculare*. Plant. Pathol. J. 21, 237–243.

Leigh, E.G., 2010. The evolution of mutualism. J. Evol. Biol. 23, 2507–2528.

Leyval, C., Joner, E.J., del Val, C., Haselwandter, K., 2002. Potential of arbuscular mycorrhizal fungi for bioremediation. In: Gianinazzi, S., Schiüepp, H., Barea, J.M., Haselwandter, K. (Eds.), Mycorrhizal Technology in Agriculture. Birkhäuser, Basel.

Li, Y., Liu, Z., Hou, H., Lei, H., Zhu, X., Li, X., et al., 2013. Arbuscular mycorrhizal fungi-enhanced resistance against *Phytophthora sojae* infection on soybean leaves is mediated by a network involving hydrogen peroxide, jasmonic acid, and the metabolism of carbon and nitrogen. Acta Physiol. Plant. 35, 3465–3475.

Li, H., Chen, X.W., Wong, M.H., 2016. Arbuscular mycorrhizal fungi reduced the ratios of inorganic/organic arsenic in rice grains. Chemosphere. 145, 224–230.

Liao, J.P., Lin, X.G., Cao, Z.H., Shi, Y.Q., Wong, M.H., 2003. Interactions between arbuscular mycorrhizae and heavy metals under sand culture experiment. Chemosphere. 50 (6), 847–853.

Linderman, R.G., 1994. Role of VAM fungi in biocontrol. In: Pfleger, F.L., Linderman, R. G. (Eds.), Mycorrhizae and Plant Health. APS, St. Paul, MN, pp. 1–26.

Lingua, G., D'Agostino, G., Massa, N., Antosiano, M., Berta, G., 2002. Mycorrhiza-induced differential response to a yellows disease in tomato. Mycorrhiza. 12, 191–198.

Liu, J., Maldonado-Mendoza, I., Lopez-Meyer, M., Cheung, F., Town, C.D., Harrison, M. J., 2007. Arbuscular mycorrhizal symbiosis is accompanied by local and systemic alterations in gene expression and an increase in disease resistance in the shoots. Plant. J. 50, 529–544.

Liu, R.J., 1995. Effect of vesicular-arbuscular mycorrhizal fungi on Verticillium wilt of cotton. Mycorrhiza. 5, 293–297.

Liu, Y., Feng, X., Gao, P., Li, Y., Christensen, M.J., Duan, T., 2018. Arbuscular mycorrhiza fungi increased the susceptibility of *Astragalus adsurgens* to powdery mildew caused by *Erysiphe pisi*. Mycology 9, 223–232.

Liu, X., Feng, Z., Zhao, Z., Zhu, H., Yao, Q., 2020. Acidic soil inhibits the functionality of arbuscular mycorrhizal fungi by reducing arbuscule formation in tomato roots. Soil. Sci. Plant. Nutr. 66, 275–284.

Lovelock, C.E., Wright, S.F., Clark, D.A., Ruess, R.W., 2004. Soil stocks of glomalin produced by arbuscular mycorrhizal fungi across a tropical rain forest landscape. J. Ecol. 92, 278–287.

Maffei, G., Miozzi, L., Fiorilli, V., Novero, M., Lanfranco, L., Accotto, G.P., 2014. The arbuscular mycorrhizal symbiosis attenuates symptom severity and reduces virus concentration in tomato infected by *Tomato yellow leaf curl Sardinia virus* (TYLCSV). Mycorrhiza. 24, 179–186.

Malhi, G.S., Kaur, M., Kaushik, P., Alyemeni, M.N., Alsahli, A.A., Ahmad, P., 2021. Arbuscular mycorrhiza in combating abiotic stresses in vegetables: an eco-friendly approach. Saudi J. Biol. Sci. 28 (2), 1465–1476.

Malusá, E., Sas-Paszt, L., Ciesielska, J., 2012. Technologies for beneficial microorganisms inocula used as biofertilizers. Sci. World J. 1–12.

Manjunatha, H.P., Jones Nirmalnath, P., Chandranath, H.T., Shiney, A., Jagadeesh, K.S., 2018. Field evalualtion of native arbuscular mycorrhizal fungi in the management of Striga in sugarcane (*Saccharum ofcinarum* L.). J. Pharm. Phytochem. 7, 2496–2500.

Marschner, H., Dell, B., 1994. Nutrient uptake and mycorrhizal symbiosis. Plant. Soil. 159, 89–102.

Martínez-Medina, A., Roldan, A., Pascual, J.A., 2011. Interaction between arbuscular mycorrhizal fungi and *Trichoderma harzianum* under conventional and low input fertilization field condition in melon crops: growth response and Fusarium wilt biocontrol. Appl. Soil. Ecol. 47, 98–105.

Masadeh, B., von Alten, H., Grunewaldt-Stoecker, G., Sikora, R.A., 2004. Biocontrol of root knot nematodes using the arbuscular mycorrhizal fungus *Glomus intraradices* and the antagonistic *Trichoderma viridae* in two tomato cultivars differing in their suitability as hosts for the nematodes. J. Plant. Dis. Protec 111, 322–333.

Miller, R.M., Jastrow, J.D., 1994. Vesicular-arbuscular mycorrhizae and biogeo-chemical cycling. In: Pfleger, F.L., Lindermann, R.G. (Eds.), Mycorrhizae and Plant Health. APS, St. Paul, MN, pp. 189–212.

Miozzi, L., Catoni, M., Fiorilli, V., Mullineaux, P.M., Accotto, G.P., Lanfranco, L., 2011. Arbuscular mycorrhizal symbiosis limits foliar transcriptional responses to viral infection and favors long-term virus accumulation. Mol. Plant-Microbe Interact. 24, 1562–1572.

Mo, Y., Wang, Y., Yang, R., Zheng, J., Liu, C., Li, H., et al., 2016. Regulation of plant growth, photosynthesis, antioxidation and osmosis by an arbuscular mycorrhizal fungus in watermelon seedlings under well-watered and drought conditions. Front. Plant. Sci. 7, 644.

Modi, A., Kanani, P., Kumar, A., 2020. Fungal endophytes-induced gene expression studies in biotic and abiotic stress management. In: Kumar, A., Radhakrishnan, E.K. (Eds.), Microbial Endophytes. Woodhead Publishing, Cambridge USA, pp. 251–271.

Mohanta, T.K., Bae, H., 2015. Functional genomics and signaling events in mycorrhizal symbiosis. J. Plant. Interact. 10 (1), 21–40.

Møller, K., Kristensen, K., Yohalem, D., Larsen, J., 2009. Biological management of gray mold in pot roses by co-inoculation of the biocontrol agent *Ulocladium atrum* and the mycorrhizal fungus *Glomus mosseae*. Biol. Control. 49, 120–125.

Mora-Romero, G.A., Cervantes-Gámez, R.G., Galindo-Flores, H., González-Ortíz, M.A., Félix-Gastélum, R., Maldonado-Mendoza, I.E., et al., 2015. Mycorrhiza-induced protection against pathogens is both genotype-specific and graft-transmissible. Symbiosis. 66, 55–64.

Muresu, R., Sulas, L., Caredda, S., 2003. Legume-Rhizobium symbiosis. Characteristics and Prospects of Inoculation. Rivista di Agronomia, Italy.

Mustafa, G., Randoux, B., Tisserant, B., Fontaine, J., Magnin-Robert, M., Lounès-Hadj Sahraoui, A., et al., 2016. Phosphorus supply, arbuscular mycorrhizal fungal species, and plant genotype impact on the protective efficacy of mycorrhizal inoculation against wheat powdery mildew. Mycorrhiza. 26, 685–697.

Mustafa, G., Khong, N.G., Tisserant, B., Randoux, B., Fontaine, J., Magnin-Robert, M., et al., 2017. Defence mechanisms associated with mycorrhiza induced resistance in wheat against powdery mildew. Func. Plant. Biol. 44 (4), 443–454.

Nair, A., Kolet, S.P., Thulasiram, H.V., Bhargava, S., 2015. Systemic jasmonic acid modulation in mycorrhizal tomato plants and its role in induced resistance against *Alternaria alternata*. Plant. Biol. 17, 625–631.

Nakatani, A.S., Mescolotti, D.L.C., Nogueira, M.A., Martines, A.M., Miyauchi, M.Y.H., Stürmer, S.L., et al., 2011. Dosage-dependent shift in the spore community of arbuscular mycorrhizal fungi following application of tannery sludge. Mycorrhiza. 21, 515–522.

Nelson, C.E., Safir, G.R., 1982. The water relations of well watered mycorrhizal and non mycorrhizal onion plants. J. Am. Soc. Mortc. Sci. 107, 271–276.

Newsham, K.K., Fitter, A.H., Watkinson, A.R., 1994. Root pathogenic and arbuscular mycorrhizal fungi determine fecundity of asymptomatic plants in the field. J. Ecol. 82, 805–814.

Nogales, A., Aguirreolea, J., María, E.S., Camprubí, A., Calvet, C., 2009. Response of mycorrhizal grapevine to *Armillaria mellea* inoculation: disease development and polyamines. Plant. Soil. 317, 177–187.

O'Bannon, J.H., Nemec, S., 1979. The response of *Citrus* lemon seedlings to a symbionts, *Glomus eutnicatus*, and a pathogen, *Radopholus similis*. J. Nematol. 11, 270–275.

Oliveira, R.S., Dodd, J.C., Castro, P.M.L., 2001. The mycorrhizal status of *Phragmites australis* in several polluted soils and sediments of an industrialised region of northern Portugal. Mycorrhiza. 10, 241–247.

Orak, A.B., Demir, S., 2011. Effects of arbuscular mycorrhizal fungus and different phosphorus doses against cotton wilt caused *Verticillium dahliae* Kleb. Plant. Pathol. J. 10 (3), 108–114.

Othira, J.O., 2012. Effectiveness of arbuscular mycorrhizal fungi in protection of maize (*Zea mays* L.) against witchweed (*Striga hermonthica* Del Benth) infestation. J. Agric. Biotechnol. Sustain. Dev. 4, 37–44.

Ozgonen, H., Erkilic, A., 2007. Growth enhancement and Phytophthora blight (*Phytophthora capsici* Leonian) control by arbuscular mycorrhizal fungal inoculation in pepper. Crop. Protec 26, 1682–1688.

Pandey, R., 2005. Field application of bio-organics in the management of *Meloidogyne incognita* in *Mentha arvensis*. Nematol. Medit. 33, 51–54.

Patel, P., Kumar, S., Modi, A., Kumar, A., 2021. Deciphering fungal endophytes combating abiotic stresses in crop plants (cereals and vegetables). Microbial Management of Plant Stresses. Woodhead Publishing, Cambridge USA, pp. 131–147.

Pathak, P., Rai, V.K., Can, H., Singh, S.K., Kumar, D., Bhardwaj, N., et al., 2022. Plant-endophyte interaction during biotic stress management. Plants 11 (17), 2203.

Pellegrino, E., Öpik, M., Bonari, E., Ercoli, L., 2015. Responses of wheat to arbuscular mycorrhizal fungi: a meta-analysis of field studies from 1975 to 2013. Soil. Biol. Biochem. 84, 210–217.

Pinochet, J., Fernandez, C., Jaimez, M., De, Tenoury, P., 1997. Micropropagated banana infected with *Meloidogyne javanica* responds to *Glomus intraradices* and phosphorus. Hort. Sci. 32, 35–49.

Piotrowski, J.S., Denich, T., Klironomos, J.N., Graham, J.M., Rillig, M.C., 2004. The effects of arbuscular mycorrhizas on soil aggregation depend on the interaction between plant and fungal species. N. Phytol. 164, 365–373.

Pirozynski, K.A., Dalpe, Y., 1989. Geological history of the Glomaceae with particular reference to mycorrhizal symbiosis. Symbiosis. 7, 1–36.

Pozo, M.J., Cordier, C., Dumas-Gaudot, E., Gianinazzi, S., Barea, J.M., Azcón-Aguilar, C., 2002. Localized versus systemic effect of arbuscular mycorrhizal fungi on defence responses to *Phytophthora* infection in tomato plants. J. Exp. Bot. 53, 525–534.

Pozo, M.J., Azcón-Aguilar, C., 2007. Unraveling mycorrhiza-induced resistance. Curr. Opin. Plant. Biol. 10 (4), 393–398.

Pozo, M.J., Jung, S.C., López-Ráez, J.A., Azcón-Aguilar, C., 2010. Impact of arbuscular mycorrhizal symbiosis on plant response to biotic stress: the role of plant defence mechanisms. Arbuscular Mycorrhizas: Physiology and Function. Springer, Dordrecht, pp. 193–207.

Read, D.J., Perez-Moreno, J., 2003. Mycorrhizas and nutrient cycling in ecosystems – a journey towards relevance? N. Phytol. 157, 475–492.

Redecker, D., Schüßler, A., Stockinger, H., Stürmer, S.L., Morton, J.B., Walker, C., 2013. An evidence-based consensus for the classification of arbuscular mycorrhizal fungi (Glomeromycota). Mycorrhiza. 23, 515–531.

Remy, W., Taylor, T.N., Hass, H., Kerp, H., 1994. Four hundred million year old vesicular arbuscular mycorrhizae. Proc. Natl. Acad. Sci. USA 91, 11841–11843.

Rillig, M.C., Sosa-Hernández, M.A., Roy, J., Aguilar-Trigueros, C.A., Vályi, K., Lehmann, A., 2016. Towards an integrated mycorrhizal technology: harnessing mycorrhiza for sustainable intensification in agriculture. Front. Plant. Sci. 7, 1625.

Rillig, M.C., 2004. Arbuscular mycorrhizae, glomalin, and soil aggregation. Can. J. Soil. Sci. 84, 355–363.

Rosendahl, C.N., Rosendahl, S., 1990. The role of vesicular arbuscular mycorrhizal fungi in controlling damping-off and growth reduction in cucumber caused by *Pythium ultimum*. Symbiosis. 9, 363–366.

Ruairidh, J.H., Sawers, J.H., Yang, S.Y., Gutjahr, C., Paszkowski, U., 2008. The molecular components of nutrient exchange in arbuscular mycorrhizal interactions. Sustainable Agriculture and Forestry. Springer, pp. 37–60.

Saldajeno, M.G.B., Hyakumachi, M., 2011. The plant growth-promoting fungus *Fusarium equiseti* and the arbuscular mycorrhizal fungus *Glomus mosseae* stimulate plant growth and reduce severity of anthracnose and damping-off diseases in cucumber (*Cucumis sativus*) seedlings. Ann. Appl. Biol. 159, 28–40.

Schönbeck, F., 1980. Endomycorrhiza in relation to plant diseases. In: Schippers, B., Gams, W. (Eds.), Soil – Borne Pathogens. Academic Press, New York, NY, pp. 271–280.

Schreiner, R.P., Bethlenfalvay, G.J., 1995. Mycorrhizal interactions in sustainable agriculture. Crit. Rev. Biotechnol. 15, 271–287.

Schultz, R.C., Colletti, J.P., Isenhart, T.M., Simkins, W.W., Mize, C.W., Thompson, M.L., 1995. Design and placement of a multi-species riparian buffer strip system. Agrofores. System. 29, 1–16.

Schüßler, A., Schwarzott, D., Walker, C., 2001. A new fungal phylum, the Glomeromycota: phylogeny and evolution. Mycol. Res. 105, 1413–1421.

Selvakumar, G., Shagol, C.C., Kang, Y., Chung, B.N., Han, S.G., Sa, T.M., 2018. Arbuscular mycorrhizal fungi spore propagation using single spore as starter inoculum and a plant host. J. Appl. Microbiol. 124, 1556–1565.

Sensoy, S., Demir, S., Turkmen, O., Erdinc, C., Savur, O.B., 2007. Responses of some different pepper (*Capsicum annuum* L.) genotypes to inoculation with two different arbuscular mycorrhizal fungi. Sci. Horti 113 (1), 92–95.

Sharma, M.P., Gaur, A., Bhatia, N.P., Adholeya, A., 1996. Growth responses and dependence of *Acacia nilotica* var. *cupriciformis* on the indigenous arbuscular mycorrhizal consortium of a marginal wasteland soil. Mycorrhiza. 6 (5), 441–446.

Shaul, O., Galili, S., Volpin, H., Ginzberg, I., Elad, Y., Chet, I., et al., 1999. Mycorrhiza induced changes in disease severity and PR protein expression tobacco leaves. Mol. Plant. Microbe Inter. 12, 1000–1007.

Sheoran, V., Sheoran, A.S., Poonia, P., 2010. Soil reclamation of abandoned mine land by revegetation: a review. Int. J. Soil. Sediment. Water 3, 13.

Shreenivasa, K.R., Krishnappa, K., Ravichandra, N.G., 2007. Interaction effects of arbuscular mycorrhizal fungus *Glomus fasciculatum* and root-knot nematode, *Meloidogyne incognita* on growth and phosphorous uptake of tomato. Karnat. J. Agric. Sci. 20, 57–61.

Siddiqui, Z.A., Mahmood, I., 1995. Biological control of *Heterodera cajani* and *Fusarium udum* by *Bacillus subtilis*, *Bradyrhizobium japonicum* and *Glomus fasciculatum* on pigeonpea. Fundam. Appl. Nematol. 18, 559–566.

Siddiqui, Z.A., Mahmood, I., 1998. Effect of a plant growth promoting bacterium, an AM fungus and soil types on the morphometrics and reproduction of *Meloidogyne javanica* on tomato. Appl. Soil. Ecol. 8, 77–84.

Siddiqui, Z.A., Singh, L.P., 2004. Effects of soil inoculants on the growth, transpiration and wilt disease of chickpea. J. Plant. Dis. Protec 111, 151–157.

Siddiqui, Z.A., Akhtar, M.S., 2006. Biological control of root-rot disease complex of chickpea by AM fungi. Arch. Phytopathol. Plant. Protec 39, 389–395.

Siddiqui, Z.A., Pichtel, J., 2008. Mycorrhizae: an overview. Mycorrhizae: Sustainable Agriculture and Forestry. Springer, pp. 1–36.

Siddiqui, Z.A., Akhtar, M.S., 2009. Effects of antagonistic fungi, plant growth-promoting rhizobacteria, and arbuscular mycorrhizal fungi alone and in combination on the reproduction of *Meloidogyne incognita* and growth of tomato. J. Gen. Plant. Pathol. 75, 144–153.

Siddiqui, Z.A., Mahmood, I., Hayat, S., 1998. Biocontrol of *Heterodera cajani* and *Fusarium udum* on pigeonpea using *Glomus mosseae*, *Paecilomyces lilacinus* and *Pseudomonas fluorescens*. Thai J. Agri. Sci. 31, 310–321.

Siddiqui, Z.A., Mahmood, I., Hayat, S., 2000. Influence of plant symbionts and potassium fertilizer on *Heterodera cajani*, crop growth and yield of pigeon pea under field condition. Indian. J. Bot. Soc. 79, 109–114.

Sieverding, E., 1991. Vesicular-Arbuscular Mycorrhiza Management in Tropical Agroecosystem. Deutshe Gesellschaft Technische Zusammenarbeit (GTZ) GmbH, Eschborn.

Singh, P.K., Singh, M., Vyas, D., 2010. Biocontrol of Fusarium wilt of chickpea using arbuscular mycorrhizal fungi and *Rhizobium leguminosorum* biovar, Caryologia., 63. pp. 349–353.

Singh, V.K., Singh, A.K., Kumar, A., 2017. Disease management of tomato through PGPB: current trends and future perspective. 3 Biotech. 7 (4), 1–10.

Singh, V.K., Singh, A.K., Singh, P.P., Kumar, A., 2018. Interaction of plant growth promoting bacteria with tomato under abiotic stress: a review. Agric. Ecosyst. Environ. 267, 129–140.

Singh, M., Srivastava, M., Kumar, A., Singh, A.K., Pandey, K.D., 2020. Endophytic bacteria in plant disease management. In: Kumar, A., Singh, K.V. (Eds.), Microbial Endophytes: Prospects for Sustainable Agriculture. Woodhead Publ, Cambridge USA, pp. 61–89.

Sipahioglu, M.H., Demir, S., Usta, M., Akkopru, A., 2009. Biological relationship of potato virus Y and arbuscular mycorrhizal fungus *Glomus intraradices* in potato. Pest. Tech. 3, 63–66.

Sitaramaiah, K., Sikora, R.A., 1982. Effect of mycorrhizal fungus *Glomus fasciculatum* on the host parasite relationship of *Rotylenchulus reniformis* in tomato. Nematologica. 28, 412–419.

Smith, S.E., Read, D.J., 1997. Mycorrhizal Symbiosis. Academic Press, San Diego, p. 607.

Smith, G.E., Kaplan, D.T., 1988. Influence of mycorrhizal fungus, phosphorus and burrowing nematode interactions on the growth of rough lemon citrus seedlings. J. Nematol. 20, 539–544.

Smith, S.E., Read, D.J., 2008. The symbiont forming arbuscular mycorrhizas. Mycorrhizal Symbiosis. Academic Press, pp. 13–41.

Smith, S.E., Jakobsen, I., Grnlund, M., Smith, F.A., 2011. Roles of arbuscular mycorrhizas in plant phosphorus nutrition: interactions between pathways of phosphorus uptake in arbuscular mycorrhizal roots have important implications for understanding and manipulating plant phosphorus acquisition. Plant. Physiol. 156, 1050–1057.

Solaiman, Z.M., Hirata, H., 1997. Effect of arbuscular mycorrhizal fungi inoculation of rice seedlings at the nursery stage upon performance in the paddy field and greenhouse. Plant. Soil. 191, 1–12.

Song, Y., Chen, D., Lu, K., Sun, Z., Zeng, R., 2015. Enhanced tomato disease resistance primed by arbuscular mycorrhizal fungus. Front. Plant. Sci. 6, 786.

Spagnoletti, F.N., Cornero, M., Chiocchio, V., Lavado, R.S., Roberts, I.N., 2020. Arbuscular mycorrhiza protects soybean plants against *Macrophomina phaseolina* even under nitrogen fertilization. Eur. J. Plant. Pathol. 156, 839–849.

St-Arnaud, M., Hamel, C., Caron, M., Fortin, J.A., 1994. Inhibition of *Pythium ultimum* in roots and growth substrate of mycorrhizal *Tagetes patula* colonized with *Glomus intraradices*. Can. J. Plant. Pathol. 16, 187–194.

Stolyarchuk, I.M., Shevchenko, T.P., Polischuk, V.P., Kripka, A.V., 2009. Virus infection course in different plant species under influence of arbuscular mycorrhiza. Microbiology. 3, 70–75.

Sulistiono, W., 2020. The role of mycorrhizae on seedlings and early growth of sugarcane. Mycorrhizal Fungi-Utilization in Agriculture and Industry. IntechOpen.

Sundaresan, P., Ubalthoose Raja, N., Gunasekaran, P., 1993. Induction and accumulation of phytoalexins in cowpea roots infected with a mycorrhizal fungus *Glomus fasciculatum* and their resistance to Fusarium wilt disease. J. Biosci. 18, 291–301.

Sylvia, D.M., Williams, S.E., 1992. Mycorrhizae and environmental stresses. In: Bethlenfalvay, G.J., Linderman, R.G. (Eds.), Mycorrhizae in Sustainable Agriculture. ASA Special Publication No. 54, Madison, WI, pp. 101–124.

Talaat, N.B., Shawky, B.T., 2011. Influence of arbuscular mycorrhizae on yield, nutrients, organic solutes, and antioxidant enzymes of two wheat cultivars under salt stress. J. Plant. Nutr. Soil. Sci. 174, 283–291.

Taylor, T.N., Remy, W., Hass, H., Kerp, H., 1995. Fossil arbuscular mycorrhizae from the early Devonian. Mycologia. 87, 560–573.

Thiem, D., Szmidt-Jaworska, A., Baum, C., Muders, K., Niedojadło, K., Hrynkiewicz, K., 2014. Interactive physiological response of potato (*Solanum tuberosum* L.) plants to fungal colonization and Potato virus Y (PVY) infection. Acta Mycol. 1, 291–303.

Thygesen, K., Larsen, J., Bodker, l, 2004. Arbuscular mycorrhizal fungi reduce development of pea root-rot caused by *Aphanomyces euteiches* using oospores as pathogen inoculum. Eur. J. Plant. Pathol. 110, 411–419.

Tisdall, J.M., 1991. Fungal hyphae and structural stability of soil. Aus. J. Soil. Res. 29, 729–743.

Turhan, P., Demir, S., 2013. The effects of some biological control agents against Black root rot disease (*Rhizoctonia solani* Kühn.) on strawberry. Turk. J. Biol. Control. 4 (2), 125–140.

Turnau, K., Orlowska, E., Ryszka, P., Zubek, S., Anielska, T., Gawronski, S., 2006. Role of AMF in phytoremediation and toxicity monitoring of heavy metal rich industrial wastes in Southern Poland. In: Twardowska, I., Allen, H.E., Häggblom, M.M., Stefaniak, S. (Eds.), Soil and Water Pollution Monitoring, Protection and Remediation. Springer, Berlin.

Turnau, K., Ryszka, P., Gianinazzi-Pearson, V., van Tuinen, D., 2001. Identification of arbuscular mycorrhizal fungi in soils and roots of plants colonizing zinc wastes in southern Poland. Mycorrhiza. 10, 169–174.

Turrini, A., Bedini, A., Loor, M.B., Santini, G., Sbrana, C., Giovannetti, M., et al., 2018. Local diversity of native arbuscular mycorrhizal symbionts differentially affects growth and nutrition of three crop plant species. Biol. Fertil. Soil. 54, 203–217.

Tylka, G.L., Hussey, R.S., Roncadori, R.W., 1991. Interactions of vesicular-arbuscular mycorrhizal fungi, phosphorus and *Heterodera glyciens* on soybean. J. Nematol. 23, 122–123.

Umesh, K.C., Krishnappa, K., Bagyaraj, D.J., 1988. Interaction of burrowing nematode, *Radopholus similis* (Cobb, 1983) Thorne, 1949, and VA mycorrhiza, *Glomus fasciculatum* (Thaxt.) Gerd. and Trappe, in banana (*Musa acuminata* Colla.). Ind. J. Nematol. 18, 6–11.

Valentine, A.J., Mortimer, P.E., Kleinert, A., Kang, Y., Benedito, V.A., 2013. Carbon metabolism and costs of arbuscular mycorrhizal associations to host roots. Symbiotic Endophytes. Soil. Biol. 37, 233–252.

van der Heijden, M.G.A., Martin, F.M., Selosse, M.A., Sanders, J.R., 2014. Mycorrhizal ecology and evolution: the past, the present, and the future. N. Phytol. 205, 1406–1423.

Verbruggen, E., Kiers, T., 2010. Evolutionary ecology of mycorrhizal functional diversity in agricultural systems. Evol. Appl. 3 (506), 547–560.

Veresoglou, S.D., Johnson, D., Mola, M., Yang, G., Rillig, M.C., 2022. Evolutionary bet-hedging inarbuscular mycorrhiza – associating angiosperms. N. Phytol. 233, 1984–1987.

Vestberg, M., Palmujoki, H., Parikka, P., Uosukainen, M., 1994. Effect of arbuscular mycorrhizas on crown rot (*Phytophthora cactorum*) in micropropagated strawberry plants. Special Issue of the Third COST 87–8.10 Joint Meeting, 17–18 September 1993, Agricultural Research Centre of Finland, Laukaa, Finland.

Vos, C., Tesfahun, A., Panis, B., De Waele, D., Elsen, A., 2012. Arbuscular mycorrhizal fungi induce systemic resistance in tomato against the sedentary nematode *Meloidogyne incognita* and the migratory nematode *Pratylenchus penetrans*. Appl. Soil. Ecol. 61, 1–6.

Waceke, J.W., Waudo, S.W., Sikora, R., 2002. Effect of inorganic phosphatic fertilizers on the efficacy of an arbuscular mycorrhizal fungus against a root-knot nematode on *Pytherum*. Int. J. Pest. Manag. 48, 307–313.

Wang, Y.-Y., Yin, Q.-S., Qu, Y., Li, G.-Z., Hao, L., 2017. Arbuscular mycorrhiza-mediated resistance in tomato against *Cladosporium fulvum*-induced mould disease. J. Phytopathol. 166, 67–74.

Weissenhorn, I., Leyval, C., Belgy, G., Berthelin, J., 1995. Arbuscular mycorrhizal contribution to heavy-metal uptake by maize (*Zea mays* L) in pot culture with contaminated soil. Mycorrhiza. 5, 245–251.

Whipps, J.M., 2004. Prospects and limitations for mycorrhizals in biocontrol of root pathogens. Can. J. Bot. 82, 1198–1227.

Wright, S.F., Upadhyaya, A., 1998. A survey of soils for aggregate stability and glomalin, a glycoprotein produced by hyphae of arbuscular mycorrhizal fungi. Plant. Soil. 198, 97–107.

Wright, S.F., Anderson, R.L., 2000. Aggregate stability and glomalin in alternative crop rotations for central Great Plains. Biol. Fert. Soil. 31, 249–253.

Yang, Y., Song, Y., Scheller, H.V., Ghosh, A., Ban, Y., Chen, H., et al., 2015. Community structure of arbuscular mycorrhizal fungi associated with *Robinia pseudoacacia* in uncontaminated and heavy metal contaminated soils. Soil. Biol. Biochem. 86, 146–158.

Yao, M., Tweddell, R., Desilets, H., 2002. Effect of two vesicular-arbuscular mycorrhizal fungi on the growth of micropropagated potato plantlets and on the extent of disease caused by *Rhizoctonia solani*. Mycorrhiza. 12, 235–242.

Zhang, T., Yang, X., Guo, R., Guo, J., 2016. Response of AM fungi spore population to elevated temperature and nitrogen addition and their influence on the plant community composition and productivity. Sci. Rep. 6, 24749.

Zhang, X., Zhang, H., Lou, X., Tang, M., 2019. Mycorrhizal and non-mycorrhizal *Medicago truncatula* roots exhibit differentially regulated NADPH oxidase and antioxidant response under Pb stress. Environ. Exp. Bot. 164, 10–19.

CHAPTER

5 Biodiversity and biotechnological applications of host-specific endophytic fungi for sustainable agriculture and allied sector

Jignesh Prajapati[1], Riya Sheth[2], Riddhi Bhatt[2], Karan Chavda[1], Zalak Solanki[2], Rakesh Rawal[1,2] and Dweipayan Goswami[3]

[1]*Department of Biochemistry & Forensic Science, University School of Sciences, Gujarat University, Ahmedabad, Gujarat, India*

[2]*Department of Life Science, University School of Sciences, Gujarat University, Ahmedabad, Gujarat, India*

[3]*Department of Microbiology & Biotechnology, University School of Sciences, Gujarat University, Ahmedabad, Gujarat, India*

5.1 Introduction

Microorganisms are small living things that are present in water, soil, plant tissue, animal body, and even in the air. Endophyte is Greek for "within the plant." Endophytes are microorganisms (bacteria/fungi) that live inside plant tissues (leaves, stems, roots, and fruits) without triggering any disease signs in the host (Schulz and Boyle, 2005). De Bary invented the term endophyte (Gr. Endon—within; phyton—plant) referring to any organisms found within plants. Numerous kinds of fungi and bacteria live inside plant tissues for the entire span of their lives or for a specific time of their cycles, colonizing plants without producing any obvious harm or morphological changes to their hosts (Stone et al., 2004). Endophytic microorganisms can be divided into facultative and obligatory groups based on how they colonize. Facultative endophytes invade plants at specific times in their life cycles, while on the other hand, obligate endophytes remain inside the plants during their whole life cycles. They often reproduce vertically between generations of plants and utilize or modify the metabolic processes and byproducts produced by plants for their own survival (Gouda et al., 2016; Hardoim et al., 2008a).

Sustainable Agricultural Practices. DOI: https://doi.org/10.1016/B978-0-443-19150-3.00003-5

Across these endophytic microorganisms, endophytic fungi (EF) garnered significant research interest since they not just offer alternative sources of cytotoxic substances like anticarcinogenic molecules and antimicrobials, but more because they enhance nutrient solubilization in the plant rhizosphere, encourage plant growth, function as biopesticides, or withstand biotic and abiotic factors (Poveda, 2021). The fungus coexists biologically with the living plant in mycelial form. All types of plants, including trees, grasses, algae, and herbaceous plants, contain EF. The interior mycota of living plants have commonly been referred to as endophyte and endophytic fungus. Practically endophytes have been isolated from all habitats on planet. EF are currently regarded as one of the key elements of biodiversity and have their role in diverse range of applications (Rajamanikyam et al., 2017). In current chapter, we have discussed the biological characteristics and biodiversity of EF along with their applications in sustainable agriculture and allied sector.

5.2 Biological characteristics

Numerous biological traits of EF are influenced by a variety of variables, including host species, host developmental stage, inoculum density, and environmental conditions (Gautam and Avasthi, 2019). A diagrammatic presentation for the biological characteristics of EF is shown in Fig. 5.1.

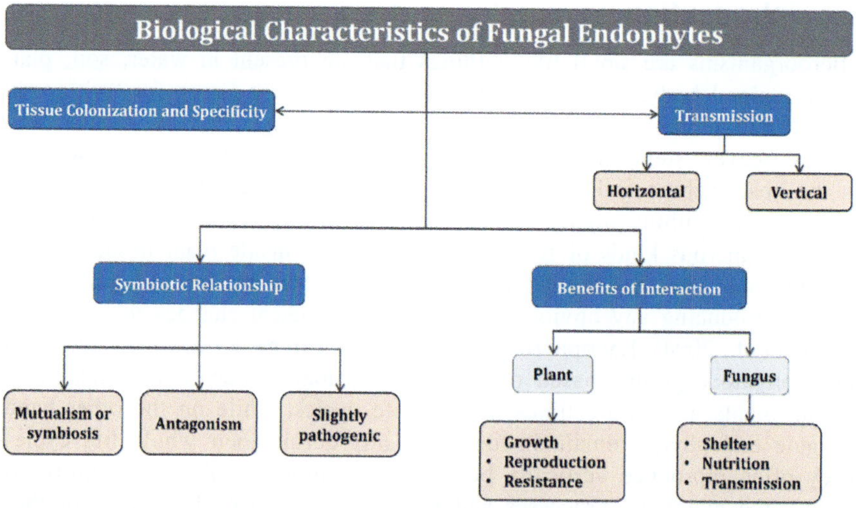

FIGURE 5.1

Diagrammatic presentation for the biological characteristics of endophytic fungi.

5.3 Biodiversity and distribution

Since the EF were found in all the plant species studied, fungi are extremely common creatures. *Sphaeria typhena*, now known as *Epichloe typhina*, was the first endophytic fungus reported by Person in 1772. However, endophytes were found to be associated with 400-million-year-old plant fossils in paleontological examinations of fossilized plants (Gautam and Avasthi, 2019).

EF can have harmful or symbiotic relationships with their plant hosts and may live on a variety of hosts or may colonize certain host plants, organs, or tissues. Both in terms of the fungus species that may live an endophytic existence and in terms of the diversity of hosts they occupy, EF are incredibly varied. Almost all classes or divisions of fungus can infect plants at any stage and age by acting as endophytes. According to diversity, the majority of EF are Ascomycota members; however, certain Basidiomycota, Zygomycota, and Oomycota taxa are also endophytic (Rajamanikyam et al., 2017; Stone et al., 2004). Fungal endophytes may typically be divided into several ecological groups based on the host organism and the tissues they occupy, the mechanisms of proliferation and transmission, the food source, and the presentation of infestation. In general, fungal endophytes can be classified into two ecological groups, Clavicipitaceous (C) and Non-Clavicipitaceous (NC) (Rodriguez et al., 2009). C-group EF transmit directly from parents to progeny by invading host seeds in the ovules of the host plant. The target tissues for their colonization are active rhizomes and stems of the hosts, but the host range is constrained to grasses. In 1988 it was reviewed that, C-group varieties are usual obligatory endophytes that secrete defensive or supportive bioactive compounds to either defend their hosts against predator viciousness or assist the hosts withstand drought (Carrol, 2012; Roberts and Lindow, 2013). The NC-group EF (Ascomycota, Basidiomycota) has a vast diversification and distribution across tropical to polar areas, specifically infects nonvascular, vascular, and woody plant groups. By creating spores or conidia, they spread sexually or asexually and aid in the induction of symbiosis by horizontal proliferation. NC-EF can exist in a dormant state till they detect the biochemical alterations brought on by host plants experiencing scars, injuries, or any climate changes, therefore they are not directly linked with the host plants (Mishra et al., 2021). In contrast to the widespread, structured, systematic, and intracellular establishment of the NC-group in roots or the rhizosphere, colonization of them in aerial parts is typically localized, constrained, and primarily intercellular manner. *Fusarium* species, *Piriformospora indica*, and dark septate mycobiota are a few examples of the mycobiomes from NC-endophytic group (Schulz et al., 2002).

Based on some investigations, the host variety of EF comprises ferns, grasses, liverworts, hornworts, grasses, shrubs, deciduous and coniferous trees, gymnosperms, angiosperms, annual/perennial herbaceous and broad-leaved plants, as well as algae. Additionally, EF are found in a wide variety of geographic environments, including tropical, continental, northern regions, mountain, marine, and xerophytic ecosystems (Alam et al., 2021; Zhang et al., 2006). Plant endophytes are part of a supervaried ecosystem with an estimated diversity of 270,000 plant species on the globe, which might contain around 1.36×10^6 distinct endophytic fungal species in addition to the 100,000 species of fungus discovered so far.

5.4 Ecology and ecological role

EF are ubiquitous and as previously mentioned, the diversification of EF in a host is known to be affected by environmental factors and the age of the host tissue. With increasing age, less EF are extracted from the host tissues (Petrini et al., 1993). Tree-colonizing EF are recognized to be host-specific, and the quality of the air also has an impact on a plant endophytic community. This demonstrates the special niche that the EF inhabit. Even though they might not be subject to the same environmental stress as Phylloplane fungi, they still ought to deal with the host defensive responses and as a result, their life habits are probably distinct from other fungi.

EF were found to inhabit a range of niches in their native ecology, which determine their functional characteristics. These functional characteristics help plants adapt to new environments, boost their resilience to abiotic and biotic stressors, and protect them from numerous viruses and pests (Card et al., 2016). They receive benefits from host plants in exchange through a variety of mechanisms, such as nutrition provision, desiccation protection, spatial organization, and sexual transfer of fungal spores to the following generation of hosts. Some endophytes have apparently been demonstrated to have a significant role in improving soil fertility and mineralization via their assistance in the biodegradation of dead and decomposing host plants. Moreover, endophytic interaction in hosts influences plant development promotion, which improves overall plant productivity and vitality through better nutrient absorption, nutrient distribution, and regulation of phytohormones engaged in the development of plants (Johnson et al., 2014).

5.5 Sheltering of endophytic fungi

The majority of EF are derived from the conditions in which their host species flourish, which include rhizobacteria, fungal spores in the environment, and insect and animal dietary habits (Sasse et al., 2018). There are commonly two transmission modes used by EF. The main method by which the offspring is affected is vertical transfer of fungus from mother plants to progeny seed (Gagic et al., 2018). When the right conditions are met, EF can penetrate recently formed plants during germination of seeds, enabling the transfer of endophytic fungus between host organism and the progeny. The fact that some EF in terrestrial tissues were being laterally transferred via spores and/or mycelium fragmentation, by biotic (herbivores or insects) or abiotic (wind or rain) spreading agents, also makes it possible for the transmission of EF among various host plants (Wiewióra et al., 2015). According to reports, external microorganisms, types of plants, soil characteristics, and plant location play a role in determining how a fungal endophytic community develops in plants (Gundale et al., 2016).

In plant roots, fungi colonize through largely random processes due to their growth habits and dispersal abilities, while bacteria follow more deterministic colonization patterns (Powell et al., 2015). Rhizosphere serve as a representative of all phytochemical constituents that reach via root system and are essential for determining the composition and aggregation of the rhizosphere microbiome (Sasse et al., 2018). The process of connection and communication starts when soil bacteria recognize the chemical signals generated by plants.

5.6 Association between endophytic fungi and host plants

Plants developed symbiosis with a wide variety of soil microbes under the motive of recruitment; some of these fungi may first adhere to the root exterior and, after that, develop structures resembling appressorium. After entering the outer layers of plant root systems, these related fungi move into and occupy the interior tissues of plants (Viterbo and Chet, 2006). Underneath the epidermal cell layers, endophytes dwell inside the interstitium of internal plant tissues. EF may alter the host's genetic makeup and phenotypic manifestation to boost resilience to pathogens and herbivores to counterbalance plant photosynthesis loss and host nitrogen metabolism alteration. EF retain asymptomatic existence and benefit their host due to this dynamic regulation. Therefore the link between EF and their host plants is the outcome of the precise control of gene products, phenotypic appearance, and metabolic activity (Yan et al., 2019). In 2005 a theory was established which stated that, asymptomatic colonization happens when endophytic infectivity and plant's defensive responses exist in balance. If the equilibrium between the host and pathogen is upset, either the fungus dies, or the plant develops a disease (Gautam and Avasthi, 2019; Schulz and Boyle, 2005).

5.7 Relevance of endophytic fungi for sustainable agriculture

It is very typical for plants and fungus to coexist; however, there are still some aspects of the EF interaction that require to be completely understood, and scientific community is making progress in this direction (Chadha et al., 2015). The incorrect and uncontrolled use of pesticides and fertilizers harms both the ecosystem and public health. Depending on the methods used for their manufacture, residues from these processes can be detected in beverages like juices and wines as well as foods such as vegetables, fruits, cereals, and grains (Zikankuba et al., 2017). The manifestation of several human health conditions in infants, teenagers, and the elderly are result of pesticide exposure and includes cancer, endocrine abnormalities, respiratory issues, and reproductive issues. Nontarget organisms are also continually influenced, including those in the marine habitat, such as plankton, crustaceans, and fish, as well as those in the terrestrial environment, such as animals, birds, and beneficial soil bacteria. In addition, excessive fertilizer usage results in the accumulation of heavy metals, algae blooms of rivers and lakes, soil contamination, pollution of groundwater and water sources, and greenhouse gas emissions (Savci, 2012). Hence, there is an unmet need to create more sustainable agriculture practice by reducing or even replacing the usage of pesticides and fertilizers.

Recent scientific study has mostly been motivated by public awareness of the need for enhancing sustainable agriculture that will assist to safeguard and mitigate future adverse environmental effects (Jaber and Enkerli, 2016). Studies have been done in this context that go further than the hypothetical use of microbes as inundate inoculation in crops as traditional biological control agents (BCAs). For instance, *Beauveria bassiana* and *Metarhizium anisopliae* are the best-known and commonly used entomopathogenic fungi in biological control programs intended to manage arthropods that might behave as insects or disease vectors (Baron et al., 2019). In this case, fungi can function as plant pathogen antagonists by using a variety of mechanisms, such as the synthesis of metabolites (antimicrobials, volatile compounds, and enzymes), involvement in competition (for area, nitrogen source, carbon source, and mineral resources), parasitism, initiation of systemic resistance by the plant, and helps for increase in plant growth, each of which has the effect of decreasing the interaction of the pathogenic organisms (Quesada Moraga, 2020). Studies on endophytes not just unveil fascinating details about ecological system and the manner that these microbes communicate with plants, but they also shed light on the variables that require to be investigated to develop environmentally friendly agricultural practices and the potential benefits of this interaction. Sustainable agriculture must be performed to increase or maintain the existing rate of food yield with limiting damage to the environment and human health (Suman et al., 2016). The next portion of the chapter explains the possible applications of endophytic organisms in crop cultivation and sale as bioproducts for agriculture, as well as the status of the research in this domain.

5.8 Applications of endophytic fungi

There is no doubt that the advancement of natural substitutes holds great promise for accomplishing more organic farming, particularly in light of the difficulties in plant biotechnology for several crops and the reality that their use is currently outlawed in many countries due to the unknowable consequences of genetic modification on human and animal and environmental health (Chadha et al., 2015). The use of EF is still just getting established, and in the following subsections, we will discuss both immediate and indirect ways that these bacteria can improve plant development as well as how they might be exploited in agricultural operations.

5.9 Direct benefits

One of the immediate benefits of working with EF is the rise in nutrient intake and plant hormones content in the plant, which are directly related to increases in plant biomass, roots and shoots growth, reproduction, and productivity (Bamisile et al., 2018).

5.9.1 Nutrient acquisition and plant growth

The EF has ability to solubilize various sources of phosphorus and transfer insufficient soil nutrients (such as nitrogen, sulfur, and phosphorus) to host plants (Kiers et al., 2011). For instance, the dark septate EF can break down substances that are good for plants, such as cellulose, starch, lipids, casamino acids, urea, pectin, and gelatin. Moreover, the combination of nitrogen-fixing root EF in leguminous plant root nodules is well known to boost plant development and health. Insect herbivory causes vascular plants to lose a major portion of their nitrogen supply. Interesting research has revealed that endophytic insect-pathogenic fungus can cycle nitrogen into plants. According to reports, *B. bassiana* can colonize a variety of plants, both naturally and artificially, which can help the plants to flourish (Jaber and Enkerli, 2016). More such significant EF are mentioned in Table 5.1. These findings showed that the active nitrogen uptake by plants during interactions between plants, endophytes, and herbivores may be crucial to the ecological nitrogen cycle (Yan et al., 2018).

5.9.2 Production of plant hormones

Phytohormones are essential for reducing the harmful effects of climate change and promoting sustainable agriculture. Key hormones including auxins, gibberellins (GAs), and cytokinins can be synthesized by fungal endophytes. Indole-3-acetic acid (IAA) is the primary auxin that is made by fungi. The primary regulators of plant growth, auxins, have a number of beneficial impacts on shoot as well as root growth, including reactions to tropism, cell division and elongation, vascular tissue differentiation, and the beginning of the root formation process (Jaroszuk-Ściseł et al., 2014). Although the metabolic pathway for IAA synthesis has not been discovered, L-tryptophan serves as the main substrate for IAA production by EF.

GAs play a key role in several plant responses, including fruit production, senescence, sexual expression, stem lengthening, and seed germination. The main end products of the mevalonic acid pathway, which produces GAs in EF, are GA1 and GA3, which are derived from GA4, GA5, and GA7. GAs are created by endophytic fungus from acetyl-CoA (Bömke and Tudzynski, 2009) (Table 5.2).

Table 5.1 Endophytic fungi and their biological role in nutrient uptake.

Endophytic fungi	Host plant	Biological role	References
Phomopsis liquidambari	Peanuts	Increase in nodulation and N_2 fixation via enhancing H_2O_2/NO-dependent signaling crosstalk	Xie et al. (2017)
	Rice	Stimulation of expression of several genes involved in N-uptake enhancing N- accumulation	Yang et al. (2014)
		Root exudates have increased concentration of soluble saccharides, total free amino acids, and organic acids	Yang et al. (2015)
Entomopathogenic fungi: Species of *Metarhizium* (*M. robertsii*) and *Beauveria bassiana*	Bean plants, wheat, switch grass	Transfer of N from insect larvae to host plant Killing of pathogenic insect larvae	Behie and Bidochka (2014)
Trichoderma asperellum	Onion (*Allium cepa*)	Phosphorous acquisition, hence reducing the usage of phosphorous fertilizers	Ortega-García et al. (2015)
Aspergillus sydowii	Maize (*Zea mays*)	Accumulation of higher amounts of P in their tissues despite of receiving lower fertilizer doses	Baron et al. (2018)

5.10 Indirect benefits

5.10.1 Production of siderophore

All living cells require the microelement iron for their growth (Rana et al., 2020). Some microorganisms, such as EF, create siderophores, which are tiny molecules with iron-chelation characteristics that attract ferric ions (Fe^{3+}) in the rhizosphere. The function of siderophores synthesized by EF is presently unclear, even after speculation that they may be related to induced systemic resistance (ISR) (Card et al., 2016). Endophytic strains of *Trichoderma koningii* ST-KKU1, *Macrophomina phaseolina* SS1L10, and *M. phaseolina* SS1R10 produced siderophores (Suebrasri et al., 2020). Additionally, siderophores produced by recombinant *Trichoderma harzianum* colonized legumes (*Phaseolus vulgaris*) are documented. The authors have hypothesized that the synthesis of siderophores by fungus had a significant contribution to boosting the sunchoke plant's growth (Eslahi et al., 2020). The PGP (plant growth promoting) siderophore ferrirubin is present in the endophyte *Talaromyces pinophilus* that inhabits strawberry trees (Aamir et al., 2019).

Table 5.2 Endophytic fungi and their role in production of plant hormones.

Endophytic fungi	Host plant	Biological role	References
Penicillium chrysogenum, *Alternaria alternate*, and Sterile hyphae (Sh-26)	*Asclepias sinaica* (medicinal plant)	Plant growth promotion by ammonia and indole-3-acetic acid (IAA) production	Fouda et al. (2015)
Paecilomyces formosus (*P. maximus*) strain LH10	Cucumber plants	Production of IAA and gibberellins (GAs)	Khan et al. (2012)
Phoma glomerata and *Penicillium* spp.	Cucumber plants	Production of IAA and GAs	Waqas et al. (2012)
Penicillium citrinum IR-3–3	Dune plants	Improve length of seedlings—Waito-c rice dwarf mutant (deficient of GA production)	Khan et al. (2008)
Trichoderma strains	Tomato seeds	Significant increase in plant height, fresh and dry matter of shoots and roots, chlorophyll content in leaves in addition to higher surface area due to high IAA production High phosphate solubilization capacity	Bader et al. (2020)
Purpureocillium lilacinum, *Purpureocillium lavendulum*, and *Metarhizium marquandii*	Soybean plants, bean plants, and maize plants	Promote growth parameters such as dry matter (shoot and roots) through IAA production Solubilize phosphate from fluorapatite	Baron et al. (2020)

5.10.2 Abiotic stress tolerance

As a result of climate change, plants are exposed to increasingly difficult environmental circumstances for their growth and maintenance. Additionally, the need for increased yields makes the situation for crops even more challenging. Realizing that it is evident that some support is needed for optimal plant growth, EF are a viable option for plant protection against both biotic and abiotic challenges in this scenario (Baron and Rigobelo, 2022).

Increased resistance to various types of abiotic stress is perhaps one of the most notable advantages that plants experience from interacting with endophytic fungus. Abiotic stressors like salt, extreme heat and cold, dryness, even harmful heavy metals can be overcome by endophytic fungus. EF can withstand abiotic stresses such as salt, extreme heat and cold, dryness, even toxic heavy metals. Abiotic stresses have a negative impact on the structure and physiology of plants because of genetic control over cell processes that result in diverse imbalances

(Egamberdieva et al., 2017). It was reviewed that plants enhance the action of antioxidant enzymes, primarily catalases along with peroxidases, during oxidative stress, which causes the production of ROS (reactive oxygen species), results in the membrane damage and the peroxidation of membrane lipids (Khan et al., 2015; Yan et al., 2019). EF impart resistance to ROS, lowering lipid peroxidation, through various unknown processes. Another important problem with abiotic stressors (drought, high temperature, and salt) on membranes is electrolyte leakage, which is connected to variations in the lipidic composition of the cellular membranes brought on by stress conditions. Alterations in the lipidic makeup of the cellular membrane can be induced by EF to stop leakage (Khan et al., 2015; Yan et al., 2019). Other than directly supporting plant growth, phytohormones also indirectly help plants by regulating how quickly they adapt to abiotic stressors. For instance, abscisic acid (ABA) causes stomata to shut, avoiding excessive water loss, and regulates the expression of genes implicated in stress responses. ABA levels are decreased by interaction with endophytic fungus. Salicylic acid (SA), a phytohormone, directly stimulates the abscisic acid-stimulated response (ASR), regulates the synthesis of PR (pathogen-related) proteins, and increases plant levels of SA through its interaction with EF. EF alter the physiological and biochemical processes in drought-tolerant plants to increase their capacity for drought tolerance. These modifications include increasing photosynthesis and plant growth by enhancing chlorophyll and gaseous exchange as well as some phytohormones, maintaining the osmotic balance by accumulating osmolytes, reducing ROS production by certain metabolic activity, and improving water efficiency for plants under drought stress (Dastogeer and Wylie, 2017). The advantages offered through these fungi, however, seem to be influenced by the host organism, host genotype, as well as environmental factors (Yan et al., 2018).

5.10.3 Biotic stress tolerance

The primary defenses against diseases, herbivores, and nematodes in response to biotic stress involve the generation of secondary metabolites as well as the induction of systemic resistance by EF (Poveda et al., 2020). The mechanisms of competition and mycoparasitism are other possibilities. Fungal endophytes activate ISR and ASR, which create compounds that are resistant to infections and provide biotic stress protection. Parasitism or competition can also occur to ward against illness and herbivory (Chitnis et al., 2020). One of the most crucial ways endophytes provide for disease prevention is induced resistance. The stimulating substance triggers the production of signaling molecules in the host, which increases the plant's resistance against pathogen invasion by ensuring that certain genes and proteins are expressed. Changes in plant metabolism make the host more important and lessen the severity of the illness. Both biotic and abiotic factors can make host plants resistant, either locally or throughout the entire plant. It has been discovered that the endophyte, *Serendipita indica*, isolated from roots aids in greater growth as well as resistance in a variety of plants (Nassimi and Taheri, 2017). The second strategy is known as mycoparasitism, in which a fungus becomes a parasite on another fungus.

Through the use of haustoria, where it destroys the plant cells, biotrophic fungi receive their nutrition from the live cells of their hosts (Kim and Vujanovic, 2016). For instance, *Thanatephorus cucumeris* is the host of *Stachybotrys elegans.* The fungus parasitizes another fungus by adhering directly to it, puncturing the hyphae, accumulating around the hyphae, and eventually producing aberrant hyphae. The third approach used by EF is competition. By competing for space and nutrients, the endophyte, which lives inside the plant body, prevents infections from growing. When compared to pathogens residing inside plant tissue, the endophyte grows and consumes nutrients quickly (Rodriguez et al., 2009). An endophyte, *Heteroconium chaetospira,* was identified from oilseed rape roots which suppressed the pathogen causing clubroot disease. Endophytes have been proposed as a BCA that really is crucial for them to aggregate at the entrance site for pathogen elimination (Card et al., 2016). The last strategy is called antibiosis, and it refers to the utilization of substances made by EF as biocontrol agents to inhibit the spread of diseases. Endophytes that have been isolated from plants have the ability to create antimicrobial substances that inhibit the growth of harmful diseases, such as flavonoids, alkaloids, peptides, phenols, steroids, terpenoids, polyketides, and VOC-specific metabolites (Lugtenberg et al., 2016). To identify and extract the precise chemical compounds generated by biocontrol agents employed for antimicrobial action, a research of possible endophytic fungus is required (Daguerre et al., 2017).

5.10.4 Activation of systemic resistance

EF help plants strengthen their defensive mechanisms, which in turn stimulates the activation of ISR and ASR pathways, which may eventually combine to enhance plant development and provide protection from infections and pests (Chadha et al., 2015). The identification of pathogen-associated molecular patterns (PAMPs) triggers the activation of systemic resistance (Poveda et al., 2020). Microbe-associated molecular patterns (MAMPs) are molecular patterns associated with microbes that are identified by plant receptors and cause a phenomenon known as MAMP-triggered immunity (MTI). Chitin and glucan found in the fungal cell wall, as well as secreted enzymes (xylanases, glucanases, and chitinases) and their byproducts, are examples of MAMPs that cause MTI (Latz et al., 2018). In addition to MAMPs, effector molecules like secondary metabolites also cause resistance and activate effector-triggered immunity (ETI). It has been shown that EF can alter their MAMPs to be identified by plants. Beneficial microbes appear to be able to get through this barrier in the case of ETI, which speeds up the colonization process (Yan et al., 2019). A complicated reaction that is mostly unstudied results from MAMPs and effectors activating ISR and ASR pathways. However, it also involves ion flux, protein phosphorylation and dephosphorylation, the production of signaling molecules (like ethylene and SA), as well as ROS and the selective expression of genes for defense responses, like the thickening of the plant cell wall, the production of PR proteins, and even cell death (Chadha et al., 2015). It has a priming effect in this situation, preparing the plant for more infections by nematodes, herbivores, or pathogenic microbes (Latz et al., 2018). Because they do not possess any harmful traits, fungal endophytes interact with their plant hosts in a balanced way. A comparison of the pathogenic *Colletotrichum incanum* with the endophytic strain *Colletotrichum tofieldiae* in *Arabidopsis thaliana* serves as a nice illustration. At the time of plant colonization, negatively selected effector protein genes in the endophytic strain are directly implicated in the pathogenic activity. Brassicaceae plants, on the other hand, showed less receptors for these proteins. As a result, species with a propensity to interact favorably with plants decreased the repertoire of these genes in their genome, whereas harmful species may use the maintenance of these genes as a possible host assault tactic (Baron and Rigobelo, 2022).

5.11 Endophytic fungi as biofertilizer

According to latest developments in the new fields of research for sustainable agricultural science, endophytic microorganisms and their consortia are a preferred option for enhancing crop yield, soil health, and fertility management (Mukherjee et al., 2021). Biofertilizers are generally microorganisms that improve soil nutritional status by making nutrients readily accessible to plants (Yadav, 2018). Fungal endophytes colonize interior plant tissues without developing disease indications and promote the plant growth that benefit both the host plant and them. As a result of this interaction, they are known as the PGP microbes (Baron and Rigobelo, 2022). The direct advantages of communication with EF are increased in nutrient uptake and numbers of phytohormones synthesis in the plant. These benefits are related to improved productivity, better root growth, plant height, reproduction, and biomass output (Bamisile et al., 2018). In contrast, endophytic invasion has indirect benefits such as induced tolerance, secondary metabolite and antimicrobial synthesis, siderophores synthesis, and resilience against biotic as well as abiotic stressors (Baron and Rigobelo, 2022).

5.12 Evaluation of endophytic diversity for sustainable agriculture: technical aspects

The first crucial step in effectively applying and exploiting endophytic fungal variety for sustainable agricultural practices is their isolation, screening, and selection in a methodical manner. Additionally, creating the most efficient inoculum is a vital step in maximizing the potential and benefits of agriculture biofertilizer. The kind of substance utilized as an active component for microbial inoculant, the manner of administration, the extended shelf life of microbial products, and the requirement that the substrate in discussion must offer stable environment are additional factors connected to such goals. The effectiveness of the microbial therapy is significantly influenced by the inoculation of seeds with agriculturally significant, secure, and environmentally persistent endophytic fungal spores (Aamir et al., 2019).

5.13 Precautions about utilizing endophytic fungi in the field of agriculture

Endophytes from different plants that promote growth may have negative impacts on people and other animals. Although most studies on endophytic microbial toxicity to humans have focused on bacterial strains, such as those from the genera *Burkholderia, Enterobacter, Pseudomonas*, and *Staphylococcus*, the use of EF is still promising. *Epichlo* (=*Neotyphodium*) spp. is the fungi having negative effects on vertebrates which endophytically colonize grasses. They are involved in the synthesis and buildup of alkaloids in plant tissues, which are poisonous to several vertebrates and several invertebrates, particularly cattle. In 2013 it was documented how to feed dairy cows ryegrass that has been endophytically colonized by two different strains of *Epichloe festucae* (=*Neotyphodium lolii*) (Finch et al., 2013). The scientists found that the milk contained fungus-produced alkaloids at concentrations that were deemed acceptable for human consumption. Despite the obvious advantages of using EF as biocontrol agents to resist diseases, certain drawbacks do occur. Inoculation with the fungus *Cryptosporiopsis* sp. can inhibit and slow the host's root development (Norway spruce seedlings) (Wani et al., 2016). This serves as a warning that the lines between an organism working as a helpful mutualist or pathogen to the same plant are accessible. Mycotoxins, which can be harmful to human health, are produced by several EF (Chitnis et al., 2020). A thorough understanding of the variables and dangers involved is required in order to prevent such detrimental consequences and enhance the application of endophytes in the biocontrol of plant diseases. Endophytes may behave as latent pathogens under certain inherent or environmental stress. Thus when endophytes are employed for biocontrol, it is important to understand the pathogenesis (Yan et al., 2019). Therefore it is important to take precautions to avoid choosing any established (plant or human) pathogen (Murphy et al., 2018).

5.14 Role of endophytic fungi in agriculture allied sectors

5.14.1 Application in environmental remediation

Heavy metal emissions were enhanced by the ongoing global industrialization of numerous contaminants. These heavy metals infiltrate into soils from water sources and industrial runoff, seriously contaminating the ecosystem and endangering human health. In addition to the often employed physicochemical techniques like filtration, chemical precipitation, and electrochemical treatment, bioremediation has newly been recognized as a viable technique to remove metals from polluted soils (Xiao et al., 2010). Living plants (particularly hyperaccumulators) are used in phytoremediation, which has considerable benefits in terms of great effectiveness and environmental friendliness. High levels of heavy metal pollution, however, can harm plant metabolisms, which limits their capacity to effectively remove the metals. Phytoremediation poses certain limitation which is the slower rate of plant development and the time it takes for biomass production. Recent studies, however, suggest that EF have a greater ability to help plants tolerate heavy metals (Khan et al., 2017). The fungi can reduce metal availability and toxicity to host plants by bio transforming metals and/or accumulating heavy metals in its hyphae from the soil around plants (Khan et al., 2017). In this situation, IAA produced by endophytic fungal strains was shown to boost the growth of plant biomass and hence enhance the removal of metals by plants. Endophytic fungus, both active and dormant, can increase phytoremediation of heavy metal-contaminated soils and absorb metals from water and soils without the need for a plant host. From the stem of *Portulaca oleracea* growing in metal-contaminated soils, *Lasiodiplodia* sp. MXSF31, an endophytic fungus, was isolated. The active fungal biomass absorbed more hazardous Cd, Pb, and Zn in contrast to the decaying matter from contaminated soils and water. According to studies, functional groups on such fungal surface, such as the cell wall's hydroxyl, amino, carbonyl, sulfonate, and benzene moieties, may have a role in the biosorption of heavy metals. It is possible that a factor impacting adsorption effectiveness is the density of these functional groups. EF could be chemically altered to become a bio sorbent to increase the biosorption effectiveness for metal particles. For instance, after being dried and used as raw biomass, the mycelium of the mangrove fungal endophyte *Fusarium* sp. was chemically altered by formaldehyde, methanol, and acetic acid to improve its attraction for uranium from wastewater (Chen et al., 2014). Additional variables include contact duration, solution pH, the ratio of solid to liquid, and the initial metal content in the contaminated samples (Yan et al., 2018).

Neotyphodium coenophialum and *Neotyphodium uncinatum*, the EF, infect *Festuca arundinacea* schreb. and *Festuca pratensis* Huds. plants, while uninfected plants are grown on petroleum-contaminated soil. The endophyte-infected plants had more root and shoot biomass than the uninfected plants produced. It demonstrates that the endophytes are crucial to phytoremediation. Since they are one of the most frequent colonists of dead plant tissues, endophytes play an important function in the environment as decomposers (Sudha et al., 2016).

5.14.2 Biosynthesis of biocatalysts

It is believed that endophytic fungus must infect and live inside of plants without exhibiting obvious symptoms. These organisms may release a range of enzymes to degrade complex molecules from the atmosphere if they get their sustenance from sources other than the plant (Suryanarayanan et al., 2012). For instance, glutaminase-free L-asparaginase is used to treat both adult and pediatric acute lymphoblastic leukemia and has been found to be produced by EF derived from trees in wet evergreen and semievergreen forests of India. EF are efficient for producing chitin-modifying enzymes (chitinases, chitin deacetylases, and chitosanases) because they are more likely to remodel their cellular structure during growth and plant infection, so this capacity can be applied in biotechnology (Suryanarayanan et al., 2012). For instance, 31 fungal endophytes found in the Western Ghats forests of southern India showed positive chitinase activity, with several isoforms of the enzyme being characterized. Numerous isolates developed chitosanases, which act on chitosan that has been acetylated to varying degrees (Govinda Rajulu et al., 2011). The fungal enzymes often have lower molecular weights, higher polydispersity, and function on substrates with less acetylation. Assessing and filtering for the variety of isoenzymes as well as the oligomers formed by these enzymes is a novel but interesting technique since these chemicals are beneficial in a range of applications, including the food and environmental sectors. Enzymes with different properties are needed for a variety of industrial applications; as a result, fungal endophytes are becoming incredibly valuable as a source of particular and effective biocatalysts (Yan et al., 2018). Fig. 5.2 depicts the representation for the various applications of EF as biocatalysts.

5.14.3 Biosynthesis of biofuel

New sustainable alternative energy sources are urgently needed because of the rising global demand for fuel and the environmental issues brought on by greenhouse gas emissions (Wu et al., 2017). Volatile organic compounds (VOCs) can be produced by EF when they thrive on plant and agricultural waste. These VOCs, often known as "mycodiesel," are predominantly hydrocarbons and other oxygenated compounds, with properties resembling those of fossil fuels and are considered to be viable fossil fuel alternatives (Wu et al., 2017). For instance, a number of species of *Trichoderma* and certain strains of *Penicillium brasilianum*, *Penicillium griseoroseum*, *Xylaria* sp. (NICl3), *Xylaria* sp. (NICL5), and *Penicillium* sp. (PAOE) produce high concentrations of a lipid matrix that may be useful as sources of potential biofuel precursors (Santos-Fo et al., 2011). Due to the interaction between endophytes and plants, lignocellulolytic enzymes produced by EF may degrade and convert plant cellulose into mycodiesel under microaerophilic conditions (Suryanarayanan et al., 2012). Under microaerophilic circumstances, fungal endophyte from the genus *Gliocladium* was able to break down plant cellulose and produce complex hydrocarbons. Without the need for hydrolytic pretreatments, this fungus was able to generate hydrocarbons with a molecular weight range of C6–C19 (including benzene, hexane, and heptane) (Ahamed and Ahring, 2011). Graphical representation for various mentioned applications of EF is displayed in Fig. 5.3.

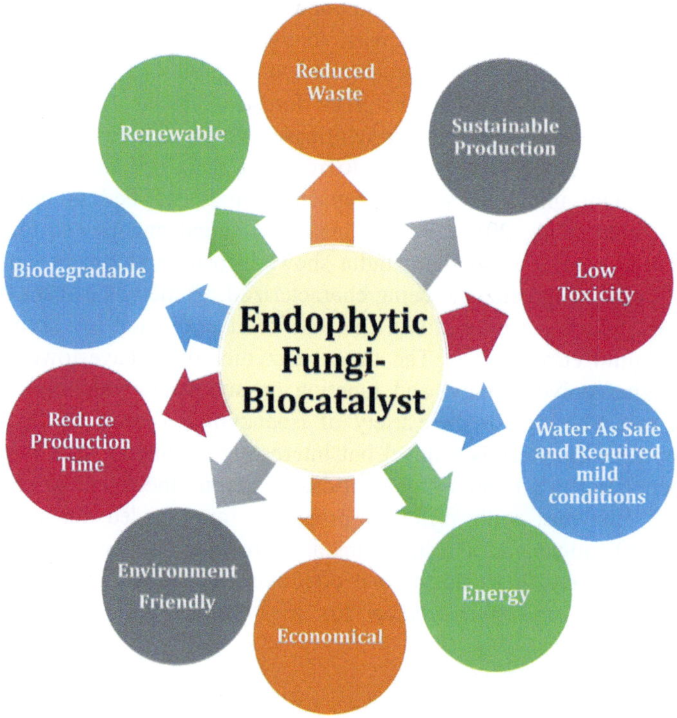

FIGURE 5.2

Representation for the various applications of endophytic fungi as biocatalysts.

5.15 Commercialization of endophytic fungi

Endophytes are being offered for sale to improve the insect and environmental resistance of agricultural crops. Commercially available EF-containing products include BioEnsureR-Corn and BioEnsureR-Rice. Under severe drought and other stressful situations, these compounds can boost crop output by up to 80%. Before planting seeds, the products are sprinkled on as they are available in liquid form. Before germination, they start a symbiotic association with seedlings and aid in enhancing plant performance during growth. Another example of a commercially product based on these organisms is the *Epichloe* endophytes, which are particularly popular in New Zealand, Australia, South America, and the United States (Johnson et al., 2013). These endophytes are helpful in agriculture because they can increase plant longevity by shielding it from biotic and abiotic stresses (Gautam and Avasthi, 2019).

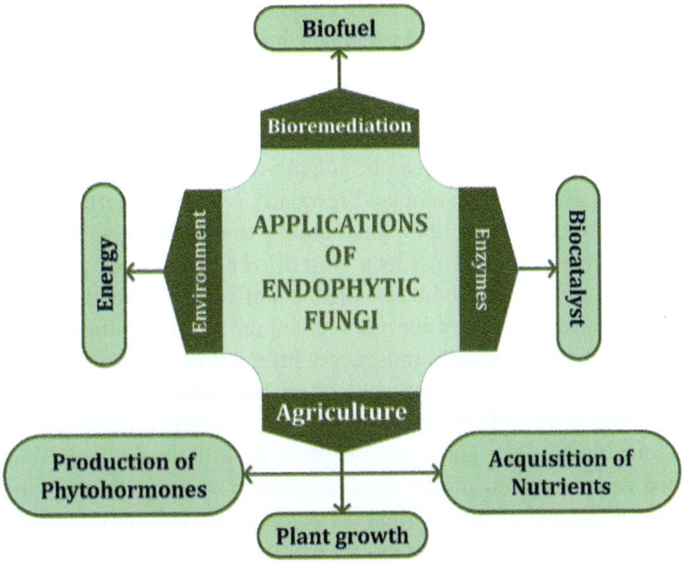

FIGURE 5.3

Applications of endophytic fungi in sustainable agriculture and allied sectors.

5.16 Conclusion and future perspectives

Agriculture is under a tremendous pressure to do something for the global food security, and the climate change is contributing to the worsening of the situation. Increasing temperature ultimately cause desired crop yields to decline and simultaneously promoting the growth of weed, pest, and disease-causing microorganisms. Hence, with the purpose of controlling weed development, insect issues, and plant diseases, farmers administer pesticides in current agricultural practice to increase the crop yield. However, if we look at the other side of the coin, pesticides are seriously endangering the environment as well as human beings. Extensive use of pesticides leads to the depletion of the soil structure and fertility as well as water quality, which has a dangerous effect on all living creatures, including animals, plants, birds, and humans. Therefore there is an unmet demand to switch to new ecofriendly approaches to maintain sustainable agriculture, a system based on ecological principles where the farming is focused on investigation of interactions between different species and their environment. An innovative strategy for sustainable farming is the use of biological resources, which reduce the emission of hazardous chemicals into the environment and support ecological sustainability. In this perspective, utilizing the benefits of beneficial microorganism and their metabolites for agricultural management and integrated nutrient and pest control becomes the viable and ecofriendly option.

Microorganisms are by far the most ubiquitous entities in the ecosystem and govern a number of important biochemical and elemental processes, including phosphate solubilization and N2 fixation, among many others. Endophytes are the microbes which are living inside plant tissues, owing to its advantages from mutualistic interactions among host plants. This gives the new direction for the use of microbes in sustainable agriculture and allied sectors. Endophytic fungus has an impact on a plant ability to develop and respond to diseases, herbivores, and environmental changes. Moreover, they have ability to produce significant secondary metabolites which lead to their use as a biofertilizer and biocontrol agent.

Even though EF are used in sustainable agriculture, much more extensive study is still required to determine their potential for application in this sector. Endophytes are like hidden gold mines, yet little is known about their biochemical makeup and how they interact with their host plants. For instance, EF are explored to produce phytohormones, but research on how other endophytic metabolites interact with plant hormone receptors is still unresolved. With today's sophisticated computational power, it is possible to examine these types of interactions using a variety of bioinformatics techniques, such as molecular docking and molecular simulation interactions. Furthermore, validation of these computation study in plant experiments can lead to new biochemical entities which can be used as phytohormone analogs in regular farming practice as well as hydroponic farming system. Exploration of other endophytic metabolites as antibacterial and antifungal compounds can decrease the load of biotic stress in current agriculture practice. Moreover, the level of biotic stress in the existing farming practices can be reduced by investigating novel endophytic metabolites as antibacterial and antifungal compounds.

EF are also used to produce mycodiesel, which is primarily hydrocarbons and other oxygenated chemicals with characteristics like fossil fuels and thought to be promising fossil fuel substitutes. But limited research exists for EF to increase the production of biofuels. A framework for developing the powerful fungal strains to produce biofuel may be provided by a thorough understanding of molecular biology and the mechanism of enzyme used for synthesis of fuel-like compounds by EF. This may be accomplished by using a variety of modern techniques, including genetic engineering, metabolomics, and proteomics. The exploration of EF, their metabolites, and in-depth study to understand their mode of action for various activity using modern technology will surely help the farmers to use EF and their products to increase the crop yields in ecofriendly sustainable agricultural practice.

References

Aamir, M., Rai, K.K., Zehra, A., Kumar, S., Yadav, M., Shukla, V., et al., 2019. Fungal endophytes: classification, diversity, ecological role, and their relevance in sustainable agriculturein:Microbial Endophytes: Prospects for Sustainable Agriculture 291–323. Available from: https://doi.org/10.1016/B978-0-12-818734-0.00012-7.

Ahamed, A., Ahring, B.K., 2011. Production of hydrocarbon compounds by endophytic fungi Gliocladium species grown on cellulose. Bioresour. Technol. 102, 9718–9722. Available from: https://doi.org/10.1016/j.biortech.2011.07.073.

Alam, B., Lǐ, J., Gě, Q., Khan, M.A., Gōng, J., Mehmood, S., et al., 2021. Endophytic fungi: from symbiosis to secondary metabolite communications or vice versa? Front. Plant. Sci. 12, 1–24. Available from: https://doi.org/10.3389/fpls.2021.791033.

Bader, A.N., Salerno, G.L., Covacevich, F., Consolo, V.F., 2020. Native *Trichoderma harzianum* strains from Argentina produce indole-3 acetic acid and phosphorus solubilization, promote growth and control wilt disease on tomato (*Solanum lycopersicum* L.). J. King Saud. Univ. - Sci. 32, 867–873. Available from: https://doi.org/10.1016/j.jksus.2019.04.002.

Bamisile, B.S., Dash, C.K., Akutse, K.S., Keppanan, R., Wang, L., 2018. Fungal endophytes: beyond herbivore management. Front. Microbiol. 9. Available from: https://doi.org/10.3389/FMICB.2018.00544/FULL.

Baron, N.C., Costa, N.T.A., Mochi, D.A., Rigobelo, E.C., 2018. First report of *Aspergillus sydowii* and *Aspergillus brasiliensis* as phosphorus solubilizers in maize. Ann. Microbiol. 68, 863–870. Available from: https://doi.org/10.1007/S13213-018-1392-5/FIGURES/2.

Baron, N.C., de Souza Pollo, A., Rigobelo, E.C., 2020. *Purpureocillium lilacinum* and *Metarhizium marquandii* as plant growth-promoting fungi. PeerJ 2020. Available from: https://doi.org/10.7717/peerj.9005.

Baron, N.C., Rigobelo, E.C., 2022. Endophytic fungi: a tool for plant growth promotion and sustainable agriculture. Mycology 13, 39–55. Available from: https://doi.org/10.1080/21501203.2021.1945699.

Baron, N.C., Rigobelo, E.C., Zied, D.C., Baron, N.C., Rigobelo, E.C., Zied, D.C., 2019. Filamentous fungi in biological control: current status and future perspectives. Chil. J. Agric. Res. 79, 307–315. Available from: https://doi.org/10.4067/S0718-58392019000200307.

Behie, S.W., Bidochka, M.J., 2014. Ubiquity of insect-derived nitrogen transfer to plants by endophytic insect-pathogenic fungi: an additional branch of the soil nitrogen cycle. Appl. Environ. Microbiol. 80, 1553–1560. Available from: https://doi.org/10.1128/AEM.03338-13.

Bömke, C., Tudzynski, B., 2009. Diversity, regulation, and evolution of the gibberellin biosynthetic pathway in fungi compared to plants and bacteria. Phytochemistry . Available from: https://doi.org/10.1016/j.phytochem.2009.05.020.

Card, S., Johnson, L., Teasdale, S., Caradus, J., 2016. Deciphering endophyte behaviour: the link between endophyte biology and efficacious biological control agents. FEMS Microbiol. Ecol. 92, 114. Available from: https://doi.org/10.1093/femsec/fiw114.

Carrol, G., 2012. Fungal endophytes in stems and leaves : from latent pathogen to mutualistic symbiont author (s): George Carroll reviewed work (s): Published by : Ecological Society of America Stable URL. Ecology 69, 2–9. http://www.jstor.org/stable/1943154.

Chadha, N., Mishra, M., Rajpal, K., Bajaj, R., Choudhary, D.K., Varma, A., 2015. An ecological role of fungal endophytes to ameliorate plants under biotic stress. Arch. Microbiol. 197, 869–881. Available from: https://doi.org/10.1007/S00203-015-1130-3.

Chen, F., Tan, N., Long, W., Yang, S.K., She, Z.G., Lin, Y.C., 2014. Enhancement of uranium(VI) biosorption by chemically modified marine-derived mangrove endophytic fungus *Fusarium* sp. #ZZF51. J. Radioanal. Nucl. Chem. 299, 193–201. Available from: https://doi.org/10.1007/S10967-013-2758-6.

Chitnis, V.R., Suryanarayanan, T.S., Nataraja, K.N., Prasad, S.R., Oelmüller, R., Shaanker, R.U., 2020. Fungal endophyte-mediated crop improvement: the way ahead. Front. Plant. Sci. 11. Available from: https://doi.org/10.3389/FPLS.2020.561007/FULL.

Daguerre, Y., Edel-Hermann, V., Steinberg, C., 2017. Fungal genes and metabolites associated with the biocontrol of soil-borne plant pathogenic fungi. Fungal Metab. 33–104. Available from: https://doi.org/10.1007/978-3-319-25001-4_27.

Dastogeer, K.M.G., Wylie, S.J., 2017. Plant-fungi association: role of fungal endophytes in improving plant tolerance to water stress. Plant-Microbe Interact. Agro-Ecological Perspect. 1, 161–176. Available from: https://doi.org/10.1007/978-981-10-5813-4_8.

Egamberdieva, D., Wirth, S.J., Alqarawi, A.A., Abd-Allah, E.F., Hashem, A., 2017. Phytohormones and beneficial microbes: essential components for plants to balance stress and fitness. Front. Microbiol. 8. Available from: https://doi.org/10.3389/FMICB.2017.02104/FULL.

Eslahi, N., Kowsari, M., Motallebi, M., Zamani, M.R., Moghadasi, Z., 2020. Influence of recombinant Trichoderma strains on growth of bean (*Phaseolus vulgaris* L) by increased root colonization and induction of root growth related genes. Sci. Hortic. 261. Available from: https://doi.org/10.1016/j.scienta.2019.108932.

Finch, S.C., Thom, E.R., Babu, J.V., Hawkes, A.D., Waugh, C.D., 2013. The evaluation of fungal endophyte toxin residues in milk. N. Z. Vet. J. 61, 11–17. Available from: https://doi.org/10.1080/00480169.2012.704626.

Fouda, A.H., Hassan, S.E.D., Eid, A.M., Ewais, E.E.D., 2015. Biotechnological applications of fungal endophytes associated with medicinal plant *Asclepias sinaica* (Bioss.). Ann. Agric. Sci. 60, 95–104. Available from: https://doi.org/10.1016/j.aoas.2015.04.001.

Gagic, M., Faville, M.J., Zhang, W., Forester, N.T., Rolston, M.P., Johnson, R.D., et al., 2018. Seed transmission of epichloë endophytes in lolium perenne is heavily influenced by host genetics. Front. Plant. Sci. 871. Available from: https://doi.org/10.3389/FPLS.2018.01580/FULL.

Gautam, A.K., Avasthi, S., 2019. Fungal endophytes: potential biocontrol agents in agriculture. Role of Plant Growth Promoting Microorganisms in Sustainable Agriculture and Nanotechnology. Elsevier, pp. 241–283. Available from: https://doi.org/10.1016/b978-0-12-817004-5.00014-2.

Gouda, S., Das, G., Sen, S.K., Shin, H.S., Patra, J.K., 2016. Endophytes: a treasure house of bioactive compounds of medicinal importance. Front. Microbiol. 7, 1538. Available from: https://doi.org/10.3389/FMICB.2016.01538/BIBTEX.

Govinda Rajulu, M.B., Thirunavukkarasu, N., Suryanarayanan, T.S., Ravishankar, J.P., El Gueddari, N.E., Moerschbacher, B.M., 2011. Chitinolytic enzymes from endophytic fungi. Fungal Divers. 47, 43–53. Available from: https://doi.org/10.1007/S13225-010-0071-Z.

Gundale, M.J., Almeida, J.P., Wallander, H., Wardle, D.A., Kardol, P., Nilsson, M.C., et al., 2016. Differences in endophyte communities of introduced trees depend on the phylogenetic relatedness of the receiving forest. J. Ecol. 104, 1219–1232. Available from: https://doi.org/10.1111/1365-2745.12595.

Hardoim, P.R., van Overbeek, L.S., Elsas, J.D.van, 2008a. Properties of bacterial endophytes and their proposed role in plant growth. Trends Microbiol. 16, 463–471. Available from: https://doi.org/10.1016/j.tim.2008.07.008.

Jaber, L.R., Enkerli, J., 2017. Fungal entomopathogens as endophytes: can they promote plant growth? Technol. Taylor Fr. 27, 28–41. Available from: https://doi.org/10.1080/09583157.2016.1243227.

Jaroszuk-Ściseł, J., Kurek, E., Trytek, M., 2014. Efficiency of indoleacetic acid, gibberellic acid and ethylene synthesized in vitro by *Fusarium culmorum* strains with different effects on cereal growth. Biologia 69, 281–292. Available from: https://doi.org/10.2478/S11756-013-0328-6/HTML.

Johnson, J.M., Alex, T., Oelmüller, R., 2014. *Piriformospora indica*: the versatile and multifunctional root endophytic fungus for enhanced yield and tolerance to biotic and abiotic stress in crop plants. J. Trop. Agric. 52, 103–122.

Johnson, L.J., De Bonth, A.C.M., Briggs, L.R., Caradus, J.R., Finch, S.C., Fleetwood, D.J., et al., 2013. The exploitation of epichloae endophytes for agricultural benefit. Fungal Divers. 60, 171–188. Available from: https://doi.org/10.1007/s13225-013-0239-4.

Khan, A., Hamayun, M., Kang, S.M., Kim, Y.H., Jung, H.Y., Lee, J.H., et al., 2012. Endophytic fungal association via gibberellins and indole acetic acid can improve plant growth under abiotic stress: an example of *Paecilomyces formosus* LHL10. BMC Microbiol. 12. Available from: https://doi.org/10.1186/1471-2180-12-3.

Khan, A.L., Hussain, J., Al-Harrasi, A., Al-Rawahi, A., Lee, I.J., 2015. Endophytic fungi: resource for gibberellins and crop abiotic stress resistance. Crit. Rev. Biotechnol. 35, 62–74. Available from: https://doi.org/10.3109/07388551.2013.800018.

Khan, A.R., Ullah, I., Waqas, M., Park, G.S., Khan, A.L., Hong, S.J., et al., 2017. Host plant growth promotion and cadmium detoxification in *Solanum nigrum*, mediated by endophytic fungi. Ecotoxicol. Environ. Saf. 136, 180–188. Available from: https://doi.org/10.1016/j.ecoenv.2016.03.014.

Khan, S.A., Hamayun, M., Yoon, H., Kim, H.Y., Suh, S.J., Hwang, S.K., et al., 2008. Plant growth promotion and *Penicillium citrinum*. BMC Microbiol. 8. Available from: https://doi.org/10.1186/1471-2180-8-231.

Kiers, E.T., Duhamel, M., Beesetty, Y., Mensah, J.A., Franken, O., Verbruggen, E., et al., 2011. Reciprocal rewards stabilize cooperation in the mycorrhizal symbiosis. Science (80-.) 333, 880–882. Available from: https://doi.org/10.1126/SCIENCE.1208473.

Kim, S.H., Vujanovic, V., 2016. Relationship between mycoparasites lifestyles and biocontrol behaviors against *Fusarium* spp. and mycotoxins production. Appl. Microbiol. Biotechnol. 100, 5257–5272. Available from: https://doi.org/10.1007/S00253-016-7539-Z.

Latz, M.A.C., Jensen, B., Colling, D.B., Jørgensen, H.J.L., 2018. Endophytic fungi as biocontrol agents: elucidating mechanisms in disease suppression. Taylor Fr. 11, 555–567. Available from: https://doi.org/10.1080/17550874.2018.1534146.

Lugtenberg, B.J.J., Caradus, J.R., Johnson, L.J., 2016. Fungal endophytes for sustainable crop production. F. Microbiol. 92, 194. Available from: https://doi.org/10.1093/femsec/fiw194. Available from: academic.oup.com.

Mishra, Y., Sharma, L., Dhiman, M., Sharma, M.M., 2021. Endophytic fungal diversity of selected medicinal plants and their bio-potential applications. Fungi Bio-Prospects Sustain. Agriculture, Environ. Nano-Technol 227–283. Available from: https://doi.org/10.1016/b978-0-12-821394-0.00010-x.

Mukherjee, A., Bhowmick, S., Yadav, S., Rashid, M.M., Chouhan, G.K., Vaishya, J.K., et al., 2021. Re-vitalizing of endophytic microbes for soil health management and plant protection. 3 Biotech. 11. Available from: https://doi.org/10.1007/S13205-021-02931-4.

Murphy, B.R., Doohan, F.M., Hodkinson, T.R., 2018. From concept to commerce: developing a successful fungal endophyte inoculant for agricultural crops. J. Fungi. Available from: https://doi.org/10.3390/jof4010024.

Nassimi, Z., Taheri, P., 2017. Endophytic fungus *Piriformospora indica* induced systemic resistance against rice sheath blight via affecting hydrogen peroxide and antioxidants. Biocontrol. Sci. Technol. 27, 252–267. Available from: https://doi.org/10.1080/09583157.2016.1277690.

Ortega-García, J.G., Montes-Belmont, R., Rodríguez-Monroy, M., Ramírez-Trujillo, J.A., Suárez-Rodríguez, R., Sepúlveda-Jiménez, G., 2015. Effect of *Trichoderma asperellum* applications and mineral fertilization on growth promotion and the content of phenolic compounds and flavonoids in onions. Sci. Hortic. 195, 8–16. Available from: https://doi.org/10.1016/j.scienta.2015.08.027.

Petrini, O., Sieber, T.N., Toti, L., Viret, O., 1993. Ecology, metabolite production, and substrate utilization in endophytic fungi. Nat. Toxins 1, 185–196. Available from: https://doi.org/10.1002/nt.2620010306.

Poveda, J., 2021. Beneficial effects of microbial volatile organic compounds (MVOCs) in plants. Appl. Soil. Ecol. 168. Available from: https://doi.org/10.1016/J.APSOIL.2021.104118.

Poveda, J., Abril-Urias, P., Escobar, C., 2020. Biological control of plant-parasitic nematodes by filamentous fungi inducers of resistance: trichoderma, mycorrhizal and endophytic fungi. Front. Microbiol. 11. Available from: https://doi.org/10.3389/FMICB.2020.00992/FULL.

Powell, J.R., Karunaratne, S., Campbell, C.D., Yao, H., Robinson, L., Singh, B.K., 2015. Deterministic processes vary during community assembly for ecologically dissimilar taxa. Nat. Commun. 6. Available from: https://doi.org/10.1038/ncomms9444.

Quesada Moraga, E., 2020. Entomopathogenic fungi as endophytes: their broader contribution to IPM and crop production. Biocontrol. Sci. Technol. Available from: https://doi.org/10.1080/09583157.2020.1771279.

Rajamanikyam, M., Vadlapudi, V., Amanchy, R., Upadhyayula, S.M., 2017. Endophytic fungi as novel resources of natural therapeutics. Braz. Arch. Biol. Technol. 60. Available from: https://doi.org/10.1590/1678-4324-2017160542.

Rana, K.L., Kour, D., Kaur, T., Devi, R., Yadav, A.N., Yadav, N., et al., 2020. Endophytic microbes: biodiversity, plant growth-promoting mechanisms and potential applications for agricultural sustainability. Antonie van. Leeuwenhoek, Int. J. Gen. Mol. Microbiol. 113, 1075–1107. Available from: https://doi.org/10.1007/S10482-020-01429-Y.

Roberts, E., Lindow, S., 2013. Loline alkaloid production by fungal endophytes of Fescue species select for particular epiphytic bacterial microflora. ISME J. 82 (8), 359–368. Available from: https://doi.org/10.1038/ismej.2013.170.

Rodriguez, R.J., White Jr, J.F., Arnold, A.E., Redman, R.S., 2009. Fungal Endophytes: Diversity and Functional Roles. Wiley Online Library, pp. 314–330. 182. Available from: https://doi.org/10.1111/j.1469-8137.2009.02773.x.

Santos-Fo, F.C., Fill, T.P., Nakamura, J., Monteiro, M.R., Rodrigues-Fo, E., 2011. Endophytic fungi as a source of biofuel precursors. J. Microbiol. Biotechnol. 21, 728–733. Available from: https://doi.org/10.4014/jmb.1010.10052.

Sasse, J., Martinoia, E., Northen, T., 2018. Feed your friends: do plant exudates shape the root microbiome? Trends Plant. Sci. Available from: https://doi.org/10.1016/j.tplants.2017.09.003.

Savci, S., 2012. An agricultural pollutant: chemical fertilizer. Int. J. Environ. Sci. Dev. 73–80. Available from: https://doi.org/10.7763/ijesd.2012.v3.191.

Schulz, B., Boyle, C., 2005. The endophytic continuum. Mycol. Res. Available from: https://doi.org/10.1017/S095375620500273X.

Schulz, B., Boyle, C., Draeger, S., Römmert, A.K., Krohn, K., 2002. Endophytic fungi: a source of novel biologically active secondary metabolites. Mycol. Res. Available from: https://doi.org/10.1017/S0953756202006342.

Stone, J.K., Polishook, J.D., White, J.F., 2004. Endophytic fungi. Biodivers. Fungi: Inventory Monit. Methods 241–270. Available from: https://doi.org/10.1016/B978-012509551-8/50015-5.

Sudha, V., Govindaraj, R., Baskar, K., Al-Dhabi, N.A., Duraipandiyan, V., 2016. Biological properties of endophytic fungi. Braz. Arch. Biol. Technol. 59, 1–7. Available from: https://doi.org/10.1590/1678-4324-2016150436.

Suebrasri, T., Harada, H., Jogloy, S., Ekprasert, J., Boonlue, S., 2020. Auxin-producing fungal endophytes promote growth of sunchoke. Rhizosphere 16, 100271. Available from: https://doi.org/10.1016/j.rhisph.2020.100271.

Suman, A., Nath Yadav, A., Verma, P., 2016. Endophytic microbes in crops: diversity and beneficial impact for sustainable agriculture. Microb. Inoculants Sustain. Agric. Product. Vol. 1 Res. Perspect. 117–143. Available from: https://doi.org/10.1007/978-81-322-2647-5_7.

Suryanarayanan, T.S., Thirunavukkarasu, N., Govindarajulu, M.B., Gopalan, V., 2012. Fungal endophytes: an untapped source of biocatalysts. Fungal Divers. 54, 19–30. Available from: https://doi.org/10.1007/S13225-012-0168-7.

Viterbo, A., Chet, I., 2006. TasHyd1, a new hydrophobin gene from the biocontrol agent Trichoderma asperellum, is involved in plant root colonization. Mol. Plant. Pathol. 7, 249–258. Available from: https://doi.org/10.1111/J.1364-3703.2006.00335.X. Wiley Online Libr.

Wani, Z.A., Mirza, D.N., Arora, P., Riyaz-Ul-Hassan, S., 2016. Molecular phylogeny, diversity, community structure, and plant growth promoting properties of fungal endophytes associated with the corms of saffron plant: an insight into the microbiome of *Crocus sativus* Linn. Fungal Biol. 120, 1509–1524. Available from: https://doi.org/10.1016/j.funbio.2016.07.011.

Waqas, M., Khan, A.L., Kamran, M., Hamayun, M., Kang, S.M., Kim, Y.H., et al., 2012. Endophytic fungi produce gibberellins and indoleacetic acid and promotes host-plant growth during stress. Molecules 17, 10754–10773. Available from: https://doi.org/10.3390/molecules170910754.

Wiewióra, B., Żurek, G., Pañka, D., 2015. Is the vertical transmission of *Neotyphodium lolii* in perennial ryegrass the only possible way to the spread of endophytes? PLoS One 10. Available from: https://doi.org/10.1371/JOURNAL.PONE.0117231.

Wu, W., Davis, R.W., Tran-Gyamfi, M.B., Kuo, A., LaButti, K., Mihaltcheva, S., et al., 2017. Characterization of four endophytic fungi as potential consolidated bioprocessing hosts for conversion of lignocellulose into advanced biofuels. Appl. Microbiol. Biotechnol. 101, 2603–2618. Available from: https://doi.org/10.1007/S00253-017-8091-1.

Xiao, X., Luo, S., Zeng, G., Wei, W., Wan, Y., Chen, L., et al., 2010. Biosorption of cadmium by endophytic fungus (EF) *Microsphaeropsis* sp. LSE10 isolated from cadmium hyperaccumulator *Solanum nigrum* L. Bioresour. Technol. 101, 1668–1674. Available from: https://doi.org/10.1016/j.biortech.2009.09.083.

Xie, X.G., Fu, W.Q., Zhang, F.M., Shi, X.M., Zeng, Y.T., Li, H., et al., 2017. The endophytic fungus phomopsis liquidambari increases nodulation and N_2 fixation in *Arachis hypogaea* by enhancing hydrogen peroxide and nitric oxide signalling. Microb. Ecol. 74, 427–440. Available from: https://doi.org/10.1007/S00248-017-0944-8.

Yadav, A.N., 2018. Biodiversity and biotechnological applications of host-specific endophytic fungi for sustainable agriculture and allied sectors acta scientific microbiology (ISSN : 2581-3226) biodiversity and biotechnological applications of host-specific endophytic Fu. Acta Sci. Microbiol. 1.

Yan, L., Zhao, H., Zhao, X., Xu, X., Di, Y., Jiang, C., et al., 2018. Production of bioproducts by endophytic fungi: chemical ecology, biotechnological applications, bottlenecks, and solutions. Appl. Microbiol. Biotechnol. 102, 6279–6298. Available from: https://doi.org/10.1007/S00253-018-9101-7.

Yan, L., Zhu, J., Zhao, X., Shi, J., Jiang, C., Shao, D., 2019. Beneficial effects of endophytic fungi colonization on plants. Appl. Microbiol. Biotechnol. 103, 3327–3340. Available from: https://doi.org/10.1007/S00253-019-09713-2.

Yang, B., Ma, H.Y., Wang, X.M., Jia, Y., Hu, J., Li, X., et al., 2014. Improvement of nitrogen accumulation and metabolism in rice (*Oryza sativa* L.) by the endophyte *Phomopsis liquidambari*. Plant. Physiol. Biochem. 82, 172–182. Available from: https://doi.org/10.1016/j.plaphy.2014.06.002.

Yang, B., Wang, X.M., Ma, H.Y., Yang, T., Jia, Y., Zhou, J., et al., 2015. Fungal endophyte *Phomopsis liquidambari* affects nitrogen transformation processes and related microorganisms in the rice rhizosphere. Front. Microbiol. 6. Available from: https://doi.org/10.3389/FMICB.2015.00982/FULL.

Zhang, H.W., Song, Y.C., Tan, R.X., 2006. Biology and chemistry of endophytes. Nat. Prod. Rep. Available from: https://doi.org/10.1039/b609472b.

Zikankuba, V.L., Mwanyika, G., Ntwenya, J.E., James, A., Lwoga Zikankuba, V., 2017. Pesticide regulations and their malpractice implications on food and environment safety. Taylor Fr. 5. Available from: https://doi.org/10.1080/23311932.2019.1601544.

CHAPTER

Plant growth-promoting bacteria: sustainable use in abiotic stress management

6

Amanda Lys dos Santos Silva[1] **and Ana Maria Queijeiro López**[2]

[1]*Institute of Biological and Health Sciences (ICBS), Microbiology Section, Federal University of Alagoas, Maceió-AL, Brazil*

[2]*Institute of Chemistry and Biotechnology (IQB) , Biochemistry Section, Federal University of Alagoas, Maceió-AL, Brazil*

6.1 Introduction

In nature, plants are colonized by countless soil microorganisms, which play important roles in maintaining plant health and productivity. When such microbes live in close contact with the plants root zone, they are called rhizosphere, which has ability to withstand changes in environmental conditions. This justifies the increased interest in understanding modifications to rhizospheric microbial diversity and community structure in response to different stresses. In a review published by Allison and Martiny (2008), the authors describe three potential impacts of a disturbance on microbial composition and/or ecosystem processes, and explain that to be considered "resistant," microbial composition must remain the same in the face of the disturbance. Alternatively, if the community changes but quickly recovers in its initial composition when the disturbing agent ceases and the original environmental conditions are restored, it is sensitive but "resilient." Finally, a community whose composition is sensitive and nonresilient can still generate process rates like the original community only if the community members are "functionally redundant."

Plant growth-promoting bacteria (PGPB) are the free-living bacteria in the soil, in addition to those that colonize the rhizosphere, and cannot only improve plant vigor, but tolerate different biotic and abiotic stresses (Modi et al., 2020; Patel et al., 2021; Singh et al., 2018). They do this through several mechanisms that act synergistically during plant development, such as production of extracellular polymeric substance (EPS), production of ACC (1-aminocyclopropane-1-carboxylate) deaminase, changes in phytohormone contents, reduction in production of ethylene, inducing plant synthesis of antioxidant enzymes, improving the absorption of essential mineral elements, decreasing the absorption of excess nutrients/heavy metals (HM), and by altering the fitness of individual plant genotypes. Combining all the mentioned characteristics, PGPB induce morphological

Sustainable Agricultural Practices. DOI: https://doi.org/10.1016/B978-0-443-19150-3.00011-4

and biochemical changes in inoculated plants, that is, they induce systemic tolerance (ITS) to abiotic stresses.

Since it is imperative to exponentially increase food crop production to sustain the increase in global population, and the susceptibility of plants to a range of abiotic stresses is one of the challenges to achieve this goal, screening for PGPB with multiple inoculants uses for the management of these crops in the field is relevant and an ecologically correct biotechnological tool.

This chapter is an update on the role of PGPBs from their screening in many environments impacted by abiotic stresses, their use in phytoremediation, and the potential challenges of commercializing these microorganisms as inoculants to improve agricultural production. We present the current state of the concepts, as well as the characterization of the mechanisms of action of PGPB that can be determined in vitro, as well as many case studies and field applications of these inoculants.

6.2 Rhizosphere

Once roots perform many functions for a plant, including anchoring and acquiring vital nutrients and water needed for growth, its development depends on the combination and concentration of mineral nutrients available in the soil that gets from decaying plants and animals. Soils also function as a system for filtering surface water, storing carbon, and maintaining atmospheric gases. However, more than a century ago, Lorenz Hiltner first proposed the idea that plant roots are not only influenced by the soil, defining them as "rhizosphere," but the soil compartment is influenced by plant roots (Hiltner, 1904). This scientist postulated that the area around a plant root releases chemicals capable of supporting the development of different microbial communities around it. Hiltner observed bacterial cells even inside the rhizodermis of healthy roots.

In the following years, the definition of rhizosphere was refined. In fact, the rhizosphere is considered one of the most complex ecosystems on Earth (Jones and Hinsinger, 2008; Raaijmakers et al., 2009), and extends beyond the physical association of roots and soil particles to a more complex volume of overlap and functionally integrated zones. It consists of a gradient in chemical, biological, and physical properties that change both radially and longitudinally along the root. It is now well recognized that the rhizosphere is not a region of definable size or shape, but instead we consider three habitats as integral parts of the rhizosphere continuum: (1) the root endosphere, which includes portions of the cortex and endodermis, where microbes and cations can occupy the "free space" between cells (apoplastic space); (2) the rhizoplane, that is, the medial zone directly adjacent to the root, including its epidermis and mucilage; and (3) the rhizospheric soil which extends from the rhizoplane to the massive soil. Such zones are known to be hot-spots for microbial diversity, containing different communities of microbial diversity compared to the bulk soil. Both plants and microorganisms obtain their nutrients from the soil and alter soil properties through organic litter deposition and metabolic activities, respectively. Although microorganisms have a series of direct effects on plants, communication between them occurs via metabolites exuded by the roots, but despite such importance, the influence of plants on microorganisms remains poorly studied (Praeg et al., 2019). The main knowledge gaps for understanding the mechanisms of plant–microbe interactions in the rhizosphere are exudates and metabolic activity (Fig. 6.1).

It is already well established that microbial activity in the rhizosphere is essential for plant functioning and directly and indirectly influences plant composition and biomass (Schnitzer et al., 2011). Colonization is not uniform, but occurs in patches along the root, covering about 15%–40% of the total root surface of the plant. One of the groups of microorganisms intensively studied for their beneficial effects is the symbiotic N_2-fixing bacteria observed since Hiltner's studies. In analogy with the symbiotic root fungi (mycorhiza), Lorenz Hiltner named the bacterial community that is strongly associated with roots as "bacteriorhiza," with this taxonomic group being the most abundant associated with roots (Mohanty et al., 2021). Other representatives of this beneficial microbiota are PGPB, biocontrol microorganisms such as parasitic fungi of other pathogens and pests, in addition to protozoa, while rhizosphere organisms that are deleterious to plant growth and health include nematodes, oomycetes, bacteria, and the pathogenic fungi (York et al., 2016). Also, a third group of microorganisms that can be found in the rhizosphere are the human pathogens.

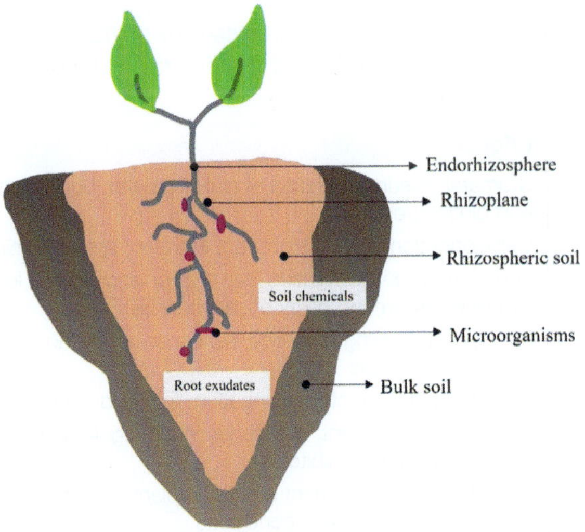

FIGURE 6.1

Schematic of a plant root section showing the bulk soil and the structure of the rhizosphere with its different zones.

Therefore plants carefully regulate the composition and activity of these microbial communities, and if under normal conditions they are all essential for plant growth, in environments subjected to some stress, particularly beneficial bacteria can improve such performance. The main objective of understanding the complexity and dynamics of interactions between soil, plant roots, and microbial communities in the rhizosphere is to seek the balance of beneficial microbes over harmful ones in this region, favoring plant productivity.

6.3 Plant growth-promoting bacteria

PGPB can live inside plant cells, attaching to specialized structures (nodules), or outside them but still stimulating plant growth (Gray and Smith, 2005). Although numerous soil bacteria have been reported as PGPB (the most studied genera include *Azospirillum*, *Bacillus*, and *Pseudomonas*), the diversity of soil and rhizosphere microbiomes is highly underestimated (Singh et al., 2020b; York et al., 2016; Kumar et al., 2016, 2017).

Furthermore, the mechanisms by which bacteria exhibit beneficial activities are often not well understood, but at least three can be identified: (1) the regulation of plant hormone signaling; (2) the repellency or overcoming of pathogenic microbial strains; and (3) increasing the bioavailability of soil nutrients. In this context, as previously mentioned, several important bacterial characteristics, such as biological nitrogen fixation, production of siderophores and phytohormones (Table 6.1),

Table 6.1 Descriptive summary of main classes of phytohormones produced by plants growth-promoting bacteria and its effects on the plant (Miura and Furumoto, 2013; Mamédio et al., 2020).

Phytohormones	Effects
Abscisic acid	Stomatal closure; inhibition of shoot growth; bud dormancy; abiotic and biotic stresses
Auxin	Cell elongation and division; root initiation; apical dominance
Cytokinin	Inhibition of root elongation; cell division stimulation; leaf expansion by cell enlargement; delay of senescence
Ethylene	Stress and ripening hormone; senescence and abscission; abiotic and biotic stresses
Gibberellin	Seed germination; stem elongation; floral induction and fruit growth
Jasmonic acid	Growth, development, and senescence regulator; pest resistance inducer (protease inhibitor)
Salicylic acid	Growth and development regulator; plant defense inducer under biotic and abiotic stress conditions

Adapted from Miura, K., Furumoto, T., 2013. Cold signaling and cold response in plants. Int. J. Mol. Sci. 14, 5312–5337; Mamédio, D., Cecato, U., Sanches, R., da Silva, S.M., Silva, D.R., et al., 2020. Do plant-growth promoting bacteria contribute to greater persistence of tropical pastures in water deficit? A review. Res. Soc. Dev. 9 (8), e523985756.

expression of ACC deaminase and phosphate solubilization, contribute to a better plant vigor. According to Backer et al. (2018), these are direct mechanisms, while the indirect ones are related to biocontrol, through antagonistic activity against phytopathogenic microorganisms, or inducing systemic plant resistance responses, or even interfering with bacterial quorum sensing (QS) systems, etc.

The efficiency of PGPB to be introduced in the management of an agricultural crop depends strongly on its ability to win the competition of autochthonous soil microorganisms, as well as to adhere and colonize the external and/or internal part of plant tissues, in addition to the age and stage of development of the plants and the resistance to all the stressors present (Massa et al., 2022).

6.4 Stress factors

Most authors use the term "plant stress" in an extremely broad sense. In the physical sciences, for example, stress refers to the ability of a force applied to a material to induce a deformation (deformation). Hans Selye (1975) adapted this term to living organisms, and soon after Lichtenthaler extended the concept of stress to plants, including their regeneration phase, when there are no more stressors, and differentiating two types of stress, that is, small doses of stressors can cause positive effects on plant performance ("positive stress" or eustress), while high doses can cause serious damage to the plant ("negative stress" or stress). Although the concept of stress is still confusing and even controversial in many scientific publications—an extensive review can be found in Lu et al. (2021), it is already clear that it refers to something that challenges homeostasis. Thus any unfavorable condition or substance that affects or blocks the metabolism, growth, or development of a plant is considered a stressor. It includes many biotic (insects, bacteria, fungi, and viruses) and abiotic (radiation, salinity, temperature, flooding, drought, and contaminants) stresses, resulting in highly negative impacts on plant survival and biomass (Verma et al., 2021). According to the source, whether natural or anthropogenic, abiotic stress can be classified as shown in Table 6.2.

According to Nagarajan and Nagarajan (2010), abiotic stresses are responsible for 50% of crop loss in the world, which justifies investments in plant stress research on two different fronts, one agronomic, to sustain agricultural and forestry productivity in a constantly changing world, and another ecological, to preserve the biodiversity of ecosystems subjected to more extreme environmental conditions induced by ongoing climate change (Bussotti and Pollastrini, 2021).

Since abiotic stresses induce the production of phytohormones such as ethylene (leading to suppression of plant growth and reduced root proliferation), and given the multiplicity and diversity of microorganisms on the planet and their importance in modulating plant physiology, it is attributed to the improvement of plant–microbe interactions in the rhizosphere not only the increase in plant resistance but also its survival to all kinds of abiotic stressors (York et al., 2016; Balestrini et al., 2022).

Table 6.2 Natural and anthropogenic abiotic stressors acting on plants (Kumar et al., 2020).

Source of abiotic stress factors	Examples
Natural	High irradiance (i.e., photoinhibition, photooxidation)
	Heat (increased temperature) and low temperature (chilling)
	Sudden and late frost
	Water shortage (desiccation problems)
	Mineral deficiency
	Long rainy periods
Anthropogenic	Herbicides, pesticides, fungicides
	Air pollutants
	Ozone and photochemical smog
	Formation of highly reactive oxygen species
	Photooxidants (e.g., peroxy-acyl-nitrates)
	Acid rain, acid fog, acid morning dew
	Acid pH of soil and water
	Mineral deficiency
	Oversupply of nitrogen
	Heavy metal load
	Overproduction of NH_4^+ in breeding stations
	Increased UV radiation and CO_2

Zinnert et al. (2013) compared the effects of natural stress to light regimes, as well as exposure to explosives, on plant physiology, noting different physiological responses in each condition. The authors highlighted that despite the relevance of mentioning that the differences between the results of the treatments could also take into account the changes in hyperspectral reflectance due to remote sensing, plants are constantly subjected to several stress factors simultaneously, and although they evolved to have several mechanisms of protection against excessive exposure to light under natural conditions, they are not adapted to withstand. In this sense, it is likely that future climate change scenarios will intensify the effects of abiotic stresses on plants, and further increase the virulence and spread of existing pathogens and pests, negatively affecting crop yields and survival (Balestrini et al., 2022), the numerous anthropogenic disturbances to which they have been exposed.

6.4.1 Temperature

The temperature around plants is a primary factor that affects the rate of their development and growth, especially since each taxonomic group of plants has specific photo-systems in which photosynthesis will occur optimally within a certain ideal temperature range (Fernandez et al., 2012). Therefore the roots are extremely sensitive to extreme temperatures, far below or far above the minimum or maximum limits of the ideal range, especially at critical moments of development. Therefore temperature becomes a strong stressor responsible for severe damage to productivity (Hatfield and Prueger, 2015), and considering that this factor is gradually increasing, due to global climate change, a technological challenge has been the search for innovative solutions, tools to protect plants or provide them to combat the effects of these changes.

The first evidence that a bacterium mitigated the effects of exposure to high temperatures on sorghum (*Sorghum bicolor* L.) seedlings (var CSV-15) was presented by Ali et al. (2009), by inoculating them with a thermotolerant strain of *Pseudomonas aeruginosa* (AKM-P6), isolated from the rhizosphere of pigeon pea (*Cajanus cajan* L.) grown under semiarid conditions in India. The authors observed that, in relation to the noninoculated control, the inoculated seedlings improved shoot development, root length, dry biomass, and survival at elevated temperatures (47°C–50°C day/30°C–33°C night) under sterile and nonsterile conditions, although no significant difference was observed between treated and untreated plants kept at room temperature. The uninoculated seedlings began to wilt after 2 days of exposure to elevated temperature and died at the end of the fifth day of the experiment, while the inoculated ones resisted for up to 15 days. In contrast, inoculation of sorghum seedlings with *P. aeruginosa* AKM-P6 influenced biochemical and physiological parameters at elevated temperatures, as indicated by the increase in the synthesis of high-molecular weight proteins in leaves, reduced membrane injury, and improved levels of cellular metabolites such as proline, chlorophyll, sugars, amino acids, and total proteins.

The same group of research achieved comparable results after inoculation of *Pseudomonas putida* AKM-P7 in wheat (*Triticum aestivum* L.), that is, there was an increase in stress tolerance due to the synthesis of high molecular weight proteins and an improvement in levels of cellular metabolites (Ali et al., 2011). The genus *Pseudomonas* belongs to the γ subclass of the Proteobacteria and includes mostly fluorescent Pseudomonads (FP) as well as a few nonfluorescent species. FP are known for its production of diverse microbial metabolites including siderophores, antibiotics, and volatile compounds, being one of the common groups of PGPB which have received much attention in recent years also as biocontrol agent (Karimzadeh et al., 2020). These findings intensified the hypothesis of the relevant role of plant–microorganisms interaction in mitigating the adverse effects of climate change on crop growth, leading to the development of microbial consortia to mitigate such effects.

Kang et al. (2015) isolated 17 bacteria from the rhizosphere of randomly selected pepper plants (*Capsicum annum* L.) and subsequently selected them for their ability to promote plant growth. *Serratia nematodiphila* PEJ1011 produced three different active gibberellins (phytohormones that control key aspects of plant growth such as germination, growth elongation, flower development, and flowering time), such as GA_4, GA_{20}, and GA_9. The authors inoculated *S. nematodiphila* PEJ1011 in pepper seedlings and found that the deleterious effects of low temperature (5°C) on them were attenuated, as well as there was an improvement significant in the plant growth parameters. Inoculation increased GA_4 production under normal and cold stress conditions.

The first study to monitor carbohydrate concentrations in seedlings of grapevine (*Vitis vinifera* L.) colonized by PGPB during exposure to low temperature (4°C) was conducted by Fernandez et al. (2012). These authors hypothesized that the induced tolerance could be explained, at least in part, by the impact of bacteria on photosynthesis or carbohydrate metabolism of this culture during the process known as "cold acclimation" (the increase in freezing tolerance during exposure to at low but not freezing temperatures), as described by Miura and Furumoto (2013). Thus a suspension of 3×10^8 colony forming units (CFU) mL^{-1} of *Burkholderia phytofirmans* was inoculated into roots of 2-week-old grapevine seedlings. After growing for 4 weeks at constant temperature of 26°C under white fluorescent light, the seedlings were submitted to a cooling treatment (4°C under 16 hours of light and 8 hours of darkness), being analyzed the photosynthetic pigments, chlorophyll fluorescence, levels of sugars, and primary metabolites. It was found that *B. phytofirmans* could modulate carbohydrate metabolism to reduce cold damage in sweet pepper seedlings exposed to low temperature stress comparable to cold acclimatization and low temperature tolerance.

Subramanian et al. (2015) also investigated the inoculation of bacterial species to understand cold acclimatization, but used a consortium formed by psychrotolerant *Pseudomonas vancouverensis* OB155 and *Pseudomonas frederiksbergensis* OS261 in tomato plants (*Solanum lycopersicum* L.). They found that these strains increased the expression of cold acclimatization genes and the antioxidant activity in tomato leaf tissues inoculated and exposed to low temperatures.

6.4.2 Water

Water is essential for the transport of nutrients and metabolic reactions of all organisms, and therefore for their survival. The volume of water required by a plant depends on its type, stage of development/age, and the amount of light it receives. Therefore water deficit negatively affects many aspects of plant physiology, including seed germination, photosynthesis, and productivity.

Recently, Karimzadeh et al. (2020) obtained 15 *Pseudomonas* spp. and subjected them to various levels of water stress, using different concentrations of aqueous solutions of polyethylene glycol (PEG-6000) (0, 200.2, 295.7, and 367.7 g/L), with equivalent osmotic potential (ψs) of 0, -5, -10, and -15 bar, respectively. The tolerance of these isolates to different salinity levels (0%, 4%, and 10% NaCl) was also evaluated, as well as their ability to solubilize phosphate and produce auxin, ACC deaminase, and siderophores. After the selection of the isolates with the desired abilities, the authors disinfected wheat seeds (*T. aestivum* L.) and immersed them in aqueous suspensions of these microorganisms (5×10^9

CFU mL^{-1}) for 45 minutes, sowing them then in pots with soil. Each pot received 10 of these healthy seeds and were taken to a greenhouse, where they were kept for 125 days under water stress conditions. The results showed that bacterial strains significantly increased soil available phosphate, growth rates (a 41.6% increase in root dry weight and an 11% increase in shoot dry weight), grain yield (a 17% increase), and phosphate uptake from the aerial part of the wheat plant as compared to the control (i.e., no bacterial strains).

In one research conducted by Vitorino et al. (2022), 56 bacteria and 21 fungi were isolated from the rhizosphere and roots of *Hymenaea courbaril* L.—a fruiting species sampled in two biomes in central-western Brazil—Pantanal and Cerrado. All of them were investigated for: (1) the biosynthesis of indole-3-acetic acid (IAA) and the enzyme ACC deaminase; (2) the solubilization of calcium phosphate, iron phosphate, aluminum phosphate; and (3) the antibiosis against the phytopathogenic fungi *Sclerotinia sclerotiorum*, *Fusarium* sp., and *Rhizoctonia* sp. The data obtained were then statistically evaluated in a correlation matrix and combined through principal component analysis (PCA), with the purpose of discriminating the best PGPB under water restriction conditions (by biopriming of *H. courbaril* L.). The authors verified that the bacteria were more effective than the fungi in this ability, being selected from Pantanal the strains of rhizobacteria HPA12 (*Paenibacillus alvei*) and HPA26 (*Lysinibacillus fusiformis*), and from Cerrado the strains HCE10 (*Bacillus thuringiensis*, rhizospheric) and HCE5 (*Bacillus cereus*, endophytic). Thus the plants were inoculated, submitted to water suppression condition, and evaluated every 15 days regarding some biometric parameters (height, stem diameter, number of nodes and leaves). At the end of the assay, the parameters of total dry mass, relative water content, and chlorophyll indices (chlorophyll a, b, and total) were determined in the leaves. The authors concluded that *B. thuringiensis* HCE10 potentially aids *H. courbaril* L. in its dispersal through water-poor soils and proved to be one PGPB that reduces the effects of water stress to be applied in agriculture.

6.4.3 Salinity

According to Egamberdieva et al. (2019), it is estimated that around 62 million hectares or 20% of the world's irrigated lands are affected by salinity. Salt stress is one of the main abiotic stresses in soil, with an ionic and osmotic effect that disturbs homeostasis, reducing the ability of crops to absorb water. This makes it difficult not only for plants to survive, but also for other life forms. A plant under saline stress undergoes several morphological, physiological, and molecular changes that impact its growth and development (Table 6.3). Therefore salinity is responsible for a significant reduction in the overall productivity of different crops.

However, several microbes that inhabit plants growing under high salinity conditions show adaptations to saline stress and thrive well under such conditions (Arora et al., 2020). This halophyte microbiome influences soil quality and fertility parameters, as seen in Fig. 6.2, emerging as an efficient biological tool to

Table 6.3 Effects of salt stress on different plant attributes.

Kind of response	Examples
Physiological	Loss of turgor
	Reduced water potential
	Decrease in stomatal conductance
	Plant growth reduction
	Leaf senescence
	Reduced photosynthesis
	Reduced CO_2 concentration
Biochemical	Plant hormones induction (ethylene, for example)
	Increased antioxidant enzymes
	Ion homeostasis
	Osmolytes accumulation
	Change in photosynthesis pathway
Molecular	Expression of stress responsive gene
	Activation of osmotic function gene

Adapted from Kumar, A., Singh, S., Gaurav, A.K., Srivastava, S., Verma, J.P., 2020. Plant growth-promoting bacteria: biological tools for the mitigation of salinity stress in plants. Front. Microbiol. 11, 1216.

mitigate the toxic effects of high salt concentrations on plants, while simultaneously remedying degraded saline soils.

Zilaie et al. (2022) evaluated the effect of two strains of halotolerant rhizobacteria, *Bacillus pumilus* HR (isolated from *Halostachys belangeriana*) and *Zhihengliuella halotolerans* SB, on improved tolerance of the halophyte *Seidlitzia rosmarinus*—a species from the flora of desert areas of Iran that is used in the diet of camels and sheep—to salinity and dust stress. The authors reported the effects of a single and coinoculation of these bacteria on the morphophysiological, biochemical, and ionomic properties of *S. rosmarinus*. The bacteria studied not only improved plant growth conditions under salt stress, but also alleviated the negative effects of dust. However, *Z. halotolerans* SB increased the seedling quality index better than *B. pumilus* HR in the two investigated stress conditions. Thus the study revealed that the application of halotolerant rhizobacteria can be an economical and ecologically sustainable method to improve the yield and forage quality in the desert halophyte.

In another study, Hou et al. (2022) isolated two strains of bacteria from the rhizosphere of one of the most important agricultural production areas in arid and semiarid regions of China, *Bacillus atrophaeus* WZYH01 and *Planococcus soli* WZYH02. Both strains can tolerate up to 1197 mM NaCl and exhibited three growth-promoting characteristics: production of IAA, abscisic acid (ABA), and exopolysaccharides and phosphate solubilization. Each bacterium was inoculated into corn seedlings in pots, where they were exposed to 5.9 dS/m of saline stress during 31 days of growth. After this time, the maize growth index (height and dry

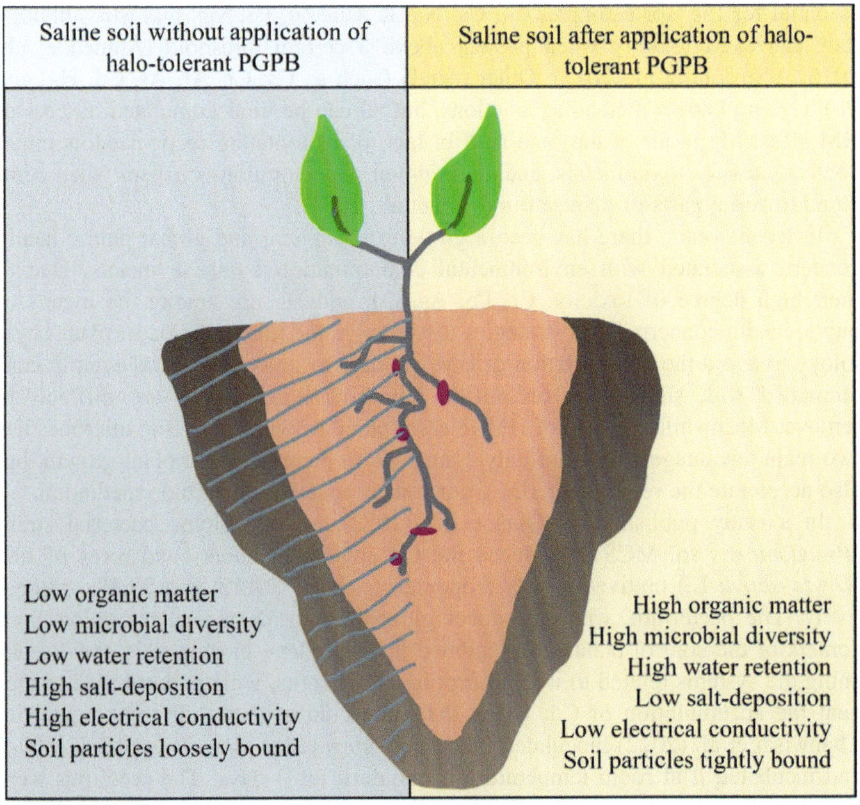

FIGURE 6.2

Effects of application (right) or not (left) of a halotolerant plant growth-promoting bacteria (PGPB) in soil with high salinity, in relation to several parameters.

biomass) was obtained. The results showed that plant height and dry mass increased significantly with the inoculation of the lines under saline stress conditions. A similar result was observed in relation to the IAA content in the leaves. In contrast, the ABA content in corn leaves was significantly decreased compared to the control treatments, concluding that *B. atrophaeus* WZYH01 and *P. soli* WZYH02 can protect corn from salt stress because of the integration of multiple physiological processes, but the effect of the *P. soli* WZYH02 strain was more prominent on the growth rate and gene expression.

6.4.4 Heavy metals

Metals are ubiquitous in the biosphere and vital to industry, infrastructure, and daily life. They constitute about 75% of the known elements and many are essential for life, for example, Co, Ca, Na, K, Cu, Zn, Fe, Mg, and Mn, although they can exert toxicity when present above a certain threshold (Ahmed et al., 2016; Aransiola et al., 2017). Other metals (such as Cs, Cr, Al, As, Cd, Hg, and Pb) have no known metabolic functions, but all can be bioaccumulated. Excess of HM affect life in air, water, and soil. In fact, plant roots are recognized as initial contact sites for toxic metals, and store higher concentrations of them when compared to aerial parts of plants (Burkhead et al., 2009).

In recent years, there has been a growing ecological and global public health concern associated with environmental contamination by these metals. Due to their high degree of toxicity, Cr, Pb, As, Cd, and Hg are among the metals of public health concern. Thus strategies that involve the manipulation of plant physiology to avoid the accumulation of HM can help to grow safe foods even in contaminated soil, since once present in the environment, they are difficult to remove. Meanwhile, the use of HM-tolerant plant growth-promoting microbes has two main advantages: they not only contribute to increased host plant growth, but also accelerate the removal of HM from contaminated soils by bioremediation.

In a study published by Qian et al. (2022), an endophytic bacterial strain (*Pseudomonas* sp. MCS15) reduced the Cd content of shoots and roots of rice (*Oryza sativa* L.) cultivated in hydroponic boxes, by 29.2% and 39.8%, respectively. The bacterium, which produces gluconic acid, decreased the chlorophyll content of the control plants and inhibited their ethylene biosynthesis, thus weakening the systems related to iron absorption in the roots, which contributed to prevent the accumulation of Cd. Using the same culture but a different organism, Thanwisai et al. (2022) inoculated rice with *Cupriavidus taiwanensis* KKU2500-3 and incubated it at room temperature in the dark for 3 days. The seedlings were then transferred to trays with mixed media containing soil, burnt rice husk, and manure. After 30 days, healthy seedlings were then selected for transplanting in the rice field using soils selected according to their total and available Cd levels, that is, no Cd, low and medium levels (no high Cd content was found). At the end, plants were collected from the field experiment to analyze the concentration of Cd. The accumulation of Cd in rice grains inoculated with *C. taiwanensis* KKU2500-3 and cultivated in soils contaminated with 2.44, 7.43, and 10.69 ppm of available Cd, decreased by 41.14%, 24.76%, and 83.20%, which resulted in Cd levels in the grains below the maximum limit indicated by the Codex Committee on Contaminants in Food (0.4 ppm), that is, safe for consumers.

Microorganisms with tolerance or resistance to Hg can contribute to the reduction and/or elimination of different forms of Hg in contaminated environments and have potential use in phytoremediation. Considering that, as mentioned earlier, the presentation of a single plant growth-promoting activity, such as the synthesis of IAA, does not guarantee the greatest success of a strain to be used as PGPB. Robas et al. (2021) proposed an index entitled "Bio-Mercury Remediation Suitability Index" (BMR-SI) to discriminate and select those strains with the highest probability of success in soil recovery and use as an inoculant. To build it, the authors considered information such as auxin production, phosphate solubilization, synthesis of siderophores and ACC deaminase, in addition to the minimum concentration of Hg capable of acting as a bactericide. In turn, González et al. (2021), selected strains of *Bacillus toyonensis*, *Pseudomonas moraviensis*, *P. baetica*, and *Brevibacterium frigoritolerans* as good candidates for new phytorhizoremediation assays in soils contaminated with Hg, using the BMR-SI proposed by Robas et al. (2021).

In recent years, several entire bacterial genomes have been sequenced to identify loci involved in metal resistance. In a study published by Ouertani et al. (2020), a bacterial strain (TL13) resistant to HM (such as chromium, copper, nickel, cobalt, and arsenic) was isolated from effluents of a Tunisian leather industry and subjected to in vitro screening for promoting activities of plant growth. After performing the complete genome sequencing of TL13, the sequence of the 16S ribosomal DNA (rDNA) genes was evaluated, and this strain was identified as *Microbacterium metallidurans* TL13. The 50% growth inhibitory concentration (IC_{50}) of this bacterium for Na_2HAsO_4, $FeSO_4$, $NiCl_2$, $CuSO_4$, $K_2Cr_2O_7$, $CoCl_2$, and $HgCl_2$ showed values of 7007, 4403, 1680, 1590, 676, 445, and 368 mg/L, respectively, that means, in growing order of toxicity of such solutions for the bacterium. As expected, the analysis of its genome revealed the presence of several genes that contribute directly or indirectly to PGPB activities, that is, involved in inorganic phosphate solubilization, ammonia assimilation, levan production, IAA synthesis, siderophores, and polyketides, for the production of glutathione peroxidase, catalase, superoxide dismutase, and thioredoxin reductase, which confer a higher tolerance to oxidative/metal stresses, for heat shock tolerance, cold shock tolerance, glycine-betaine production, and lytic enzymes (cellulase, chitinase, and proteases) production, that enable bacteria to survive biotic/abiotic stress, ensuring that such an isolate has a very high potential for use in areas to be phytoremediated (Pathak et al., 2022).

6.5 Challenges of plant growth-promoting bacteria application

As mentioned in the studies cited in previous items, the most used approach for the development of new PGPB inoculants is based on laboratory screening tests, under aseptic conditions, which screen specific mechanisms that stimulate plant growth (Massa et al., 2022). However, screening microorganisms with many PGPB functions may not necessarily promote crop growth under field conditions, while those that have minimal in vitro plant growth-promoting functions may have extra mechanisms that enable them to do so. According to Backer et al. (2018), since these other skills are less understood, they are difficult to track under laboratory conditions.

Two other points must be considered when the objective is to apply the PGPB in the field. The first is that planktonic growth of bacteria in pure culture rarely exists in natural environments. Indeed, bacteria in the environment exist in a complex and dynamic surface—associated community called a biofilm—communities of microorganisms that begin when a cell attaches to a surface or interphase and is incorporated into a self-produced extracellular matrix called EPS. The EPS is made of both bound and secreted form of lipids, polysaccharides, proteins, humic substances, nucleic acids, and water, allowing microorganisms to develop complex, three-dimensional, attached communities that are resistant to attacks that would destroy individual cells not part of the biofilm colony. Many studies have shown that biofilm-producing bacteria express fine aggregative fimbriae (curli) and cellulose—two important components of biofilm matrices (Bokranz et al., 2005; Gualdi et al., 2008). Their aggregation in biofilms has the main advantage of increasing tolerance to changes in environmental conditions (nutrients, predation, exposure to toxic chemicals) or abiotic stressors (changes in salt concentration, pH, temperature, and water content, for example). Thus PGPB biofilms are more effective under field conditions than any planktonic PGPB (Pandin et al., 2017; Backer et al., 2018). On the other hand, the endophytic colonization of PGPB is better controlled compared to the beneficial behavior of free-living soil bacteria (Kumar et al., 2015, 2020, 2021). As a result, there is a possibility that the use of PGPB is in modeling the behavior of endophytic bacteria and biofilms (Singh et al., 2017, 2020a, 2022; Timmusk et al., 2017).

Another issue is the challenge of maintaining viable PGPB (plant growth–promoting bacteria) cells during storage and application stages. These microorganisms are often used to inoculate plant materials. However, they are either used without an appropriate carrier or in quantities insufficient for efficient colonization of the rhizosphere. This is problematic, especially under field conditions, where they face competition from the resident soil micro- and macrofauna (Backer et al., 2018). Although direct application to soils allows for a higher concentration of PGPB compared to seed treatment, it is associated with a greater risk of contamination and loss of growth-promoting metabolic activities contained therein (Massa et al., 2022). As reported by Massa et al. (2022), the two main aspects that dominate the success of inoculation are the effectiveness of the bacterial isolate and the appropriate application technology, that is, in addition to the potential for application of beneficial microbes in agriculture appearing limitless, it is essential to establish a link between the laboratory and field tests. Timmusk et al. (2017), for example, published a review where they advocate the development of mathematical models based on "custom" inoculums as a way of finding their way to facilitate reproducible field application, and sustainable food production under a changing climate. The combination of different methodologies related to PGPB, such as identification of microorganisms with characteristics that stimulate plant growth, as well as seed inoculation tests under laboratory conditions and field cultivation experiments, are part of the search for modern technologies for agricultural crops. Based on studies by Backer et al. (2018) and

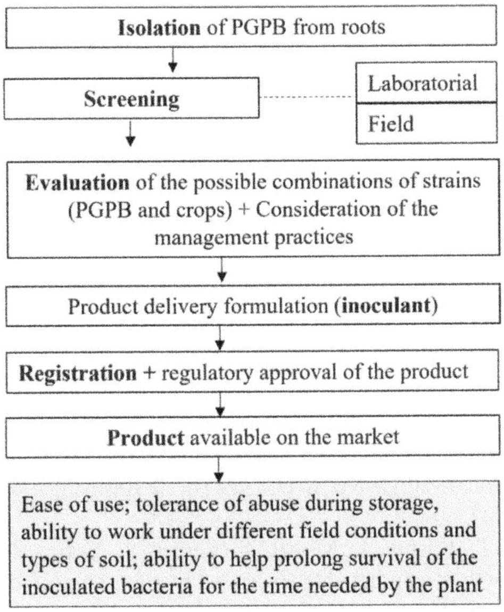

FIGURE 6.3

Flow diagram of basic steps for developing a bacterial inoculant formed by plant growing promoting bacteria (PGPB).

Massa et al. (2022), the development of PGPB-based inoculants can be summarized in a few steps, as shown in Fig. 6.3.

6.6 Conclusion

In this chapter, we explain the role of PGPBs in reducing the effects of abiotic stressors in plants, listing many studies related to the screening and evaluation of efficient strains acting alone or in consortium, as well as some of the challenges of applying inoculants in the field. Considering all the examples cited, it is evident that the proper use of this technology—inoculation of plants with effective bacterial strains—in agrosystems, can change the scenario of current agronomy and is extremely necessary due to global climate change, altering the use of land and the intensification of agriculture on previously less productive soils.

Furthermore, the possibility of combining the plant growth promotion with the cleaning of environmental pollutants, such as HMs, contributes to the greener environmental sustainability campaign. Considering that in the field plants are exposed to an unpredictable combination of different stressors, and not just one of them, the continuity of this type of research favors the choice of innovative agricultural production alternatives, combining microorganisms with different capacities and without harming the environment, increasing productivity.

References

Allison, S.D., Martiny, J.B.H., 2008. Resistance, resilience, and redundancy in microbial communities. PNAS 105, 11512–11519.

Aransiola, E.F., Ige, O.A., Ehinmitola, E.O., Layokun, S.K., 2017. Heavy metals bioremediation potential of *Klebsiella* species isolated from diesel polluted soil. Afr. J. Biotech. 16, 1098–1105.

Ali, S.K.Z., Sandhya, V., Grover, M., Rao, L.V., Venkateswarlu, B., 2011. Effect of inoculation with a thermotolerant plant growth promoting *Pseudomonas putida* strain AKMP7 on growth of wheat (*Triticum* spp.) under heat stress. J. Plant Interact 6, 239–246.

Ali, S.Z., Sandhya, V., Grover, M., Kishore, N., Rao, L.V., Venkateswarlu, B., 2009. *Pseudomonas* sp. strain AKM-P6 enhances tolerance of sorghum seedlings to elevated temperatures. Biol. Fert. Soils 46 (1), 45–55.

Ahmed, M.K., Baki, M.A., Kundu, G.K., Islam, M.S., Islam, M.M., Hossain, M.M., 2016. Human Health Risks from Heavy Metals in Fish of Buriganga River, 5. Springer Plus, Bangladesh, p. 1697.

Arora, K.N., Fatima, T., Mishra, J., Mishra, I., Verma, S., Verma, R., et al., 2020. Halotolerant plant growth promoting rhizobacteria for improving productivity and remediation of saline soils. J. Adv. Res. 11 (26), 69–82.

Bussotti, F., Pollastrini, M., 2021. Revisiting the concept of stress in forest trees at the time of global change and issues for stress monitoring. Plant Stress. 2, 100013.

Bokranz, W., Wang, X., Tschape, H., Romling, U., 2005. Expression of cellulose and curli fimbriae by *Escherichia coli* isolated from the gastrointestinal tract. J. Med. Microb. 54, 1171–1182.

Balestrini, R., Chitarra, W., Ghirardo, A., Nardini, A., Nerva, L., 2022. A stressful life: how plants cope with multiple biotic and abiotic adverse factors. Plant Stress. 5, 100095.

Backer, R., Rokem, J.S., Ilangumaran, G., Lamont, J., Praslickova, D., Ricci, E., et al., 2018. Plant growth-promoting rhizobacteria: context, mechanisms of action, and roadmap to commercialization of biostimulants for sustainable agriculture. Front. Plant Sci. 9, 1473.

Burkhead, J.L., Gogolin-Reynolds, K.A., Abdel-Ghany, S.E., Cohu, C.M., Pilon, M., 2009. Copper homeostasis. N. Phytol. 182, 799–816.

Egamberdieva, D., Wirth, S., Bellingrath-Kimura, S.D., Mishra, J., Arora, N.K., 2019. Salt-tolerant plant growth promoting rhizobacteria for enhancing crop productivity of saline soils. Front. Microb. 10, 2791.

Fernandez, O., Theocharis, A., Bordiec, S., Feil, R., Jacquens, L., Clement, C., et al., 2012. *Burkholderia phytofirmans* PsJN acclimates grapevine to cold by modulating carbohydrate metabolism. Mol. Plant Microbe Interact 25, 496–504.

Gray, E.J., Smith, D.L., 2005. Intracellular and extracellular PGPR: commonalities and distinctions in the plant-bacterium signaling processes. Soil Biol. Biochem. 37, 395–412.

González, D., Robas, M., Probanza, A., Jiménez, P.A., 2021. Selection of mercury-resistant PGPR strains using the BMRSI for bioremediation purposes. Int. J. Environ. Res. Pub. Health 18 (18), 9867.

Gualdi, L., Tagliabue, L., Bertagnoli, S., Ieranò, T., De Castro, C., Landini, P., 2008. Cellulose modulates biofilm formation by counteracting curli-mediated colonization of solid surfaces in *Escherichia coli*. Microbiology 154 (7), 2017–2024.

Hiltner, L., 1904. Uber neuere Erfahrungen und Probleme auf dem Gebiete der Bodenbakteriologie unter besonderden berucksichtigung und Brache. Arb. Dtsch. Landwirtsch. Gesellschaft 98, 59–78.

Hatfield, J.L., Prueger, J.H., 2015. Temperature extremes: effect on plant growth and development. Weather Clim. Extrem. 10 (A), 4–10.

Hou, Y., Zeng, W., Ao, C., Luo, Y., Wang, Z., Hou, M., et al., 2022. *Bacillus atrophaeus* WZYH01 and *Planococcus soli* WZYH02 improve salt tolerance of maize (*Zea mays* L.) in saline soil. Front. Plant Sci. 13, 891372.

Jones, D., Hinsinger, P., 2008. The rhizosphere: complex by design. Plant Soil 312, 1–6.

Kumar, V., Kumar, A., Pandey, K.D., Roy, B.K., 2015. Isolation and characterization of bacterial endophytes from the roots of *Cassia tora* L. Ann. Microbiol. 65 (3), 1391–1399.

Karimzadeh, J., Alikhani, H.A., Etesami, H., Pourbabaei, A.A., 2020. Improved phosphorus uptake by wheat plant (*Triticum aestivum* L.) with rhizosphere fluorescent pseudomonads strains under water deficit stress. J. Plant Growth Reg. 162–178.

Kumar, A., Singh, S., Gaurav, A.K., Srivastava, S., Verma, J.P., 2020. Plant growth-promoting bacteria: biological tools for the mitigation of salinity stress in plants. Front. Microbiol. 11, 1216.

Kumar, A., Droby, S., Singh, V.K., Singh, S.K., White, J.F., 2020. Entry, colonization, and distribution of endophytic microorganisms in plants. In: Kumar, A., Radhakrishnan, E. K. (Eds.), Microbial Endophytes: Functional Biology and Applications. Woodhead Publishing, Cambridge USA, pp. 1–33.

Kumar, A., Singh, M., Singh, P.P., Singh, S.K., Singh, P.K., Pandey, K.D., 2016. Isolation of plant growth promoting rhizobacteria and their impact on growth and curcumin content in *Curcuma longa* L. Biocatal. Agric. Biotechnol. 8, 1–7.

Kang, S.-M., Khan, A., Waqas, M., You, Y.-H., Hamayun, M., Joo, G.-J., et al., 2015. Gibberellin-producing *Serratia nematodiphila* PEJ1011 ameliorates low temperature stress in *Capsicum annum* L. Eur. J. Soil Biol. 68, 85–93.

Kumar, A., Zhimo, Y., Biasi, A., Salim, S., Feygenberg, O., Wisniewski, M., et al., 2021. Endophytic microbiome in the carposphere and its importance in fruit physiology and pathology. In: Spadaro, D., Droby, S., Gullino, M.L. (Eds.), Postharvest Pathology. Plant Pathology in the 21st Century, vol 11. Springer, Cham, pp. 73–88.

Kumar, A., Verma, H., Singh, V.K., Singh, P.P., Singh, S.K., Ansari, W.A., et al., 2017. Role of *Pseudomonas* sp. in sustainable agriculture and disease management. Agriculturally Important Microbes for Sustainable Agriculture. Springer, Singapore, pp. 195–215.

Lu, S., Wei, F., Li, G., 2021. The evolution of the concept of stress and the framework of the stress system. Cell Stress 5 (6), 76–85.

Massa, F., Defez, R., Bianco, C., 2022. Exploitation of plant growth promoting bacteria for sustainable agriculture: hierarchical approach to link laboratory and field experiments. Microorganisms 10, 865.

Miura, K., Furumoto, T., 2013. Cold signaling and cold response in plants. Int. J. Mol. Sci. 14, 5312–5337.

Modi, A., Kanani, P., Kumar, A., 2020. Fungal endophytes-induced gene expression studies in biotic and abiotic stress management. In: Kumar, A., Radhakrishnan, E.K. (Eds.), Microbial Endophytes. Woodhead Publishing, Cambridge USA, pp. 251–271.

Mohanty, P., Singh, P.K., Chakraborty, D., Mishra, S., Pattnaik, R., 2021. Insight into the role of PGPR in sustainable agriculture and environment. Front. Sustain. Food Syst. 5, 1–12.

Mamédio, D., Cecato, U., Sanches, R., da Silva, S.M., Silva, D.R., et al., 2020. Do plant-growth promoting bacteria contribute to greater persistence of tropical pastures in water deficit? A review. Res. Soc. Dev. 9 (8), e523985756.

Nagarajan, S., Nagarajan, S., 2010. Abiotic tolerance and crop improvement. In: Pareek, A., Sopory, S.K., Bohnert, H.J. (Eds.), Abiotic Stress Adaptation in Plants: Physiological, 504/Molecular Plant-Microbe Interactions Molecular and Genomic Foundation. Springer, Dordrecht, The Netherlands, pp. 1–11.

Ouertani, R., Ouertani, A., Mahjoubi, M., Bousselmi, Y., Afef, N., Hanene, C., et al., 2020. New plant growth-promoting, chromium-detoxifying *Microbacterium* species isolated from a tannery wastewater: performance and genomic insights. Front. Bioeng. Biotech. 8, 521.

Praeg, N., Pauli, H., Ilmer, P., 2019. Microbial diversity in bulk and rhizosphere soil of *Ranunculus glacialis* along a high-alpine altitudinal gradient. Front. Microb. 10, 1429.

Patel, P., Kumar, S., Modi, A., Kumar, A., 2021. Deciphering fungal endophytes combating abiotic stresses in crop plants (cereals and vegetables). Microbial Management of Plant Stresses. Woodhead Publishing, Cambridge USA, pp. 131–147.

Pandin, C., Le Coq, D., Canette, A., Aymerich, S., Briandet, R., 2017. Should the biofilm mode of life be taken into consideration for microbial biocontrol agents? Microb. Biotech. 10, 719–734.

Pathak, P., Rai, V.K., Can, H., Singh, S.K., Kumar, D., Bhardwaj, N., et al., 2022. Plant-endophyte interaction during biotic stress management. Plants 11 (17), 2203.

Qian, L., Song, F., Xia, J., Wang, R.A., 2022. Glucuronic acid-producing endophyte *Pseudomonas* sp. MCS15 reduces cadmium uptake in rice by inhibition of ethylene biosynthesis. Front. Plant Sci. 13, 876545.

Robas, M., Jiménez, P.A., González, D., Probanza, A., 2021. Bio-mercury remediation suitability index: a novel proposal that compiles the pgpr features of bacterial strains and its potential use in phytoremediation. Int. J. Environ. Res. Pub. Health 18 (8), 4213.

Raaijmakers, J.M., Paulitz, T.C., Steinberg, C., Alabouvette, C., Moënne-Loccoz, Y., 2009. The rhizosphere: a playground and battlefield for soilborne pathogens and beneficial microorganisms. Plant Soil 321, 341–361.

Selye, H., 1975. Implications of stress concept. N. Y. State J. Med. 75, 2139–2145.

Schnitzer, S., Klironomos, J., Hillerislambers, J., Kinkel, L., 2011. Soil microbes drive the classic plant diversity–productivity pattern. Ecology 92 (2), 296–303.

Singh, M., Qureshi, K.A., Jaremko, M., Rajput, M., Singh, S.K., Pandey, K.D., et al., 2022. Bioprospects of endophytic bacteria in plant growth promotion and Ag-nanoparticle biosynthesis. Plants 11 (14), 1787.

Singh, D., Singh, S.K., Singh, V.K., Gupta, A., Aamir, M., Kumar, A., 2020b. Plant growth–promoting bacteria and their role in environmental management. In: Singh, P., Kumar, A., Borthakur, A. (Eds.), Abatement of Environmental Pollutants. Elsevier, Cambridge USA, pp. 161–175.

Singh, M., Kumar, A., Singh, R., Pandey, K.D., 2017. Endophytic bacteria: a new source of bioactive compounds. 3 Biotech 7 (5), 1–14.

Singh, M., Srivastava, M., Kumar, A., Singh, A.K., Pandey, K.D., 2020a. Endophytic bacteria in plant disease management. In: Kumar, A., Singh, K.V. (Eds.), Microbial Endophytes: Prospects for Sustainable Agriculture. *Woodhead Publishing*, Cambridge USA, pp. 61–89.

Singh, V.K., Singh, A.K., Singh, P.P., Kumar, A., 2018. Interaction of plant growth promoting bacteria with tomato under abiotic stress: a review, Agric. Ecosyst. Environ., 267, pp. 129–140.

Subramanian, P., Mageswari, A., Kim, K., Lee, Y., Sa, T., 2015. Psychrotolerant endophytic *Pseudomonas* sp strains OB155 and OS261 induced chilling resistance in tomato plants (*Solanum lycopersicum* mill.) by activation of their antioxidant capacity. Mol. Plant Microbe Interact. 28, 1073–1081.

Thanwisai, L., Janket, A., Siripornadulsil, W., Siripornadulsil, S., 2022. A cadmium-tolerant bacterium (*Cupriavidus taiwanensis* KKU2500-3) reduces cadmium accumulation in grains of KDML105 rice cultivated in a contaminated paddy field. Environ. Technol. Innov. 28, 102685.

Timmusk, S., Behers, L., Muthoni, J., Muraya, A., Aronsson, A.-C., 2017. Perspectives and challenges of microbial application for crop improvement. Front. Plant. Sci. 8.

Vitorino, L.C., Rocha, A.F.S., Bessa, L.A., Lourenço, L.L., Costa, A.C., Silva, F.G., 2022. Symbiotic microorganisms affect the resilience of *Hymenaea courbaril* L., a neotropical fruit tree, to water restriction. Plant Stress 5, 100092.

York, L.M., Carminati, A., Mooney, S.J., Ritz, K., Bennett, M.J., 2016. The holistic rhizosphere: integrating zones, processes, and semantics in the soil influenced by roots. J. Exp. Bot. 67, 3629–3643.

Verma, H., Kumar, D., Kumar, V., Kumari, M., Singh, S.K., Sharma, V.K., et al., 2021. The potential application of endophytes in management of stress from drought and salinity in crop plants. Microorganisms 9 (8), 1729.

Zilaie, M.N., Arani, A.M., Etesami, H., Dinarvand, M., Dolati, A., 2022. Halotolerant plant growth-promoting rhizobacteria-mediated alleviation of salinity and dust stress and improvement of forage yield in the desert halophyte *Seidlitzia rosmarinus*. Environ. Exp. Bot. 201, 104952.

Zinnert, J., Via, S.M., Young, D.R., 2013. Distinguishing natural from anthropogenic stress in plants: physiology, fluorescence, and hyperspectral reflectance. Plant Soil 133–141.

CHAPTER

7 Plant growth–promoting rhizobacteria: their potential as biological control agents in sustainable agriculture

Younes Rezaee Danesh[1,2], Marika Pellegrini[3], Ahmet Akköprü[2], Beatrice Farda[3], Gökhan Boyno[2] and Rihab Djebaili[3]

[1]*Department of Plant Protection, Faculty of Agriculture, Urmia University, Urmia, Iran*

[2]*Department of Plant Protection, Faculty of Agriculture, Van Yuzuncu Yil University, Van, Turkey*

[3]*Department of Life, Health and Environmental Sciences, University of L'Aquila, L'Aquila, Italy*

7.1 Introduction

The activities carried out in sustainable agricultural systems are based on maintaining soil health, optimal use of water resources, and minimizing the amount of pollution in the environment, which results in increasing the amount of food production in the country. Biotic stresses caused by plant pathogens are one of the major factors in the economic losses of crop yield. Several agrochemicals are used by farmers to manage diseases caused by pathogens. However, the consumption of these compounds is decreasing significantly due to several reasons, including creating pollution in the environment, the development of resistance in the population of the pathogens, and the high costs of application for plant protection. Using of microbial agents as an alternative method for plant disease management has received special attention by researchers. In nature, there are a large number of useful microorganisms in the soil that can be used to manage plant diseases by stimulating resistance or improving plant growth (Singh et al., 2017, 2020a). One of the most important groups of these microbial agents are plant growth–promoting rhizobacteria (PGPRs), which are widely used in the production of agricultural and horticultural crops such as cereals, legumes, oil seeds, fiber plants, fruits, vegetables, medicinal plants, ornamental plants, and spices. During the past few decades, a large number of bacteria including *Alcaligenes*, *Aeromonas*, *Azotobacter*, *Arthrobacter*, *Azoarcus*, *Azospirillum*, *Acinetobacter*, *Agrobacterium*, *Aneurinibacillus*, *Bacillus*, *Beijerinckia*, *Burkholderia*, *Gluconacetobacter*, *Gluconobacter*, *Herbaspirillum*, *Paenibacillus*, *Pseudomonas*, *Rhizobium*, *Rhodococcus*, *Saccharothrix*, *Serratia*, *Thiobacillus*, and *Variovorax* are known as PGPR in plants (Annapurna et al., 2013; Crepin et al., 2012; Dobbelaere et al., 2003). These beneficial rhizobacteria have

Sustainable Agricultural Practices. DOI: https://doi.org/10.1016/B978-0-443-19150-3.00015-1

been used in sustainable agriculture as bio-fertilizers or biological control agents (Babalola, 2010; Kumar et al., 2021a). The bacteria from *Pseudomonadaceae* and *Burkholderiaceae* are the most important microbial agents colonizing the rhizosphere (Peiffer et al., 2013). The rhizobacteria can survive in the soil or seed (Suslow, 1980), then colonize the root cortex area as endophytes and distribute in the rhizosphere (Bahme and Schroth, 1987). Varoius strains of PGPR have been commercialized in laboratory, greenhouse, and field conditions as plant disease control agents, including *Agrobacterium, Azospirillum, Azotobacter, Bacillus, Burkholderia, Delftia, Paenibacillus, Pantoea, Pseudomonas, Rhizobium,* and *Serratia* (Glick, 2012; Kumar et al., 2017). Although many strains PGPRs have been isolated, it is of particular importance to identify effective strains with high host plant root colonization ability as well as their potential to inhibit plant disease and stimulate plant growth. On the other hand, understanding the PGPRs mechanism of action is also very important in disease management programs. In this chapter, the concept of plant growth–promoting bacteria and their mechanism of action in the management of plant diseases are discussed.

7.2 Definition of plant growth–promoting rhizobacteria

Plant roots release huge amounts of organic compounds such as nucleotides, amino acids, fatty acids, organic acids, and vitamins (Uren, 2007) that constitute 5%–21% of the carbon fixed by the plant (Marschner, 1995). Therefore the rhizosphere is an attractive colonization site for microorganisms. The rhizosphere microorganism concentration is 10–1000 times higher than that in bulk soil (Lugtenberg and Kamilova, 2009). One gram of soil contains 10^3–10^5 protozoa, 10^3–10^6 algae, 10^5–10^6 fungi, and 10^8–10^9 bacteria (Rughöft et al., 2016). The bacteria are the main group of soil microorganism community. They use a wide range of substances as nutrient and energy sources, and also have a huge ability to affect plant health and growth positively or negatively (Oleńska et al., 2020). Some of these bacteria, entitled PGPRs, directly or indirectly have positive effects on plant growth and health and belonged to different genera such as *Bacillus, Pseudomonas, Rhizobium, Streptomyces, Azorhizobium, Mesorhizobium, Azospirillum, Azotobacter, Bradyrhizobium, Paenibacillus, Acinetobacter, Flavobacterium, Aeromonas, Enterobacter, Agrobacterium, Allorhizobium, Pantoea, Arthrobacter, Burkholderia, Acetobacter, Caulobacter, Azoarcus, Chromobacterium, Delftia, Micrococcus, Flavobacterium, Frankia, Gluconacetobacter, Frankia, Klebsiella, Pantoea,* and *Serratia* (Basu et al., 2021; Parray et al., 2016; Vessey, 2003). PGPR were first defined by Kloepper and Schroth (1978) as organisms that, after being inoculated on seeds, could successfully colonize plant roots and positively enhance plant growth. These bacteria directly contribute plant growth and health by dissolving minerals, nitrogen fixation, and production of plant hormones such as auxins, gibberellins, and cytokinins (Pieterse et al., 2014; Saharan and Nehra, 2011).

Also, they improve indirectly plant health and growth by biocontrol activity against plant pathogens and deleterious microorganisms. PGPRs can reduce the incidence or severity of plant diseases with different mechanisms including production of antimicrobial compounds (antibiosis), production of siderophores, production of lytic enzymes, production of volatile compounds, hydrogen cyanide (HCN) production, competition for nutrients or place, and induction of systemic resistance (ISR) of plants (Lugtenberg and Kamilova, 2009; Navarro et al., 2019). An efficient plant growth–promoting rhizobacterium should have high potential to colonize host plant roots. So, the bacterium establishes itself in the plant rhizosphere, increases its population, and can show its beneficial effects. However, the beneficial effect of the bacterial strains of a particular genus and species are not always the same in all plants. More detail about biocontrol mechanisms will be discussed in the next sections of this chapter.

7.3 Plant growth–promoting rhizobacteria in biocontrol of plant diseases and their mechanisms

Nowadays, there is a growing interest in the use of natural resources to fight phytopathogens. This growing interest is linked to the limitations in the use of agrochemicals started since 2011 for environmental and human health protection (Barratt et al., 2018). Due to legislative limits and customer demand for pesticide-free foods, the major phytosanitary businesses have been making investments in the field of biocontrol (Bale et al., 2008). The increasing demand for organic products has a huge impact on the worldwide biocontrol industry (Pacios-Michelena et al., 2021).

Based on the July 2022 report of P&S Intelligence, the global market for biocontrol agents generated $4851.7 million in revenue in 2021, and is anticipated to reach $13,634.3 million by 2030, growing at a compound annual growth rate of 12.2% from 2021 to 2030 (P&S Intelligence, 2022). The market is expanding due to the growing preference for biological products over chemical-based crop protection products. The biocontrol agents are also advantageous for pest control because they are safe for the environment, do not damage people, and work all year round (P&S Intelligence, 2022). Due to the rigorous government rules on the use of chemical pesticides, Europe had the greatest market share in 2021, and the region's sector is predicted to expand in the years to come. Europe, for instance, wants to use 50% fewer dangerous pesticides by 2030 (P&S Intelligence, 2022).

Plant interactions with its microbiota are essential for their growth and development (Dastogeer et al., 2020). The plant's root system is the most active portion, as it collects nutrients and interacts with and controls its microbiome the most (Kumar et al., 2021b; Pascale et al., 2020; Zhimo et al., 2022). Numerous bacteria connected to the roots offer the plant ecological services (Xiong et al., 2021). The variety and composition of rhizosphere microbiota are dynamic due to the flexibility of plant–bacteria interactions. The factors that drive endogenous and external interactions that change the diversity and composition of the microbial community are plant type and pedoclimatic conditions (Edwards et al., 2015). The host plant, field location and conditions, crop management practices, growing season, plant genotype and developmental stage, and soil properties and nutrients have been mentioned to be important in microbiota shaping (Compant et al., 2019; Liu et al., 2017).

Rhizosphere bacteria typically belong to *Bacteroidetes*, *Proteobacteria*, and other copiotrophs lineages (Ling et al., 2022). By producing bioactive chemicals (such as antibiotics, bacteriocins, siderophores, and other inhibitory compounds) and physically blocking phytopathogen invasion, root microorganisms are crucial for nutrition intake and pathogen defense (Compant et al., 2011, 2021; Santos and Olivares, 2021). Among rhizosphere microbiota, PGPR engage in interactions with the host plant, stimulating optimal physiological and phytosanitary status (Khare et al., 2018). PGPR are essential for crop yield and plant health defense and enhancement by combining several strategies that involve direct and indirect mechanisms (dos Santos et al., 2020; Kumar et al., 2016; Singh et al., 2018, 2020b).

Direct mechanisms comprise the positive effects that PGPR have on plants (e.g., phytohormones production), soil fertility and structure (e.g., nutrients availability), biogeochemical cycling (e.g., nitrogen fixation), and microbial diversity (e.g., biodiversity enhancement) (Mokrani et al., 2020). Beneficial microbes also help plants in responding better to abiotic stresses like salinity, drought, and temperature (Ayuso-Calles et al., 2021; Pathak et al., 2022; Verma et al., 2021). For example, the presence of ACC (1-Aminocyclopropane-1-carboxylic acid) deaminase in PGPR assures the modulation of stress ethylene (Pellegrini et al., 2021) while proline production promotes the osmotic balance (Chen et al., 2007). Indirect mechanisms mainly include the phytopathogens control. The plant may acquire resistance mechanisms that are specific to the injured tissue or systemically distributed throughout the entire plant. These mechanisms include induced systemic resistance (ISR) and systemic acquired resistance (SAR) (Lahlali et al., 2022). While parasites and pathogens cause SAR, the plant's own helpful microbes cause ISR (Huang et al., 2013). Fig. 7.1 summarizes the biocontrol effects that PGPR have against phytopathogens. The following sections describe in detail all mechanisms important for phytopathogens defense and biocontrol, that are ISR, competition, antibiosis, and HCN, lytic enzymes, siderophores, and volatiles production.

7.3.1 Competition and antibiosis

The composition, diversity, and functions of a microbial community are significantly shaped by resource competition and antibiosis. Within competition, dominant bacterial species rule over other organisms that contest for the same resources (i.e., space and nutrients) thanks to interspecies competitive strategies

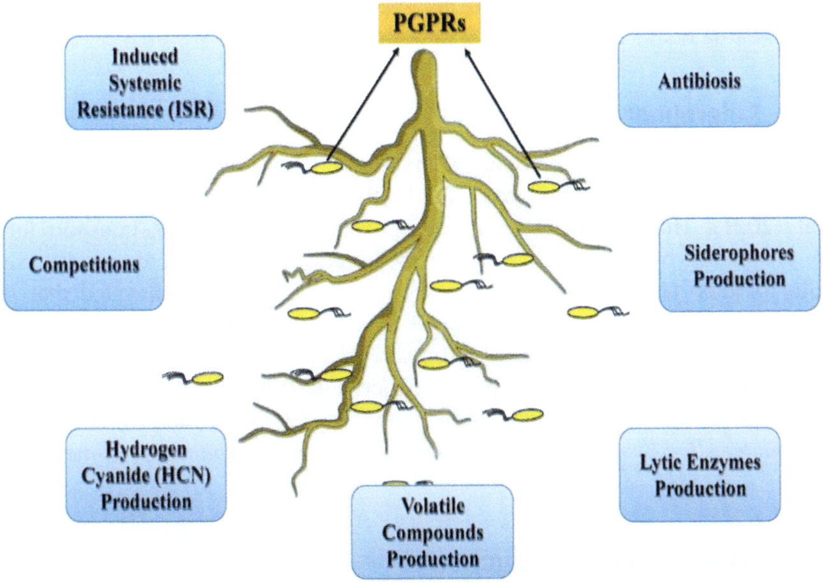

FIGURE 7.1

Schematic summary of the biocontrol mechanisms employed by plant growth–promoting rhizobacteria (PGPRs) against phytopathogens.

(Hibbing et al., 2010). The knowledge of this interaction is of relevance for phytopathogens inhibition by biocontrol agents.

The term "antibiosis" is used to indicate the interaction that arises when one microorganism produces secondary metabolites that have a negative effect on other microorganisms (Nehra et al., 2021). These secondary metabolites, mostly called antimicrobials, are environment-friendly and have no negative effects on either human or animal health (Bolívar-Anillo et al., 2020). Antibiosis toward phytopathogens is obtained using a wide range of molecules that, if present in sufficient amounts, successfully suppress phytopathogens (Peng et al., 2021). Hence, a quorum-sensing mechanism is frequently used to control the production of antimicrobials (Harriott, 2019). The significance of secondary metabolites in mediating competition in antimicrobial capacity has been confirmed by extensive in vitro and in situ research (Hansen et al., 2020; Lucke et al., 2020).

Antibiotics produced by PGPR are more efficient than others due to their biological activity (e.g., antimicrobial, anthelminthic, antiviral, cytotoxic, insecticidal, and phytotoxic) (Fernando et al., 2007). Among the most common, *Pseudomonas* spp. produce a wide scope of antifungal molecules, such as 2,4-diacetylphloroglucinol (2,4-DAPG) (Elshahat et al., 2016). *Bacillus* spp. excretes a large variety of antibiotics (e.g., bacilysin) and antimicrobial lipopeptide biosurfactants (e.g., bacillomycin). Other common sources of antimicrobial compound are actinomycetes, mainly *Streptomyces* spp., accounting for 66% of known antibiotics (Farda et al., 2022).

7.3.2 Siderophores' production

Microbial cells produce siderophores in iron-deficient environments, converting insoluble form of this metal into soluble ones by producing complexes (Terra et al., 2021). The cell surfaces of siderophore-producing PGPR transport ferric iron complexes by iron regulated outer membrane proteins, improving the availability of this metal for metabolisms. Numerous reports demonstrated that this ability is important to improve the plant iron uptake (e.g., under salinity stress) and limit phytopathogens spread (particularly fungal ones) (Mónica et al., 2022; Sayyed et al., 2013). The ability to produce siderophores is widespread in the PGPR group. Most common genera found to be siderophore producers are *Azotobacter*, *Bacillus*, *Pseudomonas*, *Rhizobium*, and *Streptomyces* (Singh et al., 2022).

7.3.3 Lytic enzymes' production

Lytic enzymes released by PGPR provide another biocontrol mechanism against phytopathogens. These enzymes are not constitutive and are produced following a complex series of signals induced by host identification (Karlsson et al., 2017). The role of some lytic enzymes, for example, chitinases, has been extensively reported to be essential in the biocontrol of fungal phytopathogens (Köhl et al., 2019). Several reports also described the activity of chitinases, lipases, and proteases for insecticidal purposes (Rakshiya et al., 2016). Among PGPR able to produce lytic enzymes, most strains belong to *Bacillus*, *Pseudomonas*, and *Streptomyces* (Veliz et al., 2017).

7.3.4 Volatile compounds' production

Several volatile organic compounds (VOCs) that PGPR secretes are excellent biocontrol agents against a wide range of phytopathogens (Pacios-Michelena et al., 2021). PGPR catabolic mechanisms such as glycolysis, proteolysis, and lipolysis are used to create different chemical forms of VOCs. Several VOCs affect the growth, differentiation, stress tolerance, and/or behavior of invertebrates, plants, or fungi (Audrain et al., 2015). Recent research has highlighted the importance of VOCs in bacterial interactions in a variety of settings, including soil, animal and plant microbiota, and biofilms, in addition to these interactions with a wide spectrum of eukaryote organisms (Audrain et al., 2015). VOCs have a direct activity on pathogenic bacterial cells and on their quorum sensing, disrupting the gene expression regulation in response to cell-population density and biofilm formation (Jones et al., 2017; Modi et al., 2020). The main PGPR genera known to be producers of VOCs are *Pseudomonas*, *Bacillus*, and *Streptomyces* (Chen et al., 2015;

Djebaili et al., 2021; Mohanty et al., 2021). VOCs produced by bacteria, including ammonia, HCN, hydrogen sulfide, and nitric oxide. Among them, HCN is common in many PGPR genera belonging to *Pseudomonas*, *Chromobacterium*, and *Rhizobium* (Sehrawat et al., 2022). Due to the sensitivity of heme groups in the target eukaryotic cells, HCN is toxic to a wide range of organisms, including plants, nematodes, insects, and fungi (Zdor, 2015).

7.3.5 Induced systemic resistance

Several studies unveiled that using PGPR is an efficient way to counteract the negative impacts of numerous environmental stresses (Bacilio et al., 2004; de-Bashan et al., 2012; Djebaili et al., 2021; Etesami and Beattie, 2017; Etesami and Maheshwari, 2018; Grover et al., 2011; Kim et al., 2012; Qin et al., 2016). It is supposed that the release of several chemicals (such as elicitors and antibiotics) and the induction of the ISR are responsible for this ability (Compant et al., 2005; Etesami and Maheshwari, 2018; Glick, 2004, 2010, 2014; Pagnani et al., 2018, 2020). Plants can respond to stresses by activating defense mechanisms that are either systemically distributed throughout the plant or restricted to the injured organ (Romera et al., 2019). The SAR and ISR are among the latter. Pathogens and pests cause SAR, while rhizosphere-dwelling beneficial microorganisms mediate ISR (Choudhary et al., 2007). Several plant species have demonstrated PGPR-induced ISR (e.g., cress, bean, carnation, cucumber, radish, tobacco, and tomato). It has been shown that ISR generated by PGPR is effective in lowering pathogenic assaults by bacteria, viruses, and fungi (van Loon et al., 1998). Some of the elicitors generated by PGPR are also used in the ISR such as lipopolysaccharides, flagella, siderophores, and antibiotics (Bakker et al., 2003; van Loon et al., 1998).

7.4 Future perspectives and conclusions

PGPRs are fascinating life forms that provide significant benefits to many host plants. These bacteria, along with their success in plant growth and production, offer an excellent opportunity to reduce or eliminate toxic agrochemicals as they biologically control pathogens throughout the growing season (Boyno et al., 2022; Calvo et al., 2014; Lamichhane et al., 2016; Morales-Cedeño et al., 2021; Santos et al., 2021). In addition, the plant growth−promoting characteristics shown by PGPRs communities, such as the solubility of nutrients including nitrogen (N), phosphorus (P), and potassium (K), boosting the plant immune system, and managing abiotic stress, also significantly contribute to plant development and production. The practice of activating, fixing, and purchasing PGPRs can significantly reduce inputs of chemical fertilizers as well as contribute to sustainable agriculture. Additionally, through systematic and focused research and/or methodology, it is necessary to assess or calculate the influence of PGPRs on root characteristics and the resulting benefits to the plants and ecosystem services. It can be desirable to employ seedling stage screening of potential PGPRs strains to produce favorable root characteristics. Direct applications of these strains in sustainable agriculture are interesting area of research.

Mechanisms used by PGPRs include recognition of chemical signals and nutrient intake, antibiosis, siderophores/lytic enzymes/volatile compounds/HCN production, and motility (Etesami and Maheshwari, 2018; Kuzyakov and Razavi, 2019; Zboralski and Filion, 2020). In addition, PGPRs compete with other microorganisms in the rhizosphere by producing antimicrobial chemicals (del Carmen Orozco-Mosqueda et al., 2018; Santoyo et al., 2012; Wu et al., 2019). To get effective results in the field, it is necessary to identify and choose the most competitive PGPRs because many of these pathways are also utilized by plant diseases. Also, unlike chemical pesticides, biocontrol agents need support even after application to be in their targeted niche. Therefore for the success of biological control, it must be ensured not only the quality of the applied biocontrol agent, but also its establishment so that it can thrive in the natural ecosystem and compete well with pathogens. Plants are thought to have evolved to internalize microbes for nutrient and defense. The relationship between PGPRs and the plant starts with root exudation in the rhizosphere and progresses through attachment, colonization in the endo-rhizosphere and/or on the rhizoplane. By releasing different metabolites, such as phytohormones and volatile organic substances, PGPRs alter root property routes and/or disrupts hormonal balance. The microbial effect on changing root properties can occur in terms of root morphological, architectural, physiological, and biotic properties. Changing external characteristics may contribute to plant development and ecosystem functioning through cumulative effects in terms of improved anchorage due to root branching (del Carmen Orozco-Mosqueda and Santoyo, 2021). As a result of enhanced microbial polysaccharide synthesis, PGPRs inoculation also enhances soil aggregation.

Finally, as only a few procedures have been presented, it would be beneficial to develop novel screening techniques for significant PGPRs functions to colonize the rhizosphere (Santoyo et al., 2019). Furthermore, it is critical to pinpoint the precise roles that some mechanisms play in promoting plant growth and rhizosphere colonization. For instance, some volatile substances have been shown to have antibiotic effects in laboratory settings, but several studies have found that they also play a significant role in the rhizosphere (Kai et al., 2016). Filling all existing gaps in the knowledge of rhizosphere colonization mechanisms by PGPRs will facilitate the choice of potential strains as bio-inoculants in the field. In addition, since pathogens behave differently in laboratory and field conditions, there is a need for the dissemination of PGPRs in field conditions to regain their biological activity and viability. Integrated participation of scientists from different disciplines, including microbiology, agriculture, biochemistry, genetics, and molecular biology, can help better understand how PGPRs work and identify applications.

References

Annapurna, K., Kumar, A., Kumar, L.V., Govindasamy, V., Bose, P., Ramadoss, D., 2013. PGPR-induced systemic resistance (ISR) in plant disease management. In: Maheshwari, D. (Ed.), Bacteria in Agrobiology: Disease Management. Springer, Berlin, pp. 405–425.

Audrain, B., Farag, M.A., Ryu, C.M., Ghigo, J.M., 2015. Role of bacterial volatile compounds in bacterial biology. FEMS Microbiol. Rev. 39, 222–233.

Ayuso-Calles, M., Flores-Félix, J.D., Rivas, R., 2021. Overview of the role of rhizobacteria in plant salt stress tolerance. Agronomy 11, 1759.

Babalola, O.O., 2010. Beneficial bacteria of agricultural importance. Biotechnol. Lett. 32, 1559–1570.

Bacilio, M., Rodriguez, H., Moreno, M., Hernandez, J.P., Bashan, Y., 2004. Mitigation of salt stress in wheat seedlings by a gfp-tagged *Azospirillum lipoferum*. Biol. Fertil. Soils 40, 188–193.

Bahme, J.B., Schroth, M.N., 1987. Spatial-temporal colonization patterns of a rhizobacterium on underground organs of potato. Phytopathology 77, 1093–1100.

Bakker, P.A.H.M., Ran, L.X., Pieterse, C.M.J., van Loon, L.C., 2003. Understanding the involvement of rhizobacteria-mediated induction of systemic resistance in biocontrol of plant diseases. Can. J. Plant Pathol. 25, 5–9.

Bale, J., van Lenteren, J., Bigler, F., 2008. Biological control and sustainable food production. Philos. Trans. R. Soc. B Biol. Sci. 363, 761–776.

Barratt, B.I.P., Moran, V.C., Bigler, F., van Lenteren, J.C., 2018. The status of biological control and recommendations for improving uptake for the future. Biocontrol 63, 155–167.

Basu, A., Prasad, P., Das, S.N., Kalam, S., Sayyed, R.Z., Reddy, M.S., El Enshasy, H., 2021. Plant growth promoting rhizobacteria (PGPR) as green bioinoculants: recent developments, constraints, and prospects. Sustainability 13, 1140.

Bolívar-Anillo, H.J., Garrido, C., Collado, I.G., 2020. Endophytic microorganisms for biocontrol of the phytopathogenic fungus *Botrytis cinerea*. Phytochem. Rev. 19, 721–740.

Boyno, G., Demir, S., Danesh, Y.R., 2022. Effects of some biological agents on the growth and biochemical parameters of tomato plants infected with *Alternaria solani* (Ellis & Martin) Sorauer. Eur. J. Plant Pathol. 162 (1), 19–29.

Calvo, P., Nelson, L., Kloepper, J.W., 2014. Agricultural uses of plant biostimulants. Plant Soil 383 (1), 3–41.

Chen, Y., Gozzi, K., Yan, F., Chai, Y., 2015. Acetic acid acts as a volatile signal to stimulate bacterial biofilm formation. mBio 6, 392.

Chen, Z., Pottosin, I.I., Cuin, T.A., Fuglsang, A.T., Tester, M., Jha, D., 2007. Root plasma membrane transporters controlling K^+/Na^+ homeostasis in salt-stressed barley. Plant Physiol. 145, 1714–1725.

Choudhary, D.K., Prakash, A., Johri, B.N., 2007. Induced systemic resistance (ISR) in plants: mechanism of action. Ind. J. Microbiol. 47, 289–297.

Compant, S., Reiter, B., Sessitsch, A., Nowak, J., Clément, C., Barka, E.A., 2005. Endophytic colonization of *Vitis vinifera* L. by plant growth-promoting bacterium *Burkholderia* sp. strain PsJN. Appl. Environ. Microbiol. 71, 1685–1693.

Compant, S., Mitter, B., Colli-Mull, J.G., Gangl, H., Sessitsch, A., 2011. Endophytes of grapevine flowers, berries, and seeds: identification of cultivable bacteria, comparison with other plant parts, and visualization of niches of colonization. Microbiol. Ecol. 62, 188–197.

Compant, S., Samad, A., Faist, H., Sessitsch, A., 2019. A review on the plant microbiome: ecology, functions, and emerging trends in microbial application. J. Adv. Res. 19, 29–37.

Compant, S., Cambon, M.C., Vacher, C., Mitter, B., Samad, A., Sessitsch, A., 2021. The plant endosphere world – bacterial life within plants. Environ. Microbiol. 23, 1812–1829.

Crepin, A., Barbey, C., Cirou, A., Tannières, M., Orange, N., Feuilloley, M., Dessaux, Y., Burini, J., Faure, D., Latour, X., 2012. Biological control of pathogen communication in the rhizosphere: a novel approach applied to potato soft rot due to *Pectobacterium atrosepticum*. Plant Soil 358, 27–37.

Dastogeer, K.M.G., Tumpa, F.H., Sultana, A., Akter, M.A., Chakraborty, A., 2020. Plant microbiome—an account of the factors that shape community composition and diversity. Curr. Plant Biol. 23, 100161.

de-Bashan, L.E., Hernandez, J.P., Bashan, Y., 2012. The potential contribution of plant growth-promoting bacteria to reduce environmental degradation—a comprehensive evaluation. Appl. Soil Ecol. 61, 171–189.

del Carmen Orozco-Mosqueda, M., Santoyo, G., 2021. Plant-microbial endophytes interactions: scrutinizing their beneficial mechanisms from genomic explorations. Cur. Plant Biol. 25, 100189.

del Carmen Orozco-Mosqueda, M., del Carmen Rocha-Granados, M., Glick, B.R., Santoyo, G., 2018. Microbiome engineering to improve biocontrol and plant growth-promoting mechanisms. Microbiol. Res. 208, 25–31.

Djebaili, R., Pellegrini, M., Ercole, C., Farda, B., Kitouni, M., Del Gallo, M., 2021. Biocontrol of soil-borne pathogens of *Solanum lycopersicum* L. and *Daucus carota* L. by plant growth-promoting actinomycetes: *in vitro* and in planta antagonistic activity. Pathogens 10, 1305.

Dobbelaere, S., Vanderleyden, J., Okon, Y., 2003. Plant growth promoting effects of diazotrophs in the rhizosphere. Crit. Rev. Plant Sci. 22, 107–149.

dos Santos, R.M., Diaz, P.A.E., Lobo, L.L.B., Rigobelo, E.C., 2020. Use of plant growth-promoting rhizobacteria in maize and sugarcane: characteristics and applications. Front. Sustain. Food Syst. 4, 136.

Edwards, J., Johnson, C., Santos-Medellín, C., Lurie, E., Podishetty, N.K., Bhatnagar, S., 2015. Structure, variation, and assembly of the root-associated microbiomes of rice. Proc. Natl. Acad. Sci. 112.

Elshahat, M.R., Ahmed, A.A., Enas, A.H., Fekria, M.S., 2016. Plant growth promoting rhizobacteria and their potential for biocontrol of phytopathogens. Afr. J. Microbiol. Res. 10, 486–504.

Etesami, H., Beattie, G.A., 2017. Plant-microbe interactions in adaptation of agricultural crops to abiotic stress conditions. Probiotics and Plant Health. Springer, pp. 163–200.

Etesami, H., Maheshwari, D.K., 2018. Use of plant growth promoting rhizobacteria (PGPRs) with multiple plant growth promoting traits in stress agriculture: action mechanisms and future prospects. Ecotoxicol. Environ. Saf. 156, 225–246.

Farda, B., Djebaili, R., Vaccarelli, I., Del Gallo, M., Pellegrini, M., 2022. Actinomycetes from caves: an overview of their diversity, biotechnological properties, and insights for their use in soil environments. Microorganisms 10.

Fernando, W.G.D., Nakkeeran, S., Zhang, Y., Savchuk, S., 2007. Biological control of *Sclerotinia sclerotiorum* (Lib.) de Bary by *Pseudomonas* and *Bacillus* species on canola petals. Crop Prot. 26, 100–107.

Glick, B.R., 2004. Bacterial ACC deaminase and the alleviation of plant stress. Adv. Appl. Microbiol. 56, 291–312.

Glick, B.R., 2010. Using soil bacteria to facilitate phytoremediation. Biotechnol. Adv. 28, 367–374.

Glick, B.R., 2012. Plant growth-promoting bacteria: mechanisms and applications. Scientifica 2012, 963401.

Glick, B.R., 2014. Bacteria with ACC deaminase can promote plant growth and help to feed the world. Microbiol. Res. 169, 30–39.

Grover, M., Ali, S.Z., Sandhya, V., Rasul, A., Venkateswarlu, B., 2011. Role of microorganisms in adaptation of agriculture crops to abiotic stresses. World J. Microbiol. Biotechnol. 27, 1231–1240.

Hansen, B.L., Pessotti, R., de, C., Fischer, M.S., Collins, A., El-Hifnawi, L., Liu, M.D., 2020. Cooperation, competition, and specialized metabolism in a simplified root nodule microbiome. mBio 11, e01917–e01920.

Harriott, M.M., 2019. Biofilms and antibiotics. Reference Module in Biomedical Sciences. Elsevier, pp. 1–11.

Hibbing, M.E., Fuqua, C., Parsek, M.R., Peterson, S.B., 2010. Bacterial competition: surviving and thriving in the microbial jungle. Nat. Rev. Microbiol. 8, 15–25.

Huang, J., Wei, Z., Tan, S., Mei, X., Yin, S., Shen, Q., 2013. The rhizosphere soil of diseased tomato plants as a source for novel microorganisms to control bacterial wilt. Appl. Soil Ecol. 72, 79–84.

Jones, S.E., Ho, L., Rees, C.A., Hill, J.E., Nodwell, J.R., Elliot, M.A., 2017. *Streptomyces* exploration is triggered by fungal interactions and volatile signals. Elife 6.

Kai, M., Effmert, U., Piechulla, B., 2016. Bacterial-plant-interactions: approaches to unravel the biological function of bacterial volatiles in the rhizosphere. Front. Microbiol. 7, 108.

Karlsson, M., Atanasova, L., Jensen, D.F., Zeilinger, S., 2017. Necrotrophic mycoparasites and their genomes. Microbiol. Spectr. 5.

Khare, E., Mishra, J., Arora, N.K., 2018. Multifaceted interactions between endophytes and plant: developments and prospects. Front. Microbiol. 9, 2732.

Kim, Y.C., Glick, B.R., Bashan, Y., Ryu, C.M., 2012. Enhancement of plant drought tolerance by microbes. Plant Responses to Drought Stress. Springer, pp. 383–413.

Kloepper, J.W., Schroth, M.N., 1978. Plant growth-promoting rhizobacteria on radishes. In: Proceedings of the Fourth International Conference on Plant Pathogenic Bacteria, vol. 2. Station de Pathologie Ve'ge'tale et de Phytobacte'riologie, INRA, Angers, France, pp. 879–882.

Köhl, J., Kolnaar, R., Ravensberg, W.J., 2019. Mode of action of microbial biological control agents against plant diseases: relevance beyond efficacy. Front. Plant Sci. 10, 845.

Kumar, A., Singh, M., Singh, P.P., Singh, S.K., Singh, P.K., Pandey, K.D., 2016. Isolation of plant growth promoting rhizobacteria and their impact on growth and curcumin content in *Curcuma longa* L. Biocatal. Agric. Biotechnol. 8, 1–7.

Kumar, A., Singh, S.K., Kant, C., Verma, H., Kumar, D., Singh, P.P., Modi, A., Droby, S., Kesawat, M.S., Alavilli, H., Bhatia, S.K., 2021a. Microbial biosurfactant: a new frontier for sustainable agriculture and pharmaceutical industries. Antioxidants 10 (9), 1472.

Kumar, A., Zhimo, Y., Biasi, A., Salim, S., Feygenberg, O., Wisniewski, M., Droby, S., 2021b. Endophytic microbiome in the carposphere and its importance in fruit physiology and pathology. In: Spadaro, D., Droby, S., Gullino, M.L. (Eds.), Postharvest Pathology. Plant Pathology in the 21st Century, vol 11. Springer, Cham, pp. 73–88.

Kumar, A., Verma, H., Singh, V.K., Singh, P.P., Singh, S.K., Ansari, W.A., Yadav, A., Singh, P.K., Pandey, K.D., 2017. Role of *Pseudomonas* sp. in sustainable agriculture and disease management. Agriculturally Important Microbes for Sustainable Agriculture. Springer, Singapore, pp. 195–215.

Kuzyakov, Y., Razavi, B.S., 2019. Rhizosphere size and shape: temporal dynamics and spatial stationarity. Soil Biol. Biochem. 135, 343–360.

Lahlali, R., Ezrari, S., Radouane, N., Kenfaoui, J., Esmaeel, Q., El Hamss, H., Belabess, Z., Barka, E.A., 2022. Biological control of plant pathogens: a global perspective. Microorganisms 10, 596.

Lamichhane, J.R., Dachbrodt-Saaydeh, S., Kudsk, P., Messéan, A., 2016. Toward a reduced reliance on conventional pesticides in European agriculture. Plant Dis. 100 (1), 10–24.

Ling, N., Wang, T., Kuzyakov, Y., 2022. Rhizosphere bacteriome structure and functions. Nat. Commun. 13, 836.

Liu, H., Carvalhais, L.C., Crawford, M., Singh, E., Dennis, P.G., Pieterse, C.M.J., 2017. Inner plant values: diversity, colonization and benefits from endophytic bacteria. Front. Microbiol. 8, 2552.

Lucke, M., Correa, M.G., Levy, A., 2020. The role of secretion systems, effectors, and secondary metabolites of beneficial rhizobacteria in interactions with plants and microbes. Front. Plant Sci. 11, 589416.

Lugtenberg, B., Kamilova, F., 2009. Plant-growth-promoting rhizobacteria. Ann. Rev. Microbiol. 63 (1), 541–556.

Marschner, H., 1995. Mineral Nutrition of Higher Plants, second ed. Academic, London.

Modi, A., Kanani, P., Kumar, A., 2020. Fungal endophytes-induced gene expression studies in biotic and abiotic stress management. In: Kumar, A., Radhakrishnan, E.K. (Eds.), Microbial Endophytes. Woodhead Publishing, Cambridge USA, pp. 251–271.

Mohanty, P., Singh, P.K., Chakraborty, D., Mishra, S., Pattnaik, R., 2021. Insight into the role of PGPR in sustainable agriculture and environment. Front. Sustain. Food Syst. 5, 667150.

Mokrani, S., Nabti, E., Cruz, C., 2020. Current advances in plant growth promoting bacteria alleviating salt stress for sustainable agriculture. Appl. Sci. 10, 7025.

Mónica, D.I.F., Wong, A.V., Rubio, P.J.S., Vaca-Paulín, R., Yañez-Ocampo, G., 2022. Exploring plant growth-promoting rhizobacteria as stress alleviators: a methodological insight. Arch. Microbiol. 204, 316.

Morales-Cedeño, L.R., del Carmen Orozco-Mosqueda, M., Loeza-Lara, P.D., Parra-Cota, F.I., de Los Santos-Villalobos, S., Santoyo, G., 2021. Plant growth-promoting bacterial endophytes as biocontrol agents of pre-and post-harvest diseases: fundamentals, methods of application and future perspectives. Microbiol. Res. 242, 126612.

Navarro, M.O.P., Barazetti, A., Niekawa, E.T.G., Dealis, M.L., Matos, J.M.S., Liuti, G., Modolon, F., Oliveira, I.M., Andreata, M., Cely, M.V.T., Andrade, G., 2019. Microbial biological control of diseases and pests by PGPR and PGPF. In: Singh, D.P., et al., (Eds.), Microbial Interventions in Agriculture and Environment. Springer Nature, Singapore.

Nehra, S., Gothwal, R.K., Varshney, A.K., Solanki, P.S., Chandra, S., Meena, P., 2021. Bio-management of *Fusarium* spp. associated with fruit crops. Fungi Bio-Prospects in Sustainable Agriculture, Environment and Nano-Technology. Elsevier, pp. 475–505.

Oleńska, E., Małek, W., Wójcik, M., Swiecicka, I., Thijs, S., Vangronsveld, J., 2020. Beneficial features of plant growth-promoting rhizobacteria for improving plant growth and health in challenging conditions: a methodical review. Sci. Total Environ. 743140682.

P&S Intelligence, 2022. Biocontrol agents market size and share analysis report by active substance (microbials, macrobials, entomopathogenic nematodes), crop type (cereals & grains, vegetables & fruits, pulses & oilseeds), target pest (arthropods, weeds, microorganisms), application (seed treatment, on-field, post-harvest) – Global Industry Growth Forecast to 2030.

Pacios-Michelena, S., Aguilar González, C.N., Alvarez-Perez, O.B., Rodriguez-Herrera, R., Chávez-González, M., Arredondo Valdés, R., 2021. Application of *Streptomyces* antimicrobial compounds for the control of phytopathogens. Front. Sustain. Food Syst. 5, 696518.

Pagnani, G., Pellegrini, M., Galieni, A., D'Egidio, S., Matteucci, F., Ricci, A., Stagnari, F., Sergi, M., Lo Sterzo, C., Pisante, M., Del Gallo, M., 2018. Plant growth-promoting rhizobacteria (PGPR) in *Cannabis sativa* 'Finola'cultivation': an alternative fertilization strategy to improve plant growth and quality characteristics. Ind. Crops Prod. 123, 75–83.

Pagnani, G., Galieni, A., Stagnari, F., Pellegrini, M., Del Gallo, M., Pisante, M., 2020. Open field inoculation with PGPR as a strategy to manage fertilization of ancient *Triticum* genotypes. Biol. Fertil. Soils 56, 111–124.

Parray, J.A., Jan, S., Kamili, A.N., Qadri, R.A., Egamberdieva, D., Ahmad, P., 2016. Current perspectives on plant growth-promoting rhizobacteria. J. Plant Growth Reg 35, 877–902.

Pascale, A., Proietti, S., Pantelides, I.S., Stringlis, I.A., 2020. Modulation of the root microbiome by plant molecules: the basis for targeted disease suppression and plant growth promotion. Front. Plant Sci. 10.

Pathak, P., Rai, V.K., Can, H., Singh, S.K., Kumar, D., Bhardwaj, N., Roychowdhury, R., de Azevedo, L.C.B., Verma, H., Kumar, A., 2022. Plant-endophyte interaction during biotic stress management. Plants 11 (17), 2203.

Peiffer, J.A., Spor, A., Koren, O., Jin, Z., Tringe, S.G., Dangl, J.L., Buckler, E.S., Ley, R. E., 2013. Diversity and heritability of the maize rhizosphere microbiome under field conditions. Proc. Nat. Acad. Sci. 110, 6548–6553.

Pellegrini, M., Pagnani, G., Rossi, M., D'Egidio, S., Del Gallo, M., Forni, C., 2021. *Daucus carota* L. seed inoculation with a consortium of bacteria improves plant growth, soil fertility status and microbial community. Appl. Sci. 11.

Peng, Y., Li, S.J., Yan, J., Tang, Y., Cheng, J.P., Gao, A.J., 2021. Research progress on phytopathogenic fungi and their role as biocontrol agents. Front. Microbiol. 12, 670135.

Pieterse, C.M.J., Zamioudis, C., Berendsen, R.L., Weller, D.M., Van Wees, S.C.M., Bakker, P.A.H.M., 2014. Induced systemic resistance by beneficial microbes. Ann. Rev. Phytopath 52 (1), 347–375.

Qin, Y., Druzhinina, I.S., Pan, X., Yuan, Z., 2016. Microbially mediated plant salt tolerance and microbiome-based solutions for saline agriculture. Biotechnol. Adv. 34, 1245–1259.

Rakshiya, Y.S., Verma, M.K., Sindhu, S.S., 2016. Efficacy of antagonistic soil bacteria in management of subterranean termites (Isoptera). Res. Environ. Life Sci. 9, 949–955.

Romera, F.J., García, M.J., Lucena, C., Martínez-Medina, A., Aparicio, M.A., Ramos, J., Alcántara, E., Angulo, M., Pérez-Vicente, R., 2019. Induced systemic resistance (ISR) and Fe deficiency responses in dicot plants. Front. Plant Sci. 10, 287.

Rughöft, S., Herrmann, M., Lazar, C.S., Cesarz, S., Levick, S.R., Trumbore, S.E., Küsel, K., 2016. Community composition and abundance of bacterial, archaeal and nitrifying populations in Savanna soils on contrasting bedrock material in Kruger national park, South Africa. Front. Microbiol. 7, 1638.

Saharan, B.S., Nehra, V., 2011. Plant growth promoting rhizobacteria: a critical review. Life Sci. Med. Res. 2011, 1–30.

Santos, L.F., Olivares, F.L., 2021. Plant microbiome structure and benefits for sustainable agriculture. Curr. Plant Biol. 26, 100198.

Santos, M.S., Rodrigues, T.F., Nogueira, M.A., Hungria, M., 2021. The challenge of combining high yields with environmentally friendly bioproducts: a review on the compatibility of pesticides with microbial inoculants. Agronomy 11 (5), 870.

Santoyo, G., Orozco-Mosqueda, M.D.C., Govindappa, M., 2012. Mechanisms of biocontrol and plant growth-promoting activity in soil bacterial species of *Bacillus* and *Pseudomonas*: a review. Biocontrol Sci. Technol. 22 (8), 855–872.

Santoyo, G., Sánchez-Yáñez, J.M., de los Santos-Villalobos, S., 2019. Methods for detecting biocontrol and plant growth-promoting traits in Rhizobacteria. Methods in Rhizosphere Biology Research. Springer, Singapore, pp. 133–149.

Sayyed, R.Z., Chincholkar, S.B., Reddy, M.S., Gangurde, N.S., Patel, P.R., 2013. Siderophore producing PGPR for crop nutrition and phytopathogen suppression. Bacteria in Agrobiology: Disease Management. Springer, Berlin, Heidelberg, pp. 449–471.

Sehrawat, A., Sindhu, S.S., Glick, B.R., 2022. Hydrogen cyanide production by soil bacteria: biological control of pests and promotion of plant growth in sustainable agriculture. Pedosphere 32, 15–38.

Singh, V.K., Singh, A.K., Kumar, A., 2017. Disease management of tomato through PGPB: current trends and future perspective. 3 Biotech 7 (4), 1–10.

Singh, V.K., Singh, A.K., Singh, P.P., Kumar, A., 2018. Interaction of plant growth promoting bacteria with tomato under abiotic stress: a review. Agric. Ecosyst. Environ. 267, 129–140.

Singh, M., Srivastava, M., Kumar, A., Singh, A.K., Pandey, K.D., 2020a. Endophytic bacteria in plant disease management. In: Kumar, A., Singh, K.V. (Eds.), Microbial Endophytes: Prospects for Sustainable Agriculture. Woodhead Publ., Cambridge USA, pp. 61–89.

Singh, D., Singh, S.K., Singh, V.K., Gupta, A., Aamir, M., Kumar, A., 2020b. Plant growth–promoting bacteria and their role in environmental management. In: Singh, P., Kumar, A., Borthakur, A. (Eds.), Abatement of Environmental Pollutants. Elsevier, Cambridge USA, pp. 161–175.

Singh, P., Chauhan, P.K., Upadhyay, S.K., Singh, R.K., Dwivedi, P., Wang, J., 2022. Mechanistic insights and potential use of siderophores producing microbes in rhizosphere for mitigation of stress in plants grown in degraded land, Front. Microbiol., 13. p. 898979.

Suslow, T.V., 1980. Increased growth and yield of sugar beets by seed treatment with selected *Pseudomonas* spp. and bacterial culture preservation in frozen or dry film of cellulose methyl ether. Ph.D. Thesis, University of California, Los Angeles.

Terra, L., Ratcliffe, N., Castro, H.C., Vicente, A.C.P., Dyson, P., 2021. Biotechnological potential of *Streptomyces* siderophores as new antibiotics. Curr. Med. Chem. 28, 1407–1421.

Uren, N.C., 2007. Types, amounts, and possible functions of compounds released into the rhizosphere by soil-grown plants. In: Pinton, R., Varanini, Z., Nannipieri, P. (Eds.), The Rhizosphere. Biochemistry and Organic Substances at the Soil-Plant Interface. CRC Press/Taylor & Francis Group, Boca Raton, FL, pp. 1–21.

Van Loon, L.C., Bakker, P.A.H.M., Pieterse, C.M.J., 1998. Systemic resistance induced by rhizosphere bacteria. Ann. Rev. Phytopathol. 36, 453–483.

Veliz, E.A., Martínez-Hidalgo, P., Hirsch, A.M., 2017. Chitinase-producing bacteria and their role in biocontrol. AIMS Microbiol. 3, 689–705.

Verma, H., Kumar, D., Kumar, V., Kumari, M., Singh, S.K., Sharma, V.K., Droby, S., Santoyo, G., White, J.F., Kumar, A., 2021. The potential application of endophytes in management of stress from drought and salinity in crop plants. Microorganisms 9 (8), 1729.

Vessey, J.K., 2003. Plant growth promoting rhizobacteria as biofertilizers. Plant Soil 255, 571–586.

Wu, Y., Zhou, J., Li, C., Ma, Y., 2019. Antifungal and plant growth promotion activity of volatile organic compounds produced by *Bacillus amyloliquefaciens*. Microbiol. Open 8 (8)e00813.

Xiong, Q., Hu, J., Wei, H., Zhang, H., Zhu, J., 2021. Relationship between plant roots, rhizosphere microorganisms, and nitrogen and its special focus on rice. Agriculture 11, 234.

Zboralski, A., Filion, M., 2020. Genetic factors involved in rhizosphere colonization by phytobeneficial *Pseudomonas* spp. Comput. Struc. Biotechnol. J. 18, 3539–3554.

Zdor, R.E., 2015. Bacterial cyanogenesis: impact on biotic interactions. J. Appl. Microbiol. 118, 267–274.

Zhimo, V.Y., Kumar, A., Biasi, A., Abdelfattah, A., Sharma, V.K., Salim, S., Feygenberg, O., Bartuv, R., Freilich, S., Whitehead, S.R., Wisniewski, M., 2022. Assembly and dynamics of the apple carposphere microbiome during fruit development and storage. Front. Microbiol. 2973.

CHAPTER

Bioprospecting of bacterial endophytes from the tuber of *Dioscorea bulbifera*

8

Sougata Ghosh[1,2], Vikas Ghattargi[3], Komal E. Kaware[4], Shivani P. Kulkarni[4] and Sirikanjana Thongmee[1]

[1]*Department of Physics, Faculty of Science, Kasetsart University, Bangkok, Thailand*
[2]*Department of Microbiology, School of Science, RK University, Rajkot, Gujarat, India*
[3]*Department of Serology and Microbiome, Mylab Discovery Solutions Pvt. Ltd., Pune, Maharashtra, India*
[4]*Department of Microbiology, Modern College of Arts, Science and Commerce, Ganeshkhind, Pune, Maharashtra, India*

8.1 Introduction

Endophytes are generally isolated from the surface-disinfested plant tissues or extracted from the internal parts of the plant. The endophytic bacteria do not visibly harm the plant (Santoyo et al., 2016). These bacterial communities are advantageous to the host plants as they not only promote plant growth but also protect from several phytopathogens. It is interesting to note that under environmental stress, communication between host plants and the bacterial endophytes are more efficient compared to the rhizospheric bacteria (Ali et al., 2012; Coutinho et al., 2015). The endophytes can promote plant growth by direct mechanisms that include acquisition of nutrients such as iron, phosphorous, and nitrogen. Modulation of the phytohormones such as auxin, cytokinin, and gibberellins apart from lowering of ethylene by synthesizing 1-aminocyclopropane-1-carboxylate (ACC) deaminase are among other direct mechanisms (Ghosh et al., 2021). On the other hand, several indirect mechanisms of plant growth promotion are also reported in bacterial endophytes that include resistance of infection by plant pathogens and lowering the associated damage to the host plants (Bhattacharya et al., 2021; Ghosh and Das, 2022; Coutinho et al., 2015; Pandya et al., 2015; Rashid et al., 2012; Saini et al., 2015). Sometimes it is thought that the rhizospheric bacteria secrete certain plant cell wall lytic enzymes to gain entry and colonize the internal tissues as endophytes as the microecosystem is ideal for survival and nutrition. The endophytic bacterial diversity is often seen as a subset of the rhizosphere and/or root-associated bacterial population (Marquez-Santacruz et al., 2010). Since rhizosphere being most competitive ecological niche for the microbes, it makes it difficult to occupy space and acquire required nutrients

Sustainable Agricultural Practices. DOI: https://doi.org/10.1016/B978-0-443-19150-3.00005-9

conveniently. Hence, some potentially beneficial or pathogenic bacteria with high colonizing capacities gain entry within plant tissues where they efficiently proliferate, obtain nutrients from the microenvironment, and modulate the growth and development of their host plant (Haas and Keel, 2003). Endophytes are generally motile and can produce polysaccharides.

Several bacterial endophytes such as *Azoarcus* sp. BH72, *Azospirillum lipoferum* 4B, *Burkholderia phytofirmans* PsJN, *Enterobacter cloacae* ENHKU01, *Gluconacetobacter diazotrophicus* PaI5, *Klebsiella pneumoniae* 342, *Pseudomonas putida* W619, *Pseudomonas stuzeri* A1501, *Serratia proteamaculans* 568, and others are reported from rice, potato, tomato, pepper, and other plants (Krause et al., 2006; Wisniewski-Dyé et al., 2011; Weilharter et al., 2011; Luo et al., 2012; Bertalan et al., 2009; Fouts et al., 2008; Taghavi et al., 2009; Yan et al., 2008).

Among various yams (*Dioscorea* species) which is a tropical tuber crop, *Dioscorea bulbifera* L. (aerial yam) has got more health benefits due to their antidiabetic, antioxidant, and anticancer applications (Ghosh et al., 2012a, 2013, 2014, 2015a). Yams are considered as source of food by more than 60 million people living in West Africa (Asiedu and Sartie, 2010). However, the poor soil quality results in low yield of yams (Diby et al., 2009). Chemical fertilizers are often insufficient for improving productivity (Ettien et al., 2014). Hence, plant growth-promoting bacteria can be employed for improving the growth, yield, and enhanced nutritional content in yams. More recently, isolation and bioprospecting of endophytes from yams have gained more attention as the tubers are considered as untapped sources of endophytic biodiversity (Rezaei et al., 2017; Takada et al., 2018). Some preliminary study attempted to report two endophytic plant growth-promoting Rhizobia strains, S-93T and S-62, from *Dioscorea* species (Ouyabe et al., 2019a; Ouyabe et al., 2019b). In view of the background, this chapter gives an elaborate account on the plant growth-promoting traits of bacterial endophytes isolated from *D. bulbifera*.

8.2 Isolation and identification of bacterial endophytes

Bacterial endophytes were isolated from the *D. bulbifera* tubers collected from the Regional Research Institute, Pune, India. The tubers were initially surface sterilized by treating with sterile distilled water, 95% ethanol and 0.1% mercuric chloride solution followed by washing thoroughly with sterile distilled water. The bulbs were then chopped and crushed aseptically using a sterile mortar and pestle followed by dilution in sterile phosphate-buffered saline (PBS) that was further plated on sterile Luria Bertani agar, standard plate count agar (SPCA), and Cystine Lactose Electrolyte Deficient (CLED) agar. All plates were incubated at 30°C for 48 hours. Based on similar colony characteristics, nine bacteria were isolated that were denoted as DBTE (*D. bulbifera* tuber endophytes) 1–9 (Kaware and Kulkarni, 2018).

Extraction of genomic DNA from the endophytic isolates was accomplished using QIAGEN blood and tissue kit (QIAGEN, USA) followed by its quantification on Nanodrop ND1000 (Thermo Scientific, USA). Identification of the isolates was carried out by initial amplification of the 16S rRNA gene using eubacteria universal primer but specific primers viz. 27F (5′-AGA GTT TGA TCM TGG CTC AG-3′) and 1492R (5′-ACG GCTACC TTG TTA CGA CTT-3′) as reported earlier (Ghattargi et al., 2018). Purified amplified polymerase chain reaction (PCR) products were recovered employing polyethylene glycol (PEG)–NaCl precipitation. ABI 3730xl DNA analyzer using the Big Dye terminator kit was used for sequencing (Applied Biosystems, Inc., Foster City, CA). The DNASTARPro, version 10 was used to assemble the obtained sequence.

In order to do phylogenetic analysis, the ChromasPro software was used to edit the generated sequence followed by comparison using the EZBioCloud Server, eventually retrieving the closely related sequences (Kim et al., 2012). Next, MEGA11 was used for further alignment. Phylogenetic trees and bootstrap analysis were performed according to Tamura et al. (2021).

The Maximum Likelihood method and Tamura–Nei model was used for deducing the evolutionary history (Tamura and Nei, 1993). Fig. 8.1 shows the phylogenetic tree associated with the highest log likelihood (−6090.16). Initial tree(s) for the heuristic search were obtained automatically by applying Neighbor-Join and BioNJ algorithms to a matrix of pairwise distances estimated using the Tamura–Nei model, and then selecting the topology with superior log likelihood value (Ghattargi et al., 2018). This analysis involved 42 nucleotide sequences. Codon positions included were first + second + third + noncoding. There were a total of 1387 positions in the final dataset. Evolutionary analyses were conducted in MEGA11 (Tamura et al., 2021). Newly generated sequences were submitted to NCBI GenBank and the accession numbers are listed in Table 8.1.

Table 8.1 Molecular identification of endophytic isolates.

Endophytic isolates	Closest phylogenetic neighbor	GenBank accession number
DBTE 1	*Alkalihalobacillus oshimensis*	MH057382
DBTE 2	*Alkalihalobacillus oshimensis*	MH057383
DBTE 3	*Alkalihalobacillus oshimensis*	MH057384
DBTE 4	*Bacillus nakamurai*	MH057385
DBTE 5	*Priestia flexus*	MH057386
DBTE 6	*Priestia aryabhattai*	MH057387
DBTE 7	*Priestia aryabhattai*	MH057388
DBTE 8	*Staphylococcus cohnii*	MH057389
DBTE 9	*Staphylococcus cohnii*	MH057390

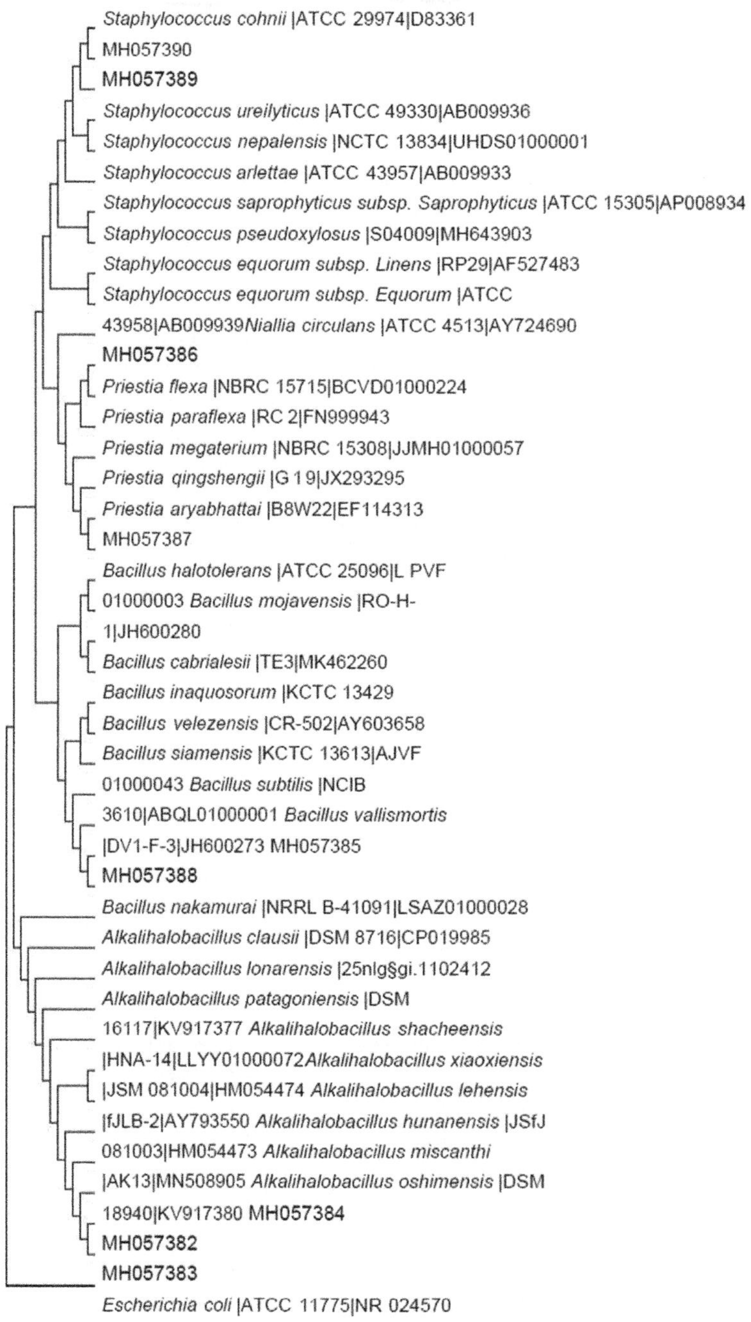

FIGURE 8.1

Phylogenetic relationship between bacterial endophytes isolated from *Dioscorea bulbifera* tubers along with the sequences from selected references strains. The analysis was conducted using the Maximum Likelihood method and Tamura–Nei model.

8.3 Enzyme activity

Endophytes can secrete lytic enzymes such as β-1,3-glucanases, lipases, cellulases, and chitinases to target the cell walls of fungal pathogens that are mostly composed of chitin, cellulase, and glucan (Bhagwat et al., 2019). *Bacillus thuringiensis* UM96 is reported to produce chitinases against gray mold causing pathogen (Martinez-Absalon et al., 2012; Martínez-Absalón et al., 2014). Enzyme activity by bacteria helps degrade complex molecules to simpler forms that can be easily taken up by the host plants resulting in promotion of plant growth and yield. A strain UM96 was reported to interact synergistically with other rhizospheric strains such as *Pseudomonas fluorescens* UM16, UM256, UM240, and UM270 for promotion of growth in corn seedlings (Rojas-Solis et al., 2016). Enzymatic activity of cellulases enables the endophytes to colonize the endosphere of plants (Menendez et al., 2015). Thus cellulolytic properties of endophytes do not cause damage to the plant tissues. Similarly, all the endophytes isolated from *D. bulbifera* tubers were screened for the production of enzymes, such as amylase, cellulase, gelatinase, lipase, chitinase, and pectinase, the results for which are summarized in Table 8.2.

A loopful of 24 hours grown cultures was spot inoculated onto the plates of corresponding media. The spots were allowed to dry followed by incubation at 30°C for 72 hours. The plates were checked every 24 hours to detect the enzyme activity. Bacterial cultures were spot inoculated on plates of nutrient agar supplemented with 1%(w/v) soluble starch followed by incubation to check for amylase production. The plates were the exposed to iodine vapors to enable the proper visualization of clearance zones around the spot. DBTE 2, 6, and 9 showed high amylase activity followed by DBTE 4 and 5 as seen in Table 8.2. Likewise, cellulase activity was determined on carboxymethyl cellulose (CMC) agar. After incubation, the plates were flooded with 1% Congo red solution and kept undisturbed for 15 minutes followed by several washes with 1 M NaCl to remove unbound excess dye. Zone of clearance around the spots of bacterial growth represented cellulase activity. DBTE 3 and 4 exhibited identical activity while DBTE 6–9 failed to produce cellulase. Gelatinase activity of the isolated endophytes was estimated on nutrient agar supplemented with 1.2% gelatine. The plates were flooded with 5 mL of mercuric chloride reagent for checking zone of clearance around the spots indicating gelatinase production. DBTE 2 and 7 exhibited superior gelatinase production. Pectinase activity was checked on media containing 5% pectin. Flooding with 1% cetrimide for 15 minutes resulted in zone of clearance around the spots indicating pectin degradation. DBTE 9 exhibited high pectinase activity followed by DBTE 4 and 8. Lipase activity of the bacterial endophytes was determined on Rhodamine B agar supplemented with 12% olive oil and 2% Rhodamine B. Although all endophytic isolates grew on the media, they failed to show pink zone of hydrolysis around the bacterial growth after incubation confirming no lipase production. The bacteria were spot inoculated on chitinolytic

Table 8.2 Plant growth-promoting rhizobacteria (PGPR) properties of endophytes isolated from *Dioscorea bulbifera* tubers (Kaware and Kulkarni, 2018).

Endophytes	Zone diameter in mm								
	DBTE 1	**DBTE 2**	**DBTE 3**	**DBTE 4**	**DBTE 5**	**DBTE 6**	**DBTE 7**	**DBTE 8**	**DBTE 9**
Amylase	7	19	12	17	15	18	9	12	18
Cellulase	9	13	14	14	16	–	–	–	–
Gelatinase	–	11	–	3	–	1	12	8	6
Pectinase	G	G	–	13	7	7	6	12	28
Lipase	G	G	G	G	G	G	G	G	G
Chitinase	–	3	3	15	34	14	04	–	18
Phosphate solubilization	G	G	G	G	G	G	G	G	G
Zinc solubilization	–	7	7	–	G	6	G	5	G
Nitrogen fixation	G	G	G	G	G	G	G	G	G

Note: *Only growth but no enzyme production (G); no growth and no enzyme production (–).*

activity medium supplemented with 1% colloidal chitin followed by incubation. The plates were then flooded with Gram's iodine solution and checked for zone of clearance around the spots. DBTE 5 exhibited highest chitinase activity followed by DBTE 9 and 4. For checking phosphate solubilizing property of the bacteria, they were spot inoculated on Pikovskya's medium followed by incubation at 30°C for 72 hours. Although all bacteria grew on the medium but they did not show any zone of clearance around the colonies indicating no phosphate solubilization. Likewise, zinc solubilization of the bacteria was checked on Pikovskya's medium with 0.12% zinc oxide. After incubation at 30°C for 72 hours, DBTE 2 and 3 exhibited identical zone of clearance around the colonies indicating positive zinc solubilization. Further, the isolates were checked for nitrogen fixation by confirming their ability to grow on nitrogen-free Ashby's medium. After incubation at 30°C for 72 hours, all isolates grew on the medium confirming their ability to fix atmospheric nitrogen.

8.4 Bioemulsifier production

Various bacterial endophytes are reported to synthesize and secrete bioemulsifiers and biosurfactants that are surface active amphiphilic compounds that can promote emulsification of two immiscible phases effectively (Kharangate-Lad and D'Souza, 2021). Several bacteria and fungi producing bioemulsifier have got wide applications in environmental bioremediation and food industries. It is important to note that the biosurfactants and bioemulsifiers are generally used interchangeably with each other, although they differ in their physicochemical properties. Bioemulsifiers can dissolve in both polar and nonpolar solvents owing to their hydrophilic and hydrophobic structural moieties. It is important to note that biosurfactants produced by Acinetobacter venetianus ATCC 31012 removed 89% of the crude oil by emulsification (Uzoigwe et al., 2015). Likewise, *Candida sphaerica* produced biosurfactants that resulted in bioremediation of iron, zinc, and lead up to 95%, 90%, and 79%, respectively. Bioaccumulation of nickel and copper by *Candida* sp. took place due to biosurfactant production (Luna et al., 2016). It is significant to note that biofilm plays a major role due to their exopolysaccharides (EPS) production that possess surfactant or emulsifying properties (Grujić et al., 2017). Emulsification of hexadecane and xylene was reported for *Vibrio* sp. that was attributed to the reduction of the surface tension between the two immiscible phases (Kharangate-Lad and Bhosle, 2014). Likewise, *Halobacillus trueperi* also emulsified hexadecane (Kharangate-Lad and Bhosle, 2015). Bioemulsifier was secreted by the *Acinetobacter* sp. that remained cell bound. It could emulsify long chain alkanes and aromatics apart from solubilization of crude oils, polyaromatic hydrocarbons (PAHs), and paraffins (Uzoigwe et al., 2015). In view of the background, the bioemulsification potential of the bacterial endophytes isolated from *D. bulbifera* was explored.

Table 8.3 Screening for bioemulsifier production using drop collapse assay (Kaware and Kulkarni, 2018).

Cultures	Time (s)			
	Petrol	Diesel	Kerosene	Engine oil
DBTE 1	11.0	3.3	4.8	5.4
DBTE 2	5.0	4.0	2.88	5.16
DBTE 3	5.1	6.91	2.16	3.1
DBTE 4	7.12	8.42	4.06	5.03
DBTE 5	5.29	5.29	3.73	4.78
DBTE 6	13.12	7.64	3.07	5.68
DBTE 7	9.1	8.3	3.46	5.08
DBTE 8	9.53	4.38	5.88	5.49
DBTE 9	11.15	5.58	3.2	6.4

The bacterial isolates were screened for the bioemulsifier production using drop collapse assay. In brief, overnight grown cultures were inoculated in minimal medium (MM) containing 1% of different oils like engine oil, petrol, diesel, and kerosene and incubated at 30°C for 48 hours under shaking condition. After centrifuging the culture broths at 10,000 rpm for 10 minutes, the supernatant was collected. Then 10 μL of this cell-free supernatant of each culture was added to the layer of oil spread on distilled water. The time taken for bursting of the drop with appearance of an oil-free clear zone indicated the presence of bioemulsifier in the cell-free broth. All of the isolated bacterial endophytes exhibited bioemulsifier production as seen in Table 8.3. The lesser time taken for drop collapse denotes stronger bioemulsification. DBTE 1 and 8 showed strong bioemulsification against diesel oil while DBTE 2–6 and DBTE 9 showed maximum activity against kerosene oil.

The isolates positive for drop collapse assay were further checked for emulsification activity where 3 mL of cell-free supernatant of each cultures was vortexed vigorously with 0.5 mL of petrol, diesel, kerosene, and engine oils separately for 2 minutes followed by incubation at 30°C for 1 hour that resulted in the phase separation. After removal of aqueous phase carefully, its absorbance was recorded at 400 nm for evaluating the emulsification activity per mL (EU/mL) as presented in Table 8.4. DBTE 9 exhibited highest emulsification activity against petrol (2.9 EU/mL) followed by DBTE 2 and DBTE 9 (2.48 and 2.38 EU/mL, respectively) against engine oil.

Various parameters like time, temperature, pH, oil concentration, and salt concentrations were optimized for high bioemulsifier production (Atre et al., 2015). The effect of time (in days) on bioemulsifier production was checked by growing the bacterial endophytes in MM broth at 30°C followed by measuring the bioemulsification activity after every 24 hours till 7 days as depicted in Fig. 8.2.

Table 8.4 Emulsification activity of the endophytes from *Dioscorea bulbifera* (Kaware and Kulkarni, 2018).

Cultures	Emulsification activity (EU/mL)			
	Petrol	**Diesel**	**Kerosene**	**Engine oil**
DBTE 1	0.49	0.38	1.11	1.22
DBTE 2	0.51	0.51	1.93	2.48
DBTE 3	1.59	1.21	0.52	1.58
DBTE 4	1.43	0.7	1.09	1.43
DBTE 5	2.9	1.02	1.73	2.18
DBTE 6	1.99	1.21	0.79	1.25
DBTE 7	1.48	0.78	1.11	1.3
DBTE 8	0.8	0.85	1.0	1.58
DBTE 9	0.78	1.04	1.62	2.38

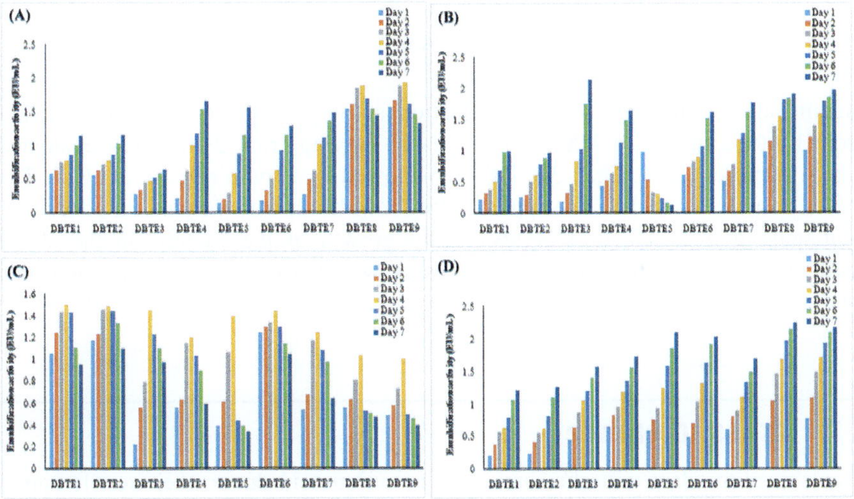

FIGURE 8.2

Time optimization for bioemulsifier production by bacterial endophytes against various oils (Kaware and Kulkarni, 2018).

Maximum activity was seen on Day 4 against petrol and kerosene while for diesel and engine oil all cultures showed maximum activity on Day 7.

The effect of different temperatures on bioemulsifier production was checked by growing the cultures at temperatures varying from 4°C to 60°C which is shown in Fig. 8.3A. Bioemulsifier production was measured after 3 days of incubation. DBTE 4, 5, and 7 showed that 45°C facilitated higher production of the

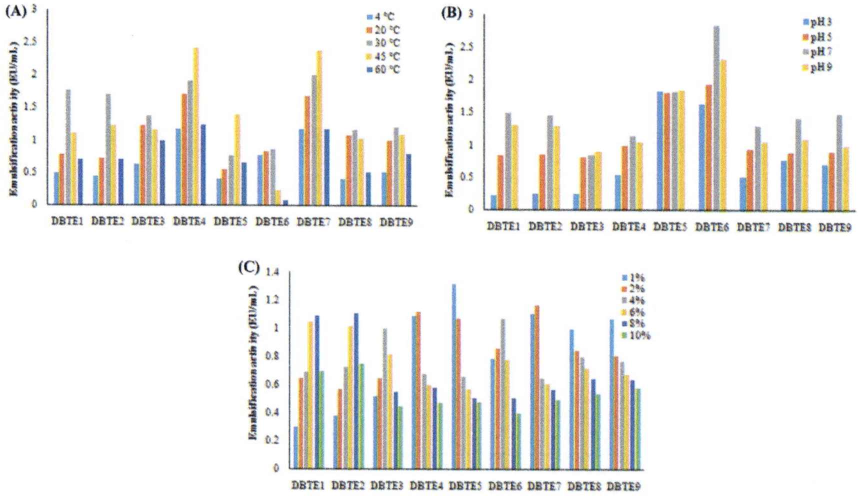

FIGURE 8.3

Optimization of various parameters for bioemulsifier production by bacterial endophytes. (A) Temperature; (B) pH; and (C) oil concentration (Kaware and Kulkarni, 2018).

bioemulsifier. Similarly, the emulsification activity of the cell-free supernatant from bacteria grown in MM broth with pH varying from 3 to 9 was evaluated which is presented in Fig. 8.3B. DBTE 6 showed maximum activity at pH 7. The effect of oil was checked by supplementing MM broth with varying oil concentration ranging from 1% to 10%(v/v). After growth of the bacteria in the above medium for 3 days, the emulsification assay was carried out; the result of which can be seen in Fig. 8.3C. DBTE 5 and 9 exhibited highest activity with 1% oil.

8.5 Stress tolerance

Plants need to adapt with the surrounding environment due to their sessile nature (Vocciante et al., 2022). Hence, endophytes play a significant role in this coevolution where their symbiotic relationship promotes efficient plant survival in the contaminated sites and adverse growth conditions (Gkorezis et al., 2016). This symbiotic system created in response to the abiotic stress is referred as a holobiont (Rohrbacher and St-Arnaud, 2016). The major abiotic stresses that plants undergo include organic and inorganic contaminants, salinity, flood, and drought. Endophytes provide a protective mechanism to plants not only against abiotic stress but also its associated susceptibility to pathogenic attack (Kong and Glick, 2017). Hence, the tolerance to various stresses such as pH, temperature, and salinity was evaluated for the bacterial endophytes of *D. bulbifera* that is presented in Table 8.5. The growth of the bacteria was checked in MM broth that was adjusted

Table 8.5 Abiotic stress tolerance by endophytes (Kaware and Kulkarni, 2018).

Cultures	pH					Temperature (°C)					NaCl concentration (%)					
	3	5	7	9	11	4	20	30	45	60	0.5	0.8	1	2	4	6
DBTE 1	−	−	+	+	−	+	+	+	+	+	+	+	+	+	+	+
DBTE 2	−	−	+	+	−	+	+	+	+	+	+	+	+	+	+	+
DBTE 3	−	+	+	+	−	+	+	+	+	+	+	+	+	+	+	−
DBTE 4	−	+	+	+	−	+	+	+	+	+	+	+	+	+	+	+
DBTE 5	−	+	+	+	+	+	+	+	+	+	+	+	+	+	+	+
DBTE 6	−	−	+	+	−	+	+	+	+	+	+	+	+	+	+	+
DBTE 7	−	+	+	+	−	+	+	+	+	+	+	+	+	+	+	+
DBTE 8	−	+	+	−	−	+	+	+	+	+	+	+	+	+	+	+
DBTE 9	−	+	+	−	−	+	+	+	+	+	+	+	+	+	+	+

to different pH ranging from 3 to 11. An acidic pH 3 inhibited growth of all bacteria while DBTE 1, 2, and 6 were only inhibited at pH 5. All of the bacterial cultures could grow luxuriously at neutral pH. It is interesting to note that DBTE 5 could only grow at pH 11 indicating its alkaliphilic nature. All bacteria were grown at temperatures varying from 4°C to 60°C and interestingly they could grow in all temperatures. This might be attributed to the robust nature of the internal tissue of the *D. bulbifera* that protects the tubers in all environmental conditions. Hence, the bacteria inhabiting the tubers might have adapted to survive and grow at all temperatures.

The growth, development, physiological processes, and cellular metabolism are adversely affected above the normal temperature range due to heat stress (Shaffique et al., 2022). This condition results in morbidity and mortality of plants (Dos Santos et al., 2022; Haider et al., 2022; Perrella et al., 2022). Moreover, irreversible changes such as wilting, leaf damage, fruit drop, blossom end rot, and bolting cause huge agricultural loss (Faizan et al., 2022; Baniwal et al., 2004). Hence, microbial endophytes can significantly combat heat stress serving as biocontrol agents following the mechanism similar to the plant growth-promoting rhizobacteria (PGPR) as illustrated in Fig. 8.4 (Shaffique et al., 2022). Endophytes can induce thermotolerance in plants which is often associated with release of EPS that contain 97% water. This keeps the plant hydrated. Such thermoregulation are often attributed to production of proline and glycine betaine (Anli et al., 2020; Shekhawat et al., 2022). Earlier studies have shown that inoculation of *Septoglomus deserticola* and *S. constrictu* resulted in protection of tomato plant seedlings under extreme heat and drought stress. These microbes not only reduced oxidative stress by decreasing reactive oxygen species (ROS) generation. The symbiotic effect improved cellular performance, stomatal conductance, and leaf water content (Meena et al., 2015). Likewise, the thermoregulatory potential of *Bacillus cereus* SA1 was reported on soybean plants which were attributed to enhancement of chlorophyll a and b, carotenoid, protein, ascorbic acid peroxidase, and superoxide dismutase levels (Bisht et al., 2020).

The salinity stress tolerance was checked by growing the bacteria in MM broth supplemented with varying sodium chloride (NaCl) concentration ranging from 0.5% to 6%(w/v). All cultures showed growth at salt concentrations from 0.5 to 4. However, the growth of DBTE 3 was inhibited at 6% NaCl concentration. Osmotic stress can severely affect the plant health due to reduction in soil water that can be addressed by molecular, cellular, and physiological alterations effectively brought about by the microflora living in symbiotic association with the host plant as seen in Fig. 8.5 (Arora et al., 2020).

Halotolerant endophytes can increase the expression of aquaporins (AQPs) genes that in turn can promote water uptake by the plants even in saline environment and under drought stress (Moshelion et al., 2015). In an earlier study, halotolerant plant growth-promoting *Azospirillum brasilense*, *Pantoea agglomerans*, and *Bacillus megaterium* induced the expression of vital genes such as, PIP2, ZmPIP1−1, and HvPIP2−1 that are involved in AQPs synthesis (Marulanda

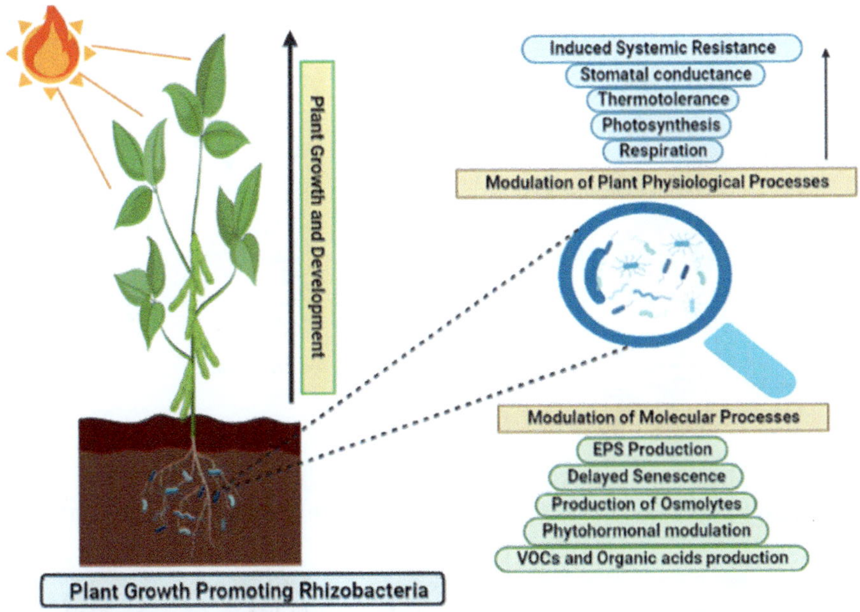

FIGURE 8.4

The illustration represents the role of plant growth-promoting rhizobacteria (PGPR) in mitigating heat stress in plants. PGPRs promote plant growth and development by modulating the physiological and molecular processes in plants. *EPS*, Exopolysaccharides; *VOCs*, volatile organic compounds.

Reprinted from Shaffique, S., Khan, M.A., Wani, S.H., Pande, A., Imran, M., Kang, S.-M., Rahim, W., Khan, S.A., Bhatta, D., Kwon, E.-H., Lee, I.-J., 2022. A review on the role of endophytes and plant growth promoting rhizobacteria in mitigating heat stress in plants. Microorganisms 10, 1286 (open access).

et al., 2010; Gond et al., 2015). These microbes can even upregulate the production of certain key enzymes such as superoxide dismutase (SOD), peroxidase, catalase (CAT), nitrate reductase (NR), glutathione reductase (GR), polyphenol oxidase (PO), guaiacol peroxidase (GP), monohydrate dehydrogenase (MDHAR), and dehydroascorbate reductase (DHAR) that can significantly ameliorate salt stress associated ROS generation and related oxidative stress (Chawla et al., 2013; Islam et al., 2016). Likewise various nonenzymatic antioxidants such as ascorbate (AsA), carotenoids, tocopherols, glutathione (GSH), and phenolics produced by the microbial endophytes can serve as defense molecules for the plant under saline conditions (Sharma et al., 2012). A heat-tolerant endophyte *Pseudomonas stutzeri* ISE12 isolated from halophyte *Salicornia europaea* induced antioxidant system and plant cell wall rearrangement in *Brassica napus* L. (Szymańska et al., 2019). Likewise, *Bradyrhizobium* and *Pseudomonas graminis* resulted in the accumulation of AsA and GSH in cowpea that attributed to its survival under salt stress (Santos et al., 2018). Additionally, hyperosmotic and

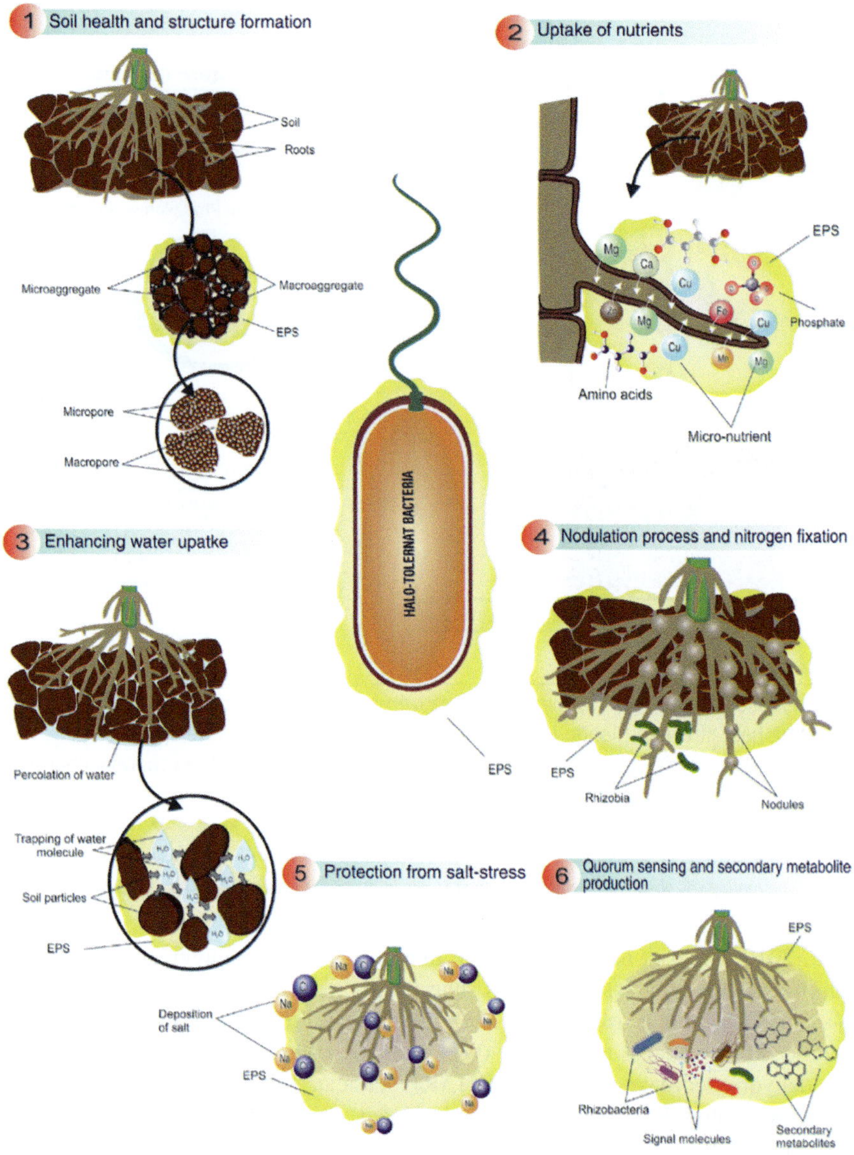

FIGURE 8.5

Benefits of biofilm formation by exopolysaccharides (EPS) producing plant growth-promoting rhizobacteria (PGPR) in improving the nutrient status and structure of saline soil suitable for plant and microbial associations.

Reprinted with permission from Arora, N.K., Fatima, T., Mishra, J., Mishra, I., Verma, S., Verma, R., Verma, M., Bhattacharya, A., Verma, P., Mishra, P., Bharti, C., 2020. Halo-tolerant plant growth promoting rhizobacteria for improving productivity and remediation of saline soils. J. Adv. Res. 26, 69–82. Copyright © 2020 The Authors. Published by Elsevier B.V. on behalf of Cairo University.

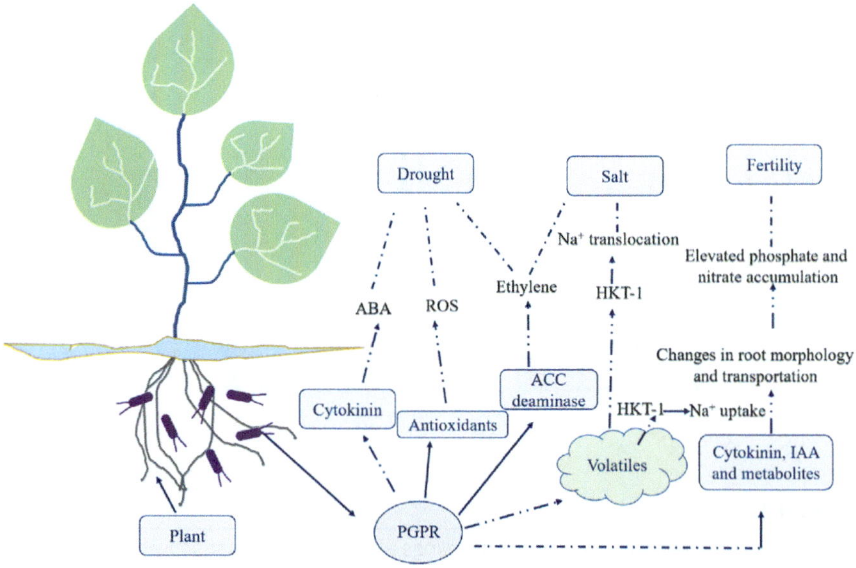

FIGURE 8.6

A diagrammatic representation of plant growth-promoting rhizobacteria (PGPR) role against various stresses. *ABA*, Abscisic acid; *ROS*, reactive oxygen species; *IAA*, indole 2–3-acetic acid; *ACC*, 1-aminocyclopropane-1-carboxylic acid.

Reprinted from Mohanty, P., Singh, P.K., Chakraborty, D., Mishra, S., Pattnaik, R., 2021. Insight into the role of PGPR in sustainable agriculture and environment. Front. Sustain. Food Syst. 5, 667150 (open access).

hyponutritive conditions may result in modulation of phytohormones and accumulation of nutrients by the endophytes that promote the growth and survival of the host plant (Sorty et al., 2016; Khan et al., 2019). Hence, the bacterial endophytes may follow a similar mechanism like PGPR for supporting the host plants to overcome abiotic stress as depicted in Fig. 8.6 (Mohanty et al., 2021).

8.6 Plant growth promotion

The effect of plant growth promotion of the isolated endophytes on the germinated seedlings of *Vigna aconitifolia*, *Vigna radiata*, *Cicer arietinum*, and *Macrotyloma uniflorum* was determined on soft agar. Overnight soaked seeds were placed onto the sterile soft agar and inoculated with 0.1 mL of endophyte culture. Tubes were incubated at room temperature for 12 days and root as well as shoot length was measured as shown in Fig. 8.7. DBTE 1 and 7 showed maximum plant growth promotion in *C. arietinum* where the length of roots was 15.5 and 15 cm, respectively. DBTE 3, 5, and 9 promoted the growth in *V. radiata* where the shoot lengths were 13.5, 16, and 15 cm, respectively.

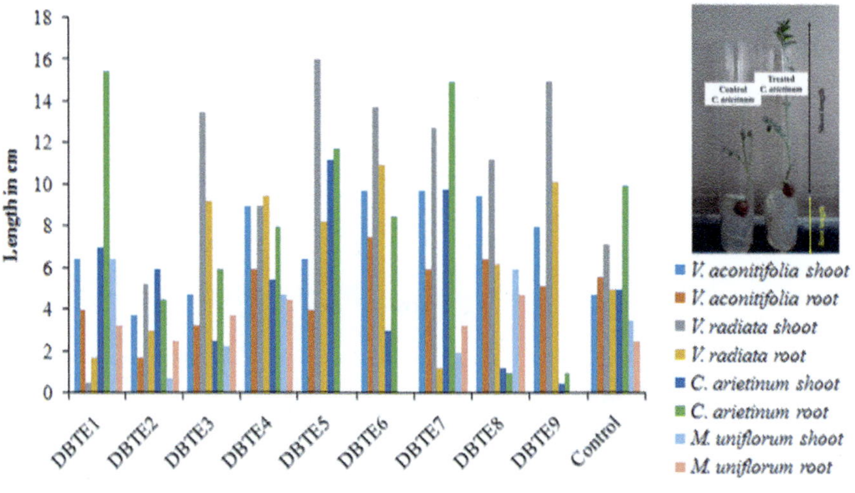

FIGURE 8.7

Plant growth promotion in germinating seedlings after 12 days. Inset showing the representative image of control (untreated) and DBTE 7 treated seedlings of *Cicer arietinum* (Kaware and Kulkarni, 2018).

8.7 Conclusion and future perspectives

Several endophytic bacteria have shown promising plant growth-promoting ability either by direct mechanism or indirectly where they can be used as potential biocontrol agents. Majority of the plant growth-promoting properties are explored in rhizospheric bacteria. Hence, there is dearth of knowledge regarding the exploration of the untapped endophytic diversity which is predominant in the tissue microenvironment of the plants. However, it is challenging to isolate the endophytes as they are critical in their nutritional requirement and abiotic parameters. Hence, growing the endophytes as culturable bacteria needs thorough optimization of the growth media components. Likewise, in order to assure all endophytes are isolated, both fungus- and bacteria-specific medium should be used. Endophytes from *D. bulbifera* can be explored for synthesis of several bioactive compounds that are often isolated from the host plant. Since the host plant is annual in nature, the endophytic flora can be grown and used for bioreactor-based large-scale production of therapeutic compounds without harming the plant.

D. bulbifera is reported to synthesize various metal nanostructures that include silver, gold, copper, platinum, palladium, and bimetallic nanoparticles (Ghosh et al., 2011, 2012b, 2015b,c,d). These nanoparticles have antimicrobial, anticancer, antioxidant, and antidiabetic activities (Ghosh et al., 2015e, 2016a,b; Shende et al., 2017; Ghosh, 2018). Hence, it would be interesting to check if the endophytes isolated from *D. bulbifera* can synthesize metal and metal oxide nanoparticles. Unlike nanoparticles synthesized by physical and chemical approaches, endophyte-mediated synthesis would be environmentally benign route as it would restrict the use of hazardous reducing and stabilizing agents (Shende et al., 2018; Bhagwat et al., 2018; Bloch et al., 2021).

Being more biocompatible, such microbially synthesized nanoparticles can be used as drug carriers for sustained release and targeted delivery to the site of disease (Ghosh et al., 2015f, 2018; Ghosh, 2019).

In conclusion, endophytes isolated from *D. bulbifera* can open a new area that can revolutionize drug development, nanomedicine, and bioprospecting. Integrative approach using genomics, proteomics, and metabolomics will help to identify and characterize the numerous endophytic genes that are dispersed throughout the chromosome that would certainly help in genetic engineering in order to develop recombinant microbes with superior plant growth-promoting traits.

Acknowledgment

Dr. Sougata Ghosh acknowledges Kasetsart University, Bangkok, Thailand for Post Doctoral Fellowship and funding under Reinventing University Program (Ref. No. 6501.0207/9219 dated September 14, 2022). Dr. Ghosh is also thankful to The Program Management Unit for Human Resources & Institutional Development and Innovation (PMU-B) for funding the proposal entitled "Developing High Performance workforce in Postdoc and Postmaster under Research Topic "Study and Development of Porous Nanofibers, Graphene Oxide Quantum Dot Nanofiber Composite, and Metal Doped Graphene Oxide Quantum Dot Nanofiber Composite for Carbon Dioxide Reduction and Capture" under the Program of National Postdoctoral and Postgraduate System approved by PMU-B Board Committees (Contract No. B13F660065).

References

Ali, S., Charles, T.C., Glick, B.R., 2012. Delay of flower senescence by bacterial endophytes expressing 1-aminocyclopropane-1-carboxylate deaminase. J. Appl. Microbiol. 113, 1139–1144.

Anli, M., Baslam, M., Tahiri, A., Raklami, A., Symanczik, S., Boutasknit, A., et al., 2020. Biofertilizers as strategies to improve photosynthetic apparatus, growth, and drought stress tolerance in the date palm. Front. Plant Sci. 11, 516818.

Arora, N.K., Fatima, T., Mishra, J., Mishra, I., Verma, S., Verma, R., et al., 2020. Halotolerant plant growth promoting rhizobacteria for improving productivity and remediation of saline soils. J. Adv. Res. 26, 69–82.

Asiedu, R., Sartie, A., 2010. Crops that feed the world 1. Yams. Food Secur. 2, 305–315.

Atre, C., Bhandari, T., Kothawade, M., Mujumdar, M., 2015. Isolation, characterization and screening of bioemulsifier producing micro-organisms from soil sample (Kaas Pathar, Satara, India) and its application. Int. J. Curr. Microbiol. App. Sci. 2, 65–81.

Baniwal, S.K., Bharti, K., Chan, K.Y., Fauth, M., Ganguli, A., Kotak, S., et al., 2004. Heat stress response in plants: a complex game with chaperones and more than twenty heat stress transcription factors. J. Biosci. 29, 471–487.

Bertalan, M., Albano, R., de Pádua, V., Rouws, L., Rojas, C., Hemerly, A., et al., 2009. Complete genome sequence of the sugarcane nitrogen-fixing endophyte *Gluconacetobacter diazotrophicus* Pal5. BMC Genomics 10, 450.

Bhagwat, T.R., Joshi, K.A., Parihar, V.S., Asok, A., Bellare, J., Ghosh, S., 2018. Biogenic copper nanoparticles from medicinal plants as novel antidiabetic nanomedicine. World J. Pharm. Res. 7 (4), 183–196.

Bhagwat, A., Collins, C.H., Dordick, J.S., 2019. Selective antimicrobial activity of cell lytic enzymes in a bacterial consortium. Appl. Microbiol. Biotechnol. 103, 7041–7054.

Bhattacharya, J., Nitnavare, R., Shankhapal, A., Ghosh, S., 2021. Microbially synthesized nanoparticles: aspect in plant disease management. In: Radhakrishnan, E.K., Kumar, A., Raveendran, A. (Eds.), Biocontrol Mechanisms of Endophytic Microorganisms. Elsevier, USA, pp. 303–325.

Bisht, N., Mishra, S.K., Chauhan, P.S., 2020. *Bacillus amyloliquefaciens* inoculation alters physiology of rice (*Oryza sativa* L. var. IR-36) through modulating carbohydrate metabolism to mitigate stress induced by nutrient starvation. Int. J. Biol. Macromol. 143, 937–951.

Bloch, K., Pardesi, K., Satriano, C., Ghosh, S., 2021. Bacteriogenic platinum nanoparticles for application in nanomedicine. Front. Chem. 9, 624344.

Chawla, S., Jain, S., Jain, V., 2013. Salinity induced oxidative stress and antioxidant system in salt-tolerant and salt-sensitive cultivars of rice (*Oryza sativa* L.). J. Plant Biochem. Biotechnol. 22, 27–34.

Coutinho, B.G., Licastro, D., Mendonc, a-Previato, L., Cámara, M., Venturi, V., 2015. Plant-influenced gene expression in the rice endophyte *Burkholderia kururiensis* M130. Mol. Plant Microbe Interact. 28, 10–21.

Diby, L.N.G., Hgaza, V.K., Tieb, T.B., Assa, A., Carsky, R., Girardin, O., et al., 2009. Productivity of yams (*Dioscorea* spp.) as affected by soil fertility. J. Anim. Plant Sci. 5, 494–506.

Dos Santos, T.B., Ribas, A.F., de Souza, S.G.H., Budzinski, I.G.F., Domingues, D.S., 2022. Physiological responses to drought, salinity, and heat stress in plants: a review. Stresses 2, 113–135.

Ettien, D.J.B., Kouadio, K.H., N'goran, K.E., Yao-Kouamé, A., Girardin, O., 2014. Improving the performance of a traditional variety of yam produced under ferralsol poor in organic matter in the forest areas of Côte d'Ivoire. Int. J. Adv. Appl. Res. 4, 76–84.

Faizan, M., Yu, F., Rajput, V.D., Minkina, T., Hayat, S., 2022. Role of brassinosteroids in protein folding under high-temperature stress. In: Khan, M.T.A., Yusuf, M., Qazi, F., Ahmad, A. (Eds.), Brassinosteroids Signalling. Springer, Singapore, pp. 259–268.

Fouts, D.E., Tyler, H.L., DeBoy, R.T., Daugherty, S., Ren, Q., Badger, J.H., et al., 2008. Complete genome sequence of the N2-fixing broad host range endophyte *Klebsiella pneumoniae* 342 and virulence predictions verified in mice. PLoS Genet. 4, e1000141.

Ghattargi, V.C., Nimonkar, Y.S., Burse, S.A., Davray, D., Kumbhare, S.V., Shetty, S.A., et al., 2018. Genomic and physiological analyses of an indigenous strain, *Enterococcus faecium* 17OM39. Funct. Integr. Genomics 18 (4), 385–399.

Ghosh, S., 2018. Copper and palladium nanostructures: a bacteriogenic approach. Appl. Microbiol. Biotechnol. 101 (18), 7693–7701.

Ghosh, S., 2019. Mesoporous silica based nano drug delivery system synthesis, characterization and applications. In: Mohapatra, S.S., Ranjan, S., Dasgupta, N., Mishra, R.K., Thomas, S. (Eds.), Nanocarriers for Drug Delivery. Elsevier Inc., Amsterdam, The Netherlands, pp. 285–317.

Ghosh, S., Das, S., 2022. Impact of climate change on microbial endophytes: novel nanoscale cell factories. In: Kumar, A., Fernando, L., Ferreira, R., Panwar, J.S. (Eds.), Microbiome Under Changing Climate: Implications and Solutions. Elsevier, USA, pp. 161–185.

Ghosh, S., Patil, S., Ahire, M., Kitture, R., Jabgunde, A., Kale, S., et al., 2011. Synthesis of gold nano-anisotrops using *Dioscorea bulbifera* tuber extract. J. Nanomater. 2011, 354793.

Ghosh, S., Derle, A., Ahire, M., More, P., Jagtap, S., Phadatare, S.D., et al., 2013. Phytochemical analysis and free radical scavenging activity of medicinal plants *Gnidia glauca* and *Dioscorea bulbifera*. PLoS One 8, e82529.

Ghosh, S., Ahire, M., Patil, S., Jabgunde, A., Bhat Dusane, M., Joshi, B.N., et al., 2012a. Antidiabetic activity of *Gnidia glauca* and *Dioscorea bulbifera*: potent amylase and glucosidase inhibitors. Evid. Based Complement. Alternat. Med. 2012, 929051.

Ghosh, S., Patil, S., Ahire, M., Kitture, R., Jabgunde, A., Kale, S., et al., 2012b. Synthesis of silver nanoparticles using *Dioscorea bulbifera* tuber extract and evaluation of its synergistic potential in combination with antimicrobial agents. Int. J. Nanomed. 7, 483–496.

Ghosh, S., More, P., Derle, A., Patil, A.B., Markad, P., Asok, A., et al., 2014. Diosgenin from *Dioscorea bulbifera*: novel hit for treatment of Type II *Diabetes mellitus* with inhibitory activity against α-amylase and α-glucosidase. PLoS One 9, e106039.

Ghosh, S., Parihar, V.S., More, P., Dhavale, D.D., Chopade, B.A., 2015a. Phytochemistry and therapeutic potential of medicinal plant: *Dioscorea bulbifera*. Med. Chem. 5, 160–172.

Ghosh, S., More, P., Nitnavare, R., Jagtap, S., Chippalkatti, R., Derle, A., et al., 2015b. Antidiabetic and antioxidant properties of copper nanoparticles synthesized by medicinal plant *Dioscorea bulbifera*. J. Nanomed. Nanotechnol. S6, 007.

Ghosh, S., Nitnavare, R., Dewle, A., Tomar, G.B., Chippalkatti, R., More, P., et al., 2015c. Novel platinum-palladium bimetallic nanoparticles synthesized by *Dioscorea bulbifera*: anticancer and antioxidant activities. Int. J. Nanomed. 10, 7477–7490.

Ghosh, S., Jagtap, S., More, P., Shete, U.J., Maheshwari, N.O., Rao, S.J., et al., 2015d. *Dioscorea bulbifera* mediated synthesis of novel $Au_{core}Ag_{shell}$ nanoparticles with potent antibiofilm and antileishmanial activity. J. Nanomater. 2015, 562938.

Ghosh, S., Parihar, V.S., Dhavale, D.D., Chopade, B.A., 2015e. Commentary on therapeutic potential of *Gnidia glauca*: a novel medicinal plant. Med. Chem. 5 (8), 351–353.

Ghosh, S., More, P., Derle, A., Kitture, R., Kale, T., Gorain, M., et al., 2015f. Diosgenin functionalized iron oxide nanoparticles as novel nanomaterial against breast cancer. J. Nanosci. Nanotechnol. 15, 9464–9472.

Ghosh, S., Chacko, M.J., Harke, A.N., Gurav, S.P., Joshi, K.A., Dhepe, A., et al., 2016a. *Barleria prionitis* leaf mediated synthesis of silver and gold nanocatalysts. J. Nanomed. Nanotechnol. 7, 4.

Ghosh, S., Gurav, S.P., Harke, A.N., Chacko, M.J., Joshi, K.A., Dhepe, A., et al., 2016b. *Dioscorea oppositifolia* mediated synthesis of gold and silver nanoparticles with catalytic activity. J. Nanomed. Nanotechnol. 7, 5.

Ghosh, S., Sanghavi, S., Sancheti, P., 2018. Metallic biomaterial for bone support and replacement. In: Balakrishnan, P., Sreekala, M.S., Thomas, S. (Eds.), Fundamental Biomaterials: Metals. Vol. 2. Woodhead Publishing Series in Biomaterials. Woodhead Publishing, pp. 139–165.

Ghosh, S., Bhagwat, T., Webster, T.J., 2021. Endophytic microbiomes and their plant growth promoting attributes for plant health. In: Yadav, A.N., Singh, J., Singh, C., Yadav, N. (Eds.), Current Trends in Microbial Biotechnology for Sustainable Agriculture, Environmental and Microbial Biotechnology. Springer, Singapore, pp. 245–278.

Gkorezis, P., Daghio, M., Franzetti, A., Van Hamme, J.D., Sillen, W., Vangronsveld, J., 2016. The interaction between plants and bacteria in the remediation of petroleum hydrocarbons: an environmental perspective. Front. Microbiol. 7, 1836.

Gond, S.K., Bergen, M.S., Torres, M.S., White Jr, J.F., 2015. Endophytic *Bacillus* spp. produce antifungal lipopeptides and induce host defence gene expression in maize. Microbiol. Res. 172, 79–87.

Grujić, S., Vasić, S., Radojević, I., Čomić, L., Ostojić, A., 2017. Comparison of the *Rhodotorula mucilaginosa* biofilm and planktonic culture on heavy metal susceptibility and removal potential. Water Air Soil Pollut. 228, 73.

Haas, D., Keel, C., 2003. Regulation of antibiotic production in root-colonizing *Pseudomonas* spp. and relevance for biological control of plant disease. Ann. Rev. Phytopathol. 41, 117–153.

Haider, S., Raza, A., Iqbal, J., Shaukat, M., Mahmood, T., 2022. Analyzing the regulatory role of heat shock transcription factors in plant heat stress tolerance: a brief appraisal. Mol. Biol. Rep. 49, 5771–5785.

Islam, F., Ali, B., Wang, J., Farooq, M.A., Gill, R.A., Ali, S., et al., 2016. Combined herbicide and saline stress differentially modulates hormonal regulation and antioxidant defense system in *Oryza sativa* cultivars. Plant. Physiol. Biochem. 107, 82–95.

Kaware, K.E., Kulkarni, S.P., 2018. Bioprospecting of bacterial endophytes from tuber of *Dioscorea bulbifera* (Unpublished Master's thesis). Savitribai Phule Pune University, Pune, India.

Khan, M.A., Asaf, S., Khan, A.L., Ullah, I., Ali, S., Kang, S.M., et al., 2019. Alleviation of salt stress response in soybean plants with the endophytic bacterial isolate *Curtobacterium* sp. SAK1. Ann. Microbiol. 69, 797–808.

Kharangate-Lad, A., Bhosle, S., 2014. Studies on bioemulsifier production by a bacterial isolate. J. Sci. Ind. Res. 73, 674–679.

Kharangate-Lad, A., Bhosle, S., 2015. Studies on the effects of zerovalent iron nanoparticles on bacteria from the mangrove ecosystem. Environ. Sci. Pollut. Res. 23, 927–931.

Kharangate-Lad, A., D'Souza, N.C., 2021. Current approaches in bioremediation of toxic contaminants by application of microbial cells; biosurfactants and bioemulsifiers of microbial origin. In: Kumar, V., Prasad, R., Kumar, M. (Eds.), Rhizobiont in Bioremediation of Hazardous Waste. Springer, Singapore, pp. 217–263.

Kim, O.S., Cho, Y.J., Lee, K., Yoon, S.H., Kim, M., Na, H., et al., 2012. Introducing EzTaxon-e: a prokaryotic 16S rRNA gene sequence database with phylotypes that represent uncultured species. Int. J. Syst. Evol. Microbiol. 62 (Pt 3), 716–721.

Kong, Z., Glick, B.R., 2017. The role of plant growth-promoting bacteria in metal phytoremediation. Adv. Microb. Physiol. 71, 97–132.

Krause, A., Ramakumar, A., Bartels, D., Battistoni, F., Bekel, T., Boch, J., et al., 2006. Complete genome of the mutualistic, N2-fixing grass endophyte *Azoarcus* sp. strain BH72. Nat. Biotechnol. 24, 1385–1391.

Luna, J.M., Santos Filho, A.S., Rufino, R.D., Sarubbo, L.A., 2016. Production of biosurfactant from *Candida bombicola* URM 3718 for environmental applications. Chem. Eng. Trans. 49, 583–588.

Luo, S., Xu, T., Chen, L., Chen, J., Rao, C., Xiao, X., et al., 2012. Endophyte-assisted promotion of biomass production and metal-uptake of energy crop sweet sorghum by plant-growth-promoting endophyte *Bacillus* sp. Appl. Microbiol. Biotechnol. 93, 1745–1753.

Marquez-Santacruz, H.A., Hernandez-Leon, R., Orozco-Mosqueda, M.C., Velazquez-Sepulveda, I., Santoyo, G., 2010. Diversity of bacterial endophytes in roots of Mexican husk tomato plants (*Physalis ixocarpa*) and their detection in the rhizosphere. Gen. Mol. Res. 9, 2372–2380.

Martinez-Absalon, S.C., Orozco-Mosqueda, M.D.C., Martinez-Pacheco, M.M., Farias-Rodriguez, R., Govindappa, M., Santoyo, G., 2012. Isolation and molecular characterization of a novel strain of *Bacillus* with antifungal activity from the sorghum rhizosphere. Genet. Mol. Res. 11, 2665–2673.

Martínez-Absalón, S., Rojas-Solís, D., Hernández-León, R., Prieto-Barajas, C., Orozco-Mosqueda, M.D.C., Peña-Cabriales, J.J., et al., 2014. Potential use and mode of action of the new strain *Bacillus thuringiensis* UM96 for the biological control of the grey mould phytopathogen *Botrytis cinerea*. Biocontrol Sci. Technol. 24, 1349–1362.

Marulanda, A., Azcón, R., Chaumont, F., Ruiz-Lozano, J.M., Aroca, R., 2010. Regulation of plasma membrane aquaporins by inoculation with a *Bacillus megaterium* strain in maize (*Zea mays* L.) plants under unstressed and salt-stressed conditions. Planta 232, 533–543.

Meena, H., Ahmed, M.A., Prakash, P., 2015. Amelioration of heat stress in wheat, *Triticum aestivum* by PGPR (*Pseudomonas aeruginosa* strain 2CpS1). Biosci. Biotechnol. Res. 8, 171–174.

Menendez, E., Garcia-Fraile, P., Rivas, R., 2015. Biotechnological applications of bacterial cellulases. AIMS Bioeng. 2, 163–182.

Mohanty, P., Singh, P.K., Chakraborty, D., Mishra, S., Pattnaik, R., 2021. Insight into the role of PGPR in sustainable agriculture and environment. Front. Sustain. Food Syst. 5, 667150.

Moshelion, M., Halperin, O., Wallach, R., Oren, R.A.M., Way, D.A., 2015. Role of aquaporins in determining transpiration and photosynthesis in water stressed plants: crop water use efficiency, growth and yield. Plant Cell Environ. 38, 1785–1793.

Ouyabe, M., Kikuno, H., Tanaka, N., Babil, P., Shiwachi, H., 2019a. Nitrogen-fixing endophytic bacteria of water yam (*Dioscorea alata* L.) in relation with fertilization practices. Trop. Agric. Dev. 63, 122–130.

Ouyabe, M., Kikuno, H., Tanaka, N., Babil, P., Shiwachi, H., 2019b. Contribution of biological nitrogen fixation in lesser yam (*Dioscorea esculenta* L.) associated with endophytic diazotrophic bacteria. Trop. Agric. Dev. 63, 131–139.

Pandya, M., Rajput, M., Rajkumar, S., 2015. Exploring plant growth promoting potential of non rhizobial root nodules endophytes of *Vigna radiata*. Microbiology 84, 80–89.

Perrella, G., Bäurle, I., van Zanten, M., 2022. Epigenetic regulation of thermomorphogenesis and heat stress tolerance. New Phytol. 234, 1144–1160.

Rashid, S., Charles, T.C., Glick, B.R., 2012. Isolation and characterization of new plant growth-promoting bacterial endophytes. Appl. Soil. Ecol. 61, 217–224.

Rezaei, A.Q., Kikuno, H., Babil, P., Tanaka, N., Park, B.J., Onjo, M., et al., 2017. Nitrogen-fixing bacteria is involved with the lesser yam (*Dioscorea esculenta* L.) growth under low fertile soil condition. Trop. Agric. Dev. 61, 40–47.

Rohrbacher, F., St-Arnaud, M., 2016. Root exudation: the ecological driver of hydrocarbon rhizoremediation. Agronomy 6, 19.

Rojas-Solis, D., Hernandez-Pacheco, C.E., Santoyo, G., 2016. Evaluation of *Bacillus* and *Pseudomonas* to colonize the rhizosphere and their effect on growth promotion in tomato (*Physalis ixocarpa* Brot. ex Horm.). Rev. Chapingo Ser. Hortic. 22, 45–57.

Saini, R., Dudeja, S.S., Giri, R., Kumar, V., 2015. Isolation, characterization, and evaluation of bacterial root and nodule endophytes from chickpea cultivated in Northern India. J. Basic Microbiol. 55, 74–81.

Santos, A.D.A., Silveira, J.A.G.D., Bonifacio, A., Rodrigues, A.C., Figueiredo, M.D., Santos, V.B., 2018. Antioxidant response of cowpea co-inoculated with plant growth promoting bacteria under salt stress. Braz. J. Microbiol. 49, 513–521.

Santoyo, G., Moreno-Hagelsieb, G., Orozco-Mosqueda, M.C., Glick, B.R., 2016. Plant growth-promoting bacterial endophytes. Microbiol. Res. 183, 92–99.

Shaffique, S., Khan, M.A., Wani, S.H., Pande, A., Imran, M., Kang, S.-M., et al., 2022. A review on the role of endophytes and plant growth promoting rhizobacteria in mitigating heat stress in plants. Microorganisms 10, 1286.

Sharma, P., Jha, A.B., Dubey, R.S., Pessarakli, M., 2012. Reactive oxygen species, oxidative damage, and antioxidative defense mechanism in plants under stressful conditions. J. Bot. 184, 57–67.

Shekhawat, K., Almeida-Trapp, M., García-Ramírez, G.X., Hirt, H., 2022. Beat the heat: plant- and microbe-mediated strategies for crop thermotolerance. Trends Plant Sci. 27, 802–813.

Shende, S., Joshi, K.A., Kulkarni, A.S., Shinde, V.S., Parihar, V.S., Kitture, R., et al., 2017. *Litchi chinensis* peel: a novel source for synthesis of gold and silver nanocatalysts. Glob. J. Nanomed. 3 (1), 555603.

Shende, S., Joshi, K.A., Kulkarni, A.S., Charolkar, C., Shinde, V.S., Parihar, V.S., et al., 2018. *Platanus orientalis* leaf mediated rapid synthesis of catalytic gold and silver nanoparticles. J. Nanomed. Nanotechnol. 9, 2.

Sorty, A.M., Meena, K.K., Choudhary, K., Bitla, U.M., Minhas, P.S., Krishnani, K.K., 2016. Effect of plant growth promoting bacteria associated with halophytic weed (*Psoralea corylifolia* L) on germination and seedling growth of wheat under saline conditions. Appl. Biochem. 180, 872–882.

Szymańska, S., Dabrowska, G.B., Tyburski, J., Niedojadło, K., Piernik, A., Hrynkiewicz, K., 2019. Boosting the *Brassica napus* L. tolerance to salinity by the halotolerant strain *Pseudomonas stutzeri* ISE12. Environ. Exp. Bot. 163, 55–68.

Taghavi, S., Garafola, C., Monchy, S., Newman, L., Hoffman, A., Weyens, N., et al., 2009. Genome survey and characterization of endophytic bacteria exhibiting a beneficial effect on growth and development of poplar trees. Appl. Environ. Microbiol. 75, 748–757.

Takada, K., Kikuno, H., Babil, P., Shiwachi, H., 2018. Analysis of the source of nitrogen during water yam (*Dioscorea alata* L.) growth using $\delta^{15}N$ observations. Trop. Agric. Dev. 62, 124–131.

Tamura, K., Nei, M., 1993. Estimation of the number of nucleotide substitutions in the control region of mitochondrial DNA in humans and chimpanzees. Mol. Biol. Evol. 10 (3), 512–526.

Tamura, K., Stecher, G., Kumar, S., 2021. MEGA11: molecular evolutionary genetics analysis version 11. Mol. Biol. Evol. 38 (7), 3022–3027.

Uzoigwe, C., Burgess, J.G., Ennis, C.J., Rahman, P.K., 2015. Bioemulsifiers are not biosurfactants and require different screening approaches. Front. Microbiol. 6, 245–251.

Vocciante, M., Grifoni, M., Fusini, D., Petruzzelli, G., Franchi, E., 2022. The role of plant growth-promoting rhizobacteria (PGPR) in mitigating plant's environmental stresses. Appl. Sci. 12, 1231.

Weilharter, A., Mitter, B., Shin, M.V., Chain, P.S., Nowak, J., Sessitsch, A., 2011. Complete genome sequence of the plant growth-promoting endophyte *Burkholderia phytofirmans* strain PsJN. J. Bacteriol. 193, 3383–3384.

Wisniewski-Dyé, F., Borziak, K., Khalsa-Moyers, G., Alexandre, G., Sukharnikov, L.O., Wuichet, K., et al., 2011. *Azospirillum* genomes reveal transition of bacteria from aquatic to terrestrial environments. PLoS Genet. 7, e1002430.

Yan, Y., Yang, J., Dou, Y., Chen, M., Ping, S., Peng, J., et al., 2008. Nitrogen fixation island and rhizosphere competence traits in the genome of root-associated *Pseudomonas stutzeri* A1501. Proc. Natl. Acad. Sci. USA 105, 7564–7569.

CHAPTER

9 Green synthesis of nanomaterials and their applications in sustainable agriculture

Beatriz Montaño-Leyva[1], Jonathan M. Sanchez-Silva[2], Luis G. Hernández-Montiel[3], Paloma P. Casas-Junco[4], Tomás Rivas-García[5], Jesús E. Reyna-Ochoa[6], Francisco J. Blancas-Benitez[6], Cristina Moreno-Hernández[6], Ulises M. López-García[6], Ana Y. Flores-Ramírez[6] and Ramsés R. González-Estrada[6]

[1]*Department of Research and Postgraduate in Food, University of Sonora, Hermosillo, Sonora, Mexico*

[2]*Postgraduate Study and Research Center, Faculty of Chemical Sciences, Autonomous University of San Luis Potosí, San Luis Potosí, Mexico*

[3]*Agriculture Program in Arid Zones, Northwest Biological Research Center, La Paz, Baja California Sur, Mexico*

[4]*Food Technology Unit, Research and Postgraduate Secretariat, Autonomous University of Nayarit, Tepic, Nayarit, Mexico*

[5]*National Council of Humanities, Sciences, and Technologies (CONAHCYT), Chapingo Autonomous University, Texcoco, Estado de Mexico, Mexico*

[6]*Division of Research and Postgraduate Studies, National Technological Institute of Mexico/ Technological Institute of Tepic, Tepic, Nayarit, Mexico*

9.1 Introduction

Nowadays, the food waste is a problem worldwide in the food industry caused by several factors. According to the Food and Agriculture Organization of the United Nations (FAO), more than 1.3 billion metric tons per year of food is wasted throughout the supply chain due to problems in postharvest techniques, during the storage, transport facilities, market, and even in consumer's home (FAO, 2019). The use and application of nanomaterials can impact on the food industry positively due to their ability to protect the food products (Nile et al., 2020). Nanostructured material has enormous potential to provide innovative improvements to a food product to meet the demands of today's consumers with positive effects on the supply and bioavailability of nutrients (satiety, weight control, pleasant sensations from eating, health, among others) (Yildirim et al., 2018; Ojeda et al., 2019). The United States National Nanotechnology Initiative (NNI) defined nanomaterials (NMs) as materials with at least one dimension in the scale

Sustainable Agricultural Practices. DOI: https://doi.org/10.1016/B978-0-443-19150-3.00009-6

of 1–100 nm. They can be classified into three classes of food materials: (1) organic NMs (lipids, proteins, and nutritional attributes); (2) inorganic NMs (metal and metal oxide); (3) combined organic and inorganic (antimicrobials, antioxidants, permeability, and rigidity regulators) (Peters et al., 2017). These nanoparticles are designed and synthesized to have composition and physicochemical properties that lead to specific desire functional attributes for customer products such as food packaging, additives, and food preservation (Prajitha et al., 2019; Dudefoi et al., 2017). At present, there are numerous possibilities for their future application in the food industry (Yong-Jin et al., 2008). Interestingly, NMs can be used not only as individual treatments but also combined with other substances. The objective of this chapter was to summarize recent information about the techniques, properties of NMs as well as their diverse applications for fruit and vegetables production.

9.2 Impact of fruit and vegetable consumption

In recent years, consumers demand food with clean labels, that is, the lifestyle of the consumer drove the importance for climate change, welfare, animal welfare, and the production systems used in the development of new foods. Consumer demands are looking for safety foods with greater benefits to health. The changes in the awareness of the relationship between food and health create a market with health-enhancing properties. At the same time, people are changing their dietary habits and introducing options such as vegan, vegetarian, organic or flexitarian with the use of natural origin ingredients more than synthetic (Rosenthal et al., 2021).

Numerous epidemiological studies have shown an inverse association between fruit and vegetables consumption and chronic diseases including different types of cancer, cardiovascular, and neurodegenerative diseases (Gan et al., 2015; Boeing et al., 2012; Yahia et al., 2011). These studies have shown mounting evidence that people who avoid fruit and vegetables completely, or consume very little, are indeed at increased risk of these diseases. Therefore interest in the health benefits of fruit and vegetable consumption is increasing. Moreover, interest in understanding the type, number, and mode of action of the different components in fruits and vegetables that confer health benefits is also increasing. Fruits and vegetables have historically been considered rich sources of some essential dietary micronutrients and of fibers, and more recently they have been recognized as important sources for a wide array of phytochemicals that individually, or in combination, may benefit health (Ruiz Rodríguez et al., 2021; Fernández-Ochoa et al., 2022). Naturally occurring compounds such as phytochemicals, which possess anticarcinogenic and other beneficial properties, are referred to as chemo preventive agents, being classified as blocking and suppressive agents. The blocking agents are based on their antioxidant activity and the capacity to scavenge free radicals. Resolution of the potential protective roles of specific antioxidants and other constituents of fruits and vegetables deserves major attention. Evidence indicates that for the effect of fruit and vegetables consumption on health, the whole may be more than the sum of the parts. Individual components appear to act synergistically, in that the influence of at least some of them is additive.

The global average for vegetables (based on availability and not including vegetable oils) and fruits consumption is 2.6% and 2.7% of total daily energy intake, respectively. Thus it is argued that increasing intake from 400 to 800 g/day of fruits and vegetables is a public health strategy of considerable importance for individuals and communities worldwide. Vegetable consumption is highest in North Africa, the Middle East, parts of Asia, the United States, Cuba, and southern Europe. On the other hand, fruit intakes are highest in some parts of Africa, the Middle East, southern Europe, and Oceania, and lowest in other parts of Africa and Asia (Marmot et al., 2007). The World Health Organization (WHO) recommends a daily intake of more than 400 g per person daily, and health authorities worldwide promote high consumption of fruits and vegetables (Yahia, 2017). Many of the putative chemoprotective phytochemicals in fruits and vegetables are colored (due to different pigments). The guidelines are based on selecting one serving daily of fruits and vegetables from each of seven color classes (red, yellow–green, red–purple, orange, orange–yellow, green, white–green) so that a variety of phytochemicals is consumed.

9.3 Impact of agrochemicals' use on food production

Agrochemicals are a group of chemicals which are classified as pesticides, chemical fertilizers, growth-promoters, soil stimulants, feed additives, and veterinary drugs by their agronomic purposes (Gill et al., 2022). During the last century, agrochemicals play an essential role in agricultural production systems worldwide by enhancing food production, maintaining soil and plant nutrition, and plant health (Meena et al., 2020a). Since the beginning of the green revolution, agrochemicals were used to increase food production for an increasing global demand (Tilman et al., 2011). Nonetheless, the expected world population by 2050 will be of 9.3 billion which will lead to an increasing demand of food, and hence, an increasing use of agrochemicals (Boedeker et al., 2020). The growing population will lead to an increase of agrochemical's demand, however, their indiscriminate use has led to environmental and health problems (Dhananjayan et al., 2020). Among all agrochemicals, pesticides and fertilizers are the most used products which turns consequently on residual contamination (Mahmood et al., 2016).

Agrochemicals are also called crop protection products, and include herbicides, fungicides, bactericides, insecticides, and nematicides, by their target organism against their action; and from all the mentioned agrochemicals, pesticides are the most used products (Fig. 9.1) (Zhang et al., 2018). Nonetheless,

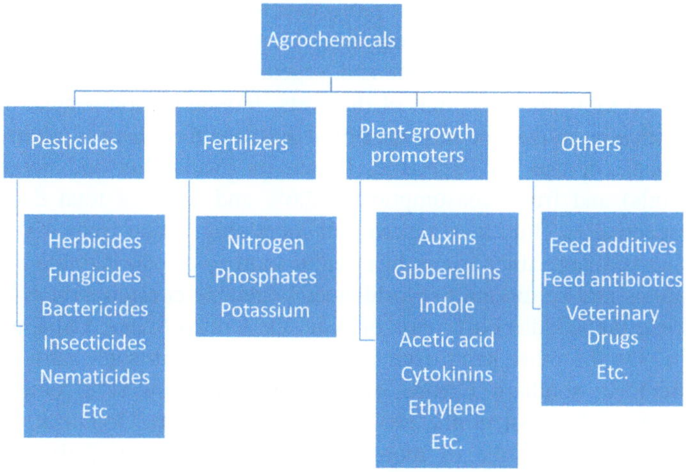

FIGURE 9.1

General classification of agrochemicals by their function.

most of the pesticides are residually accumulated by relying on variables such as edaphic factors and soil microbial diversity (Mandal et al., 2020). It has been documented that the indiscriminate use of pesticides has led to an inhibition or kills the microbial diversity in soil (Malik et al., 2017). The residual of some fungicides such as captan tends to decrease living organisms and to affect the legume–rhizobium association (Kyei-Boahen et al., 2011). Fungicides have a negative impact on several biochemical processes in soil by inhibiting enzymatic activity (Carr et al., 2005). For example, benomyl, captan, mancozeb, trifloxystrobin, and thiram fungicides inhibited phosphomonoesterase, urease, dehydrogenase, phosphatase enzymatic activity (Meena et al., 2020a).

The absorption and penetration of pesticides inside crops, vegetables, and fruits is too variable and depends on the chemical functional group and environmental conditions, thus ranging from hours to years (Yigit and Velioglu, 2019). Foods are subjected to many processes such as baking, blanching, pasteurization, peeling, firing, and washing in order to prolong shelf-life, increase diversity, enhance nutritive value and flavor, and reduce pesticide residues (Zhao et al., 2019). However, in some cases the metabolites formed during the processing of food are more toxic than the pesticides, and can led to an increase of residual too (Duan et al., 2022). "The effect of processing on residue concentration is determined by calculating the processing factor (PF) which is the ratio of the pesticide residue concentration (mg kg^{-1}) in the processed product to the pesticide residue concentration (mg kg^{-1}) in the untreated product" (Yigit and Velioglu, 2019). A PF greater than one indicates increased pesticide concentration in comparison to raw material, while a PF lower than one indicates a reduced pesticide concentration (Aguilera et al., 2014).

The pesticides are being applied for long-periods and in an indiscriminate manner, as consequence, several reports have documented the toxic effects of nontarget organisms such as amphibians, arthropods, fishes, humans, and others (Nicolopoulou-Stamati et al., 2016). In general, the toxic exposure through occupational or unintentional poisoning has led to an acute toxic human effect (Tarannum et al., 2019). The toxic effects severity depends on person exposed during application, length, and frequency of contact time, mode of application, the agrochemical used and, its mode of action (Elahi et al., 2019). The agricultural workers are the most exposed population group, but general population is also affected by residually contamination of food principally and in minor proportion by air, and water contamination (Hendges et al., 2019). It has been documented that chronic exposure to pesticides can lead to DNA damage and by consequence cause neurodegenerative disorders, respiratory diseases, diabetes, and cancer (Dhananjayan et al., 2020).

9.4 Nanotechnology in food production

Food consumption has augmented with population growth. Therefore it is necessary to reduce food waste and increase food production. The excessive use of synthetic agrochemicals and synthetic materials to increase crop yields and food preservation has led to the degradation of agroecosystems, resistance to pesticides, and environmental contamination (Reddy and Chhabra, 2022). For this reason, it is important to implement sustainable agriculture and develop new technologies to increase production, shelf-life, and food quality (Barrett, 2021). In recent years, the study of nanotechnology to be used in sustainable agriculture has increased (Neme et al., 2021; Abobatta, 2018). In recent years, the study of nanotechnology to be used in sustainable agriculture has increased because it can greatly improve the production, quality, and shelf-life of food products.

Nanotechnology is the scientific technique in which the materials produced or utilized have a smaller structure than 100 nm (Bumbudsanpharoke et al., 2015; Arora et al., 2022). Food production and preservation are one of the greatest current challenges in the food industry. In this sense, the application of nanotechnology has potential benefits, which include improvement in the quality and safety of agricultural inputs, processing, and higher quality and nutritional value of the final product (Fig. 9.2) (Neme et al., 2021).

Nanotechnology is one of the promising technologies to improve agricultural productivity. Among the most important applications is the encapsulation of active compounds in nanofertilizers, herbicides, and pesticides (Arora et al., 2022). Nano formulations differ from conventional ones by the controlled release of the active ingredient, which reduces the loss of chemical formulations into the soil, water, and the environment, reducing pollution (Naderi and Abedi, 2012). It has been observed that nano fertilizers improve plant nutrition and stress

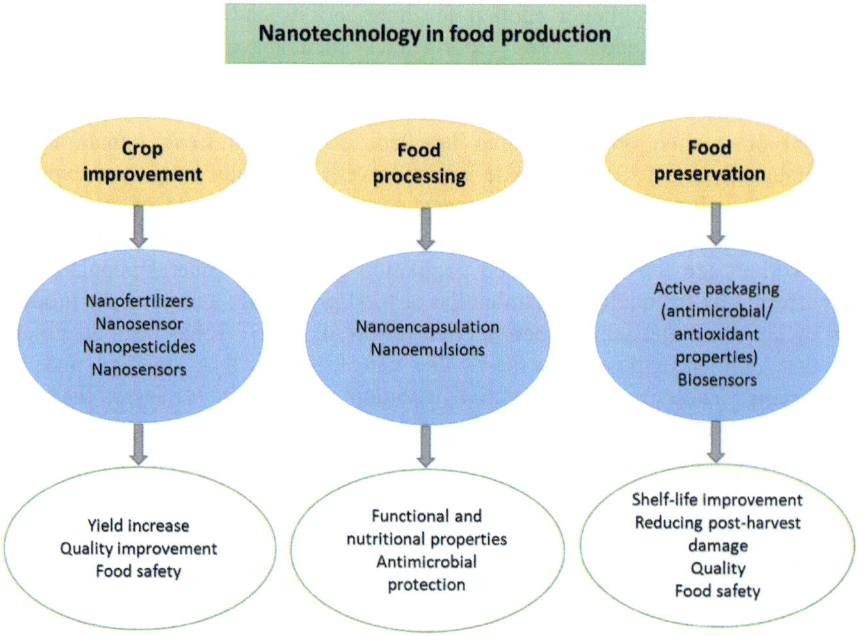

FIGURE 9.2

Scope of nanotechnology in food production.

tolerance, minimize leaching, and increase quality parameters and crop yield (Solanki et al., 2015; Duhan et al., 2017). Nanosensors facilitate growth monitoring, plant diseases, and help combat pathogens in crops. Nanosensors have advantages compared to conventional sensors, they present high sensitivity, high surface-to-volume ratio, fast and reliable response, and high selectivity (Duhan et al., 2017; Rai et al., 2012).

Nanotechnology can also be of great benefit in food processing, it can improve textures, flavors, aromas, color, nutritional properties, safety, and antimicrobial protection (Neme et al., 2021; Ashfaq et al., 2022). Also, the shelf-life of perishable foods such as fruits and vegetables can be effectively improved using nanotechnology. Fruits and vegetables are perishable, they have less shelf-life because of their high oxygen and water vapor transmission rate. Therefore they cannot be stored for long periods (Ashfaq et al., 2022; Xing et al., 2019). NMs and coating added with nanoparticles play an important role in food preservation. Thanks to nanotechnology, a wide variety of food packaging with improved physicochemical and antimicrobial properties can be manufactured, helping to extend the shelf-life of the product. The enhancement in these properties is due to nanoscale dispersion and interfacial interactions in the polymeric matrix because of their large surface area and high surface energy compared to materials on the macroscale (Arfat et al., 2014). Nanoparticles can modify the physical and mechanical properties of the polymer matrix, improving its resistance, flexibility, and durability. In addition, functional nanoparticles can improve barrier properties, durability, and thermal stability (Bumbudsanpharoke et al., 2015; González-Estrada et al., 2021). In active packaging, some nanoparticles such as oxygen and ethylene scavengers, antioxidants, and antimicrobial agents are intentionally added into the polymeric matrix in order to prevent oxidation, microorganisms, moisture gain, etc., therefore, improving the shelf-life of food (Ashfaq et al., 2022; Gaikwad et al., 2020). Nanosensors have been one of the most successful outcomes of nanotechnology, especially for highly perishable foods. Nanosensors can monitor the conditions of food products and their quality through indicators of time, temperature, pH, and of freshness (Fuertes et al., 2016).

Nanotechnology can significantly improve food production and preservation. Nanotechnology cannot only improve agriculture and food production in developed countries but could have a significant impact on rural populations and marginalized areas in developing countries (Prasad and Kochhar, 2014). However, nanotechnology also has some hazards that should be considered before using it in our daily lives. The reduced size of nanoparticles and increased surface area causes migration from the package into the food and may cause potential health risks for the consumer (Ashfaq et al., 2022; Naseer et al., 2018). Therefore more research is required to assess the benefits and potential risks of nanotechnology.

9.5 Green synthesis of nanomaterials

In recent years, research in nanoscience has increased considerably, and in a short time, devices and materials at the nanoscale have become part of our lives. In this way, the emergence of nanotechnology through the development of new nanodevices and NMs has opened the doors to disciplines such as biotechnology and agriculture, with new applications of great relevance for these areas. In particular, nanotechnology today is beginning to play an essential role in agriculture due to its great potential to improve the efficiency of agricultural inputs, production, and food safety, through the use of nanodevices and NMs such as nanosensors, nanofertilizers, nanopesticides, and nano formulations (Singh et al., 2021). However, it is vital to study and understand the phytotoxicity and possible ecological risks of these NMs for their potential application (Singh et al., 2021; Vurro et al., 2019). In this regard, synthesis methods are of great importance. There are mainly two strategies for synthesizing nanomaterials, the top-down approach and the bottom-up approach (Fig. 9.3).

On the one hand, the top-down approach involves breaking down a larger structure into smaller nanometer scale pieces through physical processes. In contrast, in the bottom-up approach, the opposite occurs; the material is synthesized from the atomic level to form the nanostructure using chemical or biological methods (Ahmed et al., 2022; Yazdanian et al., 2022; Khan, 2020; Noah and

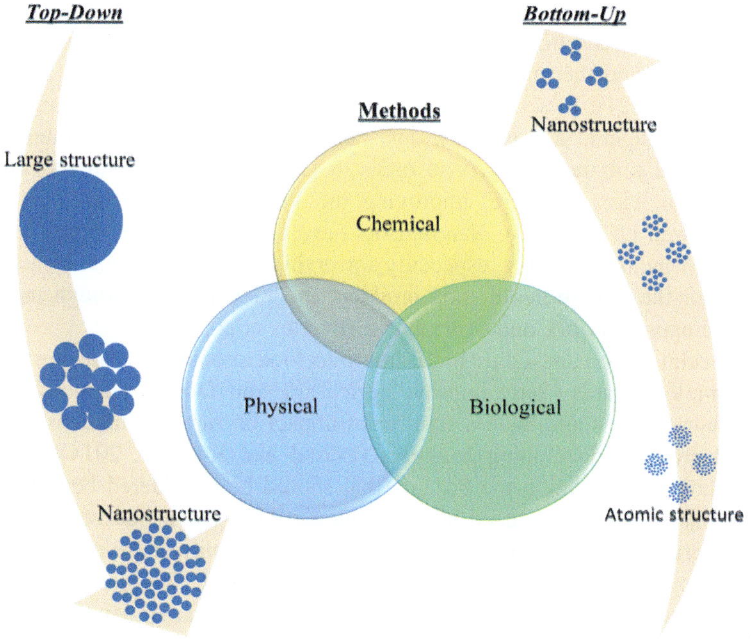

FIGURE 9.3

Different approaches to the synthesis of nanomaterials.

Ndangili, 2022). In particular, physical and chemical strategies are mainly used to synthesize nanomaterials (Khan, 2020; Noah and Ndangili, 2022). Physical methods include lithography, spray pyrolysis, physical vapor deposition, photoirradiation, ultrasonication, crushing, grinding, attrition, and ball milling (Ahmed et al., 2022; Khan, 2020; Noah and Ndangili, 2022; Zhang et al., 2020a). Chemical methods include chemical vapor deposition, hydrothermal and solvothermal methods, sol-gel, coprecipitation, chemical and photochemical reduction, thermal decomposition, microwave-assisted synthesis, ultrasound-assisted synthesis, electrochemical synthesis, reverse micelle method, and gas phase, among others (Ahmed et al., 2022; Khan, 2020; Noah and Ndangili, 2022; Zhang et al., 2020a). However, although these methods are characterized by uniform size distribution, homogeneity, and low energy consumption during the synthesis process, most of these methods are associated with problems such as high energy demand in production, high costs, and low yields; in addition to hazardous reaction conditions and the use of aggressive chemicals that can cause carcinogenicity, toxicity, and environmental toxicity, which pose a severe threat to health and the environment (Noah and Ndangili, 2022; Zhang et al., 2020a; Bandeira et al., 2020). These toxicity problems are rather prominent due to the use during the synthesis process of hazardous substances such as reducing agents, organic solvents, and stabilizers, which are used to avoid the agglomeration of colloids, but because of their toxicity and chemical contamination could limit the use of NMs obtained in agriculture and other areas (Yazdanian et al., 2022; Noah and Ndangili, 2022; Zhang et al., 2020a; Hernández-Díaz et al., 2021; Mittal et al., 2020; Prasad et al., 2017). This situation has led to the development of new economically viable, reliable, clean, biologically appropriate, and environmentally friendly strategies for the synthesis of NMs (Noah and Ndangili, 2022; Zhang et al., 2020a), such as biological methodologies that are novel and attractive since they use components such as organisms, microorganisms, metabolites or phytochemicals, which allow the synthesis of NMs in a nontoxic and environmentally friendly way (Ahmed et al., 2022; Noah and Ndangili, 2022; Zhang et al., 2020a).

These methodologies are known as *green synthesis strategies*, which collectively in the field of nanotechnology is known as *green nanotechnology*, and consist of producing NMs through the bio reduction of metal ions (Noah and Ndangili, 2022) through the use of bacteria (Fang et al., 2019), fungi (Abdelghany et al., 2018), yeasts (Olobayotan and Akin-Osanaiye, 2019), viruses (Taghavi Fardood et al., 2019), actinomycetes (Hassan et al., 2019), biopolymers (Lomelí-Marroquín et al., 2019), photosynthesis (Hamedi and Shojaosadati, 2019), micro and macro algae (Bhuyar et al., 2020), and plants (Abinaya et al., 2021). In addition, strategies have been reported in the literature incorporating physical or chemical methods in biological methodologies that help improve the obtaining of NMs while maintaining the appellative of green synthesis (Lomelí-Marroquín et al., 2019; Abinaya et al., 2021).

Of all these strategies, those involving the use of microorganisms have been most widely explored; however, synthesis using plant extracts has been gaining popularity due to their content of phenolic acids, flavonoids, polysaccharides, and other compounds that provide NMs with properties of interest for their application in the agricultural field (Ahmed et al., 2022; Abinaya et al., 2021). At the same time, solvent-free synthesis strategies are being investigated to obtain NMs that meet the objectives of green synthesis and have the desired properties to be applied in agriculture (Ahmed et al., 2022). In this way, green nanotechnology offers an important solution for the agricultural area through multiple tools for the conversion of biological systems to green methodologies for the synthesis of NMs, providing an attractive solution to inhibit any adverse effects produced by chemical and physical methods, and the possible rejection of the application of NMs, avoiding any accompanying toxicity and the harmful impact on the environment, thus reducing the risk of nanotechnology.

9.6 Properties of nanomaterials used in food production

Nanotechnology applied to food production involves the synthesis, characterization, and application of nanoscale structures and materials (1–100 nm) (Onyeaka et al., 2022) with the aim of developing sustainable agriculture focused on economic development, environmental protection, and food security (Sun et al., 2021). The application of NMs in the food production incorporates the use of materials that can be divided into six groups: metal NMs, metal oxide NMs, carbon NMs, nanoclays/surface-modified nanoclays, nanoencapsulateds y nanobiocomposites (Tuantranont et al., 2020). These NMs have the capacity to enhance the development of sustainable agriculture, as well as address the common problems, such as pest control, weed control, soil depletion, low nutritional value of food, and excessive use of traditional agrochemicals, which have affected biodiversity and threatened human health through the food chain (Lindsey et al., 2020; Singhal et al., 2022). In this regard, NMs have gained great attention for use in sustainable agriculture due to their physical and chemical properties. Some of the most important properties of NMs include hydrophobicity, stability, phytotoxicity and antimicrobial, antibacterial, and antioxidant activities (Xu et al., 2021; Ameen et al., 2021) which will be discussed below. It is important to note that the properties of the NMs depend on the synthesis techniques and the relevance of each of these properties is influenced by the application in sustainable agriculture.

The main route of entry of NMs to plants is through the soil; therefore, NMs are immersed in a physicochemical soil environment, which also possesses a mixture of minerals, organic matter, biota, water and air, as well as soil properties such as pH and ionic strength (Xu et al., 2021). Principally clay soils have the capacity to exchange anions and/or cations by adsorption, therefore, these sites can bind to NMs causing heteroaggregation (Zhang et al., 2020b). In the same way, the organic matter (e.g., fulvic acid and humic acid) of the soil has the characteristic of having hydrophobic and hydrophilic fractions, so it has the capacity to coat the NMs and this can cause an electrostatic stabilization or flocculation/aggregation modifying the surface and behavior of the NMs (Degenkolb et al., 2019; Luo et al., 2018). For that reason, it is essential to know the hydrophobicity of the NM and its interaction with various ions and organic compounds (Luo et al., 2018; Badawy et al., 2010) to evaluate if there is any transformation of the NMs before reaching the plant.

The stability of NMs is generally studied with the pH of the system with which it is interacting and especially with metallic NMs, this is due to the speciation of NMs, for example, a ZnO nanomaterial in acidic conditions ($pH < 6.0$) will release Zn^{+2}, however, if it is in basic conditions it can form a $Zn(OH)_2$ layer (Tang et al., 2002), this totally changes the chemical species with which it is required at the beginning. In addition, each of the NMs has a different isoelectric point (IEP), this property indicates the pH at which the NM has a neutral surface charge, and this value can vary from an acidic to a basic region (Zhang et al., 2020b). Several studies have been carried out where the concentration of antioxidants and polyphenols, such as quercetin and flavonoids in the plant, are one of the factors that control and improve the stability of nanoparticles due to their chelating effect (Nasrollahzadeh et al., 2020).

Although various NMs have been synthesized and their potential against plant pathogens has been studied, the exact mode of action of NMs is still not fully understood (Hajipour et al., 2012). Various mechanisms of action against phytopathogens such as reactive oxygen species (ROS) production, antioxidant depletion, damage and/or loss of membrane potential, interference with nutrient uptake, and genotoxicity have been reported (Lemire et al., 2013). The adhesion of NMs with the cell membrane occurs due to electrostatic attractions between the negative charge of the membrane of microbes and the positive charges of the NM, this causes a morphological alteration of the membrane and consequently causes disruption of permeability, damage to cellular structures, and finally their cell death (Ali et al., 2020). Therefore here we see the importance of knowing the surface charge of the nanomaterial from the IEP value.

This toxicity of NMs toward phytopathogens is beneficial to plants, however, it is important to highlight the phytotoxicity of NMs, that is, the ability to inhibit seed germination, reduce root, seedling and shoot growth, delay flowering, and decrease yield (Li et al., 2015; Sanzari et al., 2019). To evaluate the phytotoxicity of NMs, it is necessary to evaluate germination, cell cultures, and genetic effects on a plant, and detection techniques such as ion-coupled plasma-mass spectroscopy (ICE-Ms), photothermal, and photoacoustic have been used (Khodakovskaya et al., 2012). Dietz and Herth (2011) established five main modes in which NMs interact with organisms: (1) metal ions released by NMs in solution produce a chemical effect; (2) spherical and hard NMs produce mechanical effects; (3) the surface of NMs produce catalytic effects; (4) the surface of NMs can bind to proteins by noncovalent or covalent mechanisms and can cause oxidative effects; and (5) NMs can change the chemical environment, especially pH. Further research aimed at assessing phytotoxicity as a key property of NMs is imperative.

The antimicrobial and antibacterial activity of NMs is directly dependent on the size, generally, smaller NMs have high stability and enhanced microbial activity (Roy et al., 2019). The great area/volume ratio of NMs provides a larger area of intracellular interaction and penetration (Raza et al., 2016). For the NM to have effective antimicrobial activity, its size should be less than 50 nm, and if the NMs have a size range between 10 and 15 nm they will possess superior antimicrobial activity (Roy et al., 2019). In one study, a series of colloidal Ag nanoparticles were synthesized using the saccharides: glucose, galactose, maltose, and lactose; the nanoparticles that had a size greater than 50 nm exhibited the lowest antimicrobial activity (Panáček et al., 2006), several studies have been developed demonstrating the size−antimicrobial activity relationship (Li et al., 2013; Ivask et al., 2014). It is important to note that some NMs are considered safe and effective antibacterial agents, as they are highly toxic to bacteria, but not toxic to animal cells (Klueh et al., 2000; Marambio-Jones and Hoek, 2010), however, more research is needed.

It is known that environmental stress can trigger an overproduction of ROS, such as superoxide radical ($O_2^{\bullet -}$), hydroxyl radical ($^{\bullet}OH$), and hydrogen peroxide (H_2O_2), which generates oxidative stress, damages DNA, proteins, cell membrane lipids, and suppresses plant growth (Gill and Tuteja, 2010). Generally, the plant antioxidant defense system can scavenge excess ROS and counteract oxidative damage, however, the prolonged and intensified time of environmental stress causes the plant defense system to be inefficient (Liu et al., 2021; Das and Roychoudhury, 2014). Recently, it has been shown that the antioxidant activity of metal oxide NMs exhibits tremendous advantages against environmental stresses caused to a plant (Lu et al., 2020; Wu et al., 2017; Liu et al., 2019; Zhao et al., 2020). For example, foliar sprayed Mn_3O_4 NMs (1 mg per plant) have been found to successfully counteract oxidative stress in cucumber plants, thus this NM acts as a nanoregulator that helps to counteract undesirable effects of oxidative stress (Lu et al., 2020). When NMs are applied at low doses (<20 ppm), they exhibit antioxidant activity that could scavenge ROS under environmental stress, whereas, when the concentration is increased NMs can exhibit oxidative activity that can induce phytotoxicity (Liu et al., 2021).

9.7 Application of green nanomaterials in fruit and vegetable production

The perishable nature of fruit and vegetables and biotic factors contribute to reducing their shelf-life and increasing postharvest losses. The application of NMs can reduce the loss of quality and the proliferation of microorganisms, increasing the shelf-life (Tables 9.1 and 9.2). In this sense, the use of NMs has represented in recent years an efficient alternative that could reduce postharvest losses in fruit and vegetables. Different NMs have been applied to vegetables, including chemical compounds such as titanium dioxide (TiO_2), zinc oxide (ZnO), copper (Cu), and silver (Ag). These compounds have been approved for use at regulated concentrations and are considered nontoxic for human consumption. The metals are easily incorporated into a polymeric matrix that is applied as NM to the product preserving the quality and reducing the growth of microorganisms (Ismail, 2021).

The powdery mildew disease on pepper *Capsicum annuum* L. was controlled by the application of MgO and ZnO nanoparticles. The concentration of 200 mg/L of MgO NPs reduces the disease severity by 83.12% with respect to the control treatment. The MgO NPs were more efficient than the ZnO NPs treatment at the same concentrations. Scanning electron microscopy (SEM) showed morphological damages in the pathogen after 48 hours of exposure to MgO NPs and ZnO NPs at 200 mg/L. The SEM micrographs showed loss of cell integrity, deformation, and collapsing of conidiophores and conidia after the nanoparticle application (Ismail, 2021).

Nanoparticles of titanium dioxide (TiO_2 NPs), silver (Ag NPs), and selenium (Se NPs) were applied in vitro to control *Alternaria alternata* isolated from rotten tomatoes. TiO_2 NPs and Se NPs at 100 ppm reduced the mycelial growth of *A. alternata* by 90%. The in vivo application presented a similar effect, the TiO_2 NPs and the Se NPs showed higher efficiency than the Ag NPs. The control percentage of tomatoes infected by *A. alternata* was 59% and 57%, respectively

Table 9.1 Characterization of nanomaterials applied at different cultivars.

Nanomaterial	Abbreviation	Characterization	Cultivar application	References
Copper nanoparticles from eucalyptus leaf extracts	CuNP-E	Average size from 10 to 130 nm, with triangular shaped and nanoparticles poly-dispers	Chili	Iliger et al. (2021)
Copper chitosan nanoparticles	Cu-CT NPs	Size 368.3 ± 3.2 nm, hydrodynamic diameter	Tomato	Meena et al. (2020b)
Graphene oxide nanosheet-decorated copper oxide nanoparticles	rGO-CuO NPs	Three different sizes 5, 20, and 50 nm in average	Tomato and pepper plant	Al-Naamani et al. (2018)
Zinc oxide nanoparticles	ZnO NPs	Particle size about 80–100 nm	Carrots	Xu et al. (2017)
Chitosan-zinc oxide nanocomposite	CS-ZnONP	Particle size about 10–50 nm	Pepper	Asgari-Targhi et al. (2021)
Carboxymethylcellulose (CMC) spray-coated Cu^{2+} intercalated montmorillonite (MMT) nanocomposite	CMC spray-coated Cu – MMT	Cu particles intercalated in MMT layers along with the CMC	Potato	Zhu et al. (2019)

Table 9.2 Impact of nanomaterials application on several fruits.

Fruit	Nanomaterial	Relevant results	References
Lemon (*Citrus limon* L. *obs*)	Chitosan-clay nanocomposite	The TSS values ranging from 8.8 to 10.6, TA from 8.83 to 10.67, firmness from 260.13 to 252.98 N, while the punch forces ranging from 27.64 to 20.88 (NSD)	Taghinezhad and Ebadollahi (2017)
Tangerine (*Citrus tangerine Hort. ex Tanaka*)	Chitosan/montmorillonite (CS/MMT)	Treated fruits with CS/MMT-0.5, the weight loss values ranging from 0.47 to 1.47 for 3 and 11 days, respectively, the ascorbic acid content from 48.06 to 57.16, TTS from 9.0 to 9.0 for 0 and 11 days, respectively	Xu et al. (2018)
Italia and Benitaka table grapes (*V. vinifera* L.)	Chitosan/silica nanocomposite (CSN)	In vitro tests with the use of CSN at 1% the gray mold disease was reduced in artificial and natural infections of grapes. On quality parameters, TSS values ranging from 10.7 to 11.9 and 11.9 to 13, TA from 0.89 to 0.96 and 0.97 to 0.99	Youssef et al. (2019)
Cavendish bananas (*M. acuminata* L.)	Chitosan nanoparticles (CS)	The color of the treated fruit was green until Day 4 and turned to yellow at Day 5, retarding the maturation of fruits compared to control	Esyanti et al. (2019)
Cavendish bananas (*M. acuminata* L.)	Ajwain-magnesium nanoparticles Ajwain-nickel nanoparticles and silver-neemnanoparticles	In vitro studies revealed that the application of NEEM-AG NPs at 0.1%, the inhibition of spore germination was complete (100%)	Jagana et al. (2017)
Red Delicious (*M. domestica Bork*)	Nano-calcium	Fruits treated with nano-Ca at 1.5% exhibited TSS values ranging from 13.6 to 17.7, TA from 4.6 to 3.1, firmness reach up 44.83 to 33.6 N.	Ranjbar et al. (2018)
Red Delicious (*M. domestica Bork*)	ZnO nanoparticles	ZnO NPs with sizes <50 nm possess a significant antifungal activity against *Penicillium expansum*	Sardella et al. (2017)

(*Continued*)

Table 9.2 Impact of nanomaterials application on several fruits. *Continued*

Fruit	Nanomaterial	Relevant results	References
Guava (*P. guajava* L.)	Microencapsulated starch/ascorbic acid	In microcapsules of 6.25% the ascorbic acid the TSS values ranging from 10.5 to 11.6, TA from 0.7 to 0.72, pH reaches up 3.70 to 3.78, respiration rate from 12 to 20, firmness from 360 to 290 N	Martínez-Ortiz et al. (2019)

(El-Gazzar and Ismail, 2020). Chitosan–nano-ZnO composite films applied to cherry tomatoes showed antibacterial activity against *Alicyclobacillus acidoterrestris*, *Staphylococcus aureus, Escherichia coli*, and *Salmonella*. The application of the film to tomatoes reduces each bacterial 2 logs CFU/mL concerning control treatment after 15 days of storage. Quality parameters such as respiratory intensity, soluble solid content, and color were controlled after the storage and prolonged the shelf-life of tomatoes (Li et al., 2021).

According to Meena et al. (2020b), tomatoes after 21 days of storage do not present microbial development with the application of 0.4% of Cu-chitosan NPs. SEM tomato surface analysis showed the presence of the NM, this effect was related to the microbial decay on tomatoes during storage. In green tomato, the application of a film based on Nano-SiOx/chitosan preserves quality, the effect observed was the retarded weight loss and softness, delayed the degradation of organic acids and soluble solids. *E. coli* and *S. aureus* growth was efficiently controlled with the application of the nano-SiOx/chitosan solution. The inhibition rate for each bacterium was 42 y 37% respectively. The nanocomposite nano-SiOx/chitosan was more efficient than the application of chitosan solution without the silicon oxides with a bacterium control of 30% and 19% respectability (Zhu et al., 2019).

To extend the shelf-life of cucumber, Helal et al. (2021) applied a combination of titanium dioxide nanoparticles (1%), chitosan (1%), and sodium tripolyphosphate (2%). After 21 days, the visual appearance of cucumbers stored at 10°C preserved high quality compared with the control treatment. The application individual of chitosan preserves 14 days the quality. The essential oil and leaf extract have been incorporated as a component of different NMs. Postharvest quality on cucumber was preserved by the application of a nanostructured coating of chitosan with cinnamon essential oil (CEO).

In avocado, Chávez et al. (2018) applied a nanocomposite edible coating based to chitosan and pepper tree essential oil. The nanocomposite inhibited the development of the fungus *Colletotrichum gloeosporioides* during postharvest storage. In similar study, chitosan-thyme essential oil nanoparticles (CTSTEO-NPs) reduce 60% the incidence of the pathogen *C. gloeosporioides* respect to control fruits.

9.8 Conclusions and future perspectives

The production of fruits and vegetables is very significant worldwide. However, a very large number of tons is lost due to biotic and abiotic factors. Therefore it is necessary to apply diverse technologies in pre- and postharvest stages. Nanotechnology is a discipline that is revolutionizing several areas of knowledge, among them is agriculture, specifically in the production of several fruits and vegetables. NPs have been a promising tool, increasing the functionality and efficiency of chemical pesticides and other substances from biological or natural origin. Nanotechnology in agriculture sector represents a suitable alternative for controlling important postharvest pathogens of several fruits as well as for quality maintenance. The results of several investigations of the application of NPs have been very promising, however, it is necessary to evaluate its commercial application and consider also the preferences of consumers, legislations, and regulations of each country, toxicity values, and impact to the environment.

References

Abdelghany, T.M., Al-Rajhi, A.M.H., Al Abboud, M.A., Alawlaqi, M.M., Ganash Magdah, A., Helmy, E.A.M., et al., 2018. Recent advances in green synthesis of silver nanoparticles and their applications: about future directions. A review. Bionanoscience 8, 5–16.

Abinaya, S., Kavitha, H.P., Prakash, M., Muthukrishnaraj, A., 2021. Green synthesis of magnesium oxide nanoparticles and its applications: a review. Sustain. Chem. Pharm. 19, 100368.

Abobatta, W.F., 2018. Nanotechnology application in agriculture. Acta Sci. Agric. 2.

Aguilera, A., Valverde, A., Camacho, F., Boulaid, M., García-Fuentes, L., 2014. Household processing factors of acrinathrin, fipronil, kresoxim-methyl and pyridaben residues in green beans. Food Control 35, 146–152. Available from: https://doi.org/10.1016/J.FOODCONT.2013.06.038.

Ahmed, S.F., Mofijur, M., Rafa, N., Chowdhury, A.T., Chowdhury, S., Nahrin, M., et al., 2022. Green approaches in synthesising nanomaterials for environmental nanobioremediation: technological advancements, applications, benefits and challenges. Environ. Res. 204, 111967.

Ali, M.A., Ahmed, T., Wu, W., Hossain, A., Hafeez, R., Masum, M.M.I., et al., 2020. Advancements in plant and microbe-based synthesis of metallic nanoparticles and their antimicrobial activity against plant pathogens. Nanomaterials 10. Available from: https://doi.org/10.3390/nano10061146.

Al-Naamani, L., Dutta, J., Dobretsov, S., 2018. Nanocomposite zinc oxide-chitosan coatings on polyethylene films for extending storage life of okra (*Abelmoschus esculentus*). Nanomaterials 8, 479.

Ameen, F., Alsamhary, K., Alabdullatif, J.A., ALNadhari, S., 2021. A review on metal-based nanoparticles and their toxicity to beneficial soil bacteria and fungi. Ecotoxicol. Environ. Saf. 213, 112027. Available from: https://doi.org/10.1016/J.ECOENV.2021.112027.

Arfat, Y.A., Benjakul, S., Prodpran, T., Sumpavapol, P., Songtipya, P., 2014. Properties and antimicrobial activity of fish protein isolate/fish skin gelatin film containing basil leaf essential oil and zinc oxide nanoparticles. Food Hydrocoll. 41, 265–273.

Arora, S., Murmu, G., Mukherjee, K., Saha, S., Maity, D., 2022. A comprehensive overview of nanotechnology in sustainable agriculture. J. Biotechnol.

Asgari-Targhi, G., Iranbakhsh, A., Ardebili, Z.O., Tooski, A.H., 2021. Synthesis and characterization of chitosan encapsulated zinc oxide (ZnO) nanocomposite and its biological assessment in pepper (*Capsicum annuum*) as an elicitor for in vitro tissue culture applications. Int. J. Biol. Macromol. 189, 170–182.

Ashfaq, A., Khursheed, N., Fatima, S., Anjum, Z., Younis, K., 2022. Application of nanotechnology in food packaging: pros and cons. J. Agric. Food Res. 7, 100270.

Badawy, A., Luxton, T., Silva, R., Scheckel, K., Suidan, M., Tolaymat, T., 2010. Impact of environmental conditions (pH, ionic strength, and electrolyte type) on the surface charge and aggregation of silver nanoparticles suspensions. Environ. Sci. Technol. 44, 1260–1266. Available from: https://doi.org/10.1021/es902240k.

Bandeira, M., Giovanela, M., Roesch-Ely, M., Devine, D.M., da Silva Crespo, J., 2020. Green synthesis of zinc oxide nanoparticles: a review of the synthesis methodology and mechanism of formation. Sustain. Chem. Pharm. 15, 100223.

Barrett, C.B., 2021. Overcoming global food security challenges through science and solidarity. Am. J. Agric. Econ. 103, 422–447.

Bhuyar, P., Rahim, M.H.A., Sundararaju, S., Ramaraj, R., Maniam, G.P., Govindan, N., 2020. Synthesis of silver nanoparticles using marine macroalgae *Padina* sp. and its antibacterial activity towards pathogenic bacteria. Beni Suef Univ. J. Basic. Appl. Sci. 9, 1–15.

Boedeker, W., Watts, M., Clausing, P., Marquez, E., 2020. The global distribution of acute unintentional pesticide poisoning: estimations based on a systematic review. BMC Public Health 20, 1–19. Available from: https://doi.org/10.1186/S12889-020-09939-0.

Boeing, H., Bechthold, A., Bub, A., 2012. Bioactive food as dietary interventions for cardiovascular disease. Eur. J. Nutr. 51, 637–663.

Bumbudsanpharoke, N., Choi, J., Ko, S., 2015. Applications of nanomaterials in food packaging. J. Nanosci. Nanotechnol. 15, 6357–6372.

Carr, J.F., Gregory, S.T., Dahlberg, A.E., 2005. Severity of the streptomycin resistance and streptomycin dependence phenotypes of ribosomal protein S12 of *Thermus thermophilus* depends on the identity of highly conserved amino acid residues. J. Bacteriol. 187, 3548–3550. Available from: https://doi.org/10.1128/JB.187.10.3548-3550.2005/ASSET/2C7F509F-B012-401C-B361-C758B59901AF/ASSETS/GRAPHIC/ZJB0100546980002.JPEG.

Chávez, M., Mireya Esbeiddy, G.E.R.R., Maribel, A.R., Gutie, P.P., 2018. Effect of pepper tree (*Schinus molle*) essential oil-loaded chitosan bio-nanocomposites on postharvest control of *Colletotrichum gloeosporioides* and quality evaluations in avocado (*Persea americana*) cv. Hass . Available from: https://doi.org/10.1007/s10068-018-0410-5.

Das, K., Roychoudhury, A., 2014. Reactive oxygen species (ROS) and response of antioxidants as ROS-scavengers during environmental stress in plants. Front. Environ. Sci. 2. Available from: https://doi.org/10.3389/fenvs.2014.00053.

Degenkolb, L., Kaupenjohann, M., Klitzke, S., 2019. The variable fate of Ag and TiO_2 nanoparticles in natural soil solutions—sorption of organic matter and nanoparticle stability. Water Air Soil. Pollut. 230, 62. Available from: https://doi.org/10.1007/s11270-019-4123-z.

Dhananjayan, V., Jayanthi, P., Jayakumar, S., Ravichandran, B., 2020. Agrochemicals impact on ecosystem and bio-monitoring. Resour. Use Efficiency Agric. 349–388. Available from: https://doi.org/10.1007/978-981-15-6953-1_11.

Dietz, K.J., Herth, S., 2011. Plant nanotoxicology. Trends Plant Sci. 16, 582–589. Available from: https://doi.org/10.1016/J.TPLANTS.2011.08.003.

Duan, L., An, X., Pan, X., Li, R., Wang, K., Guo, L., et al., 2022. Residual levels of five pesticides in peanut oil processing and chips frying. J. Sci. Food Agric. 102, 2494–2499. Available from: https://doi.org/10.1002/JSFA.11590.

Dudefoi, W., Terrisse, H., Richard-Plouet, M., Gautron, E., Popa, F., Humbert, B., et al., 2017. Criteria to define a more relevant reference sample of titanium dioxide in the context of food: a multiscale approach. Food Addit. Contam. Part. A 1–13. Available from: https://doi.org/10.1080/19440049.2017.1284346.

Duhan, J.S., Kumar, R., Kumar, N., Kaur, P., Nehra, K., Duhan, S., 2017. Nanotechnology: the new perspective in precision agriculture. Biotechnol. Rep. 15, 11–23.

Elahi, E., Weijun, C., Zhang, H., Nazeer, M., 2019. Agricultural intensification and damages to human health in relation to agrochemicals: application of artificial intelligence. Land. Use Policy 83, 461–474. Available from: https://doi.org/10.1016/J.LANDUSEPOL.2019.02.023.

El-Gazzar, N., Ismail, A.M., 2020. The potential use of titanium, silver and selenium nanoparticles in controlling leaf blight of tomato caused by *Alternaria alternata*. Biocatal. Agric. Biotechnol. 27, 101708.

FAO, 2019. The State of Food and Agriculture 2019. Moving forward on food loss and waste reduction. Rome.

Esyanti, R.R., Zaskia, H., Amalia, A., Nugrahapraja, D.H., 2019. Chitosan nanoparticle-based coating as post-harvest technology in banana, J. Phys. Conf. Ser., 1204. Institute of Physics Publishing. Available from: https://doi.org/10.1088/1742-6596/1204/1/012109.

Fang, X., Wang, Y., Wang, Z., Jiang, Z., Dong, M., 2019. Microorganism assisted synthesized nanoparticles for catalytic applications. Energies 12, 190.

Fernández-Ochoa, Á., de la Luz Cádiz-Gurrea, M., Fernández-Moreno, P., Rojas-García, A., Arráez-Román, D., Segura-Carretero, A., 2022. Recent analytical approaches for the study of bioavailability and metabolism of bioactive phenolic compounds. Molecules 27, 1–22. Available from: https://doi.org/10.3390/molecules27030777.

Fuertes, G., Soto, I., Vargas, M., Valencia, A., Sabattin, J., Carrasco, R., 2016. Nanosensors for a monitoring system in intelligent and active packaging. J. Sens. 2016.

Gaikwad, K.K., Singh, S., Negi, Y.S., 2020. Ethylene scavengers for active packaging of fresh food produce. Environ. Chem. Lett. 18, 269–284.

Gan, Y., Tong, X., Li, L., Cao, S., Yin, X., Gao, C., et al., 2015. Consumption of fruit and vegetable and risk of coronary heart disease: a meta-analysis of prospective cohort studies. Int. J. Cardiol. 183, 129–137.

Gill, S.S., Tuteja, N., 2010. Reactive oxygen species and antioxidant machinery in abiotic stress tolerance in crop plants. Plant. Physiol. Biochem. 48, 909–930. Available from: https://doi.org/10.1016/J.PLAPHY.2010.08.016.

Gill, R., Naeem, M., Ansari, A.A., Chhikara, A., Francisco, J., Bremont, J., et al., 2022. Agrochemicals in soil and environment: conclusions and future perspectives. Agrochem. Soil. Environ. 609–612. Available from: https://doi.org/10.1007/978-981-16-9310-6_26.

González-Estrada, R.R., Blancas-Benitez, F.J., Montaño-Leyva, B., Zambrano-Zaragoza, M. L., Aguirre-Güitrón, L., Moreno-Hernández, C., et al., 2021. Zinc nanomaterials: a safe tool for postharvest disease management. In: Abd-Elsalam, K.A. (Ed.), Zinc-Based Nanostructures for Environmental and Agricultural Applications. Elsevier, pp. 243–265. Available from: https://doi.org/10.1016/B978-0-12-822836-4.00018-5.

Hajipour, M.J., Fromm, K.M., Akbar Ashkarran, A., Jimenez de Aberasturi, D., Larramendi, I. R., de, Rojo, T., et al., 2012. Antibacterial properties of nanoparticles. Trends Biotechnol. 30, 499–511. Available from: https://doi.org/10.1016/J.TIBTECH.2012.06.004.

Hamedi, S., Shojaosadati, S.A., 2019. Rapid and green synthesis of silver nanoparticles using *Diospyros* lotus extract: evaluation of their biological and catalytic activities. Polyhedron 171, 172–180.

Hassan, S.E.-D., Fouda, A., Radwan, A.A., Salem, S.S., Barghoth, M.G., Awad, M.A., et al., 2019. Endophytic actinomycetes *Streptomyces* spp. mediated biosynthesis of copper oxide nanoparticles as a promising tool for biotechnological applications. J. Biol. Inorg. Chem. 24, 377–393.

Helal, M., Sami, R., Khojah, E., Elhakem, A., Benajiba, N., Al-Mushhin, A.A.M., et al., 2021. Evaluating the coating process of titanium dioxide nanoparticles and sodium tripolyphosphate on cucumbers under chilling condition to extend the shelf-life. Sci. Rep. 11, 1–8.

Hendges, C., Schiller, A., da, P., Manfrin, J., Macedo, E.K., Gonçalves, A.C., et al., 2019. Human intoxication by agrochemicals in the region of South Brazil between 1999 and 2014. J. Environ. Sci. Health B 54, 219–225. Available from: https://doi.org/10.1080/03601234.2018.1550300.

Hernández-Díaz, J.A., Garza-García, J.J.O., Zamudio-Ojeda, A., León-Morales, J.M., López-Velázquez, J.C., García-Morales, S., 2021. Plant-mediated synthesis of nanoparticles and their antimicrobial activity against phytopathogens. J. Sci. Food Agric. 101, 1270–1287.

Iliger, K.S., Sofi, T.A., Bhat, N.A., Ahanger, F.A., Sekhar, J.C., Elhendi, A.Z., et al., 2021. Copper nanoparticles: green synthesis and managing fruit rot disease of chilli caused by *Colletotrichum capsici*. Saudi J. Biol. Sci. 28, 1477–1486.

Ismail, A.M., 2021. The antifungal effect of MgO and ZnO nanoparticles against powdery mildew disease of pepper (*Capsicum annuum* L.) under greenhouse conditions. Egypt. J. Agric. Res.

Ivask, A., Kurvet, I., Kasemets, K., Blinova, I., Aruoja, V., Suppi, S., et al., 2014. Size-dependent toxicity of silver nanoparticles to bacteria, yeast, algae, crustaceans and mammalian cells in vitro. PLoS One 9, e102108. Available from: https://doi.org/10.1371/journal.pone.0102108.

Jagana, D.R., Hegde, Y., Lella, R., 2017. Green nanoparticles - a novel approach for the management of banana anthracnose caused by *Colletotrichum musae*. Int. J. Curr. Microbiol. Appl. Sci. 6, 1749–1756. Available from: https://doi.org/10.20546/ijcmas.2017.610.211.

Khan, F.A., 2020. Synthesis of nanomaterials: methods and technology. Applications of Nanomaterials in Human Health. Springer, pp. 15–21.

Khodakovskaya, M.v, de Silva, K., Biris, A.S., Dervishi, E., Villagarcia, H., 2012. Carbon nanotubes induce growth enhancement of tobacco cells. ACS Nano 6, 2128–2135. Available from: https://doi.org/10.1021/nn204643g.

Klueh, U., Wagner, V., Kelly, S., Johnson, A., Bryers, J., 2000. Efficacy of silver-coated fabric to prevent bacterial colonization and subsequent device-based biofilm formation. J. Biomed. Mater. Res. 53, 621–631. Available from: https://doi.org/10.1002/1097-4636(2000)53:63.0.CO;2-Q.

Kyei-Boahen, S., Slinkard, A.E., Walley, F.L., 2011. Rhizobial survival and nodulation of chickpea as influenced by fungicide seed treatment. Can. J. Microbiol. 47, 585–589.

Lemire, J.A., Harrison, J.J., Turner, R.J., 2013. Antimicrobial activity of metals: mechanisms, molecular targets and applications. Nat. Rev. Microbiol. 11, 371–384. Available from: https://doi.org/10.1038/nrmicro3028.

Li, J., Rong, K., Zhao, H., Li, F., Lu, Z., Chen, R., 2013. Highly selective antibacterial activities of silver nanoparticles against *Bacillus subtilis*. J. Nanosci. Nanotechnol. 13, 6806–6813. Available from: https://doi.org/10.1166/jnn.2013.7781.

Li, K.E., Chang, Z.Y., Shen, C.X., Yao, N., 2015. Toxicity of nanomaterials to plants. Nanotechnology and Plant Sciences: Nanoparticles and Their Impact on Plants. Springer. Available from: https://doi.org/10.1007/978-3-319-14502-0_6.

Li, Y., Zhou, Y., Wang, Z., Cai, R., Yue, T., Cui, L., 2021. Preparation and characterization of chitosan–nano-ZnO composite films for preservation of cherry tomatoes. Foods 10, 3135.

Lindsey, A.P.J., Murugan, S., Renitta, R.E., 2020. Microbial disease management in agriculture: current status and future prospects. Biocatal. Agric. Biotechnol. 23, 101468. Available from: https://doi.org/10.1016/J.BCAB.2019.101468.

Liu, Y., Yue, L., Wang, Z., Xing, B., 2019. Processes and mechanisms of photosynthesis augmented by engineered nanomaterials. Environ. Chem. 16, 430–445.

Liu, Y., Xiao, Z., Chen, F., Yue, L., Zou, H., Lyu, J., et al., 2021. Metallic oxide nanomaterials act as antioxidant nanozymes in higher plants: trends, meta-analysis, and prospect. Sci. Total Environ. 780, 146578. Available from: https://doi.org/10.1016/J.SCITOTENV.2021.146578.

Lomelí-Marroquín, D., Cruz, D.M., Nieto-Argüello, A., Crua, A.V., Chen, J., Torres-Castro, A., et al., 2019. Starch-mediated synthesis of mono-and bimetallic silver/gold nanoparticles as antimicrobial and anticancer agents. Int. J. Nanomed. 14, 2171.

Lu, L., Huang, M., Huang, Y., Corvini, P.F.-X., Ji, R., Zhao, L., 2020. Mn3O4 nanozymes boost endogenous antioxidant metabolites in cucumber (*Cucumis sativus*) plant and enhance resistance to salinity stress. Environ. Sci. Nano 7, 1692–1703. Available from: https://doi.org/10.1039/D0EN00214C.

Luo, M., Huang, Y., Zhu, M., Tang, Y.N., Ren, T., Ren, J., et al., 2018. Properties of different natural organic matter influence the adsorption and aggregation behavior of TiO_2 nanoparticles. J. Saudi Chem. Soc. 22, 146–154. Available from: https://doi.org/10.1016/J.JSCS.2016.01.007.

Mahmood, I., Imadi, S.R., Shazadi, K., Gul, A., Hakeem, K.R., 2016. Effects of pesticides on environment. Plant, Soil and Microbes: Volume 1: Implications in Crop Science. Springer International, pp. 253–269. Available from: https://doi.org/10.1007/978-3-319-27455-3_13/FIGURES/2.

Malik, Z., Ahmad, M., Abassi, G.H., Dawood, M., Hussain, A., Jamil, M., 2017. Agrochemicals and soil microbes: interaction for soil health. Xenobiotics in the Soil Environment. Springer, pp. 139–152. Available from: https://doi.org/10.1007/978-3-319-47744-2_11.

Mandal, A., Sarkar, B., Mandal, S., Vithanage, M., Patra, A.K., Manna, M.C., 2020. Impact of agrochemicals on soil health. Agrochem. Detect. Treat. Remed. 161–187. Available from: https://doi.org/10.1016/B978-0-08-103017-2.00007-6.

Marambio-Jones, C., Hoek, E.M.V., 2010. A review of the antibacterial effects of silver nanomaterials and potential implications for human health and the environment. J. Nanopart. Res. 12, 1531–1551. Available from: https://doi.org/10.1007/s11051-010-9900-y.

Marmot, M., Atinmo, T., Byers, T., Chen, J., Hirohata, T., Jackson, A., et al., 2007. Food, Nutrition, Physical Activity and the Prevention of Cancer: A Global Perspective .

Martínez-Ortiz, M.A., Palma-Rodríguez, H.M., Montalvo-González, E., Sáyago-Ayerdi, S. G., Utrilla-Coello, R., Vargas-Torres, A., 2019. Effect of using microencapsulated ascorbic acid in coatings based on resistant starch chayotextle on the quality of guava fruit. Sci. Hortic. 256. Available from: https://doi.org/10.1016/j.scienta.2019.108604.

Meena, R.S., Kumar, S., Datta, R., Lal, R., Vijayakumar, V., Brtnicky, M., et al., 2020a. Impact of agrochemicals on soil microbiota and management: a review. Land 9, 34. Available from: https://doi.org/10.3390/LAND9020034.

Meena, M., Pilania, S., Pal, A., Mandhania, S., Bhushan, B., Kumar, S., et al., 2020b. Cu-chitosan nano-net improves keeping quality of tomato by modulating physio-biochemical responses. Sci. Rep. 10, 1–11.

Mittal, D., Kaur, G., Singh, P., Yadav, K., Ali, S.A., 2020. Nanoparticle-based sustainable agriculture and food science: recent advances and future outlook. Front. Nanotechnol. 2, 10.

Naderi, M.R., Abedi, A., 2012. Application of nanotechnology in agriculture and refinement of environmental pollutants. J. Nanotechnol. 11, 18–26.

Naseer, B., Srivastava, G., Qadri, O.S., Faridi, S.A., Islam, R.U., Younis, K., 2018. Importance and health hazards of nanoparticles used in the food industry. Nanotechnol. Rev. 7, 623–641.

Nasrollahzadeh, M., Sajjadi, M., Dadashi, J., Ghafuri, H., 2020. Pd-based nanoparticles: plant-assisted biosynthesis, characterization, mechanism, stability, catalytic and antimicrobial activities. Adv. Colloid Interface Sci. 276, 102103. Available from: https://doi.org/10.1016/J.CIS.2020.102103.

Neme, K., Nafady, A., Uddin, S., Tola, Y.B., 2021. Application of nanotechnology in agriculture, postharvest loss reduction and food processing: food security implication and challenges. Heliyon 7, e08539.

Nicolopoulou-Stamati, P., Maipas, S., Kotampasi, C., Stamatis, P., Hens, L., 2016. Chemical pesticides and human health: the urgent need for a new concept in agriculture. Front. Public Health 4, 148. Available from: https://doi.org/10.3389/FPUBH.2016.00148/BIBTEX.

Nile, S.H., Baskar, V., Selvaraj, D., Nile, A., Xiao, J., Kai, G., 2020. Nanotechnologies in food science: applications, recent trends, and future perspectives. Nanomicro Lett. 12. Available from: https://doi.org/10.1007/s40820-020-0383-9.

Noah, N.M., Ndangili, P.M., 2022. Green synthesis of nanomaterials from sustainable materials for biosensors and drug delivery. Sens. Int. 3, 100166.

Ojeda, G.A., Arias Gorman, A.M., Sgroppo, S.C., 2019. La nanotecnologia y su aplicación en alimentos. Mundo Nano Rev. Interdisciplinaria En. Nanociencias y. Nanotecnología 12, 1. Available from: https://doi.org/10.22201/ceiich.24485691e.2019.23.67747.

Olobayotan, I., Akin-Osanaiye, B., 2019. Biosynthesis of silver nanoparticles using baker's yeast, *Saccharomyces cerevisiae* and its antibacterial activities. Access. Microbiol. 1, 526.

Onyeaka, H., Passaretti, P., Miri, T., Al-Sharify, Z.T., 2022. The safety of nanomaterials in food production and packaging. Curr. Res. Food Sci. 5, 763–774. Available from: https://doi.org/10.1016/J.CRFS.2022.04.005.

Panáček, A., Kvítek, L., Prucek, R., Kolář, M., Večeřová, R., Pizúrová, N., et al., 2006. Silver colloid nanoparticles: synthesis, characterization, and their antibacterial activity. J. Phys. Chem. B 110, 16248–16253. Available from: https://doi.org/10.1021/jp063826h.

Peters, R., Brandhoff, P., Weigel, S., Marvin, H., Bouwmeester, H., Aschberger, K., et al., 2017. Inventory of nanotechnology applications in the agricultural, feed and food sector. EFSA Support. Publ. 11. Available from: https://doi.org/10.2903/sp.efsa.2014.en-621.

Prajitha, N., Athira, S.S., Mohanan, P.V., 2019. Bio-interactions and risks of engineered nanoparticles. Environ. Res. 172, 98–108. Available from: https://doi.org/10.1016/J.ENVRES.2019.02.003.

Prasad, P., Kochhar, A., 2014. Active packaging in food industry: a review. IOSR J. Environ. Sci. Toxicol. Food Technol. 8, 1–7.

Prasad, R., Bhattacharyya, A., Nguyen, Q.D., 2017. Nanotechnology in sustainable agriculture: recent developments, challenges, and perspectives. Front. Microbiol. 8, 1014.

Rai, V., Acharya, S., Dey, N., 2012. Implications of nanobiosensors in agriculture. J. Biomater. Nanobiotechnol.

Ranjbar, S., Rahemi, M., Ramezanian, A., 2018. Comparison of nano-calcium and calcium chloride spray on postharvest quality and cell wall enzymes activity in apple cv. *Red delicious*. Sci. Hortic. 240, 57–64. Available from: https://doi.org/10.1016/j.scienta.2018.05.035.

Raza, M.A., Kanwal, Z., Rauf, A., Sabri, A., Riaz, S., Naseem, S., 2016. Size- and shape-dependent antibacterial studies of silver nanoparticles synthesized by wet chemical routes. Nanomaterials 6, 74. Available from: https://doi.org/10.3390/nano6040074.

Reddy, S.S., Chhabra, V., 2022. Nanotechnology: its scope in agriculture, J. Phys. Conf. Ser., 2267. IOP Publishing, p. 012112.

Rosenthal, A., Guedes, A.M.M., dos Santos, K.M.O., Deliza, R., 2021. Healthy food innovation in sustainable food system 4.0: integration of entrepreneurship, research, and education. Curr. Opin. Food Sci. 42, 215–223.

Roy, A., Bulut, O., Some, S., Mandal, A.K., Yilmaz, M.D., 2019. Green synthesis of silver nanoparticles: biomolecule-nanoparticle organizations targeting antimicrobial activity. RSC Adv. 9, 2673–2702. Available from: https://doi.org/10.1039/C8RA08982E.

Ruiz Rodríguez, L.G., Zamora Gasga, V.M., Pescuma, M., Van Nieuwenhove, C., Mozzi, F., Sánchez Burgos, J.A., 2021. Fruits and fruit by-products as sources of bioactive compounds. Benefits and trends of lactic acid fermentation in the development of novel fruit-based functional beverages. Food Res. Int. 140, 109854. Available from: https://doi.org/10.1016/j.foodres.2020.109854.

Sanzari, I., Leone, A., Ambrosone, A., 2019. Nanotechnology in plant science: to make a long story short. Front. Bioeng. Biotechnol. 7.

Sardella, D., Gatt, R., Valdramidis, V.P., 2017. Physiological effects and mode of action of ZnO nanoparticles against postharvest fungal contaminants. Food Res. Int. 101, 274–279. Available from: https://doi.org/10.1016/j.foodres.2017.08.019.

Singh, H., Sharma, A., Bhardwaj, S.K., Arya, S.K., Bhardwaj, N., Khatri, M., 2021. Recent advances in the applications of nano-agrochemicals for sustainable agricultural development. Environ. Sci. Process. Impacts 23, 213–239.

Singhal, J., Verma, S., Kumar, S., 2022. The physio-chemical properties and applications of 2D nanomaterials in agricultural and environmental sustainability. Sci. Total Environ. 837, 155669. Available from: https://doi.org/10.1016/J.SCITOTENV.2022.155669.

Solanki, P., Bhargava, A., Chhipa, H., Jain, N., Panwar, J., 2015. Nano-fertilizers and their smart delivery system. Nanotechnologies in Food and Agriculture. Springer, pp. 81–101.

Sun, D.W., Huang, L., Pu, H., Ma, J., 2021. Introducing reticular chemistry into agrochemistry. Chem. Soc. Rev. 50. Available from: https://doi.org/10.1039/c9cs00829b.

Taghavi Fardood, S., Ramazani, A., Moradnia, F., Afshari, Z., Ganjkhanlu, S., Yekke Zare, F., 2019. Green synthesis of ZnO nanoparticles via Sol-gel method and investigation of its application in solvent-free synthesis of 12-aryl-tetrahydrobenzo [α] xanthene-11-one derivatives under microwave irradiation. Chem. Methodol. 3, 632–642.

Taghinezhad, E., Ebadollahi, A., 2017. Potential application of chitosan-clay coating on some quality properties of lemon during storage. CIGR J. 19.

Tang, F., Uchikoshi, T., Sakka, Y., 2002. Electrophoretic deposition behavior of aqueous nanosized zinc oxide suspensions. J. Am. Ceram. Soc. 85, 2161–2165. Available from: https://doi.org/10.1111/j.1151-2916.2002.tb00428.x.

Tarannum, N., Singh, M., Hawaldar, R., 2019. Pesticides as an occupational hazard facts and figures. Handbook of Research on the Adverse Effects of Pesticide Pollution in Aquatic Ecosystems. IGI Global. Available from: https://doi.org/10.4018/978-1-5225-6111-8.CH012.

Tilman, D., Balzer, C., Hill, J., Befort, B.L., 2011. Global food demand and the sustainable intensification of agriculture. Proc. Natl Acad. Sci. USA 108, 20260–20264. Available from: https://doi.org/10.1073/PNAS.1116437108.

Tuantranont, A., Wisitsoraat, A., Kiatpathomchai, W., Khumwan, P., Karuwan, C., Primpray, V., 2020. Nanomaterials in agricultural and food applications. Handbook of Nanotechnology Applications: Environment, Energy, Agriculture and Medicine. Available from: https://doi.org/10.1016/B978-0-12-821506-7.00016-8.

Vurro, M., Miguel-Rojas, C., Pérez-de-Luque, A., 2019. Safe nanotechnologies for increasing the effectiveness of environmentally friendly natural agrochemicals. Pest. Manag. Sci. 75, 2403–2412.

Wu, H., Tito, N., Giraldo, J.P., 2017. Anionic cerium oxide nanoparticles protect plant photosynthesis from abiotic stress by scavenging reactive oxygen species. ACS Nano 11, 11283–11297. Available from: https://doi.org/10.1021/acsnano.7b05723.

Xing, Y., Li, W., Wang, Q., Li, X., Xu, Q., Guo, X., et al., 2019. Antimicrobial nanoparticles incorporated in edible coatings and films for the preservation of fruits and vegetables. Molecules 24, 1–30.

Xu, J., Zhang, M., Bhandari, B., Kachele, R., 2017. ZnO nanoparticles combined radio frequency heating: a novel method to control microorganism and improve product quality of prepared carrots. Innov. Food Sci. Emerg. Technol. 44, 46–53.

Xu, D., Qin, H., Ren, D., 2018. Prolonged preservation of tangerine fruits using chitosan/montmorillonite composite coating. Postharvest Biol. Technol. 143, 50–57. Available from: https://doi.org/10.1016/j.postharvbio.2018.04.013.

Xu, L., Zhu, Z., Sun, D.W., 2021. Bioinspired nanomodification strategies: moving from chemical-based agrosystems to sustainable agriculture. ACS Nano 15. Available from: https://doi.org/10.1021/acsnano.1c03948.

Yahia, E.M., 2017. The contribution of fruit and vegetable consumption to human health. In: De La Rosa, L.A., Alvarez-Parrilla, E., González-Aguilar, G.A. (Eds.), Fruit and Vegetable Phytochemicals. Wiley, pp. 3–51.

Yahia, E.M., Ornelas-Paz, J.D.J., Gonzalez-Aguilar, G.A., 2011. Nutritional and health-promoting properties of tropical and subtropical fruits. Postharvest Biology and Technology of Tropical and Subtropical Fruits. Elsevier, pp. 21–78.

Yazdanian, M., Rostamzadeh, P., Rahbar, M., Alam, M., Abbasi, K., Tahmasebi, E., et al., 2022. The potential application of green-synthesized metal nanoparticles in dentistry: a comprehensive review. Bioinorg. Chem. Appl. 2022.

Yigit, N., Velioglu, Y.S., 2019. Effects of processing and storage on pesticide residues in foods. Crit. Rev. Food Sci. Nutr. 60, 3622–3641. Available from: https://doi.org/10.1080/10408398.2019.1702501.

Yildirim, S., Röcker, B., Pettersen, M.K., Nilsen-Nygaard, J., Ayhan, Z., Rutkaite, R., et al., 2018. Active packaging applications for food. Compr. Rev. Food Sci. Food Saf. 17, 165–199. Available from: https://doi.org/10.1111/1541-4337.12322.

Yong-Jin, C., Chul-Jin, K., Namsoo, K., Chong-Tai, K., Bosoon, P., 2008. Some cases in applications of nanotechnology to food and agricultural systems. Biochip J. 2, 183–185.

Youssef, K., de Oliveira, A.G., Tischer, C.A., Hussain, I., Roberto, S.R., 2019. Synergistic effect of a novel chitosan/silica nanocomposites-based formulation against gray mold of table grapes and its possible mode of action. Int. J. Biol. Macromol. 141, 247–258. Available from: https://doi.org/10.1016/j.ijbiomac.2019.08.249.

Zhang, L., Yan, C., Guo, Q., Zhang, J., Ruiz-Menjivar, J., 2018. The impact of agricultural chemical inputs on environment: global evidence from informetrics analysis and visualization. Int. J. Low-Carbon Technol. 13, 338–352. Available from: https://doi.org/10.1093/IJLCT/CTY039.

Zhang, D., Ma, X., Gu, Y., Huang, H., Zhang, G., 2020a. Green synthesis of metallic nanoparticles and their potential applications to treat cancer. Front. Chem. 8, 799.

Zhang, P., Guo, Z., Zhang, Z., Fu, H., White, J.C., Lynch, I., 2020b. Nanomaterial transformation in the soil–plant system: implications for food safety and application in agriculture. Small 16. Available from: https://doi.org/10.1002/smll.202000705.

Zhao, F., Liu, J., Han, B., Luo, J., 2019. Investigation and validation of detection of storage stability of difenoconazole residue in Mango. J. Food Qual. 2019. Available from: https://doi.org/10.1155/2019/5641643.

Zhao, L., Lu, L., Wang, A., Zhang, H., Huang, M., Wu, H., et al., 2020. Nanobiotechnology in agriculture: use of nanomaterials to promote plant growth and stress tolerance. J. Agric. Food Chem. 68, 1935–1947. Available from: https://doi.org/10.1021/acs.jafc.9b06615.

Zhu, Y., Li, D., Belwal, T., Li, L., Chen, H., Xu, T., et al., 2019. Effect of nano-SiOx/chitosan complex coating on the physicochemical characteristics and preservation performance of green tomato. Molecules 24, 4552.

CHAPTER

Cyanobacterial biodiversity and their potential application in sustainable agriculture

10

Savita Singh[1], Shivani Singh[1], Rahul Prasad Singh[2] and Ajay Kumar[3]

[1]*Department of Botany, Babu Shivnath Agrawal College, Mathura, Uttar Pradesh, India*

[2]*Department of Botany, Amity Institute of Biotechnology, Amity University, Noida, Uttar Pradesh, India*

[3]*Amity Institute of Biotechnology, Amity University, Noida, Uttar Pradesh, India*

10.1 Introduction

The constantly increasing global human population has increased the stress of feeding and meeting the needs of the population. The choice of better environmentally friendly management strategies for boosting soil fertility and agricultural practices promises good long-term food security in light of limited land and a growing population. Environmental attributes are harmed by farming practices that use artificial fertilizers and pesticides. Microalgae, including cyanobacteria (blue–green algae), emerged as potential candidates for their use in the development of environmentally friendly and sustainable agricultural practices (Pathak et al., 2018). In their capacity as a natural biofertilizer, cyanobacteria contribute significantly to the preservation of soil structure through polysaccharide-mediated soil aggregation, improved soil fertility, reclamation of atmospheric nitrogen (N), expansion of soil pores through the production of adhesive substances, growth promotion through the excretion of growth hormones (auxin and gibberellins), expansion of water-holding capacity, reduction of soil salinity, and expansion of soil phosphate (Chamizo et al., 2018; Singh and Dhar, 2010; Múnera-Porras et al., 2020). The study of cyanobacteria with advantageous effects in the area of sustainable agriculture development should receive a lot of attention.

The human population of our planet is projected to reach ~9.7 billion by 2050, and majority of this increased population would be contributed by developing countries of Asia and Africa (DESA UN, 2015). To overcome this challenge, United Nations Foundation has taken a positive lead in this direction and launched the sustainable development goals (SDGs), which are a shared plan all across the globe to end extreme poverty, reduce inequality, and protect the planet

Sustainable Agricultural Practices. DOI: https://doi.org/10.1016/B978-0-443-19150-3.00002-3

by 2030. Zero hunger, affordable and clean energy, climate action, and protecting life on land and below water are some of the sustainable goals put forward by the United Nations. The productivity gains resulted from the "Green Revolution" have essentially reached a plateau, and feeding the increased global population is further challenged by limited availability of agriculture land. Conventional crop production methods cannot meet the demands of an ever-increasing population. There are possibly two good reasons for this: limited land and the overuse of environmentally harmful pesticides and insecticides. Around 16–17 kg N/t of N fertilizer, generally urea, are used to meet the crop's demands (Singh et al., 2016a,b,c). To meet the demands of such a large population, the range of nonconventional measures involving sustainable agricultural practices and sustainable alternatives can be used to enhance the food productivity per hectare of agricultural land (Parry et al., 2011; Pathak et al., 2018).

In particular, secondary metabolites, biofertilizers, bioenergy, bioprocessing, biopesticide manufacturing, waste treatment, and bioremediation are key applications of microbial biotechnology (Du et al., 2007; Mohammadi and Sohrabi, 2012). Soil preservation, crop selection, pest control, and processing are all part of sustainable agriculture. By creating new transgenic plants, microorganisms, and animals through the use of biotechnology and sustainable farming practices, productivity could be increased (Singh, 2000). Biotechnology has made a substantial contribution to sustainable agriculture during the past 10 years through the creation of stress resistance in microorganisms and plants, bioremediation of polluted soils, improved nitrogen fixation, and improved nutrient uptake efficiency (Singh, 2000). It is still up for dispute whether or not it is appropriate to grow genetically modified organisms outdoors.

Algae and cyanobacteria are one of the most diverse groups of organisms that occupy a major proportion of the tree of life (Hug et al., 2016). Cyanobacteria are gram-negative prokaryotic oxygenic photoautotrophs that evolved during the Precambrian epoch (4.5 billion years ago) and transformed the earth from an anoxic to an oxic environment (Catherine et al., 2019). Their morphology ranges from simple unicellular forms to branched filamentous forms (Rippka et al., 1979). Cyanobacteria occur in all types of habitats: fresh water, marine, acidic and alkaline soil, extreme habitats like ice and hot springs (Sanchez-Baracaldo et al., 2005). They are easy to grow and have biomass and product yields higher than traditional crops (Dismukes et al., 2008). They can be grown at large scale using nonarable land and nonpotable water. In this new area of industrial biotechnology, they are serving as bio-manufacturing platforms to produce food, feed, clean energy, biodegradable plastics as well as pharmaceuticals and cosmetic products. With shrinking land area for agriculture and living space, the necessity to look for better sustainable alternative becomes inevitable. Thus to channelize the idea of sustainability in agriculture cyanobacterial farming and its biotechnological application may serve promising alternatives. This chapter will be focusing on understanding cyanobacterial diversity and its relation to sustainable agriculture.

10.2 Cyanobacterial biodiversity across the globe

Cyanobacteria are the most common type of creature on the planet. They are autotrophic, found in a diverse environment, and have a global distribution ranging from temperate to tropical and Arctic to Antarctic regions (Jungblut et al., 2010). From an evolutionary perspective, these organisms have a very long evolutionary history and have been credited with making the Earth oxygenic due to photosynthetic oxygen evolution. Therefore the cyanobacteria exhibit considerable plasticity and thrive in a host of stressful environmental conditions. They are a diverse group of prokaryotes found in a wide array of aquatic and terrestrial habitats and are well adapted to hostile environments. Cyanobacteria are ubiquitous in nature and found commonly in lakes, ponds, springs, wetlands, streams, and rivers, and they play a major role in the nitrogen, carbon, and oxygen dynamics of many aquatic environments. They not only have wide range of habitat but also have a range of organization that range from unicellular, to filamentous, to colonial. Cyanobacteria not only persist as free living but also form symbiotic associations like *Richelia intracellularis* lives as endosymbionts in large diatoms (Flores et al., 2022; Gomez et al., 2005). Cyanobacteria are also endosymbiotic in fungi (lichens), bryophytes (*Anthoceros*), pteridophytes (*Azolla*), cycads (coralloid roots), angiosperms (*Gunnera*), and certain marine sponges and corals. It is also believed that the chloroplasts in plants and algae derived from an endosymbiosis in which a cyanobacterium was engulfed and retained within a colorless eukaryotic cell (Martin et al., 2015; Gray, 2017). Cyanobacteria are prokaryotic and lack internal organelles, a discrete nucleus and the histone proteins associated with eukaryotic chromosomes unlike eukaryotes and like all eubacteria, their cell walls contain peptidoglycan, not cellulose (like many algae and all plants). They show a wide variation in morphology. Their morphology ranges from simple unicellular forms to complex filamentous forms.

Cyanobacteria classify into *Chroococcales* (consisting of unicellular forms and aggregation of unicells in mucilage and colonial forms), *Chamaesiphonales* (attached forms), *Pleurocapsales* (pseudoparenchymatous, pseudofilamentous, and sarcinoid forms), *Nostocales* (unbranched heterocystous filamentous forms only), and *Stigonematales* (branched heterocystous forms) is the widely accepted classical system (Desikachary, 1973; Anand et al., 2019). Morphology remained an important method for classification of cyanobacteria (Anagnostidis and Komárek, 1985), however, with the advent of new biotechnological techniques focus has shifted to taxonomy based on fatty acid composition (chemo taxonomy) (Vandamme et al., 1996), DNA finger printing (Neilan et al., 1997; Casamatta et al., 2003), and 16s rRNA sequencing (Woese, 1987). Polyphasic approach which incorporates both genotypic and phenotypic properties is being preferred in recent times (Stackebrandt et al., 2002). Cyanobacteria have played a huge role in determining the course of evolution and ecological changes throughout Earth's history. In the late Proterozoic or the early Cambrian period, cyanobacteria began to take up residence within certain eukaryote cells, this event is called endosymbiosis, for the origin of the eukaryotes. They have the potential to fix atmospheric nitrogen, so that could be used as a biofertilizer for the cultivation of economically important crops such as rice and beans.

10.3 Sustainable agriculture and cyanobacteria

With increasing population, the pressure on land for more agricultural produce is increasingly high (Guihéneuf et al., 2016). The coherence of supply and demand is difficult to be met unless some sustainable options are undertaken. In this lieu, cyanobacteria and microalgae present promising means to meet agricultural need in a sustainable manner. The biogeochemical cycles of carbon, nitrogen, and oxygen are profoundly influenced by cyanobacteria. Cyanobacteria have the ability to fix nitrogen, which helps them thrive and survive in environments with little or no combined nitrogen. This characteristic of cyanobacteria makes them crucial for agriculture and industry as biofertilizers (Singh, 1961, 2014; Vaishampayan et al., 2001; Singh et al., 2016a,b,c). They can grow easily using simple nutrients using photosynthetically active radiation (PAR) producing varied products with high protein, carbohydrate, and lipid content (Pisciotta et al., 2010), thus decreasing the load on traditional agriculture. A variety of products are obtained such as food, feed, pharma, cosmetics, bio-plastics, and clean energy sources (Rittmann, 2008).

The capacity of some cyanobacteria to form symbiotic relationships could be used to create consortiums of microorganisms for use in the bioremediation of polluted soils or aquatic systems. Cyanobacterial strains that are native to a region and have adapted to its climatic conditions can endure in moist soils. This has a considerable impact on such soils' nutritional condition, structural stability, and crop productivity. The exopolysaccharides (EPS) in cyanobacteria provide 25% of their total biomass. Cyanobacterial activities take place in the upper crust of the soil, and EPS acts as a binding agent for soil particles. The ability of the EPS to keep soil particles together causes soil aggregation, the buildup of organic matter, and an increase in the top soil's capacity to hold water. This increase in soil moisture and organic matter can help rhizobacteria, which promote growth of plant-growth promoting rhizobacteria (PGPR). Thus chemical and physical properties of soil are positively affected by cyanobacterial growth, PGPRs and EPS-producing cyanobacteria may help to rehabilitate and reclaim infertile soils (Zulpa et al., 2003; Paul and Nair, 2008). The PGPR and cyanobacteria collaboration promotes soil fertility and nutrient uptake, which in turn promotes plant development. Additionally, this consortium increases plants' resistance to environmental stressors including drought and salinity (Prasanna et al., 2012; Singh, 2014). Before submitting an application for a consortium in field conditions, community structure and variety of cyanobacteria should be thoroughly studied, especially in relation to environmental variables and ecosystem activities. The following sections present various cyanobacteria qualities that can be used to create environmentally friendly sustainable agricultural practices.

10.3.1 Unique nitrogen fixation and phosphate uptake ability

Though nitrogen occupies 78% of earth's atmosphere, still it remains a limiting nutrient for plants. Nitrogen is an essential element contained in many biomolecules necessary to sustain life (Ferguson, 1998). This necessitated the development of alternate, sustainable, and cost-effective biologically available nitrogen resources which can fulfill the nitrogen demand of agriculture in sustainable manner. Nitrogen fixation is a very expensive energy requiring process for a cell (16 ATP is required). Agriculture is heavily dependent on use of urea and potash for increasing the yield of crop plants (Rosenblueth et al., 2018). Providing nitrogen to plants without ill effects of pesticides and insecticides and fertilizers poses a daunting task for agriculture. For this purpose, biological systems have been identified which can fix atmospheric dinitrogen (Vaishampayan et al., 2001). Biological nitrogen fixation contributes $\sim 2 \times 10^2$ Mt of nitrogen annually (Guerrero et al., 1981). According to Metting (1988), the total nitrogen fixation can be ~ 90 kg N/ha/y. One of the most special attributes with which some of the heterocystous cyanobacteria are bestowed is the ability to fix atmospheric nitrogen. The free-living cyanobacteria fix <10 kg N/ha/Y, however, annually $\sim 10-30$ kg N/ha is fixed by dense mats of cyanobacteria (Aiyer et al., 1972; Watanabe et al., 1977). Therefore cyanobacteria constitute an important component of naturally available biofertilizers (Prasanna et al., 2013). Rice production in tropical countries mainly depends on biological N_2 fixation by cyanobacteria which are a natural component of paddy fields (Vaishampayan et al., 2001). In these cultivated agriculture systems, annually ~ 32 Tg of nitrogen is fixed by biological nitrogen fixers (Singh et al., 2016a,b,c), and cyanobacteria add about 20–30 kg fixed nitrogen/ha along with organic matter to the paddy fields (Issa et al., 2014).

Nitrogen fixation takes place in specialized cells known as heterocyst (Bothe et al., 2010). Heterocysts have thickened cell envelope to create micro-aerobic condition for oxygen labile nitrogenase enzyme (Awai and Wolk, 2007). Several heterocystous cyanobacterial genera make symbiotic associations with different photosynthetic and nonphotosynthetic organisms such as algae, fungi, diatoms, bryophytes, hornworts, liverworts, mosses, pteridophytes, gymnosperms, and angiosperms (Rai et al., 2000; Sarma et al., 2016). Other than cyanobacteria, only leguminous plants having *Bacteroides* and some bacteria have the ability to fix nitrogen. South Asian countries were the first to understand and use cyanobacteria like *Anaebaena*, *Nostoc*, pteridophyte *Azolla* in the rice fields (Iniesta-Pallarés et al., 2021; Hasan, 2015; Vijayan and Ray, 2015; Pham et al., 2017).

Phosphorus is the second most important macronutrient required by the plants, next to nitrogen. Phosphorous is also one of the limiting nutrients as it readily gets oxidized to various forms which cannot be utilized by plants. Cyanobacteria have the ability to solubilize insoluble $(Ca)_3(PO4)_2$ (tricalcium diphosphate); $FePO_4$ (ferric orthophosphate); $AlPO_4$ (aluminum phosphate), and $Ca_5(PO4)_3(OH)$ (hydroxylapatite) present in soils, sediments, or in pure cultures (Afkairin et al., 2021; Vaishampayan et al., 2001; Yadav et al., 2021). Application of phosphate solubilizing microbes such as *Calothrix braunii* by inoculating in soil may be an efficient way to convert the insoluble P compounds to plant available P form, resulting in better plant growth, crop yield, and quality (Sharma et al., 2013; Kalayu, 2019).

10.3.2 Plant growth promoting activity

The increase in soil moisture and organic content can support the survival and growth of PGPR. Thus cyanobacterial growth positively alters the chemical and physical property of soils, and PGPRs along with EPS-producing cyanobacteria may contribute to an improvement and reclamation of infertile soils (Flaibani et al., 1989; Zulpa et al., 2003; Paul and Nair, 2008). Cyanobacteria release a number of compounds such as growth regulators (amino acids, auxins, gibberellins, cytokinins, and abscisic acid), various metabolites, and EPS which have been found to promote plant growth (Abdel-Raouf et al., 2012). Free-living and symbiotic species of cyanobacteria like *Nostoc*, *Chlorogloeopsis*, *Calothrix*, *Plectonema*, *Gloeothece*, *Anabaena*, *Cylindrospermum*, and *Anabaenopsis* have been reported to produce IAA. Many among these species have been analyzed to produce indole-3-acetic acid in association with wheat rhizosphere (Natarajan et al., 2012). Hashtroudi et al. (2013) determined and quantified three auxins (IAA, APA and IBA) in two heterocystous cyanobacteria, that is, *Anabaena vaginicola* and *Nostoc calcicola*. Some cyanobacterial species (*Anabaena* sp. Ck1 and *Chroococcidiopsis* sp. Ck4) were shown to release both cytokinin and IAA, and inoculation of wheat with these species resulted in enhancement of seed germination, shoot length, tillering, number of lateral roots, spike length, and grain weight of plant (Hussain and Hasnain, 2011). The consortium of PGPR and cyanobacteria increases the plant growth by improving the soil fertility and nutrient utilization. In addition, this consortium also enhances the tolerance of plants against environmental stresses such as drought and salinity (Prasanna et al., 2012; Singh, 2014).

10.3.3 Biodegradation of pesticides and insecticides

In order to increase crop production, use of pesticides and insecticides has increased enormously. This has resulted in adverse impact on the ecosystem and polluting it. Some cyanobacterial strains have been reported with the remarkable tolerance against pesticides and many of them also degrade and utilize the pesticides as a nutrient (Tiwari et al., 2017a,b). These abilities of cyanobacteria can be used in the bioremediation technologies for decontamination of pesticides from the contaminated sites. Cyanobacterial species like *Nostoc linckia*, *Nostoc muscorum*, *Oscillatoria animalis*, and *Phormidium foveolarum* degraded methyl parathion, an organophosphorus insecticide (Fioravante et al., 2010). Similarly, *N. muscorum* and *Anabaena fertilissima* degraded monocrotophos, malathion, dichlorovos, and phosphomidon (Singh et al., 2013c; Singh et al., 2011). Species like *Oscillatoria, Synechococcus, Nodularia, Nostoc, Cyanothece*, and *Anabaena cylindrica* possessed the ability to degrade lindane either individually or in combination (El-Bestawy et al., 2007). *Anabaena* sp., *Nostoc* sp., *Lyngbya* sp., *Spirulina* sp., and *Microcystis* sp. are helpful in removing the contamination of glyphosate herbicide from agricultural sites (Forlani et al., 2008; Lipok et al., 2009). *Oscillatoria-Gammaproteobacteria* consortium has been found helpful in degradation of phenanthrene, dibenzothiophene, and n-octadecane (Abed and Köster, 2005).

10.3.4 Biocontrol agents

Biological control through the use of microorganisms is an attractive approach for disease control without negative impact of synthetic fertilizers. Disease outbreaks are major constraints to sustainable agricultural practices; hence, there is an urgent need to exploit the role of microorganisms in achieving the cost-effective, long-term durable food security without any side effects on environment. Fungi and bacteria are the chief biocontrol agents against plant pathogens particularly soilborne fungi. Although the use of cyanobacteria as biofertilizers is well known for a long time, recently, attention has been focused on the role of cyanobacterial secondary metabolites in phytopathogen control (Kulik, 1995) and their potential application in crop protection (Singh, 2014). The use of cyanobacterial inoculants as antagonists of phytopathogens provides a promising alternative to chemical fertilizers and pesticides which are detrimental for the health of environment. Cyanobacterial bioactive compounds span over a range of biological activities and chemical structures including fatty acids, alkaloids, peptides, polypeptides, amides, indoles, lipopeptides, and other organic chemicals (Yadav et al., 2018; Yadav et al., 2022a,b). These metabolites have a role in allelopathy that involves the use of inhibition potential of biologically active metabolites secreted by one species to the growth of other sympatric species. A number of studies have shown that cyanobacteria application as inoculants may be helpful in mitigating many of plant pathogens causing damage to the crop/vegetable produce. Cyanobacteria isolates have been found effective for the control of bacterial wilt disease and promote growth of chili plant (Yanti et al., 2019). Studies suggested that the allelopathic activity of cyanobacteria is because of phenol and polysaccharide contents (Mohamed et al., 2011). Tiwari and Kaur (2014) have found that *Spirulina platensis* showed more allelopathic activity as compared with other cyanobacterial strains like *Anabaena variabilis* and *Synechococcus elongatus* because of higher phenol and polysaccharide concentration. In another study, it was found that cyanobacteria (*S. platensis*) and bacteria (*Pseudomonas fluorescens*) showed a good potential to control powdery mildew of chamomile and improve the growth and yield of blossoms as well as oil yield (Ghebrial et al., 2019).

10.3.5 Restoration of soil fertility

Apart from serving as biofertilizer, biocontrol agent, and other fields mentioned above, cyanobacteria also play a key role in improving soil fertility and quality thus indirectly contributing in crop improvement. Reclamation of organically poor soils with cyanobacterial application has been suggested. Singh (1961) reported the biofertilizer potential as well as the ability of the cyanobacteria to reclaim the usar soils. Singh and Singh (1987) observed an increase in the nitrogen availability of rice plants due to the application of cyanobacteria. Soils from usar regions are found to be unfriendly to the crops because of high compaction, low fertility, and water deficiency (Nisha et al., 2007). Consequence of such soil characteristics leads to poor aeration, water infiltration, soil erosion, and poor diversity of microflora. Such deprived physicochemical attribute of soil hinders plant growth and productivity. Cyanobacteria have great potential as biofertilizers due to their ability to fix atmospheric nitrogen either free living or forming symbiotic association (El-Zeky et al., 2005). Cyanobacterial strains as natural biofertilizer for paddy cultivation has been reported in China, Philippines, Egypt, and India (Ladha and Reddy, 2003). The filaments of cyanobacteria form a network on the soil and help in binding of soil particles. The EPS produced by cyanobacteria have the ability to reclaim and help soil particles bind together and help in the improvement of soil moisture content (Flaibani et al., 1989; Mazor et al., 1996; Singh et al., 2019; Yadav et al., 2022a; Yadav et al., 2022b). The EPS in cyanobacteria play a major role in protecting cells from various stresses in several habitats (Singh et al., 2016a,b,c; Yadav et al., 2022a; Yadav et al., 2022b). Arora et al. (2010) studied the effect of cyanobacterial EPS on salt stress alleviation and seed germination in wheat, maize, and rice. Reclamation of the desert soils has been reported in the presence of EPS from cyanobacteria (Flaibani et al., 1989). EPS layer on soil shows different morphological forms such as capsules surrounding the cells and soil grains, slime surrounding the cyanobacterial colony (De Philippis and Vincenzini, 2003; Yadav et al., 2022a; Yadav et al., 2022b). Alteration in soil moisture content affects EPS synthesis as soil-water content directly influences metabolic activity in cyanobacteria (Satoh et al., 2002). Altered environmental conditions could affect the EPS synthesis that further influences the nutrient-capturing ability of cyanobacteria in the topsoil. In addition, the amount of EPS produced can be correlated with the soil particle aggregation and soil stability (Wolfaardt et al., 1999).

10.4 Conclusion and future prospects

There is a huge possibility of cyanobacteria meeting the aims of sustainable agriculture directly as well as indirectly. Cyanobacteria are excellent biomanufacturing platform producing a large number of commercially important products. Humanity is at the vicinity of a severe shortage of food, water, and space caused by global warming and modern agriculture practices. Hence, there is an urgent need to use cyanobacteria to meet out increasing demand of food as they are adopted to grow at environmental extremes and stressed environment. Cyanobacterial soil farming can help in developing sustainable agricultural practice by the generation of biomass that can be used for the sustainable production of food supplements, nutraceuticals, biofertilizers, and feeds for increasing populations. The overall study stated that cyanobacteria biomass can be utilized for improving the quality of food products, physicochemical properties of soil, controlling soilborne diseases, added organic matter, release growth-promoting substances, solubilize the insoluble phosphates, use as nutraceuticals and also apply in pharmaceuticals. Hence, biofertilizers prepared from cyanobacteria are economical and environment-friendly.

References

Abdel-Raouf, N., Al-Homaidan, A.A., Ibraheem, I.B.M., 2012. Agricultural importance of algae. Afr. J. Biotechnol. 11 (54), 11648–11658.

Abed, R.M.M., Köster, J., 2005. The direct role of aerobic heterotrophic bacteria associated with cyanobacteria in the degradation of oil compounds. Int. Biodeterior. Biodegrad. 55, 29–37.

Afkairin, A., Ippolito, J.A., Stromberger, M., Davis, J.G., 2021. Solubilization of organic phosphorus sources by cyanobacteria and a commercially available bacterial consortium. Appl. Soil. Ecol. 162. Available from: https://doi.org/10.1016/j.apsoil.2021.103900.

Aiyer, R.S., Sulahudean, S., Venkataraman, G.S., 1972. Long-term algalization field trial with high yielding rice varieties. Ind. J. Agric. Sci. 42, 380–383.

Anagnostidis, K., Komárek, J., 1985. Modern approach to the classification system of cyanophytes 1- introduction. Arch. für Hydrobiologie 71, 291–302.

Anand, N., Thajuddin, N., Dadheech, P.K., 2019. Cyanobacterial taxonomy: morphometry to molecular studies. Cyanobacteria. Available from: https://doi.org/10.1016/B978-0-12-814667-5.00003-9.

Arora, M., Kaushik, A., Rani, N., Kaushik, C.P., 2010. Effect of cyanobacterial exopolysaccharides on salt stress alleviation and seed germination. J. Environ. Biol. 31 (5), 701–704.

Awai, K., Wolk, C., 2007. Identification of the glycosyl transferase required for synthesis of the principal glycolipid characteristic of heterocysts of *Anabaena* sp strain PCC 7120. FEMS Microbiol. Lett. 266, 98–102.

Bothe, H., Schmitz, O., Yates, O.G., Newton, W.E., 2010. Nitrogen fixation and hydrogen metabolism in cyanobacteria. Microbiol. Mol. Biol. Rev. 74 (4), 529–551.

Casamatta, D.A., Vis, M.L., Sheath, R.G., 2003. Cryptic species in cyanobacterial systematics, a case study of *Phormidium retzii* (Oscillatoriales) using RAPD molecular markers and 16S rDNA sequence data. Aquat. Bot. 77, 295–309.

Chamizo, S., Mugnai, G., Rossi, F., Certini, G., De Philippis, R., 2018. Cyanobacteria inoculation improves soil stability and fertility on different textured soils: gaining insights for applicability in soil restoration. Front. Environ. Sci. 6, 49.

De Morais, M.G., Costa, J.A.V., 2007. Carbon dioxide fixation by *Chlorella kessleri*, *C. vulgaris*, *Scenedesmus obliquus* and *Spirulina* sp. cultivated in flasks and vertical tubular photobioreactors. Biotechnol. Lett. 29, 1349–1352. Available from: https://doi.org/10.1007/s10529-007-9394-6.

De Philippis, R., Vincenzini, M., 2003. Outermost polysaccharidic investments of cyanobacteria: nature, significance and possible applications. Recent Res. Dev. Microbiol. 7, 13–22.

Demoulin, C.F., Lara, Y.J., Cornet, L., François, C., Baurain, D., Wilmotte, A., et al., 2019. Cyanobacteria evolution: insight from the fossil record. Free Radic. Biol. Med. 140, 206–223. Available from: https://doi.org/10.1016/j.freeradbiomed.2019.05.007.

DESA UN, 2015. World population prospects: the 2015 revision, key findings and advance tables. Working Paper No ESA/P/WP. 241, United Nations, Department of Economic and Social Affairs, Population Division, New York, NY.

Desikachary, T.V., 1973. Status of classical taxonomy. In: Carr, N.G., Whitton, B.A. (Eds.), The Biology of Blue-Green Algae. Blackwell Scientific Publications, Oxford, pp. 473–481.

Dismukes, G.C., Carrieri, D., Bennette, N., Ananyev, G.M., Posewitz, M.C., 2008. Aquatic phototrophs: efficient alternatives to land-based crops for biofuels. Curr. Opin. Biotechnol. 19, 235–240.

El-Bestawy, E.A., Abd El-Salam, A.Z., Mansy, H.A.R., 2007. Potential use of environmental cyanobacterial species in bioremediation of lindane contaminated effluents, Int. Biodeteriorat. Biodegrad., 59. pp. 180–192. Available from: http://www.sciencedirect.com/science/article/pii/S0964830506002083.

El-Zeky, M., El-Shahat, R., Metwaly, G.S., Elham, M.A., 2005. Using Cyanobacteria or Azolla as alternative nitrogen sources for rice production. J. Agric. Mansoura Univ. 30 (9), 5567–5577.

Ferguson, S.J., 1998. Nitrogen cycle enzymology. Curr. Opin. Chem. Biol. 2, 182–193.

Fioravante, I.A., Barbosa, F.A.R., Augustic, R., Magalhães, S.M.S., 2010. Removal of methyl parathion by cyanobacteria *Microcystis novacekii* under culture conditions. J. Environ. Monit. 12, 1302–1306. Available from: https://doi.org/10.1039/b923288e.

Flaibani, A., Olsen, Y., Painter, T.J., 1989. Polysaccharides in desert reclamation: composition of exocellular proteoglycan complexes produced by filamentous blue-green and unicellular green edaphic algae. Carbohydr. Res. 190 (2), 235–248. Available from: https://doi.org/10.1016/0008-6215(89)84128-X.

Flores, E., Romanovicz, D.K., Nieves-Morión, M., Foster, R.A., Villareal, T.A., 2022. Adaptation to an intracellular lifestyle by a nitrogen-fixing, heterocyst-forming cyanobacterial endosymbiont of a diatom. Front. Microbiol. 13, 1–13.

Forlani, G., Pavan, M., Gramek, M., Kafarski, P., Lipok, J., 2008. Biochemical bases for a widespread tolerance of cyanobacteria to the phosphonate herbicide glyphosate. Plant. Cell Physiol. 49, 443–456.

Ghebrial, E.W.R., El-Abeid, S.E., Aref, E.M., 2019. Bio-control of chamomile powdery mildew using cyanobacteria and some antagonistic microorganisms. Egypt. J. Phytopathol. 47 (1), 53–77.

Gomez, F., Furuya, K., Takeda, S., 2005. Distribution of the cyanobacterium *Richelia intracellularis* as an epiphyte of the diatom *Chaetoceros compressus* in the western Pacific Ocean. J. Plankton Res. 27 (4), 323–330.

Gray, M.W., 2017. Lynn Margulis and the endosymbiont hypothesis: 50 years later. Mol. Biol. Cell 28, 1285–1287.

Guerrero, M.G., Vega, J.M., Losada, M., 1981. The assimilatory nitratereducing system and its regulation. Annu. Rev. Plant Physiol. 32, 168–204. Available from: https://doi.org/10.1146/annurev.pp.32.060181.001125.

Guihéneuf, F., Khan, A., Tran, L.S.P., 2016. Genetic engineering: a promising tool to engender physiological, biochemical, and molecular stress resilience in green microalgae. Front. Plant Sci. 7, 400. Available from: https://doi.org/10.3389/fpls.2016.00400.

Hasan, M., 2015. Investigation on the nitrogen fixing cyanobacteria (BGA) in rice fields of north-west region of Bangladesh. III: filamentous (heterocystous). J. Environ. Sci. Nat. Resour. 6, 253–259.

Hashtroudi, M.S., Ghassempour, A., Riahi, H., Shariatmadari, Z., Khanjir, M., 2013. Endogenous auxins in plant growth-promoting cyanobacteria *Anabaena vaginicola* and *Nostoc calcicola*. J. Appl. Phycol. 25, 379–386. Available from: http://link.springer.com/article/10.1007/s10811-012-9872-7.

Hug, L.A., Baker, B.J., Anantharaman, K., Brown, C.T., 2016. A new view of the tree of life. Nat. Microbiol. 5, 16048. Available from: https://doi.org/10.1038/nmicrobiol.2016.48.

Hussain, A., Hasnain, S., 2011. Phytostimulation and biofertilization in wheat by cyanobacteria. J. Ind. Microbiol. Biotech. 38, 85–92.

Iniesta-Pallarés, M., Álvarez, C., Gordillo-Cantón, F.M., Ramírez-Moncayo, C., Alves-Martínez, P., Molina-Heredia, F.P., et al., 2021. Sustaining rice production through biofertilization with N2-fixing cyanobacteria. Appl. Sci. 11, 4628. Available from: https://doi.org/10.3390/app11104628.

Issa, A.A., Abd-Alla, M.H., Ohyama, T., 2014. Nitrogen fixing cyanobacteria: future prospect. In: Ohyama, T. (Ed.), Advances in Biology and Ecology of Nitrogen Fixation. InTech. Available from: https://doi.org/10.5772/56995.

Jungblut, A.D., Lovejoy, C., Vincent, W.F., 2010. Global distribution of cyanobacterial ecotypes in the cold biosphere. ISME J. 4 (2), 191–202.

Kalayu, G., 2019. Phosphate solubilizing microorganisms: promising approach as biofertilizers. Int. J. Agron.

Kulik, M.M., 1995. The potential for using cyanobacteria (blue-green algae) and algae in the biological control of plant pathogenic bacteria and fungi. Eur. J. Plant Pathol. 101 (6), 585–599.

Ladha, J.K., Reddy, P.M., 2003. Nitrogen fixation in rice systems: state of knowledge and future prospects. Plant Soil 252 (1), 151–167.

Lipok, J., Wieczorek, D., Jewginski, M., Kafarski, P., 2009. Prospects of in vivo 31P NMR method in glyphosate degradation studies in whole cell system. Enzyme Microb. Technol. 44, 11–16.

Martin, W.F., Garg, S., Zimorski, V., 2015. Endosymbiotic theories for eukaryote origin. Phil. Trans. R. Soc. B370, 20140330. Available from: https://doi.org/10.1098/rstb.2014.033.

Mazor, G., Kidron, G.J., Vanshak, A., Abeliovich, A., 1996. The role of cyanobacterial exopolysaccharides in structuring desert microbial crusts. FEMS Microbiol. Ecol. 21, 121–130.

Metting, B., 1988. "Microalgae in agriculture. In: Borowitzka, M.A., Borowitzka, L.J. (Eds.), Micro-Algal Biotechnology. Cambridge University Press, Cambridge, pp. 288–304.

Mohamed, E.H., Mostafa, E., Metwally, A.M., Abd, A.I., Mona, M.I., 2011. Antagonistic activity of some fungi and cyanobacteria species. Int. J. Plant Pathol. 2, 101–114.

Mohammadi, K., Sohrabi, Y., 2012. Bacterial biofertilizers for sustainable crop production: a review. J. Agric. Biol. Sci. 7, 307–316. Available from: https://doi.org/10.1186/1475-2859-13-66.

Múnera-Porras, L.M., García-Londoño, S., Ríos-Osorio, L.A., 2020. Action mechanisms of plant growth promoting cyanobacteria in crops in situ: a systematic review of literature. Int. J. Agron.

Natarajan, C., Prasanna, R., Gupta, V., Dureja, P., Nain, L., 2012. Dissecting the fungicidal activity of *Calothrix elenkinii* using chemical analyses and microscopy. Appl. Biochem. Microbiol 48, 51–57. Available from: https://doi.org/10.1134/S0003683812010115, http://link.spr.inger.com/article/.

Neilan, B.A., Jacobs, D., del Dot, T., Blackall, L.L., Hawkins, P.R., Cox, P.T., et al., 1997. rRNA sequences and evolutionary relationships among toxic and nontoxic cyanobacteria of the genus *Microcystis*. Int. J. Syst. Bacteriol. 47, 693–697.

Parry, M.A.J., Reynolds, M., Salvucci, M.E., Raines, C., Andralojc, P.J., Zhu, X.G., et al., 2011. Raising yield potential of wheat. II. Increasing photosynthetic capacity and efficiency. J. Exp. Bot. 62, 453–467. Available from: https://doi.org/10.1093/jxb/erq304.

Pathak, J., Maurya, P.K., Singh, S.P., Häder, D.P., Sinha, R.P., 2018. Cyanobacterial farming for environment friendly sustainable agriculture practices: innovations and perspectives. Front. Environ. Sci. 6, 7.

Paul, D., Nair, S., 2008. Stress adaptations in a plant growth promoting rhizobacterium (PGPR) with increasing salinity in the coastal agricultural soils. J. Basic Microbiol. 48, 378–384. Available from: https://doi.org/10.1002/jobm.200700365.

Pham, H.T., Nguyen, L.T., Duong, T.A., Bui, D.T., Doan, Q.T., Nguyen, H.T., et al., 2017. Diversity and bioactivities of nostocacean cyanobacteria isolated from paddy soil in Vietnam. Syst. Appl. Microbiol. 40, 470–481.

Pisciotta, J.M., Zou, Y., Baskakov, I.V., 2010. Light-dependent electrogenic activity of cyanobacteria. PLoS ONE 5, e10821. Available from: https://doi.org/10.1371/journal.pone.0010821.

Prasanna, R., Joshi, M., Rana, A., Shivay, Y.S., Nain, L., 2012. Influence of co-inoculation of bacteria-cyanobacteria on crop yield and C-N sequestration in soil under rice crop. World J. Microbiol. Biotechnol. 28, 1223–1235. Available from: https://doi.org/10.1007/s11274-011-0926-9.

Prasanna, R., Sharma, E., Sharma, P., Kumar, A., Kumar, R., Gupta, V., et al., 2013. Soil fertility and establishment potential of inoculated cyanobacteria in rice crop grown under nonflooded conditions. Paddy Water Environ. 11, 175–183. Available from: https://doi.org/10.1007/s10333-011-0302-2.

Rai, A.N., Söderbäck, E., Bergman, B., 2000. Cyanobacterium-plant symbioses. N. Phytol. 147, 449–481. Available from: https://doi.org/10.1046/j.1469-8137.2000.00720.x.

Rippka, R., Deruelles, J., Waterbury, J.B., Herdman, Stanier, R.Y., 1979. Generic assignments, strain histories and properties of pure cultures of cyanobacteria. J. Gen. Microbiol. 111, 1–61.

Rittmann, B.E., 2008. Opportunities for renewable bioenergy using microorganisms. Biotechnol. Bioeng. 100, 203–212. Available from: https://doi.org/10.1002/bit.21875.

Rosenblueth, M., Ormeño-Orrillo, E., López-López, A., Rogel, M.A., Reyes, H.B.J., Martínez-Romero, J.C., 2018. Nitrogen fixation in cereals. Front. Microbiol. 9, 1794. Available from: https://doi.org/10.3389/fmicb.2018.01794.

Sanchez-Baracaldo, P., Hayes, P.K., Blank, C.E., 2005. Morphological and habitat evolution in the Cyanobacteria using a compartmentalization approach. Geobiology 3, 145–165.

Sarma, M.K., Kaushik, S., Goswami, P., 2016. Cyanobacteria: a metabolic power house for harvesting solar energy to produce bio-electricity and biofuels. Biomass Bioenerg. 90, 187–201. Available from: https://doi.org/10.1016/j.biombioe.2016.03.043.

Satoh, K., Hirai, M., Nishio, J., Yamaji, T., Kashino, Y., Koike, H., 2002. Recovery of photosynthetic systems during rewetting is quite rapid in a terrestrial cyanobacterium, *Nostoc commune*. Plant Cell Physiol. 43 (2), 170–176.

Sharma, S.B., Sayyed, R.Z., Trivedi, M.H., Gobi, T.A., 2013. Phosphate solubilizing microbes: sustainable approach for managing phosphorus deficiency in agricultural soils. SpringerPlus 2, 587.

Singh, A.L., Singh, P.K., 1987. Influence of *Azolla* management on the growth, yield of rice and soil fertility. II. N and P contents of plants and soil. Plant Soil 102, 49–54.

Singh, D.P., Khattar, J.I.S., Kaur, M., Kaur, G., Gupta, M., Singh, Y., 2013c. Anilofos tolerance and its mineralization by the cyanobacterium *Synechocyslls* sp. strain PUPCCC 64. PLoS ONE 8 (1), 1–10. Available from: https://doi.org/10.1371/journal.pone.0053445.

Singh, D.P., Khattar, J.I.S., Nadda, J., Singh, Y., Garg, A., Kaur, N., et al., 2011. Chlorpyrifos degradation by the cyanobacterium *Synechocystis* sp.strain PUPCCC 64. Environ. Sci. Pollut. Res. 18, 1351–1359. Available from: http://link.springer.com/article/10.1007%2Fs11356-011-0472-x.

Singh, N.K., Dhar, D.W., Tabassum, R., 2016a. Role of cyanobacteria in crop protection. Proc. Natl. Acad. Sci. India Sect. B Boil. Sci 86, 1–8.

Singh, R.N., 1961. Role of Blue-Green Algae in Nitrogen Economy of Indian Agriculture. Indian Council of Agricultural Research, New Delhi.

Singh, S., Kant, C., Yadav, R.K., Reddy, Y.P., Abraham, G., 2019. Cyanobacterial exopolysaccharides: composition, biosynthesis, and biotechnological applications. Cyanobacteria. Available from: https://doi.org/10.1016/B978-0-12-814667-5.00017-9.

Singh, S., Verma, E., Niveshika, Tiwari, B., Mishra, A.K., 2016b. Exopolysaccharide production in *Anabaena* sp. PCC 7120 under different $CaCl_2$ regimes. Physiol. Mol. Biol. Plants 22 (4), 557–566.

Singh, J.S., 2014. Cyanobacteria: a vital bio-agent in eco-restoration of degraded lands and sustainable agriculture. Clim. Change Environ. Sustain. 2, 133–137.

Singh, J.S., Kumar, A., Rai, A.N., Singh, D.P., 2016c. Cyanobacteria: a precious bioresource in agriculture, ecosystem, and environmental sustainability. Front. Microbiol. 7 (529), 00529. Available from: https://doi.org/10.3389/fmicb.2016.

Singh, N.K., Dhar, D.W., 2010. Cyanobacterial reclamation of salt-affected soil. Genetic Engineering, Biofertilisation, Soil Quality and Organic Farming. Springer, Dordrecht, pp. 243–275.

Singh, R.B., 2000. Biotechnology, biodiversity and sustainable agriculturea contradiction? In: Mendoza, E.M.T. (Ed.), Regional Conference in Agricultural Biotechnology Proceedings: Biotechnology Research and Policy-Needs and Priorities in the Context of Southeast Asia's Agricultural Activities. Bangkok: SEARCA (SEAMEO)/FAO/APSA.

Stackebrandt, E., Frederiksen, W., Garrity, G.M., 2002. Report of the ad hoc committee for the reevaluation of the species definition in bacteriology. Int. J. Syst. Evol. Microbiol. 52, 1043–1047.

Tiwari, B., Chakraborty, S., Srivastava, A.K., Mishra, A.K., 2017b. Biodegradation and rapid removal of methyl parathion by the paddy field cyanobacterium *Fischerella* sp. Algal Res. 25, 285–296.

Tiwari, B., Singh, S., Chakraborty, S., Verma, E., Mishra, A.K., 2017a. Sequential role of biosorption and biodegradation in rapid removal, degradation and utilization of methyl parathion as a phosphate source by a new cyanobacterial isolate *Scytonema* sp. BHUS-5. Int. J. Phytorem. 19, 884–893.

Tiwari, A., Kaur, A., 2014. Allelopathic impact of cyanobacteria on pathogenic fungi. Int. J. Pure Appl. Biosci. 2 (3), 63–70.

Vaishampayan, A., Sinha, R.P., Häder, D.-P., Dey, T., Gupta, A.K., Bhan, U., et al., 2001. Cyanobacterial biofertilizers in rice agriculture. Bot. Rev. 67, 453–516. Available from: https://doi.org/10.1007/BF02857893, http://link.springer.com/article/10.1007/BF02857893.

Vandamme, P., Pot, B., Gillis, M., deVos, P., Kersters, K., Swings, J., 1996. Polyphasic taxonomy, a consensus approach to bacterial systematics. Microbiol. Rev. 60, 407–438.

Vijayan, D., Ray, J.G., 2015. Ecology and diversity of cyanobacteria in Kuttanadu Paddy Wetlands, Kerala, India. Am. J. Plant Sci. 6, 2924–2938.

Watanabe, I., Espianas, C.R., Berja, N.S., Alimagno, B.V., 1977. Utilization of the Azolla-Anabaena complex as a nitrogen fertilizer for rice. IRRI Res. Pap. Ser. 11, 1–15.

Woese, C.R., 1987. Bacterial evolution. Microbiol. Rev. 51, 221–271.

Wolfaardt, G.M., Lawrence, J.R., Korber, D.R., 1999. Function of EPS. Microbial Extracellular Polymeric Substances. Springer, Berlin, Heidelberg, pp. 171–200.

Yadav, P., Gupta, R.K., Singh, R.P., Yadav, P.K., Patel, A.K., Pandey, K.D., 2021. Role of cyanobacteria in green remediation. Sustainable Environmental Clean-up. Elsevier, pp. 187–210.

Yadav, P., Singh, R.P., Gupta, R.K., 2022a. Role of cyanobacteria in germination and growth of paddy seedlings. Int. J. Phytol. Res. 2 (3), 11–18.

Yadav, P., Singh, R.P., Patel, A.K., Pandey, K.D., Gupta, R.K., 2022b. Cyanobacteria as a biocontrol agent. Microbial Biocontrol: Food Security and Post Harvest Management. Springer, Cham, pp. 167–185.

Yadav, S., Rai, R., Shrivastava, A.K., Singh, P.K., Sen, S., Chatterjee, A., et al., 2018. Cyanobacterial biodiversity and biotechnology: a promising approach for crop improvement. Crop Improvement Through Microbial Biotechnology. Elsevier, pp. 195–219.

Yanti, Y., Hamid, H., Syarif, Z., 2019. Screening of indigenous rhizospheric cyanobacteria as potential growth promotor and biocontrol of *Ralstonia syzygii* subsp. *indonesiensis* on Chili. Int. J. Environ. Agric. Biotechnol. 4 (6), 1665–1672.

Zulpa, G., Zaccaro, M.C., Boccazzi, F., Parada, J.L., Storni, M., 2003. Bioactivity of intra and extracellular substances from cyanobacteria and lactic acid bacteria on 'wood blue stain' fungi. Biol. Contr. 27, 345–348. Available from: https://doi.org/10.1016/S1049-9644(03)00015-X.

CHAPTER

Nanotechnology in the agricultural sector

11

Syeda Summiya
Institute of Plant Science, University of Sindh, Jamshoro, Pakistan

11.1 Introduction

The world population is increasing day by day. It is estimated that at the end of 2050, it will reach about 6 billion (Baig et al., 2021). The rapidly increasing populations depend on water, food, and energy for their food security. To fulfill the whole population's requirements, more food production is necessary, means more consumption of water, land, natural resources, and energy (Glenn and Florescu, 2016). This is the major challenge for the world's agriculture sector of developing countries. Agriculture is the backbone of their national economy. About 60% depend on agriculture and their livelihood (Ali et al., 2018). The socioeconomic issues of developing countries directly affect agricultural domains such as soil, reduction of agricultural land, commodity dependence, malnutrition, and poverty (Ditta, 2012).

Pakistan is an agricultural country and grows the world's best quality rice, wheat, sugarcane, cotton, mango, and oranges. It is estimated that Pakistan exports 1.7 million tons of mangoes among 50 countries of the world. Key crops such as cotton, rice, wheat, sugarcane, and maize contribute 19% of GDP in the economy which provides 39% labor force and 4%−11% of GDP contributing to minor crops and livestock, respectively (Khan et al., 2022a), but Pakistan's agriculture sector would be turned into a threat for national food security because of climate change, global warming, scarcity of water, and lack of food stock availability (Munir et al., 2021).

To overcome the agricultural crisis, many technologies are used in all developing countries including Pakistan. Nanotechnology is an emerging sustainable approach for the agriculture sector (Gondal and Tayyiba, 2022). Recent publications indicated that developed countries such as Germany, China, the United States, France, Brazil, Korea, and India show a great interest in the incorporation of nanotechnology with agriculture (Fadiji et al., 2022). Nanotechnology is the novel application of controlled manipulation of matter at an atomic and molecular level producing nanoscale particles of size approximately 1−100 nm (Babur and Shah) with the greater surface, volume ratio, magnetic, electrical and physical strength quantum size, and its optical characteristics. Due to the unique properties

Sustainable Agricultural Practices. DOI: https://doi.org/10.1016/B978-0-443-19150-3.00014-X

of nanoparticles, these are widely used in various agriculture and other industries as well (Hazarika et al., 2022). Nanotechnology plays a key role to change the entire scenario of the agricultural sector because it offers sustainable innovative development in the production of crops and their improvement in every domain of agriculture and is also eco-friendly (Francesconi et al., 2022).

Nanotechnology also answers the failure of the traditional system of agriculture in a sustainable way. It provides the strategies to control pests against pathogens, different organic materials inoculate with control plants pathogens and pests for the production of fresh agro foods at the commercial level (Balestra et al., 2022). Food quality, soil improvement, food distribution, and all essential aspects of agriculture can be modified and improved by nanotechnology (Ashraf et al., 2021). For soil fertility and improvement during the cultivation of crops, various kinds of hydrogel and nano clay-polymer composites are used for the soil moisture content, and water-holding capacity hence engineered nanoparticles have a significant role in crop protection by pesticides and residual detection (Kim et al., 2018).

To get sustainable achievements by nanotechnology, climate-oriented farming will be developed restoring infertile soil, and introducing drought-resistant crops with help of nano-based technologies associated with the agriculture sector. It also provides solutions to unresolved issues related to agriculture (Kim et al., 2018).

The purpose of this chapter is to discuss the essential applications of nanotechnology in the overall field of agriculture, which is the current need of our agriculture sector, and also depicts the future trends of nanotechnology. It also explores the enhancement of food security, as well as the challenges associated with its use in the agricultural and food systems.

11.2 Characteristics of nanoparticles

Nanotechnology deals with nanoparticles which are prepared by different materials at molecular or atomic level. On desired application, there are various types of nanoparticles being synthesized such as composites, carbon based, polymer, metal, metal oxide, and fabricated based (Jeevanandam et al., 2022). Nanoparticles have three dimensions with size range about 1–100 nm. Nanoparticles are classified into physical and chemical characteristics. Physical properties such as shape, size, and electrical charge chemical ones are core, shell, and stability as shown in Fig. 11.1. There are various shapes of nanoparticles such as rods, rod-like spherical, needle-like, plate-like, cubes, sheets, cylinders, ellipsoids, spheres (Sukhanova et al., 2018).

Nanoparticles are synthesized by physical, chemical, and biological methods. The physical methods are lithography, high energy irradiation, laser ablation, and all these physical methods required expensive equipment, large surface area, high

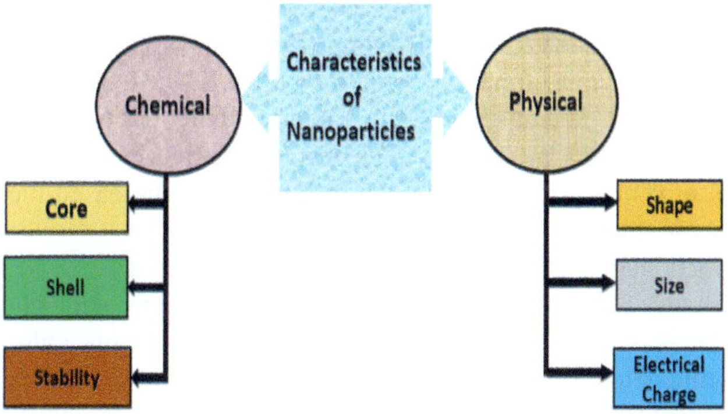

FIGURE 11.1

Illustrating physical and chemical properties of nanoparticles.

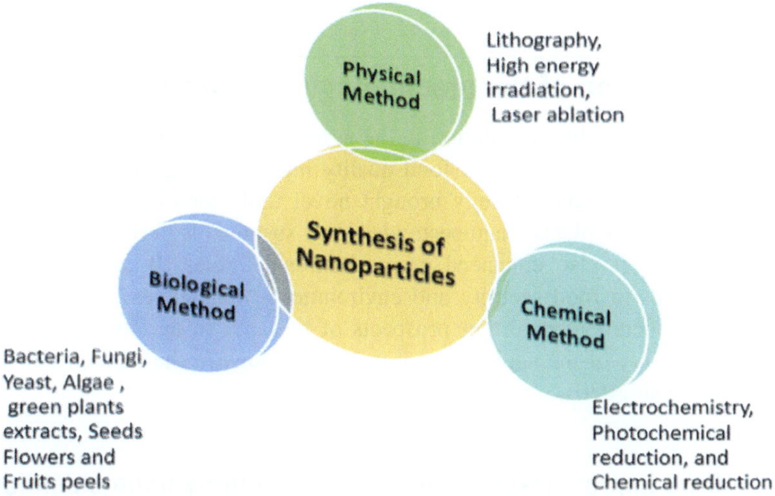

FIGURE 11.2

Showing different techniques and raw materials for nanoparticles fabrication (Pandit et al., 2022).

temperature and pressure. In chemical methods, electrochemistry, photochemical reduction, and chemical reduction are involved. Biological methods include bacteria, fungi, yeast, algae, and green plants extracts, seeds, flowers, and fruits peels also. Fig. 11.2 shows different methods and raw materials for nanoparticles fabrication (Soni et al., 2018).

There are different nanoparticles that scientist used in their research for the improvement and high yield production. A research was conducted during the year of 2016-17 effect on the genomic DNA on common bean. It was indicated that the concentrations of 20 and 30 ppm (ZnO-NPs) increased the number of branches, leaves per plant, fresh dry weight of branches leaves, and yield as well (Salama et al., 2022). Scientist used silica nanoparticles for reduction of salinity stress on basil (*Ocimum basilicum*). The results of their research demonstrated that application of nanoparticles reduced stress and also increased the physiological and morphological traits (Kalteh et al., 2018).

It is reported that MgO NPs, CuO NPs, SiO_2 NPs, and ZnO NPs efficiently prevent the fungi (*Sclerotinia sclerotioum*) attack on beans plant under different conditions. These nanoparticles also increased the pod beans chlorophyll total phenols ascorbic acids as well (Abdel-Halim and El-Ghanam, 2019). Silver nanoparticles at concentration of 40 mg/L significantly increased the physiological and biochemical parameters of fenugreek (*Trigonella foenum-graecum*) plant and also enhanced yield and seeds as well. It also increased protein, carbohydrates, phenolics, tannin, and flavonoids (Sadak, 2019).

11.3 Application of nanotechnology in agriculture

Nanotechnology brought a revolution in the agro-food sector. Possible application of nanotechnology provides better food quality inputs and significantly increases crop productivity. Nanotechnology brought novel tools for the diagnosis of disease management in plants, nanopesticides, and sustainable fertilizers for better growth of crops. It also enhanced crop nutritional values. It also ensures the global challenges in food security and environmental challenges as well. Recent literature also documented further prospects of research in agriculture. Fig. 11.3 shows the summary of the application of nanotechnology in the agricultural sector (Acharya and Pal, 2020).

11.3.1 Crop improvement and crop production by nanotechnology

Food is a fundamental need for well beings and global food security is a challenge all over the world. To fulfill the food needs of the overall population, intensive farming is practiced by people. It caused a disturbance in the ecological cycle, destruction of soil fertility, and decline of agricultural yields. According to a research, it is estimated that about 40% of the soils of the world are degraded and lose fertility to the intensive practice of crop growing (Kale and Gawade, 2016). The application of fertilizers increases soil fertility and crop yield (Dubey and Mailapalli, 2016). Macro and micronutrients are important for crop growth and soil fertility. Fertilizers and essential nutrient supplements improve crop and soil health as well (Li et al., 2018). Various nutrients deficiency problems of soil such as leaching or losses, ammonia volatilization and fertilizer burns different stages of plant growth. Nutrients are required especially during the initial stage of

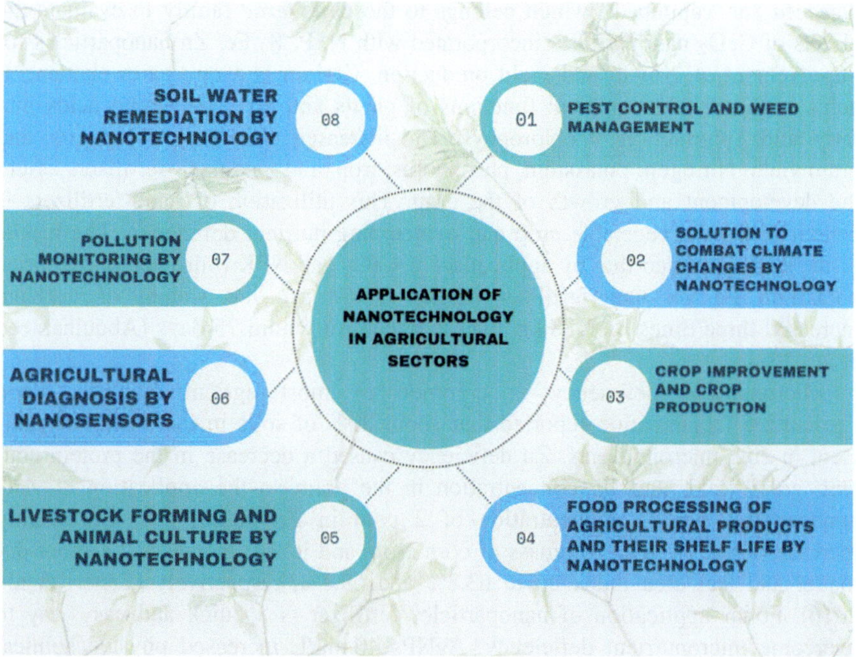

FIGURE 11.3

Contributions of nanotechnology in agriculture sectors.

plant growth. For example, plant absorbs applied nutrients: up to 40%–60% nitrogen, 15%–20% phosphorus, and 50%–60% potassium (Chinta et al., 2020). Due to volatilization, leaching and fixation of nutrients and imbalanced nutrients not only decrease the crop yield and losses but also increase contamination of soil, air, and water and also decrease the natural fertility of topmost soil (Le et al., 2021b).

To overcome soil and crop productivity problems, nano-fertilizers and agrochemicals are the best solutions as compared to conventional fertilizers. It is reported that conventional fertilizers are 50%–70% reduced by evaporation leach down or degradation (Yang et al., 2016). Nano-fertilizers have great potential to stop nutrient loss and provide an adequate amount of nutrients for the maximum growth and yield of the crop. Nano-fertilizer delivery system not only provides desired micronutrients to the soil but also decreases loss and fixation leaching of nutrients it also ensure bioavailability of nutrients for the soil (Polman et al., 2021). Nanoparticles influence tremendous effects on the overall growth of plants such as seed germination, increment of root and shoot length, and increased fruit yield (War et al., 2020).

Scientists continuously work on nutrients management for the improvement of growth and yield of crop. Current study was conducted on cabbage *Brassica oleracea* var. capitata L. which belongs to the *Cruciferae* family to evaluate the effects of CeO_2 nanoparticles incorporated with N, P, K, Fe, Zn nanoparticles for improvement of growth and yield production. Cerium is a rare earth element. It helps in the physicochemical functions of plants and converts amino acids into fatty acids, production of chlorophyll, and increased plant height, biomass, and grain yield. Nitrogen, potassium, phosphorus, iron, and zinc nano-fertilizers assure the development and growth of the crop. This utilization of nano-fertilizers is beneficial for the vegetable crop and overcoming nutrient deficiency. The results of the study revealed that by application of CeO_2 and NPK (nitrogen, phosphorus, potassium Fe Zn nano-fertilizer height (16.67 cm), the weight of cabbage increased three times by 1.88 kg/plant than control within 75 days (Abdulhameed et al., 2021).

Micronutrient deficiency is common in most agricultural soil. Food Agriculture Organization reported that about 30% of soils in the world are deficient in zinc micronutrients. Zn deficiency caused a decrease in the protein content, grain yield, and Zn concentration in the grain so the application of ZnO nanoparticles with a concentration of 2 ppm in Zn deficit soil significantly increases grain yield dry biomass of corn crop, and it also increases the shoot dry matter and leaf area index up to 63.8% and 69.7%, respectively (Taheri et al., 2016). Foliar application of nanoparticles fertilizer is a quick and easy way to overcome micronutrient deficiency. AgNPs 40 mg/L increased physicochemical parameters of fenugreek (*T. foenum-graecum*) such as the number of leaves, shoot length shoot dry weight, chlorophyll a, b carotenoids, indole acetic acid (IAA), carbohydrates, flavonoids, proteins, phenolics, and tannins. It also increased yield traits number of seeds pod^{-1}, number of pod plant^{-1}, seed index, and weight of seed plant^{-1}. AgNPs also increased the antioxidant properties and yield of vegetables (Sadak, 2019).

Silicon is an essential element for the higher plants application of silicon nanoparticles at the rate of 200 mg/kg during salinity stress and water deficit conduction on cucumber. It increased growth and productivity of cucumber plant: height up to 156% chlorophyll fruit yield. Nutrient uptake such as nitrogen 30%, potassium root 52%, stem 75%, leaf 41%, silicon 51% root, 57% stem, 8% leaf, and 78% fruit. Sodium uptake was also reduced by 38% root, 77% stem, and 38% in the leaf; potassium-sodium ratio also increased by 735% stem, 149% root, and 127% leaf. The application of SiNPs in cucumber plants also tolerates abiotic stress and homeostasis by the regulation of osmotic pressure with the control of stomatal opening (Alsaeedi et al., 2019). Various nanoparticles influence plant growth, macro-micronutrients deficiency in plants or soil, and biomass yield production (Table 11.1).

11.3.2 Pest control weeds management by nanotechnology

Crop diseases are challenges for the agriculture sector. Its control and management at its early stage is very important, so nanoparticles cooperated pesticides

Table 11.1 Uses of different nanoparticles for crop improvement and production.

S. no.	Nanoparticles	Plant	Recommended concentration	Effects	References
1.	Carbon nanoparticles (CNPs)	Corn (*Zea mays* L.)	200 mg CNPs kg^{-1}	Plant height, biomass yield, nutrients uptake, nutrients use efficiency, availability of phosphorus and nitrogen	Zhao et al. (2021)
2.	Calcium nanoparticles (CaNPs)	Drumstick tree (*Moringa oleifera* L.)	100 mg/L	Reduced salinity, immobilized heavy metals, increased germination (%), vigor indices, relative water contents, lengths of roots and shoots	Azeez et al. (2021)
3.	Titanium dioxide nanoparticles TiO_2 NPs + calcium phosphate ($Ca_3(PO_4)_2$)	Wheat (*Triticum aestivum* L.) Pakistan-13 and Zincol-16	40 ppm, 20 ppm	Increased root, shoot length, fresh weight, dry weight, chlorophyll content, relative water content, membrane stability index, and osmolyte content, nutrients uptake potassium, phosphorus, nitrogen, plant biomass, improved cellular process, seed germination reduced drought stress	Mustafa et al. (2021)
4.	Calcium carbonate ($CaCO_3$) and calcium oxide (CaO) nanoparticles	Pure green gram (*Vigna radiata*)	250 ppm	Increased higher radicle growth, seed germination, root shoot length, vigor index	Vijai Anand et al. (2021)
5.	Silicon dioxide (SiO_2) nanoparticles	Tomato (*Solanum lycopersici* L.)	0.20 g/L	Increased plant growth, chlorophyll, carotenoids, proline, superoxide dismutase (SOD), catalase (CAT), ascorbate peroxidase (APX), phenylalanine ammonia lyase (PAL)	Parveen and Siddiqui (2022)
6.	Zinc oxide (ZnO) nanoparticles	Ridge gourd seeds (*Luffa acutangula* L.)	10 ppm	Increased photosynthetic pigments, proline, relative water content, total sugars, proteins, indole acetic acid, mitigate heavy metal toxicity	Tanveer et al. (2022)
7.	Copper oxide (CuO) nanoparticles	Willow (Salix)	500 mg/kg	Decrease activity of peroxidase and polyphenol oxidase, fungal community, regulate wetland ecology	Qu et al. (2022)

(*Continued*)

Table 11.1 Uses of different nanoparticles for crop improvement and production. *Continued*

S. no.	Nanoparticles	Plant	Recommended concentration	Effects	References
8.	Iron oxide (Fe_3O_2) nanoparticles	Wheat (*T. aestivum* L.)	1.2 mM	Increased chlorophyll a, b, carotenoids, proline, SOD, peroxidase, ascorbate peroxidase, soluble sugars, Fe content in root and shoot, reduced lipid peroxidation, electrolyte leakage,	Noor et al. (2022)
9.	Manganese oxide nanoparticles (MnO_2-NPs)	Seeds of common bean (*Phaseolus vulgaris* L., cv Nebraska)	30–40 ppm	Increased growth and yield, chemical quality of leaves and seeds, genomic DNA, genes encoding protein also significantly affected	Salama et al. (2022)
10.	Magnesium oxide (MgO)	Sunflower (*Helianthus annuus* L.)	0.25 g/L	Increased relative water content, chlorophyll, carotenoids, soluble carbohydrates, antioxidant enzymatic activity under drought stress, decreased electrolyte leakage and malondialdehyde content, improved grain yield and oil (%)	Fatemi et al. (2022)

are useful for the treatment of diseases in plants (Camara et al., 2019). Nanotechnology provides biologically and chemically synthesized agrochemicals and pesticides which effectively show improvement in crop diseases due to nanoparticles properties such as induction, high solubility, a high surface area, small size particles high mobility, low toxicity, and elimintion of the organic solvent as compared to conventional pesticides (Manjunatha et al., 2016). Crops are lost by pests and weeds and for the safety and improvement of crops, fertilizers and pesticides are widely used. Continuously usage of agrochemicals is a disadvantage for crop, human beings, and the environment. The novelty in formulation of nano-based pesticides are safe for the environment. A sufficient amount of nano pesticide is an active material to respond to abiotic and biotic stress and also triggers the controlled released mechanisms (Camara et al., 2019).

Conventional pesticides only kill the upper parts of weeds. When favorable condition returns, weeds again grow but nano-based pesticides target the receptors of ground parts of weeds. Nanopesticides also solve the problems of environmental pollution which are caused by the usage of agrochemicals. Nanomaterials also prevent the loss of crops from fungal, bacterial, and viral diseases. Various nanoparticles were developed from infections or diseases such as cobalt, nickel, and ferritin (Acharya and Pal, 2020). Rhizomes and tubers of perennial weeds were significantly killed by nanoherbicides with the use of H_2O_2 at 300 mL/m^2 followed by pendimethalin at 0.75 kg/ha + ZnO nanoparticles at the rate of 500 ppm/m (Vimalrajiv et al., 2018). Pascoli et al. conducted an experiment in which they used neem tree oil extract nanoparticles. The results showed that biosynthesized nanopesticides significantly protect the crop from pesticides and it is eco-friendly as well (Pascoli et al., 2019). Another application of the eco-friendly delivery system for pesticides is that mesoporous nitrogen-rich polymers also were used. The results of the research revealed that MeSiNPs@bPEI significantly not only prevents pesticides but also improved the soil by providing Si and N as well (Plohl et al., 2021). Citrus canker is a common disease caused by *Xanthomonas citri* bacteria that attacked on citrus fruits. It caused economic loss. To control this bacterial disease, Zinkicide is used which is formulated from ZnO nanoparticles. It has low toxicity, high surface area, and move as translaminar. It also prevents the grapefruit trees from citrus scab (*Elsinoe fawcettii*) and melanosis (*Diaporthe citri*) (Graham et al., 2016). A study reported that 30%–35% of citrus fruits are lost by citrus black rot disease which is caused by Alternaria core. Conventional fungicides are not effective to control black rot of citrus while the application of green-synthesized ZnO and CuO nanoparticles used as fungicides at the rate of 80–100 mg/mL completely inhibits the growth of fungal hyphae is effective to control black rot disease of citrus fruits (Sardar et al., 2022).

Scab or Fusarium head blight is a typical cereal crop disease in humid and warm areas caused by *Fusarium graminearum*. It caused a great loss of wheat barley and contaminated grains of human and animal feed. Applying chitosan (CS) and chitosan nanoparticles as pesticides at a concentration of 1000–5000 ppm during the growth of a plant two or three times on a crop inhibits the growth of mycelium of *F. graminearum*. CS and chitosan nanoparticles are useful for the biological control of plant pathogenic fungi (Kheiri et al., 2016).

Ralstonia solanacearum is a bacterial disease of the *Solanaceae* family. In case of loss of potato and tomato crops, application of 256 μmol/L AgNPs with chitosan nanoparticles inhibits the growth of pathogenic plant bacteria in tomato plants. It is a promising and sustainable alternative for the control of bacterial wilt disease to help in the management of yield losses (Santiago et al., 2019). *Fusarium culmorum* and *Aspergillus niger* are fungal pathogens that cause an economic loss of maize, onions, mangoes, grapes, or tomatoes. Applications of 100 μg/mL green-synthesized copper oxide nanoparticles control the growth of fungal pathogen (Vanathi et al., 2016). Table 11.2 presents performance of nanopesticides and weedicides.

11.3.3 Agricultural diagnosis by nanosensors

Nanosensor brought the revolution in the agricultural farming and provided environmental-efficient and cost-effective techniques (Kashyap et al., 2019). Emerging data suggested that nanosensors might be utilized in conjunction with biomarkers to quickly, precisely, and accurately extract information about the development of early infection and disease in field conditions (Khater et al., 2017; Ray et al., 2017). As illustrated in Fig. 11.4, nanosensors can be categorized depending on the objects they can detect, the substances that make them up, and the signals they employ to communicate information. According to recent findings, field-applied nanosensors are a great tool for making crop cultivation safer and more environmentally friendly by reducing the costly usage of agrochemicals needed to maintain crop growth and health (Jasrotia et al., 2018).

It helps in the monitoring of plant health metabolism and signaling changes in pathways without disturbing the plant's usual process; monitoring of humidity and temperature of the soil, detection of microbial and pathogenic diseases in plants, and identification of heavy metal pollutants in soil or water. Nanosensors include thin film nanoparticles nanomaterials from polymers, nanoscale wires which detect high sensitivity, and carbon nanotubes that identify the surface area. These sensors also detect the conductance and semiconducting exposed to certain chemicals (Johnson et al., 2021). It is employed in the sensing of pathogens, postharvest management of vegetables and flowers, reclamation of salt-affected soils, fertilizer moisture, and pH of soil. All these employed applications of nanosensors help in nutrients management, and protect against pathogens, prevent loss of nutrients, and enhanced the yield of the crop. Microbial attacks are also reduced by oxide-based nanosensors (Kaushal and Wani, 2017). It also helps in the detection of seed viability, the number of nutrients required by the plants, and shelf life of fruits and vegetables (Dar et al., 2020).

Nanoparticles as a sensing element offer enhanced detection limits in detecting many kinds of bacterial, fungal, and viral infections (Baac et al., 2006; Boonham et al., 2008; Chartuprayoon et al., 2010; Yao et al., 2009). Generally,

Table 11.2 Uses of different nanoparticles in nano pesticides and fungicides.

S. no.	Nano pesticides/ fungicides	Plant	Recommended concentration	Effects	References
1.	KIT-6 mesoporous silica nanoparticles	Wheat (*Triticum aestivum* L.) and barley (*Hordeum vulgare* L.)	125 mg/L	Reduced oviposition and nymphal population	Alizadeh et al. (2022)
2.	Fungicides of ZnO nanoparticles	Grapefruit	1.0 mg/mL	Inhibit the fungal growth	M. Ali et al. (2022)
3.	Tribenuron-methyl-based Zein nanoparticles	Wheat seeds (cv. Pishgam)	15 mL	Reduced growth of weed	Heydari et al. (2021)
4.	Zinc oxide (ZnO) + Streptomycin bacteria	Potato (*Solanum tuberosum* L.)	75 ppm	Prevents against bacterial disease (*Streptomyces scabies*) of potato crop	Tahir et al. (2021)
5.	Silver nanoparticles (Ag NPs)	Sugar beet (*Beta vulgaris* L.)	100 ppm	Prevent bacterial soft rot disease	Ghazy et al. (2021)
6.	Silicon dioxide nanoparticles (SiO_2 NPs)	Egg plant seeds (*Solanum melongena* L.)	0.20 mg/mL	Prevent the growth of fungal, bacterial, and pathogens (Phomopsis vexans, bacterium *Ralstonia solanacearum*, and *Meloidogyne incognita*)	Khan et al. (2022b)
7.	Silica green-synthesized nanoparticles (SiNPs)	Corn (*Zea maize* L.)	10 ppm	Control the infection of seed-borne fungus (*Harpophora maydis*)	El-Shabrawy (2021)

(*Continued*)

Table 11.2 Uses of different nanoparticles in nano pesticides and fungicides. *Continued*

S. no.	Nano pesticides/ fungicides	Plant	Recommended concentration	Effects	References
8.	Copper oxide nanoparticles (CuO NPs)	Potato (*S. tuberosum* L.)	150, 200, 250 μL/L	Controlling black scurf disease of potato caused by *Rhizoctonia solani*	El-Shewy (2019)
9.	Titanium oxide nanoparticles (TiO_2 NPs)	Wheat (*T. aestivum* L.)	40 mg/L	Reduced the fungus infection of *Bipolaris sorokiniana*	Satti et al. (2021)
10.	Silver nanoparticles (Ag NPs) used as biofungicides	Wheat (*T. aectivum* L.)	50 ppm	Control pathogenic effect of fungi (*Sclerotium rolfsii*)	Desai et al. (2021)

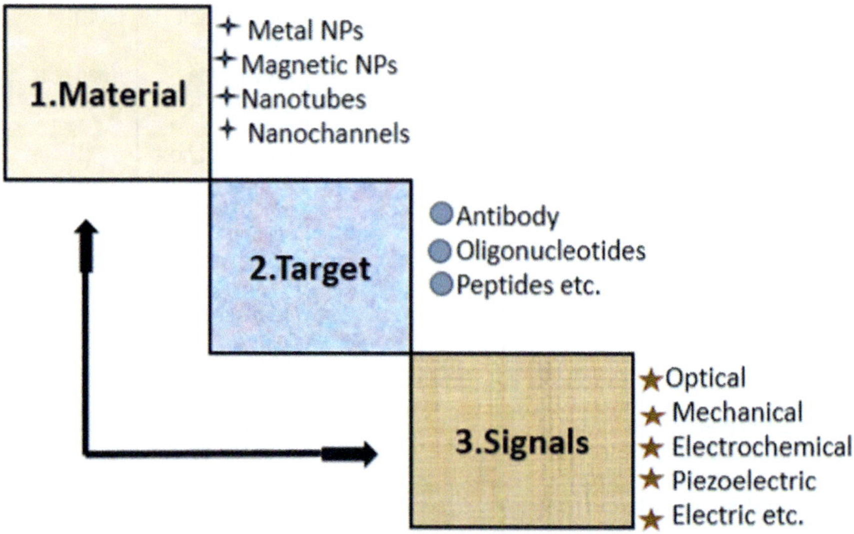

FIGURE 11.4

Basic constituents of a nanosensor (Sharma et al., 2021).

nanoparticles alone or in combination are applied in the construction of diagnostic tools, which either detect pathogens directly or detect suggestive substances implicated in progressing disease. To validate the presence of the Karnal bunt (*Tilletia indica*) infection in wheat, Singh et al. (2010) developed an immunosensor based on gold nanoparticles and the surface plasmon resonance (SPR) technique. In addition to SPR, the use of the quartz crystal microbalance (QCM) technology for identifying the presence of viruses in plants has also been described. Various nanoparticles are used in biosensors such as Au, Ag, Cu, Co, Fe_3O_4, MgO, TiO_2, ZnO, CNT, dendrimers, grapheme, polymeric, magnetic nanoparticles (Sachan et al., 2022). It is reported that fluoro-immuno assay-based nano-bio sensor cadmium with telluride quantum dot nanoparticle (CdTe QD) had detected the herbicide 2,4-D (Saha and Bhaduri, 2021). Lau et al. reported that the Au NPs-DNA probe biosensors, their results showed successful identification of *Pseudomonas syringe Arabidopsis thaliana* in infected plant samples before diseases symptom appear (Lau et al., 2017). Similarly, the study reported that fluorescent silica NPs combined with antibody nanosensors detect bacterial disease of *Solanaceae* plants caused by Xanthomonas axonopodis pv. vesicatoria (Sanzari and Leone, 2019; Yao et al., 2009). Further studies demonstrate that gold-based nanosensors detected Karnal bunt disease in wheat plants. Gold and copper nanoparticles, modified into electrode nanosensors, can detect pathogenic fungi in oil seeds and also monitor the hormone level of salicylic acid (Shang et al., 2019). For fast detection of viral infections caused by the Odontoglossum ringspot virus (ORSV) and Cymbidium mosaic virus (CymMV) in ornamental crops, antibody-functionalized gold nanorods have been used. Orchid leaf infections with ORSV and CymMV had a limit of detection (LOD) values of 48 and 42 pg/mL, respectively (Lin et al., 2014).

In order to identify the presence of the fungus *S. sclerotiorum* in an oilseed crop, Wang et al. (2010) developed an electrochemical sensor that includes a modified gold electrode with copper (Cu) nanoparticles. For the quick and accurate detection of plant viruses, Zhang et al. (2013) used the notion of combining upconversion nanoparticles (UCNPs) as biolabeling with immunomagnetic separation.

Potato bacterial disease is detected by the application of nanosensor which is prepared by Ag nanoparticles single-stranded oligonucleotide probe with a colorimetric method that can detect the DNA of *R. solanacearum* from the sample of wilt bacterial disease of potato (Khaledian et al., 2017). Platinum nanoparticle-mediated antibody immunoglobulin G (h-IgG) nanosensor was applied for the detection of plant-associated bacteria (PAB) such as thuringiensis and *Bacillus subtilis*. The samples were collected from the soil and roots of carrots, these bacteria are used in biotechnology for the enhancement of growth where the soil has low fertility (Ahmad et al., 2012).

Due to its low cost, ease of use, ease of downsizing low, high sensitivity, and capacity for immediate data processing, the electrochemical nanosensor (ECN) is a novel and effective method to identify agricultural infections at an early stage.

Due to its capacity to directly translate a biological event into an electrical signal, ECN offers a fantastic technique to evaluate any biological material (Tran and Le, 2015).

The piezoelectric nanosensor (PZN) opens up new possibilities for real-time monitoring of pathogenic infection in plant systems since it has the potential to convert mechanical energy, ultrasonic vibrations, and biofluid hydraulic energy of any biological event into electric energy. Piezoelectric sensors primarily monitor and calculate changes in mass caused by biomolecular interactions between antibodies and their corresponding antigenic determinants (Byrne, 2009). For instance, by calculating the change in frequency of the QCM, shift in the resonance frequency of quartz crystal owing to mass change has been recorded.

Table 11.3 presents recent research on nanosensors in agriculture field.

11.3.4 Soil water remediation by nanotechnology

Soil and water are important components for the life of living beings but due to urbanization, anthropogenic activities, and rapid industrialization makes it contaminated by the release of chemicals and other things (Kristanti et al., 2021). According to the World Health Organization (WHO), less drinkable water is available and approximately 1 billion people worldwide lack access to it. As a result, water management has become a significant problem for both society and public authorities (Yadav, 2021). Soil and water pollution not only threaten food security but also causes serious diseases in human beings as well (Lu et al., 2015). Soil and water contamination leave long-term effects on living beings. Main factors which cause the water soil contamination are agricultural pollutants (pesticides, industrial water synthetic fertilizers, herbicides), pathogenic and microorganisms, sewage water ditch, storm drain, chemical industries' leftover chemicals, saltwater intrusion, and sediment pollution (Chaudhry and Malik, 2017). Water and soil are mostly polluted by heavy metal accumulation, mostly reputed heavy metals such as chromium (Cr), copper (Cu), zinc (Zn), cadmium (Cd), and arsenic (As), nickel (Ni), mercury (Hg), and lead (Pb). These heavy metals are emitted from electronic waste, agricultural discharge, and industrial wastage. These heavy metals cause dangerous diseases in humans such as skin lesions, kidney and liver damage, various types of cancers, and birth defects (Chatterjee and Abraham, 2022). Soil and water contamination rate increases day by day in developed and developing countries. To overcome contamination, there are several technologies that are used such as physical and chemical or phytoremediation but nanotechnology remediation is the recent technique. It is the removal of toxic substances from the water and soil by the implementation of nanoscale application (Acharya et al., 2022).

A study was conducted on industrial wastewater that is contaminated by heavy metals such as Ni (II), Cd (II), Co (II), and Cu (II), which was significantly remediated by magnetic Fe_3O_4 nanoparticles with the use of hydrothermal technique (Habila et al., 2017). Another similar study reported that bimetallic zero-valent

Table 11.3 Uses of different nanoparticles in nanosensors for detection of soil water and food issues.

S. no.	Sample (plant/soil/ water/ food)	Nanosensors	Detection	References
1.	Soil (nitrogen status high, medium, low)	BSA/NiR/CNT-α-Fe_2O_3/SPE	Nitrate content	Kundu et al. (2022)
2.	Water	Aminophenol, phenol, polymer, polyaniline (PANI)	Organic residue, pH	Chajanovsky et al. (2021)
3.	Tobacco protoplast	Plant hormone fluorescent based	Detection of signaling molecule for plant growth and parasitism	Chesterfield et al. (2020)
4.	Tomato and maize crop	Metal-oxide based	Gases identified as markers for water stress	Fabbri et al. (2020)
5.	Tomato plants	E-nose sensor system	Diagnosis aphids attack at early stage	Cui et al. (2019)
6.	Wheat flour	Smartphone-based colorimetric (Fe_3O_4@Au)	Detection of glutathione which prevent damage to cellular reactive oxygen species	Huanan et al. (2022)
7.	Arabidopsis and rice extract	Aptamer-based nanopore thin film	Detection of salicylic acid (SA) compound regulate immune response	Chen et al. (2019)
8.	Soil	Bismuth–silver bimetallic alumina (Al_2O_3 + Bi)	Detection of heavy metals (Pd, Pt, Rh)	Van der Horst et al. (2015)
9.	Fern, rice spinach	DNA-SWNT nanoconstructs	Detection of arsenic (Ar) in plant tissues	Lew et al. (2021)
10.	Soyabean tobacco	Polyphenol-based near infrared (NIR) fluorescent single-wall carbon nanotubes (SWCNTs)	Detect plant chemical communication and accelerate phenotyping and identification of crop plants that are more tolerant to pathogen	Nißler et al. (2022)

silver/iron nanoparticles significantly remediate the toxic waste of groundwater and also degrade the industrial color dye bromophenol blue (Gallo et al., 2019). A study repoted that contamination of water by lead (Pb) effectly remediated through zero-valent Sn nanoparticles (zero-valent Sn NPS) (Mahmoud and Abdelwahab, 2021). Biao et al. conducted a similar study on heavy metal-contaminated

Ni^{2+}, Cu^{2+}, Cd^{2+}, and Pb^{2+} in an aqueous solution. They used green-synthesized gold nanoparticles. The results show that gold nanoparticles had a high rate of removal of heavy metal dye from an aqueous solution of water (Biao et al., 2018). Dendrimers (repeatedly branched molecules) are polymeric nano adsorbents that can be used to remove organics and heavy metals. Heavy metals can be absorbed by the specially shaped outside branches, whereas organic substances can be adsorbed by the interior hydrophobic shells (Khajeh et al., 2013). By creating a combination of chitosan-dendrimer nanostructure, Sadeghi-Kiakhani et al. (2013) generated a very effective bioadsorbent for the removal of anionic chemicals like dye from textile effluent. The bioadsorbent is nontoxic, biodegradable, and biocompatible. They obtain up to 99% clearance rates for some colors (Gehrke et al., 2015).

In a different study, arsenic heavy metal contamination in drinking water was remedied by utilizing nano-alumina disseminated in chitosan-grafted polyacrylamide (Saha and Sarkar, 2012). A study for the clean-up of water from pharmaceuticals also reported the use of zinc oxide nanoparticle photocatalysts (Choina et al., 2015). It was also found that super paramagnetic maghemite (γ-Fe_2O_3) nanoparticles, which were produced via flame spray pyrolysis, helped to repair the lead- and copper-contaminated water (Rajput et al., 2017). To clean-up the contaminated water, heterostructured HfO_2/TiO_2 spherical nanoparticles have been reported (Laishram et al., 2018). Eco-friendly bio-metallic silver and iron nanoparticles were also mentioned in another paper for cleaning up contaminated water (Xin-gang et al., 2019). Vanadium oxide nanoparticles activity proved as effective cleaning agent for methylene blue-contaminated water (Saghi et al., 2020). Multiwalled carbon nanotubes functionalized with a solution of nitric and sulfuric acids have been used to purge water of organic contaminants such as dioxins, polyaromatic hydrocarbons, and polychlorinated biphenyls (Salipira et al., 2007). They examined how cyclodextrin polyurethanes polymerized with carbon nanotubes degraded the model chemical trichloroethylene (TCE), a known endocrine disruptor. Serious concerns have also been raised by the prevalence of antibiotics in aquatic environments. Nonporous activated carbon and nonporous graphite are used in traditional procedures. Vanadium oxide nanoparticles have also been shown to be the utilization of single- and multiwalled carbon nanotubes have a superior adsorbent to solution distribution coefficient and adsorption affinity for tetracycline compared to these methods. The primary causes of the high affinity are strong adsorption interactions (Mehndiratta et al., 2013).

Table 11.4 shows recent research on soil and water remediation by nanotechnology.

11.3.5 Pollution monitoring by nanotechnology

The term "pollution" has many definitions, one being "the presence of a substance in the environment whose chemical composition or quantity prevents the functioning of natural processes and produces undesirable environmental and

Table 11.4 Uses of different nanoparticles for polluted soil and water remediation.

S. no.	Soil/water	Nanoparticles	Remediation method	Removal pollutants	References
1.	Contaminated Water	Zn-doped aluminum oxide (Zn: Al_2O_3) incorporated polysulfone (PSf)	Combustion	Pb (II), As (V)	Sherugar et al. (2021)
2.	Potable water sample	Nano zero-valent iron-kaolin clay (nZVI-kaol)	Sodium borohydrate reduction method	Removed As_2O_3	Selvan et al. (2022)
3.	Contaminated soil	Iron nanoparticles (nFe)	Green synthesis	Removed arsenic (As)	Wu et al. (2019)
4.	Groundwater	MOS-FeNPs (*M. oleifera* seeds + iron NPs), MOL-FeNPs (*M. oleifera* leaf + iron NPs)	Green synthesis	Removed nitrate from water, also treated coagulant and antibacterial activities	Katata-Seru et al. (2018)
5.	Wastewater	Mesoporous silica nanoparticles (MSNs) + amine, sulfonic groups, fabricate mixed matrix polysulfone (PSU)	Chemical	Removed heavy metal $Cd(NO_3)^{2 \cdot}4H_2O$, Zn $(NO_3)^2$	Alotaibi et al. (2021)
6.	Drinking water	Reduced graphene oxide (rGO) + cerium-silver mixed oxide (Ce/Ag) bimetallic oxide	Chemical precipitation	Adsorb fluoride fluoridation of drinking water	Taneja et al. (2021)
7.	Wastewater	Graphene oxide-manganese oxide (GO-MnO_2) + sulfonated polyether sulfone (SPES) ultrafiltration (UF)	Membrane phase inversion	Filtration of heavy metals Cu^{2+}, Zn^{2+}, and Ni^{2+} ions	Ibrahim et al. (2022)
8.	Contaminated soil and groundwater	Selenium nanoparticles (SeNPs) + selenite24 reducing bacterium *Citrobacter freundii* Y9	Biological	Immobilized Hg from water also remediate soil from Hg	Wang et al. (2018)
9.	Soil (fipronil contaminated agent)	Brassica-AgNps, Ipomoea-AgNps, Camellia-AgNps, and Plantago-AgNps	Green synthesis	Soil remediate by AgNps with plants roots absorption	Romeh (2018)
10.	Cu-contaminated soil	Fe_3O_4 nanoparticles (MagnNPs)	Precipitated method	Enhancing the plant protection system against reactive oxygen species, scavenge Cu and decrease Cu mobility in the soil without hampering Cu bioavailability	Demangeat et al. (2021)

health effects" (Mehndiratta et al., 2013). The environment is contaminated day by day due to anthropogenic activities and poisonous chemicals. Nanotechnology presented a solution to decrease the environmental pollutants. Toxic pollutants are detected by certain nanotechnologies such as nanorobotics, nanodevices, nanosensors, nanofillers, etc. Scientists make sure to develop low-cost and extra-responsive nanotechnology which monitors environmental pollution and reduce its harmful effects on living beings (Srivastava and Mishra, 2022). Natural components of the environment as biotic and abiotic effects of pollution caused the disturbance of the biogeochemical cycle. This disturbance can be detected by nanosensors or nano biosensors (Lahir, 2022). Several induced factors are responsible for the destruction of agroecosystems and environmental pollution such as salinization, eutrophication, and nutrients cycle (Alcon et al., 2022). These polluted agents enter our ecosystem and contaminate soil, air, and water causing different biological health issues in human such as nervous, immunity, metabolic, endocrine (hormonal disorders), and other diseases as well (Ambaye et al., 2022).

Nanotechnology presented recent research on nanosensors in which *Escherichia* coli B40 is incorporated with gold nanoparticles and fiber optic probes that monitor the heavy metal (Cd^{2+}, Hg^{2+}) contamination pollution in water (Halkare et al., 2019). The *Enterobacteriaceae* family includes the main bacterial contaminant. The detection of *E. coli* bacteria by mannose-encapsulated gold nanoparticles was made by Lin et al. (2002), since mannose has an affinity for the kind I *E. coli* pili. Mannose-encapsulated nanoparticles were found to have a greater affinity for *E. coli* in contrast to free mannose. Nanoparticle-based sensors can be used to detect biological organisms, inorganic pollutants, or both. Researchers have suggested employing porous silicon-based semiconductor nanostructures for the detection of organic pollutants (De Stefano et al., 2005). Porous silicon displays the photoluminescence phenomenon, which is extinguished in the presence of an organic or inorganic molecule. Pesticide concentrations as low as 1 ppm could be detected with this method (De Stefano et al., 2005). Titanium dioxide nanoparticles have been utilized to create self-cleaning surfaces that lessen pollution because they have photocatalytic properties. In addition, it is a strong oxidizing agent when exposed to ultra violate (UV) light. As a result, volatile organic compounds, nitrous oxides, and other pollutants are broken down into less dangerous species (Parkin and Palgrave, 2005). Groundwater mostly contaminated by certain causes such as organic and inorganic pollutants, industrial dyes, heavy metals, steel, leather, and chemicals. Gold nanoparticles have unique properties such as high surface energy, high biocompatibility, high surface-to-volume ratio, and electron transfer. It detects Cr^{3+} and monitoring contamination of groundwater pollution as well (Thangadurai et al., 2022). A similar study reported that guanidine thiocyanate (GT)-functionalized gold nanoparticles (AuNPs) detected the Cd^{2+} in polluted water by changing the color with different characterization techniques (Bhamore et al., 2021). A study reported a colorimetric detection method using Cu and Ag nanoparticles which detected and monitor the organophosphorus compound which is released into the environment by pesticides. In this method, sensing mechanism presence of Ag and

Cu, nanoparticles detected organophosphorus in the sample by changing the color (Faghiri et al., 2021).

According to studies by Kanel et al. (2005), Ponder et al. (2000), Shipley et al. (2011), iron nanoparticles are the most widely utilized remediation tool. As an electron donor with reducing characteristics, iron plays a crucial part in the removal of environmental contaminants. Its potent reluctant properties allow for the use of it in the clean-up of any contamination that can be broken down through reduction. Iron nanoparticles were used to discuss the degradation of polyhalogenated organic molecules and heavy metals (Karn et al., 2009). Iron nanoparticles are converted into ferric/ferrous ions and the halogenated organic pollutant is decreased during the reaction. Additionally, iron nanoparticles' flexible deployment capabilities enable them to effectively remove a variety of contaminants from a site (Fang et al., 2012). They have also reported on the effective breakdown of pesticides utilizing magnetic nanoparticles, such as 2, 4-dichlorophenoxyacetic acid (2,4-D). Iron nanoparticles can be used to remove contaminants from the surface, such as petrochemical compounds, as well as from below the surface, such as pesticides, organic solvents, fertilizers, and heavy metals. Nanotechnology improved and monitored environmental pollution with its unique technology. Table 11.5 shows the recent research.

11.3.6 Food processing of agricultural products and their shelf life by nanotechnology

Procedures and practices for food packaging guarantee that the food's quality is maintained and that it is suitable for eating. By removing oxygen and other gases that could cause food spoiling, packaging provides physical protection, keeping food products safe from outside influence, temperature changes, and microbial infection (Hamad et al., 2018). The application of nanotechnology opens the doors to advancement in agriculture and crop yield productivity. In the field of agriculture Traditional crop growing approaches replace by advance innovation in it with the passage of time because it is the need of the modern world to mitigate nutritional deficiency and global hunger (Singhal et al., 2022). Nanotechnology made rapid changes in food sciences to improve storage, processing, packaging, and extension in shelf life to maintain the freshness of food for long-distance transportation. Nanotechnology also ensures good food quality and reduces deterioration to detect the presence of pathogens in food material. It also reduces loss of crops postharvesting and horticulture so the food industry extends throughout the world (Ningthoujam et al., 2022). By using biodegradable materials for packaging, nanotechnology has implemented some specialized areas under packaging that have decreased environmental pollution (Kang et al., 2007). Nanotechnology innovation with food also builds the trust of customers in the market because it ensures quality and authenticity of food with low cost (Areche et al., 2022). Nanofood is frequently linked to the enhancement of color and flavor, storage, preservation, pathogen

Table 11.5 Uses of different nanoparticles for monitoring pollution of water, soil, and environment.

S. no.	Nano-chemical/ products	Monitor pollutant	Characterization technique	Detection	References
1.	Bi_2O_3@Bi	Industrial wastewater	SEM, FTIR, XRD, TGA,	Ultra low, higher sensitivity, current density of 11.2 mA/cm^2, potential current performance 0.85 V vs SCE	Munde et al. (2021)
2.	Carbon nanotube (SWCNT)	Air quality	m-Raman, X-ray photoemission spectroscopy, XPS, AFM	Assess ammonia in environment, water vapor, polluted gas sensing	Rigoni et al. (2014)
3.	Titanium nanotubes	Hydrazine in irrigated water	–	Diffusion coefficient 5.4×10^{-8} cm^2/s sensitivity 288.3 μA/ mmol/L LOD of 0.15 μmol/ L satisfactory performance of assessing hydrazine in water	Emran et al. (2020)
4.	Nanocomposite carbon black-gold nanoparticles (CBNP-AuNP-SPE)	Organic, inorganic heavy metal in river water and in soil	Plasmonic surface resonance (SPR)	Hg^{2+} determination by quick acidification, extraction protocol (HCl + ultrasound)	Cinti et al. (2016)
5.	Pencil graphite electrode nanocompositeCuO/ ZnO, Fe_2O_3/ZnO	Urea cycle water and soil	–	Linear range of sensing Fe-ZnO 0.8 μg/ mL, 4 μg/mL Fe_2O_3/ZnO perform efficiently	Dhinasekaran et al. (2022)
6.	Silver-coated gold nanostars (AuSs@Ag) anodized aluminum oxide (AAO)	Micro-plastic in water	TEM, EDS, FE-SEM, micro Raman spectroscopy	Polystyrene MPs (0.4 μm) at 1000 cm^{-1}, LOD 0.005%	Lê et al. (2021)
7.	Microfluidic paper-based + Au NPs	Hg air, fish, water	TEM	Quantification of Hg (II) in air food water of LOD 2.5 ppb and 3.0 ppb	Shariati and Khayatian (2020)

8.	Ag_2O-NPs/Au electrode	Organic pollutant 4-nitrotoluene environment and water quality assurance	–	Sensitivity 15.33 $\mu A\ (\mu M)^{-1}\ cm^{-2}$ detection limit 62.3 nM linear response ranges 0.6–5.9 mm, 37–175 mm	Chakraborty et al. (2022)
9.	Silver (Ag), gold (Au), platinum (Pt)	Heavy metals in lake water	UV-vis absorption spectroscopy, DLS, FTIR, XRD, TEM, SPR	Hg^{2+} concentration in water 4.73 10–6 ppm	Muthivhi (2017)
10.	Gold (AuNPs) + polyvinylidene fluoride (PVDF)	Pollutants of water and air	UV-vis, SERS, SEM	Quantitative detection of pH and pollutants, gas molecules and smoke of cigarette	Sun et al. (2021)

detection, antimicrobial capabilities, and intelligent packaging, and it can transform the entire food industry chain (Acevedo-Fani et al., 2017). Certain agents which play an active role in food fortification are nanoshells. Nanocapsules are used to enhance flavors and aroma (Ranjan et al., 2022). Nanosensors help in the detection of both gases released during food rotting as well as any small changes in food color. Additionally, nanosensors are more effective than traditional ways of sensing due to their great sensitivity and selectivity to these changes (Mannino and Scampicchio, 2007). The gas sensors are constructed from palladium, platinum, and gold. Nanoparticles based on gold can identify the milk toxin, Aflatoxin B1 (Mao et al., 2006). In some instances, the packaging is constructed from DNA and single-walled carbon nanotubes, which significantly boost the sensors' sensitivity. The identification of pesticides on vegetables and fruits is another application for nanosensors in agriculture. Nanosensors have additionally been utilized to detect carcinogens in food items (Meetoo, 2011).

Different valuable nanoparticles are used as color additives in food items such as TiO_2 and SiO_2 (Fadiji et al., 2022). A research study highlighted that ZnO nanoparticles have potentially inhibited growth of *Listeria monocytogenes* and *E. coli* (food-borne pathogens) and for the packaging of minced fish paste (Shankar et al., 2018). Alelwani et al. reported Ag nanoparticles, which were synthesized by goat colostrum, significantly inhibit the growth of pseudomonas *Aeruginosa* and *Aspergillus flavus* and reduced their antimicrobial effects in food items (Alelwani et al., 2022). Nanoemulsions are used in the manufacturing of sweeteners, flavored oils, salad dressing beverages, and other food products. The process involves the release of many flavors together with a variety of stimulants, including heat, pH, ultrasonic waves, and others (Kumar, 2000). Antimicrobial substances found in nanoemulsions are well recognized to be more beneficial against Gram-positive bacteria. Because of this, nanoemulsions have been employed to clean food packaging materials (Y. Wang et al., 2012). Nanoemulsions composed of soy, nonionic, or tributyl phosphate have been used to inhibit microbial growth and lessen the degree of food degradation (Sanguansri and Augustin, 2006). Sharifan et al. studied the shelf life of tomatoes and also the factors which affect quality, water loss, color change, and microbial spoilage. Their research finding proved that ZnO nanoparticles significantly increased shelf life of tomatoes and limited the spreading microbial production up to 47%, enhanced lycopene content of tomatoes 6% while the Zn content increased in tomatoes by 17% (Sharifan et al., 2021). Below mentioned data explain further, recent researches has been shown in Table 11.6.

11.4 Solution to combat climate change by nanotechnology

Climate change is a global problem. It is a nonrenewable process threatening to overall living beings on earth. Its climate is mostly affected by greenhouse

Table 11.6 Uses of different nanoparticles for food preservation and packaging.

S. no.	Applied nanoparticles	Food item	Method	Characterization techniques	Results	References
1.	Polyethylene terephthalate (PET) + TiO_2, ZnO	Mayonnaise sauce	Twin-screw extrusion	FE-SEM	PET/TiO_2 more effective against antibacterial, fungal activities compared PET/ZnO	Kohannia et al. (2021)
2.	Butterfly pea (*Clitoria ternatea*) anthocyanin + ZnO	Shrimp	Hydrothermal	FTIR	Antibacterial activity against *E. coli*, *L. monocytogenes* color indicator maintains freshness	Kim et al. (2022)
3.	Cassava peel mediated (AgNPs) + gelatin biopolymer matrix	Sapodilla (*Manilkara zapota*) fruit	Biological method	UV-vis spectrum, HR-TEM, XRD, XRS	Antimicrobial increase shelf life	Kim et al. (2022)
4.	Zinc oxide (ZnO NPs)	Cucumber and tomato	Biological method	UV-vis spectroscopy	Reduced weight loss, delay diseases, extend shelf life	Javad et al. (2022)
5.	Silver (Ag) ZnO NPs	Apple and lemon	Green synthesis method	SEM, EDX, XRD, FTIR, PL	Reduced pathogenic attack, increased shelf life, reduced weight loss	Zafar and Iqbal (2022)
6.	Zinc oxide nanoparticlesZnO NPs + chitosan/gum arabic (CH/GA)	Banana	Hydrothermal	SEM, XRD	Antibacterial against *Staphylococcus aureus*, *E. coli*, and *B. subtilis*, maintain freshness 17 days	La et al. (2021)
7.	ZnO NPs + polysaccharides (chitosan and gum arabic)	Avocados	Hydrothermal	SEM, XRD, FTIR	Maintain brightness, weight; firmness, reducing sugar content	Le et al. (2021a)
8.	TiO_2 NPs + litchi peel extract (LPE) + chitosan (CS)	Water cored Fuji apples	Microwave + green synthesis	SEM, FTIR, TGA, chromatogram TIC, DTG	Reduced respiration rate, weight loss, decay of fruit, inhibit senescence enhance phenolic compounds, antioxidant capacity, increase viscosity	Liu et al. (2021)
9.	$AgNO_3$ reduced (RP3-AgNPs)	Cherry, tomato	Green synthesis	UV-vis spectroscopy, XRD, FTIR, EDX, TGA	Increased antibacterial properties against *S. aureus* and *E. coli*	L. Wang et al. (2022)
10.	Silver particles ($AgNO_3$ NPs)	Fruits and vegetables	Green synthesis	TEM, SEM	Antibacterial activity against *E. coli* and *S. aureus*	Xing et al. (2021)

emissions and anthropogenic activities of human beings. The subset of climate change is the extinction of species, habitats change of animals, extreme weather events, water body pollution, rising temperature, sea level rise up wind storm. Nanotechnology offers sustainable solutions to combat climate change of earth (Devi and Chaturvedi, 2021). Nanotechnology provides a variety of ways to mitigate changes in the climate of the earth such as industrial wastewater treatment, drinking water treatment by hydroxyapatite-decorated activated carbon (HAp/AC) nanocomposite (Kallem et al., 2022). Nanotechnology removes the contaminants from the water such as heavy metals like lead, cadmium, mercury, and other biological toxins which cause dangerous diseases and cancers (Tom, 2021). Innovation with green nanotechnology also prevents environmental degradation and cleans the environment and helps to combat climate change emergency (Aithal and Aithal, 2021). Recent comprehensive data are presented in Table 11.7.

11.5 Future perspective

Sustainable agriculture is an important factor for ecosystem of earth because it depends on the maintenance of biotic, abiotic, food chain, and energy balances. Nano-chemicals should be used with new technology and modernization techniques for maximum production in agriculture (Prasad et al., 2017).

Nanotechnology supports the economic programs of agriculture. It also mitigates the obstacles of agri-farming products according to life cycle of plant and soil (Younis et al., 2021). Nanotechnology also gave new trends: robotics and drone-based technology that enable the farmers to produce high quality seeds and fruits and economically important crops for the farm management and sustainable food production (Husaini and Khurshid, 2021). Nanotechnology also provides solution for the absorption of nutrients for targeted nutrients deficient plant and helps absorption of toxic chemicals tolerance crops in the stress condition. Nanotools such as nanoparticles, nanocapsules, and carbon nanotubes promote selective targeted delivery of plant nutrients and absorption of toxic chemicals and provide tolerance to crops against environmental stress (Mishra et al., 2021).

1. Nanotechnology helps in fulfilling 17 goals of United Nation developmental program by 2030 such as erasing hunger by increasing the production of crop for the whole population of the world. And most of them provide strategies and solutions to climate change.
2. Application of nanotechnology in agricultural crops increases plant function and metabolism and removes deficiencies of crop, either macro or micronutrients. It also solves problem of various levels of phytotoxicity cellular induction of nanomaterial.
3. Nanotechnology investigates on heavy metal pollution resources and provides remedies and detection techniques.

Table 11.7 Uses of different nanoparticles for the solution of climate change.

S. no.	Nanoparticle	Raw material	Characterization techniques	Beneficial effects on environment	References
1.	Silver-doped zinc oxide nanoparticles and zinc oxide (Ag-ZnO) NPs + (ZnO NPs)	Cumin seeds + Pongamia seeds	XRD, FTIR, SEM, TEM, EDS, SAED	Ag-ZnO NPs reduced CO	Sam Sukumar et al. (2022)
2.	Zinc oxide green-synthesized (ZnO-GS NPs)	Cow dung + discarded cooking oil	FTIR, XRD, UV-vis spectroscopy, FE-SEM	Significantly reduced greenhouse gases hydrocarbons, CO NO	Kavalli et al. (2022)
3.	Zinc oxide nanoparticles (ZnO-NPs)	Red seaweed (*Pterocladia capillacea*)	UV-vis spectroscopy, FTIR, SEM, XRD, BET	Removing Ismate violet IV2R dye from sea water	Mansour et al. (2022)
4.	Silver oxide nanoparticles (AgO NPs)	Seed oil of *Zanthoxylum armatum*	XRD, SEM, EDX, NMR, FTIR	Production of biodiesel maintain CO_2 in the environment	Ahmad, Asif, et al. (2022a)
5.	Copper oxide nanoparticles (CuO NPs)	Citrus medica + leaf extract of *Portulaca oleracea*	XRD, SEM, EDX, FTIR, (GC–MS)	Production of renewable biodiesel to maintain CO_2 in the environment	Chia et al. (2022)
6.	Calcium oxide nanoparticles (CaO NPs)	Nonedible seed oil of *Monotheca buxifolia*	XRD, SEM, EDX (GC/MS)	Production of eco-friendly biodiesel	Ahmad, Elnaggar, et al. (2022b)
7.	Graphene oxide nanoparticles (GO NPs)	Nonedible feedstock. *Ailanthus altissima* oil (Tree of heaven)	SEM, TEM	Produced eco-friendly biodiesel also reduced in CO and unburned hydrocarbons emissions in environment slightly increases CO_2 and NO emissions	Hoseini et al. (2018)

(*Continued*)

Table 11.7 Uses of different nanoparticles for the solution of climate change. *Continued*

S. no.	Nanoparticle	Raw material	Characterization techniques	Beneficial effects on environment	References
8.	Iron oxide nanoparticles (FeO NPs)	Leaf extract of *Azadirachta indica* + cattle manure	UV-vis spectroscopy, TEM, PSA, SPR	Produced eco- friendly biogas and methane	Singh et al. (2022)
9.	Zinc oxide nanoparticles (ZnO NPs) + olive pomace	Wheat leaf extract	TEM	Production of eco-friendly biogas	Amirante et al. (2018)
10.	Alumina and titanium dioxide nanoparticles (Al + TiO_2 NPs)	Titanium tetra isopropoxide + aluminum nitrate + citric acid	SEM	CO, HC, and smoke emissions also brake-specific fuel consumption (BSFC) decreased	Srinivasan et al. (2021)

11.6 Conclusion

Nanotechnology is a promising approach for this modern era. There are variety of user-friendly applications of nanotechnology in agriculture because it fulfills all the demands of modern era and provides wide range of solutions in all fields of agriculture such as crop production, crop protection, animal husbandry, biosensors, which are used in plant diseases diagnosis, its control management, and also food sciences. It also changes traditional trends of research in agriculture and also provides solutions. Recent research articles suggested that if nanotechnology apply in crop management pest and weed control farmers gain maximum economic benefits crop diseases also treated by it. Crop production, crop yield protection, not only feed the livestock human and overall population, it also maintains life cycle of our ecosystem by solving issues related to environmental pollution and climate change. There are a lot of challenges such as environment and ecosystem for the human security which can be resolved by nanotechnology.

References

Abdel-Halim, K.Y., El-Ghanam, A.A., 2019. Antifungal potent of some metallic nanoparticles against Sclerotinia sclerotiorum on common bean plants: an emphasis for biochemical alterations and metal accumulation. Acad. J. Life Sci. 5 (11), 93–106.

Abdulhameed, M.F., Taha, A.A., Ismail, R.A., 2021. Improvement of cabbage growth and yield by nanofertilizers and nanoparticles. Environ. Nanotechnol. Monitor. Manag. 15, 100437.

Acevedo-Fani, A., Soliva-Fortuny, R., Martín-Belloso, O., 2017. Nanostructured emulsions and nanolaminates for delivery of active ingredients: improving food safety and functionality. Trends Food Sci. Technol. 60, 12–22.

Acharya, A., Pal, P.K., 2020. Agriculture nanotechnology: translating research outcome to field applications by influencing environmental sustainability. NanoImpact 19, 100232.

Acharya, B., Islam, S.M.I., Alam, M.S., 2022. Efficiency of nanotechnology in environmental pollution control by soil and groundwater remediation. 6th International Conference on Civil Engineering for Sustainable Development (ICCESD2022).

Ahmad, F., Siddiqui, M.A., Babalola, O.O., Wu, H.-F., 2012. Biofunctionalization of nanoparticle assisted mass spectrometry as biosensors for rapid detection of plant associated bacteria. Biosen. Bioelectron. 35 (1), 235–242.

Ahmad, M., Asif, S., Klemeš, J.J., Mubashir, M., Bokhari, A., Sultana, S., et al., 2022a. Conversion of the toxic and hazardous Zanthoxylum armatum seed oil into methyl ester using green and recyclable silver oxide nanoparticles. Fuel 310, 122296.

Ahmad, M., Elnaggar, A.Y., Teong, L.K., Sultana, S., Zafar, M., Munir, M., et al., 2022b. Sustainable and eco-friendly synthesis of biodiesel from novel and non-edible seed oil of Monotheca buxifolia using green nano-catalyst of calcium oxide. Energy Convers. Manag. 13, 100142.

Aithal, S., Aithal, P.S., 2021. Green and eco-friendly nanotechnology—concepts and industrial prospects. Int. J. Manag. Technol. Social Sci. 6 (1), 1–31.

Alcon, F., Zabala, J.A., Martínez-Paz, J.M., 2022. Assessment of social demand heterogeneity to inform agricultural diffuse pollution mitigation policies. Ecol. Econ. 191, 107216.

Alelwani, W., Taj, M.B., Algheshairy, R.M., Alnajeebi, A.M., Alharbi, H.F., Bannunah, A. M., et al., 2022. Synthesis and potential of bio fabricated silver nanoparticles for use as functional material against foodborne pathogens. Chem. Afr. 1–17.

Ali, S., Shafique, O., Mahmood, T., Hanif, M.A., Ahmed, I., Khan, B., 2018. A review about perspectives of nanotechnology in agriculture. Pak. J. Agric. Res. 30 (2), 116–121.

Ali, M., Wang, X., Haroon, U., Chaudhary, H.J., Kamal, A., Ali, Q., et al., 2022. Antifungal activity of zinc nitrate derived nano Zno fungicide synthesized from Trachyspermum ammi to control fruit rot disease of grapefruit. Ecotoxicol. Environ. Saf. 233, 113311.

Alizadeh, M., Sheikhi-Garjan, A., Ma'mani, L., Hosseini Salekdeh, G., Bandehagh, A., 2022. Ethology of Sunn-pest oviposition in interaction with deltamethrin loaded on mesoporous silica nanoparticles as a nanopesticide. Chem. Biol. Technol. Agric. 9 (1), 1–13.

Alotaibi, A.A., Shukla, A.K., Mrad, M.H., Alswieleh, A.M., Alotaibi, K.M., 2021. Fabrication of polysulfone-surface functionalized mesoporous silica nanocomposite membranes for removal of heavy metal ions from wastewater. Membranes 11 (12), 935.

Alsaeedi, A., El-Ramady, H., Alshaal, T., El-Garawany, M., Elhawat, N., Al-Otaibi, A., 2019. Silica nanoparticles boost growth and productivity of cucumber under water deficit and salinity stresses by balancing nutrients uptake. Plant Physiol. Biochem. 139, 1–10.

Ambaye, T.G., Chebbi, A., Formicola, F., Prasad, S., Gomez, F.H., Franzetti, A., et al., 2022. Remediation of soil polluted with petroleum hydrocarbons, and their reuse for agriculture: recent progress, challenges, and perspectives. Chemosphere 133572.

Amirante, R., Demastro, G., Distaso, E., Hassaan, M., Mormando, A., Pantaleo, A., et al., 2018. Effects of ultrasound and green synthesis ZnO nanoparticles on biogas production from olive pomace. Energy Proc. 148, 940–947.

Areche, F.O., Flores, D.D.C., Rodriguez, A.R., Huaman, J.T., Solano, M.A.Q., Jagdale, Y. D., et al., 2022. Nanoparticles designed for shelf-life extension and food safety. Nanotechnology Interventions in Food Packaging and Shelf Life. CRC Press, pp. 245–280.

Ashraf, S.A., Siddiqui, A.J., Abd Elmoneim, O.E., Khan, M.I., Patel, M., Alreshidi, M., et al., 2021. Innovations in nanoscience for the sustainable development of food and agriculture with implications on health and environment. Sci. Total Environ. 768, 144990.

Azeez, L., Lateef, A., Adetoro, R.O., Adeleke, A.E., 2021. Responses of *Moringa oleifera* to alteration in soil properties induced by calcium nanoparticles (CaNPs) on mineral absorption, physiological indices and photosynthetic indicators. Beni-Suef Univ. J. Basic Sci. 10 (1), 1–15.

Baac, H., Hajós, J.P., Lee, J., Kim, D., Kim, S.J., Shuler, M.L., 2006. Antibody-based surface plasmon resonance detection of intact viral pathogen. Biotechnol. Bioeng. 94 (4), 815–819.

Babur, M.N., & Shah, S.A. Q. Future of nanotechnology in Pakistan: a paradigm shift. J. Biol. Allied Health Sci.

Baig, N., Kammakakam, I., Falath, W., 2021. Nanomaterials: a review of synthesis methods, properties, recent progress, and challenges. Mater. Adv. 2 (6), 1821–1871.

Balestra, G.M., di Lorenzo, V., Fortunati, E., 2022. Organic antimicrobial nanomaterials and reducing copper use in sustainable plant protection. Nanotechnology-Based Sustainable Alternatives for the Management of Plant Diseases. Elsevier, pp. 179–209.

Bhamore, J.R., Gul, A.R., Kailasa, S.K., Kim, K.-W., Lee, J.S., Park, H., 2021. Functionalization of gold nanoparticles using guanidine thiocyanate for sensitive and selective visual detection of Cd2 + . Sens. Actuat. B Chem. 334, 129685.

Biao, L., Tan, S., Meng, Q., Gao, J., Zhang, X., Liu, Z., et al., 2018. Green synthesis, characterization and application of proanthocyanidins-functionalized gold nanoparticles. Nanomaterials 8 (1), 53.

Boonham, N., Glover, R., Tomlinson, J., Mumford, R., 2008. Exploiting generic platform technologies for the detection and identification of plant pathogens. Eur. J. Plant Pathol. 355–363.

Byrne, B.J.S., Gilmartin, S.E., O'Kennedy, N., 2009. Antibodybased sensors: principles, problems and potential for detection of pathogens and associated toxins. Sensors 9, 4407–4445.

Camara, M.C., Campos, E.V.R., Monteiro, R.A., do Espirito Santo Pereira, A., de Freitas Proença, P.L., Fraceto, L.F., 2019. Development of stimuli-responsive nano-based pesticides: emerging opportunities for agriculture. J. Nanobiotechnology 17 (1), 1–19.

Chajanovsky, I., Cohen, S., Shtenberg, G., Suckeveriene, R.Y., 2021. Development and characterization of integrated nano-sensors for organic residues and pH field detection. Sensors 21 (17), 5842.

Chakraborty, U., Garg, P., Bhanjana, G., Kaur, G., Kaushik, A., Chaudhary, G.R., 2022. Spherical silver oxide nanoparticles for fabrication of electrochemical sensor for efficient 4-nitrotoluene detection and assessment of their antimicrobial activity. Sci. Total Environ. 808, 152179.

Chartuprayoon, N., Rheem, Y., Chen, W., Myung, N., 2010. Detection of plant pathogen using LPNE grown single conducting polymer nanoribbon. Paper presented at the ECS Meeting Abstracts.

Chatterjee, A., Abraham, J., 2022. Heavy metal remediation through nanoparticles. Bioremediation. CRC Press, pp. 235–247.

Chaudhry, F.N., Malik, M., 2017. Factors affecting water pollution: a review. J. Ecosyst. Ecography 7 (225), 1–3.

Chen, C., Feng, S., Zhou, M., Ji, C., Que, L., Wang, W., 2019. Development of a structure-switching aptamer-based nanosensor for salicylic acid detection. Bionsens. Bioelectron. 140, 111342.

Chesterfield, R.J., Whitfield, J.H., Pouvreau, B., Cao, D., Alexandrov, K., Beveridge, C.A., et al., 2020. Rational design of novel fluorescent enzyme biosensors for direct detection of strigolactones. ACS Synth. Biol. 9 (8), 2107–2118.

Chia, S.R., Ahmad, M., Sultana, S., Zafar, M., Asif, S., Bokhari, A., et al., 2022. Green synthesis of biodiesel from citrus medica seed oil using green nanoparticles of copper oxide. Fuel 323, 124285.

Chinta, Y.D., Uchida, Y., Araki, H., 2020. Availability of nitrogen supply from cover crops during residual decomposition by soil microorganisms and its utilization by lettuce (*Lactuca sativa* L.). Sci. Hortic. 270, 109415.

Choina, J., Bagabas, A., Fischer, C., Flechsig, G.-U., Kosslick, H., Alshammari, A., et al., 2015. The influence of the textural properties of ZnO nanoparticles on adsorption and photocatalytic remediation of water from pharmaceuticals. Catal. Today 241, 47–54.

Cinti, S., Santella, F., Moscone, D., Arduini, F., 2016. Hg2 + detection using a disposable and miniaturized screen-printed electrode modified with nanocomposite carbon black and gold nanoparticles. Environ. Sci. Pollut. Res. Int. 23 (9), 8192–8199.

Cui, S., Inocente, E.A.A., Acosta, N., Keener, H.M., Zhu, H., Ling, P.P., 2019. Development of fast e-nose system for early-stage diagnosis of aphid-stressed tomato plants. Sensors 19 (16), 3480.

Dar, F.A., Qazi, G., Pirzadah, T.B., 2020. Nano-biosensors: NextGen diagnostic tools in agriculture. Nanobiotechnology in Agriculture. Springer, pp. 129–144.

De Stefano, L., Moretti, L., Rendina, I., Rotiroti, L., 2005. Pesticides detection in water and humic solutions using porous silicon technology. Sens. Actuat. B Chem. 111, 522–525.

Demangeat, E., Pédrot, M., Dia, A., Bouhnik-Le-Coz, M., Roperch, P., Compaoré, G., et al., 2021. Investigating the remediation potential of iron oxide nanoparticles in Cu-polluted soil–plant systems: coupled geochemical, geophysical and biological approaches. Nanoscale Adv. 3 (7), 2017–2029.

Desai, P., Jha, A., Markande, A., Patel, J., 2021. Silver nanoparticles as a fungicide against soil-borne Sclerotium rolfsii: a case study for wheat plants. Biobased Nanotechnology for Green Applications. Springer, pp. 513–542.

Devi, K.P., Chaturvedi, H., 2021. An overview of nanotechnology in water treatment applications and combating climate change. Water Conservation in the Era of Global Climate Change. pp. 191–212.

Dhinasekaran, D., Soundharraj, P., Jagannathan, M., Rajendran, A.R., Rajendran, S., 2022. Hybrid ZnO nanostructures modified graphite electrode as an efficient urea sensor for environmental pollution monitoring. Chemosphere 296, 133918.

Ditta, A., 2012. How helpful is nanotechnology in agriculture? Adv. Nat. Sci. Nanosci. Nanotechnol. 3 (3), 033002.

Dubey, A., Mailapalli, D.R., 2016. Nanofertilisers, nanopesticides, nanosensors of pest and nanotoxicity in agriculture. Sustainable Agriculture Reviews. Springer, pp. 307–330.

El-Shabrawy, E., 2021. Use silica nanoparticles in controlling late wilt disease in maize caused by *Harpophora maydis*. Egypt. J. Appl. Sci. 36, 1–19.

El-Shewy, E.S., 2019. The efficacy of copper oxide, tri-calcium phosphate and silicon dioxide nanoparticles in controlling black scurf disease of potato. Ann. Agric. Sci. Moshtohor 57 (1), 129–138.

Emran, K.M., Ali, S.M., Alanazi, H.E., 2020. Novel hydrazine sensors based on Pd electrodeposited on highly dispersed lanthanide-doped TiO2 nanotubes. J. Electroanal. Chem. 856, 113661.

Fabbri, B., Valt, M., Parretta, C., Gherardi, S., Gaiardo, A., Malagù, C., 2020. Correlation of gaseous emissions to water stress in tomato and maize crops: from field to laboratory and back. Sens. Actuat. B Chem. 303, 127227.

Fadiji, A.E., Mthiyane, D.M.N., Onwudiwe, D.C., Babalola, O.O., 2022. Harnessing the known and unknown impact of nanotechnology on enhancing food security and reducing postharvest losses: constraints and future prospects. Agronomy 12 (7), 1657.

Faghiri, F., Hajjami, M., Ghorbani, F., 2021. Development of a sensing system based on coupling magnetic solid phase extraction and colorimetric detection for determination of organophosphorus pesticides in fruit extract and environmental sample. Sens. Actuat. B Chem. 343, 130157.

Fang, G., Si, Y., Tian, C., Zhang, G., Zhou, D., 2012. Degradation of 2, 4-D in soils by Fe3O4 nanoparticles combined with stimulating indigenous microbes. Environ. Sci. Pollut. Res. Int. 19 (3), 784–793.

Fatemi, A., Moaveni, P., Daneshian, J., Mozafari, H., Ghaffari, M., 2022. Magnesium nanoparticles improve grain yield, oil percentage, physiological, and biochemical traits of sunflower (*Helianthus annuus* L.) under drought stress. J. Agric. Sci. Technol. 24 (3), 665–678.

Francesconi, S., Schiavi, D., di Lorenzo, V., Balestra, G.M., 2022. Inorganic nanomaterials usable in plant protection strategies. Nanotechnology-Based Sustainable Alternatives for the Management of Plant Diseases. Elsevier, pp. 211–231.

Gallo, A., Bianco, C., Tosco, T., Tiraferri, A., Sethi, R., 2019. Synthesis of eco-compatible bimetallic silver/iron nanoparticles for water remediation and reactivity assessment on bromophenol blue. J. Clean. Prod. 211, 1367–1374.

Gehrke, I., Geiser, A., Somborn-Schulz, A., 2015. Innovations in nanotechnology for water treatment. Nanotechnol. Sci. Appl. 8, 1.

Ghazy, N.A., El-Hafez, A., Omnia, A., El-Bakery, A., El-Geddawy, D.I., 2021. Impact of silver nanoparticles and two biological treatments to control soft rot disease in sugar beet (*Beta vulgaris* L). Egypt. J. Biol. Pest Control 31 (1), 1–12.

Glenn, J.C., Florescu, E., 2016. The millennium project team, 2015-16 state of the future. J. Socialomics 5 (3), 1–6.

Gondal, A.H., Tayyiba, L., 2022. Prospects of using nanotechnology in agricultural growth, environment and industrial food products. Rev. Agric. Sci. 10, 68–81.

Graham, J.H., Johnson, E., Myers, M.E., Young, M., Rajasekaran, P., Das, S., et al., 2016. Potential of nano-formulated zinc oxide for control of citrus canker on grapefruit trees. Plant Dis. 100 (12), 2442–2447.

Habila, M.A., ALOthman, Z.A., El-Toni, A.M., Labis, J.P., Khan, A., Al-Marghany, A., et al., 2017. One-step carbon coating and polyacrylamide functionalization of Fe3O4 nanoparticles for enhancing magnetic adsorptive-remediation of heavy metals. Molecules 22 (12), 2074.

Halkare, P., Punjabi, N., Wangchuk, J., Nair, A., Kondabagil, K., Mukherji, S., 2019. Bacteria functionalized gold nanoparticle matrix based fiber-optic sensor for monitoring heavy metal pollution in water. Sens. Actuat. B Chem. 281, 643–651.

Hamad, A.F., Han, J.-H., Kim, B.-C., Rather, I.A., 2018. The intertwine of nanotechnology with the food industry. Saudi J. Biol. Sci. 25 (1), 27–30.

Hazarika, A., Yadav, M., Yadav, D.K., Yadav, H.S., 2022. An overview of the role of nanoparticles in sustainable agriculture. Biocatal. Agric. Biotechnol. 102399.

Heydari, M., Yousefi, A.R., Nikfarjam, N., Rahdar, A., Kyzas, G.Z., Bilal, M., 2021. Plant-based nanoparticles prepared from protein containing tribenuron-methyl: fabrication, characterization, and application. Chem. Biol. Technol. Agric. 8 (1), 1–11.

Hoseini, S., Najafi, G., Ghobadian, B., Mamat, R., Ebadi, M., Yusaf, T., 2018. Novel environmentally friendly fuel: the effects of nanographene oxide additives on the performance and emission characteristics of diesel engines fuelled with *Ailanthus altissima* biodiesel. Renew. Energy 125, 283–294.

Huanan, G., Qiaoyan, W., Shuping, L., 2022. A smartphone-integrated dual-mode nanosensor based on Fe3O4@ Au for rapid and highly selective detection of glutathione. Spectrochim. Acta A. Mol. Biomol. Spectrosc. 271, 120866.

Husaini, A.M., Khurshid, A., 2021. Nanotechnology and robotics: the twin drivers of agriculture in future. Agricultural Biotechnology: Latest Research and Trends. Springer, pp. 553–571.

Ibrahim, Y., Wadi, V.S., Ouda, M., Naddeo, V., Banat, F., Hasan, S.W., 2022. Highly selective heavy metal ions membranes combining sulfonated polyethersulfone and self-assembled manganese oxide nanosheets on positively functionalized graphene oxide nanosheets. Chem. Eng. J. 428, 131267.

Jasrotia, P., Kashyap, P.L., Bhardwaj, A.K., Kumar, S., Singh, G., 2018. Scope and applications of nanotechnology for wheat production: a review of recent advances. J. Cereal Res. 10 (1), 1–14.

Javad, S., Azam, N., Ghaffar, N., Jabeen, K., Ahmad, A., 2022. Effect of zinc oxide nanoparticles on shelf life of commonly used vegetables. Paper presented at the Proceedings of the Bulgarian Academy of Sciences.

Jeevanandam, J., Krishnan, S., Hii, Y.S., Pan, S., Chan, Y.S., Acquah, C., et al., 2022. Synthesis approach-dependent antiviral properties of silver nanoparticles and nanocomposites. J. Nanostruct. Chem. 1–23.

Johnson, M.S., Sajeev, S., Nair, R.S., 2021. Role of nanosensors in agriculture. Paper presented at the 2021 International Conference on Computational Intelligence and Knowledge Economy (ICCIKE).

Kale, A.P., Gawade, S.N., 2016. Studies on nanoparticle induced nutrient use efficiency of fertilizer and crop productivity. Green Chem. Technol. Lett. 2, 88–92.

Kallem, P., Ouda, M., Bharath, G., Hasan, S.W., Banat, F., 2022. Enhanced water permeability and fouling resistance properties of ultrafiltration membranes incorporated with hydroxyapatite decorated orange-peel-derived activated carbon nanocomposites. Chemosphere 286, 131799.

Kalteh, M., Alipour, Z.T., Ashraf, S., Marashi Aliabadi, M., Falah Nosratabadi, A., 2018. Effect of silica nanoparticles on basil (*Ocimum basilicum*) under salinity stress. J. Chem. Health Risks 4 (3).

Kanel, S.R., Manning, B., Charlet, L., Choi, H., 2005. Removal of arsenic (III) from groundwater by nanoscale zero-valent iron. Environ. Sci. Technol. 39 (5), 1291–1298.

Kang, S., Pinault, M., Pfefferle, L.D., Elimelech, M., 2007. Single-walled carbon nanotubes exhibit strong antimicrobial activity. Langmuir 23 (17), 8670–8673.

Karn, B., Kuiken, T., Otto, M., 2009. Nanotechnology and in situ remediation: a review of the benefits and potential risks. Environ. Health Perspect. 117 (12), 1813–1831.

Kashyap, P.L., Kumar, S., Jasrotia, P., Singh, D., Singh, G.P., 2019. Nanosensors for plant disease diagnosis: current understanding and future perspectives. Nanoscience for Sustainable Agriculture. Springer, pp. 189–205.

Katata-Seru, L., Moremedi, T., Aremu, O.S., Bahadur, I., 2018. Green synthesis of iron nanoparticles using *Moringa oleifera* extracts and their applications: removal of nitrate from water and antibacterial activity against *Escherichia coli*. J. Mol. Liquids 256, 296–304.

Kaushal, M., Wani, S.P., 2017. Nanosensors: frontiers in precision agriculture. Nanotechnology. Springer, pp. 279–291.

Kavalli, K., Hebbar, G.S., Shubha, J.P., Adil, S.F., Khan, M., Hatshan, M.R., et al., 2022. Green synthesized ZnO nanoparticles as biodiesel blends and their effect on the performance and emission of greenhouse gases. Molecules 27 (9), 2845.

Khajeh, M., Laurent, S., Dastafkan, K., 2013. Nanoadsorbents: classification, preparation, and applications (with emphasis on aqueous media). Chem. Rev. 113 (10), 7728–7768.

Khaledian, S., Nikkhah, M., Shams-bakhsh, M., Hoseinzadeh, S., 2017. A sensitive biosensor based on gold nanoparticles to detect *Ralstonia solanacearum* in soil. J. Gen. Plant Pathol. 83 (4), 231–239.

Khan, H.H., Malik, M.N., Konečná, Z., Chofreh, A.G., Goni, F.A., Klemeš, J., 2022a. Blockchain technology for agricultural supply chains during the COVID-19 pandemic: benefits and cleaner solutions. J. Clean. Prod. 347, 131268.

Khan, M., Siddiqui, Z.A., Parveen, A., Khan, A.A., Moon, I.S., Alam, M., 2022b. Elucidating the role of silicon dioxide and titanium dioxide nanoparticles in mitigating the disease of the eggplant caused by *Phomopsis vexans*, *Ralstonia solanacearum*, and root-knot nematode *Meloidogyne incognita*. Nanotechnol. Rev. 11 (1), 1606–1619.

Khater, M., De La Escosura-Muñiz, A., Merkoçi, A., 2017. Biosensors for plant pathogen detection. Biosens. Bioelectron. 93, 72–86.

Kheiri, A., Jorf, S.M., Malihipour, A., Saremi, H., Nikkhah, M., 2016. Application of chitosan and chitosan nanoparticles for the control of Fusarium head blight of wheat (*Fusarium graminearum*) in vitro and greenhouse. Int. J. Biol. Macromol. 93, 1261–1272.

Kim, D.Y., Kadam, A., Shinde, S., Saratale, R.G., Patra, J., Ghodake, G., 2018. Recent developments in nanotechnology transforming the agricultural sector: a transition replete with opportunities. J. Sci. Food Agric. 98 (3), 849–864.

Kim, H.-J., Roy, S., Rhim, J.-W., 2022. Gelatin/agar-based color-indicator film integrated with *Clitoria ternatea* flower anthocyanin and zinc oxide nanoparticles for monitoring freshness of shrimp. Food Hydrocoll. 124, 107294.

Kohannia, N., Beigmohammadi, F., Ramzani Ghara, A., Nayebzadeh, K., 2021. Effect of polyethylene terephthalate incorporated with titanium dioxide and zinc oxide nanoparticles on shelf-life extension of mayonnaise sauce. J. Food Process. Preserv. 45 (5), e15453.

Kristanti, R.A., Liong, R.M.Y., Hadibarata, T.J.T.A., Pollution, S., 2021. Soil remediation applications of nanotechnology. Trop. Aquat. Soil Pollut. 1 (1), 35–45.

Kumar, M.N.R., 2000. A review of chitin and chitosan applications. React Funct. Polym. 46 (1), 1–27.

Kundu, M., Krishnan, P., Chobhe, K.A., Manjaiah, K., Pant, R., Chawla, G., 2022. Fabrication of electrochemical nanosensor for detection of nitrate content in soil extract. J. Soil Sci. Plant Nutr. 1–16.

La, D.D., Nguyen-Tri, P., Le, K.H., Nguyen, P.T., Nguyen, M.D.-B., Vo, A.T., et al., 2021. Effects of antibacterial ZnO nanoparticles on the performance of a chitosan/gum arabic edible coating for post-harvest banana preservation. Prog. Org. Coat. 151, 106057.

Lahir, Y., 2022. Nanotechnology and its implications in the environmental biotechnology. Emerging Trends in Environmental Biotechnology. CRC Press, pp. 81–95.

Laishram, D., Shejale, K.P., Gupta, R., Sharma, R.K., 2018. Heterostructured HfO2/TiO2 spherical nanoparticles for visible photocatalytic water remediation. Mater. Lett. 231, 225–228.

Lau, H.Y., Wu, H., Wee, E.J., Trau, M., Wang, Y., Botella, J.R., 2017. Specific and sensitive isothermal electrochemical biosensor for plant pathogen DNA detection with colloidal gold nanoparticles as probes. Sci. Rep. 7 (1), 1–7.

Le, K.H., Nguyen, M.D.-B., Dai Tran, L., Thi, H.P.N., Van Tran, C., Van Tran, K., et al., 2021a. A novel antimicrobial ZnO nanoparticles-added polysaccharide edible coating for the preservation of postharvest avocado under ambient conditions. Prog. Org. Coat. 158, 106339.

Le, V.S., Herrmann, L., Hudek, L., Nguyen, T.B., Bräu, L., Lesueur, D., 2021b. How application of agricultural waste can enhance soil health in soils acidified by tea cultivation: a review. Environ. Chem. Lett. 1–27.

Lê, Q.T., Ly, N.H., Kim, M.-K., Lim, S.H., Son, S.J., Zoh, K.-D., et al., 2021. Nanostructured Raman substrates for the sensitive detection of submicrometer-sized plastic pollutants in water. J. Hazard. Mater. 402, 123499.

Lew, T.T.S., Park, M., Cui, J., Strano, M.S., 2021. Plant nanobionic sensors for arsenic detection. Adv. Mater. 33 (1), 2005683.

Li, C., Li, Y., Li, Y., Fu, G., 2018. Cultivation techniques and nutrient management strategies to improve productivity of rain-fed maize in semi-arid regions. Agric. Water Manag. 210, 149–157.

Lin, C.-C., Yeh, Y.-C., Yang, C.-Y., Chen, C.-L., Chen, G.-F., Chen, C.-C., et al., 2002. Selective binding of mannose-encapsulated gold nanoparticles to type 1 pili in *Escherichia coli.* J. Am. Chem. Soc. 124 (14), 3508–3509.

Lin, H.-Y., Huang, C.-H., Lu, S.-H., Kuo, I.-T., Chau, L.-K., 2014. Direct detection of orchid viruses using nanorod-based fiber optic particle plasmon resonance immunosensor. Biosens. Bioelectron. 51, 371–378.

Liu, Z., Du, M., Liu, H., Zhang, K., Xu, X., Liu, K., et al., 2021. Chitosan films incorporating litchi peel extract and titanium dioxide nanoparticles and their application as coatings on watercored apples. Prog. Org. Coat. 151, 106103.

Lu, Y., Song, S., Wang, R., Liu, Z., Meng, J., Sweetman, A.J., et al., 2015. Impacts of soil and water pollution on food safety and health risks in China. Environ. Int. 77, 5–15.

Mahmoud, M.E., Abdelwahab, M.S., 2021. One-step synthesis of zero-valent Sn nanoparticles and potential microwave remediation of lead from water. Mater. Res. Bull. 134, 111090.

Manjunatha, S., Biradar, D., Aladakatti, Y.R., 2016. Nanotechnology and its applications in agriculture: a review. J. Farm Sci. 29 (1), 1–13.

Mannino, S., Scampicchio, M., 2007. Nanotechnology and food quality control. Vet. Res. Commun. 31 (1), 149–151.

Mansour, A.T., Alprol, A.E., Khedawy, M., Abualnaja, K.M., Shalaby, T.A., Rayan, G., et al., 2022. Green synthesis of zinc oxide nanoparticles using red seaweed for the elimination of organic toxic dye from an aqueous solution. Materials 15 (15), 5169.

Mao, X., Huang, J., Fai Leung, M., Du, Z., Ma, L., Huang, Z., 2006. Novel core-shell nanoparticles and their application in high-capacity immobilization of enzymes. Biotechnology 135 (3), 229–239.

Meetoo, D.D., 2011. Nanotechnology and the food sector: from the farm to the table. Emir. J. Food Agric. 387–403.

Mehndiratta, P., Jain, A., Srivastava, S., Gupta, N., 2013. Environmental pollution and nanotechnology. Environ. Pollut. 2 (2), 49.

Mishra, M., Pandey, A.K., Pandey, K., Dixit, S., Zohra, F., Seth, A., et al., 2021. Exploring the potential of nanotechnology in agriculture: current research and future prospects. Compos. Mater. 223–241.

Munde, A.V., Mulik, B.B., Dighole, R.P., Dhawale, S.C., Sable, L.S., Avhale, A.T., et al., 2021. Bi2O3@ Bi nanoparticles for ultrasensitive electrochemical determination of thiourea: monitoring towards environmental pollutants. Electrochim. Acta 394, 139111.

Munir, M.U., Ahmad, A., Hopmans, J.W., Belgacem, A.O., Baig, M.B., 2021. Water scarcity threats to National Food Security of Pakistan—issues, implications, and way forward. Emerging Challenges to Food Production and Security in Asia, Middle East, and Africa. Springer, pp. 241–266.

Mustafa, H., Ilyas, N., Akhtar, N., Raja, N.I., Zainab, T., Shah, T., 2021. Biosynthesis and characterization of titanium dioxide nanoparticles and its effects along with calcium phosphate on physicochemical attributes of wheat under drought stress. Ecotoxicol. Environ. Saf. 223, 112519.

Muthivhi, R., 2017. *Green Synthesis of Maltose-Reduced Silver, Gold and Platinum Nanoparticles for Colorimetric Sensing of Heavy and Alkali/Alkaline Earth Metals in Aqueous Solution:*. University of Johannesburg, South Africa.

Ningthoujam, R., Jena, B., Pattanayak, S., Dash, S., Panda, M.K., Behera, R.K., et al., 2022. Nanotechnology in food science. Bio-Nano Interface. Springer, pp. 59–73.

Nißler, R., Müller, A.T., Dohrman, F., Kurth, L., Li, H., Cosio, E.G., et al., 2022. Detection and imaging of the plant pathogen response by near-infrared fluorescent polyphenol sensors. Angew. Chem. Int. Ed. Engl. 61 (2), e202108373.

Noor, R., Yasmin, H., Ilyas, N., Nosheen, A., Hassan, M.N., Mumtaz, S., et al., 2022. Comparative analysis of iron oxide nanoparticles synthesized from ginger (*Zingiber officinale*) and cumin seeds (*Cuminum cyminum*) to induce resistance in wheat against drought stress. Chemosphere 292, 133201.

Pandit, C., Roy, A., Ghotekar, S., Khusro, A., Islam, M.N., Emran, T.B., et al., 2022. Biological agents for synthesis of nanoparticles and their applications. J. King Saud Univ. Sci. 34 (3), 101869. In this issue.

Parkin, I.P., Palgrave, R.G., 2005. Self-cleaning coatings. J. Mater. Chem. 15 (17), 1689–1695.

Parveen, A., Siddiqui, Z.A., 2022. Impact of silicon dioxide nanoparticles on growth, photosynthetic pigments, proline, activities of defense enzymes and some bacterial and fungal pathogens of tomato. Vegetos 35 (1), 83–93.

Pascoli, M., Jacques, M.T., Agarrayua, D.A., Avila, D.S., Lima, R., Fraceto, L.F., 2019. Neem oil based nanopesticide as an environmentally-friendly formulation for applications in sustainable agriculture: an ecotoxicological perspective. Sci. Total Environ. 677, 57–67.

Plohl, O., Gyergyek, S., Zemljič, L.F.J.M., Materials, M., 2021. Mesoporous silica nanoparticles modified with N-rich polymer as a potentially environmentally-friendly delivery system for pesticides. Microporous Mesoporous Mater. 310, 110663.

Polman, E.M., Gruter, G.-J.M., Parsons, J.R., Tietema, A., 2021. Comparison of the aerobic biodegradation of biopolymers and the corresponding bioplastics: a review. Sci. Total Environ. 753, 141953.

Ponder, S. Darab, J.G., Mallouk, T.E., 2000. Remediation of Cr(VI) and Pb(II) aqueous solutions using supported, nanoscale zero-valent iron. Environ. Sci. Technol. 34, 2564-2569.

Prasad, R., Bhattacharyya, A., Nguyen, Q.D., 2017. Nanotechnology in sustainable agriculture: recent developments, challenges, and perspectives. Front. Microbiol. 8, 1014.

Qu, H., Ma, C., Xing, W., Xue, L., Liu, H., White, J.C., et al., 2022. Effects of copper oxide nanoparticles on Salix growth, soil enzyme activity and microbial community composition in a wetland mesocosm. J. Hazard. Mater. 424, 127676.

Rajput, S., Singh, L.P., Pittman Jr, C.U., Mohan, D., 2017. Lead (Pb2 +) and copper (Cu2 +) remediation from water using superparamagnetic maghemite (γ-Fe2O3) nanoparticles synthesized by Flame Spray Pyrolysis (FSP). J. Colloid Interface Sci. 492, 176–190.

Ranjan, A., Arora, J., Chauhan, A., Kumari, A., Rajput, V.D., Sushkova, S., et al., 2022. Applications and implications of nanoparticles in food industries. The Role of Nanoparticles in Plant Nutrition under Soil Pollution. Springer, pp. 223–243.

Ray, M., Ray, A., Dash, S., Mishra, A., Achary, K.G., Nayak, S., 2017. Fungal disease detection in plants: traditional assays, novel diagnostic techniques and biosensors. Bioelectronics 87, 708–723.

Rigoni, F., Drera, G., Pagliara, S., Goldoni, A., Sangaletti, L., 2014. High sensitivity, moisture selective, ammonia gas sensors based on single-walled carbon nanotubes functionalized with indium tin oxide nanoparticles. Carbon 80, 356–363.

Romeh, A.A.A., 2018. Green silver nanoparticles for enhancing the phytoremediation of soil and water contaminated by fipronil and degradation products. Water Air Soil Pollut. 229 (5), 1–13.

Sachan, R., Sahni, T., Gangwar, K., Tiwari, A., Sachan, K., 2022. Role of nanobiosensors for advancement of agriculture, Recent Innovative Approaches in Agricultural Science, 193. Springer.

Sadak, M.S., 2019. Impact of silver nanoparticles on plant growth, some biochemical aspects, and yield of fenugreek plant (*Trigonella foenum-graecum*). Bull. National Res. Centre 43 (1), 1–6.

Sadeghi-Kiakhani, M., Arami, M., Gharanjig, K., 2013. Dye removal from colored-textile wastewater using chitosan-PPI dendrimer hybrid as a biopolymer: optimization, kinetic, and isotherm studies. J. Appl. Polym. Sci. 127 (4), 2607–2619.

Saghi, M.H., Qasemi, M., Alidadi, H., Alahabadi, A., Rastegar, A., Kowsari, M.H., et al., 2020. Vanadium oxide nanoparticles for methylene blue water remediation: exploring the effect of physicochemical parameters by process modeling. J. Mol. Liquids 318, 114046.

Saha, A., Bhaduri, D., 2021. Nano-biosensors: applications in agriculture and allied fields. Biotica Res. Today 3 (12), 1106–1109.

Saha, S., Sarkar, P., 2012. Arsenic remediation from drinking water by synthesized nano-alumina dispersed in chitosan-grafted polyacrylamide. J. Hazard. Mater. 227, 68–78.

Salama, D.M., Abd El-Aziz, M., Osman, S.A., Abd Elwahed, M.S., Shaaban, E., 2022. Foliar spraying of MnO2-NPs and its effect on vegetative growth, production, genomic stability, and chemical quality of the common dry bean. Arab J. Basic Appl. Sci. 29 (1), 26–39.

Salipira, K., Mamba, B., Krause, R., Malefetse, T., Durbach, S., 2007. Carbon nanotubes and cyclodextrin polymers for removing organic pollutants from water. Environ. Chem. Lett. 5 (1), 13–17.

Sam Sukumar, R., Maddula, M.R., Gopala Krishna, A., 2022. Experimental investigations with zinc oxide and silver doped zinc oxide nanoparticles for performance and emissions study of biodiesel blends in diesel engine. Int. J. Ambient Energy 43 (1), 3020–3028.

Sanguansri, P., Augustin, M.A., 2006. Nanoscale materials development—a food industry perspective. Trends Food Sci. Technol. 17 (10), 547–556.

Santiago, T.R., Bonatto, C.C., Rossato, M., Lopes, C.A., Lopes, C.A., G Mizubuti, E.S., 2019. Green synthesis of silver nanoparticles using tomato leaf extract and their entrapment in chitosan nanoparticles to control bacterial wilt. Agriculture 99 (9), 4248–4259.

Sanzari, I., Leone, A., Ambrosone, A., 2019. Nanotechnology in plant science: to make a long story short. Front. Bioeng. Biotechnol. 7, 120.

Sardar, M., Ahmed, W., Al Ayoubi, S., Nisa, S., Bibi, Y., Sabir, M., et al., 2022. Fungicidal synergistic effect of biogenically synthesized zinc oxide and copper oxide nanoparticles against *Alternaria citri* causing citrus black rot disease. Saudi J. Biol. Sci. 29 (1), 88–95.

Satti, S.H., Raja, N.I., Javed, B., Akram, A., Mashwani, Z.-u-R., Ahmad, M.S., et al., 2021. Titanium dioxide nanoparticles elicited agro-morphological and physicochemical modifications in wheat plants to control Bipolaris sorokiniana. PLoS One 16 (2), e0246880.

Selvan, B.K., Thiyagarajan, K., Das, S., Jaya, N., Jabasingh, S.A., Saravanan, P., et al., 2022. Synthesis and characterization of nano zerovalent iron-kaolin clay (nZVI-Kaol) composite polyethersulfone (PES) membrane for the efficacious As2O3 removal from potable water samples. Chemosphere 288, 132405.

Shang, Y., Hasan, M.K., Ahammed, G.J., Li, M., Yin, H., Zhou, J., 2019. Applications of nanotechnology in plant growth and crop protection: a review. Molecules 24 (14), 2558.

Shankar, S., Wang, L.-F., Rhim, J.-W., 2018. Incorporation of zinc oxide nanoparticles improved the mechanical, water vapor barrier, UV-light barrier, and antibacterial properties of PLA-based nanocomposite films. Mater. Sci. Eng. C mater. Biol. Appl. 93, 289–298.

Shariati, S., Khayatian, G., 2020. Microfluidic paper-based analytical device using gold nanoparticles modified with N, N′-bis (2-hydroxyethyl) dithiooxamide for detection of Hg (ii) in air, fish and water samples. New J. Chem. 44 (43), 18662–18667.

Sharifan, H., Noori, A., Bagheri, M., Moore, J.M., 2021. Postharvest spraying of zinc oxide nanoparticles enhances shelf life qualities and zinc concentration of tomato fruits. Crop Pasture Sci.

Sharma, P., Pandey, V., Murlidhar Sharma, M.M., Patra, A., Singh, B., Mehta, S., et al., 2021. A review on biosensors and nanosensors application in agroecosystems. Nanoscale Res. Lett. 16 (136), 22–24. In this issue.

Sherugar, P., Naik, N.S., Padaki, M., Nayak, V., Gangadharan, A., Nadig, A.R., et al., 2021. Fabrication of zinc doped aluminium oxide/polysulfone mixed matrix membranes for enhanced antifouling property and heavy metal removal. Chemosphere 275, 130024.

Shipley, H.J., Engates, K.E., Guettner, A.M., 2011. Study of iron oxide nanoparticles in soil for remediation of arsenic. J. Nanoparticle Res. 13 (6), 2387–2397.

Singh, S., Singh, M., Agrawal, V.V., Kumar, A., 2010. An attempt to develop surface plasmon resonance based immunosensor for Karnal bunt (*Tilletia indica*) diagnosis based on the experience of nano-gold based lateral flow immuno-dipstick test. Thin Solid Films 519 (3), 1156–1159.

Singh, D., Malik, K., Sindhu, M., Kumari, N., Rani, V., Mehta, S., et al., 2022. Biostimulation of anaerobic digestion using iron oxide nanoparticles (IONPs) for increasing biogas production from cattle manure. Nanomaterials 12 (3), 497.

Singhal, J., Verma, S., Kumar, S., 2022. The physio-chemical properties and applications of 2D nanomaterials in agricultural and environmental sustainability. Sci. Total Environ. 155669.

Soni, M., Mehta, P., Soni, A., Goswami, G.K., 2018. Green nanoparticles: synthesis and applications. IOSR J. Biotechnol. Biochem. 4 (3), 78–83.

Srinivasan, S.K., Kuppusamy, R., Krishnan, P., 2021. Effect of nanoparticle-blended biodiesel mixtures on diesel engine performance, emission, and combustion characteristics. Environ. Sci. Pollut. Res. Int. 28 (29), 39210–39226.

Srivastava, N., Mishra, G., 2022. Nanotechnology for environmental pollution detection and remedies. Miniaturized Analytical Devices: Materials and Technology. Wiley, pp. 279–294.

Sukhanova, A., Bozrova, S., Sokolov, P., Berestovoy, M., Karaulov, A., Nabiev, I., 2018. Dependence of nanoparticle toxicity on their physical and chemical properties. Nanoscale Res. Lett. 13 (1), 1–21.

Sun, J., Zhang, Z., Liu, C., Dai, X., Zhou, W., Jiang, K., 2021. Continuous in situ portable SERS analysis of pollutants in water and air by a highly sensitive gold nanoparticle-decorated PVDF substrate. Anal. Bioanal. Chem. 413 (21), 5469–5482.

Taheri, M., Qarache, H.A., Qarache, A.A., Yoosefi, M., 2016. The effects of zinc-oxide nanoparticles on growth parameters of corn (SC704). STEM Fellowship J. 1 (2), 17–20.

Tahir, M.H., Khalid, I., Akram, W., Sattar, A., 2021. In vitro evaluation of different nanoparticles and green synthesized nano pesticides against *Streptomyces scabies* causing the disease "common scab of potato.". Life Sci. J. 18 (10).

Taneja, L., Raghav, S., Kochar, C., Yadav, P.K., Tripathy, S.S., 2021. Effective remediation of fluoride from drinking water using cerium-silver oxide composite incorporated with reduced graphene oxide. J. Water Process Eng. 44, 102369.

Tanveer, Y., Yasmin, H., Nosheen, A., Ali, S., Ahmad, A., 2022. Ameliorative effects of plant growth promoting bacteria, zinc oxide nanoparticles and oxalic acid on *Luffa acutangula* grown on arsenic enriched soil. Environ. Pollut. 300, 118889.

Thangadurai, D.T., Manjubaashini, N., Nataraj, D., 2022. Surface-functionalized gold nanoparticles for environmental remediation. Nanotechnology for Environmental Remediation. pp. 163–182.

Tom, A.P., 2021. Nanotechnology for sustainable water treatment—a review. Mater. Today Proc.

Tran, M.V., Le, P.M.L., 2015. Nanoflake manganese oxide and nickel-manganese oxide synthesized by electrodeposition for electrochemical capacitor. J. Nanomater. 2015.

Van der Horst, C., Silwana, B., Iwuoha, E., Somerset, V., 2015. Bismuth–silver bimetallic nanosensor application for the voltammetric analysis of dust and soil samples. J. Electroanal. Chem. 752, 1–11.

Vanathi, P., Rajiv, P., Sivaraj, R., 2016. Synthesis and characterization of Eichhornia-mediated copper oxide nanoparticles and assessing their antifungal activity against plant pathogens. Bull. Mater. Sci. 39 (5), 1165–1170.

Vijai Anand, K., Reshma, M., Kannan, M., Muthamil Selvan, S., Chaturvedi, S., Shalan, A.E., et al., 2021. Preparation and characterization of calcium oxide nanoparticles from marine molluscan shell waste as nutrient source for plant growth. J. Nanostruct. Chem. 11 (3), 409–422.

Vimalrajiv, B., Chinnamuthu, C., Subramanian, E., Senthil, K., 2018. Effect of nanoparticles in combination with pendimethalin and hydrogen peroxide on growth parameters and nodulation of blackgram (*Vigna mungo* L.). Int. J. Chem. Studies 6 (3), 2816–2819.

Wang, Z., Wei, F., Liu, S.-Y., Xu, Q., Huang, J.-Y., Dong, X.-Y., et al., 2010. Electrocatalytic oxidation of phytohormone salicylic acid at copper nanoparticles-modified gold electrode and its detection in oilseed rape infected with fungal pathogen *Sclerotinia sclerotiorum*. Talanta 80 (3), 1277–1281.

Wang, Y., Zhang, Q., Zhang, C.-l, Li, P., 2012. Characterisation and cooperative antimicrobial properties of chitosan/nano-ZnO composite nanofibrous membranes. Food Chem. 132 (1), 419–427.

Wang, X., Song, W., Qian, H., Zhang, D., Pan, X., Gadd, G.M., 2018. Stabilizing interaction of exopolymers with nano-Se and impact on mercury immobilization in soil and groundwater. Environ. Sci. Nano 5 (2), 456–466.

Wang, L., Tian, Y., Zhang, P., Li, C., Chen, J., 2022. Polysaccharide isolated from *Rosa roxburghii* Tratt fruit as a stabilizing and reducing agent for the synthesis of silver nanoparticles: antibacterial and preservative properties. J. Food Measurement Charact. 16 (2), 1241–1251.

War, J.M., Fazili, M.A., Mushtaq, W., Wani, A.H., Bhat, M.Y., 2020. Role of nanotechnology in crop improvement. Nanobiotechnology in Agriculture. Springer, pp. 63–97.

Wu, Z., Su, X., Lin, Z., Owens, G., Chen, Z., 2019. Mechanism of As (V) removal by green synthesized iron nanoparticles. J. Hazard. Mater. 379, 120811.

Xing, Y., Liao, X., Liu, X., Li, W., Huang, R., Tang, J., et al., 2021. Characterization and antimicrobial activity of silver nanoparticles synthesized with the peel extract of mango. Materials 14 (19), 5878.

Xin-gang, Z., Yuan-feng, Z., Yan-bin, L., 2019. The spillovers of foreign direct investment and the convergence of energy intensity. J. Clean. Prod. 206, 611–621.

Yadav, A.N., 2021. Nanotechnology for agro-environmental sustainability. J. Appl. Biol. Biotechnol. 9 (4), 1.

Yang, H., Xu, M., Koide, R.T., Liu, Q., Dai, Y., Liu, L., 2016. Effects of ditch-buried straw return on water percolation, nitrogen leaching and crop yields in a rice–wheat rotation system. Agriculture 96 (4), 1141–1149.

Yao, K.S., Li, S., Tzeng, K., Cheng, T.C., Chang, C.Y., Chiu, C., et al., 2009. Fluorescence silica nanoprobe as a biomarker for rapid detection of plant pathogens. Paper presented at the Advanced Materials Research.

Younis, S.A., Kim, K.-H., Shaheen, S.M., Antoniadis, V., Tsang, Y.F., Rinklebe, J., 2021. Advancements of nanotechnologies in crop promotion and soil fertility: benefits, life cycle assessment, and legislation policies. Renew. Sustain. Energy Rev. 152, 111686.

Zafar, M., Iqbal, T., 2022. Green synthesis of silver and zinc oxide nanoparticles for novel application to enhance shelf life of fruits. Biomass Convers. Biorefinery 1–16.

Zhang, M., Chen, W., Chen, X., Zhang, Y., Lin, X., Wu, Z., et al., 2013. Multiplex immunoassays of plant viruses based on functionalized upconversion nanoparticles coupled with immunomagnetic separation. J. Nanomater. 2013.

Zhao, F., Xin, X., Cao, Y., Su, D., Ji, P., Zhu, Z., et al., 2021. Use of carbon nanoparticles to improve soil fertility, crop growth and nutrient uptake by corn (*Zea mays* L.). Nanomaterials 11 (10), 2717.

CHAPTER

Microbial consortia application in the sustainable agricultural practices

12

Minakshi Rajput[1], Sudhanshu Mishra[2], Akanksha Pandey[3], Neha Basera[1], Vibhuti Rana[1] and Monika Singh[1]

[1]*Department of Biotechnology, School of Applied and Life Sciences, Uttaranchal University, Dehradun, Uttarakhand, India*

[2]*School of Biosciences, Apeejay Stya University, Sohna, Haryana, India*

[3]*Department of Botany and Microbiology, Gurukula Kangri (Deemed to be University), Haridwar, Uttarakhand, India*

12.1 Introduction

Microbial communities are found almost everywhere in nature and are helpful in bioeconomy and many bio green sectors. Such microbial communities are made up of organisms that can communicate with each other (chemically or physically) and are collectively more resilient to environmental challenges, with less metabolic burden due to the exchange of resources, division of labor, and have expanded metabolic capabilities compared to single species (McCarty and Ledesma-Amaro, 2019).

In a synergistic interaction, two or more compatible microbes from different species make a consortium. In some circumstances, a combination of various strains from the same species might display improved behavior and also be regarded as a consortium. Microbial consortia enhance the favorable conditions for the growth of plants compared to individual strains as they have several biological controls as well as plant growth promotion mechanisms. The use of these microbial consortia is an efficient technique for reducing pests, phytopathogenic infections, salinity, nitrogen uptake, and drought in agricultural crops. Correspondingly, some bacterial consortia have the capabilities to chelate iron, produce phytohormones, fix nitrogen, convert inaccessible nutrients into assimilable forms, and produce phytohormones, each of which is crucial for maintaining soil quality and health. These attributes can also assist to mitigate the negative effects of some conventional, intensive agricultural practices (Santoyo et al., 2021). Bacterial consortia can be classified

Sustainable Agricultural Practices. DOI: https://doi.org/10.1016/B978-0-443-19150-3.00001-1

as either simple or complicated. The methods used to produce a large population of bacteria that would subsequently be combined with other strains to create inoculants varied, with each strain being cultivated alone or in combination with other species in a medium that is suitable for all plant growth promoting bacteria (PGPB) species (Bashan et al., 2020).

Plant growth-promoting microorganisms, that are nonpathogenic such as *Bacillus*, *Pseudomonas*, *Serratia*, *Azotobacter*, and *Azospirillum* sp. can improve the nutrient uptake and assimilation by plant, nutrient availability within the soil. These microorganisms also support nitrogen cycling so considered significant for plant growth promotion due to synergistic plant–microbe interactions (Singh et al., 2019; Woo and Pepe, 2018). Although often used plant growth-promoting microbes (PGPM) with fungus origins is less well-known in the literature. Arbuscular mycorrhizal fungus (AMF), vesicular arbuscular mycorrhiza (VAM) (mycorrhizal fungi) such as *Funneliformis, Gigaspora, Laccaria*, and *Rhizophagus* (Glomus) are obligatory biotrophs of roots, capable of establishing mutualistic relationship with most of the plants, are the best-documented example. These microbes increase the nutrient and water absorption ability of vascular plants, play a vital role in carbon exchange, and reduce the negative impacts of abiotic as well as biotic stressors (Pringle et al., 2009). Microbial interactions play key role in the metabolism of cellulose, pectin, lignin, and hemicellulose by the help of lignocellulose degrading enzymes and these enzymes are present in the cell wall of dietary fibers of major vegetables and fruits. To improve lignocellulolytic enzyme activity, several consortia were designed diversely and complexly using fungi such as *Aspergillus tubingensis*, *Aspergillus niger*, *Penicillium decumbens, Trichoderma reesei*. However, when a cellulolytic capacity having microbial community was added to the consortia to enhance the number of active microbial strains, it universally improves the β-glucosidase efficacy of that consortia. Several research on the consortia showed that the bacteria present in the microbial consortia play vital role in lignocellulolytic enzyme activity than the fungi. One fungal and bacterial genera in the consortia may interact with *T. reesei* and are potential members of a devised synergistic microbial consortium. Such devised microbial consortia may potentially be applied to effectively and economically degrade lignocelluloses (Hu et al., 2017; Toushik et al., 2017). In a study, plant growth promotion by natural microbial consortia was assessed with some bacterial strains such as *Azospirillum brasilense*, *Pseudomonas putida*, *Acinetobacter* sp., and *Sphingomonas* sp. All species were interacted with six varieties of maize studied with single as well as consortia application. The microbial consortia efficiently colonized the rhizospheric region of two different varieties. Microbial colonization by the consortium was higher when compared with single-strain colonization. The plant growth promotion activity was also high in the consortium-inoculated maize when compared with single strain as well as control. This study also concluded that the microbial consortium offers a safety and eco-friendly alternative of chemical fertilization and is also useful for sustainable agriculture (Molina-Romero et al., 2021).

12.2 Role of synthetic or artificial microbial consortia in plant growth and crop production

The evolution of synthetic microbial consortia creates the link between synthetic biology and microbiology. Synthetic microbial consortia (SMC) are produced by coculturing with two or three microbes of same group or different groups under certain environmental condition and having several applications such as development and growth of the plant and disease management. Beneficial microorganisms such as beneficial endophytic bacteria and rhizobacteria (PGPB and PGPR), actinobacteria, fungi, and microalgae can interact with plants and enhance the content of bioactive compounds of human interest (Gianinazzi et al., 2010; Glick, 2012). Hence, using SMC may be beneficial for sustainable agriculture (Bhardwaj et al., 2014).

SMC are made up of various microbial species that have been combined to provide the host plant with cumulative benefits and increased chances of adaption to the soil environment (Kaur et al., 2022). The Horizon 2020 SIMBA project (sustainable innovation of microbiome applications in the food systems), aimed to identify compatible microbial single-strain inoculants with proven plant growth capability to be used for the construction of SMC as microbial inoculants for several grain and vegetable crops (maize, wheat, tomato, potato). The identification of beneficial microbes carried out through the survey of several of articles and project reports on PGPM under different environmental stress conditions. Synthetic microbial consortia were designed on the basis of at least one plant growth-promoting character such as fixation of N_2, phosphorus solubilization, and siderophore production (Kong et al., 2018; Tabacchioni et al., 2021).

Microorganisms have been recognized to perform several functions that are helpful in crop production system. In this era, artificial microbial consortia offers a unique opportunity to improve plant health and crop production via elevating the natural secondary metabolites of the plant (Großkopf and Soyer, 2014; De Souza et al., 2015). SMC could enhance some major functions of the plant beneficial microbial genome and emerges as a key factor to improve crop performance in terms of metabolites accumulation, growth, and yield. Yin et al. (2022) formed 10 SMC using 14 different bacterial genera isolated from wheat rhizosphere and used them against fungal phytopathogen *Rhizoctonia solani*. Among all, the seven SMCs prevent *R. solani* infection and protect plant from infections such as collar and root rot, damping off, wirestem diseases etc. and help plants to flourish in a good way. They also found that nine bacterial consortia impacted on the root growth of Arabidopsis.

The bacteria–fungi consortia also play the key role as bacteria–bacteria consortia for the establishment of novel synthetic microbial communities, when these SMC applied on the host plant might enhance the growth and metabolite accumulation (Ahmad et al., 2011; Berg et al., 2014; Du Jardin, 2015; Lugtenberg, 2015), while SMC using different microbes such as bacteria and fungi may activate novel effect of plant growth promotion which are not obtained by using single microbial species (Finkel, 2020). More recently, Carrion et al. (2019) found that the *Flavobacterium* and *Chitinophaga* combination frequently prevented sugar beet from infection by the fungus *R. solani*. According to Hori et al. (2021), the genera *Fusarium* and *Curvularia*, which are fungal pairs, have a more detrimental impact on plants than single strains. Additionally, the interaction of several species can result in emergent community features, in which the activities of microbial communities go beyond simple individual additivity (Yin et al., 2022).

L-ascorbic acid is a nonenzymatic antioxidant majorly found in citrus fruits. It is commonly called as vitamin C and plays a key role in plant protection from oxidative damages caused by various abiotic stresses such as salt stress. Salt stress is a major factor that creates hindrance in crop growth and development. To alleviate this stress, the L-ascorbic acid content of plants is a strong factor. 2-Keto-L-gulonic acid (2KGA) is used as a precursor molecule in the industries to produce ascorbic acid as well as an ascorbic acid degradation product in plants. Gao et al. (2021) reported that the potential of using 2KGA to enhance crop resistance to salt stress in the crop Chinese cabbage (*Brassica campestris* ssp. chinensis), in terms of leaf and root biomass, plant metabolites, and enzymes were significantly improved after 2KGA application. Whereas the hydrogen peroxide and malondialdehyde (lipid peroxidation marker) contents were significantly decreased. These results revealed that 2KGA applications from exogenic sources can reduce the effect of salinity on plant development, and the promotion of ascorbic acid synthesis may represent a critical fundamental mechanism. His findings have some future prospective and gave significant results for application of 2KGA in sustainable agriculture.

At present, the establishment of SMC have great attention in the field of synthetic biology. This biology can offer insights into the design and construction of SMC using various microbial strains. An example is the high-efficiency production of 2KGA, Wang et al. (2022) designed a SMC on the basis of transcriptomics analyses, named as *Saccharomyces cerevisiae* and *Ketogulonigenium vulgare*. In this SMC, *K. vulgare* was designed as key strain for the production of 2KGA and *S. cerevisiae* designed as a helper strain. Comparative transcriptomic analysis was performed on an engineered *S. cerevisiae* (VTC2) with the wild strain *S. cerevisiae* BY4741. The results showed that the upregulated genes in engineered strain VTC2 were mainly involved in several metabolic pathways such as glycolysis, Kreb's cycle, biosynthesis of nucleotides and amino acids, vitamins, and proteolysis catalyzing enzymes, when compared with wild strain. Furthermore, ascorbic acid produced by engineered *S. cerevisiae* VTC2 could further relieve the oxidative stress in the environment to increase the production of 2KGA, and have great advantage when working with *K. vulgare* strain. Thus the SMC "VTC2-*K. vulgare*" showed the significant increase in the accumulation of 2KGA when compared with wild strain. The findings revealed that the oxidative stress in the synthetic microbial consortium was efficiently reduced. Thus systems analysis confirmed a favorable symbiotic relationship between microorganisms, providing guidance for further engineering synthetic consortia (Wang et al., 2022).

The effects of SMC comprising with four different bacterial genera such as *Bacillus* sp., *Delftia* sp., *Enterobacter* sp., *Achromobacter* sp. were assessed tomato plant development and mineral acquisition under the salinity and normal soil conditions. The results revealed that synthetic bacterial consortia-inoculated seedlings grown under saline soil significantly increased morphological parameters of the tomato plant such as root, shoot, leaves dry weight, number of leaves, shoot, root lengths, secondary roots, and pigment content as compared to un-inoculated control. More secondary roots are useful to absorb more nutritional elements and transport them to the different plant parts; thus, it helps to increase the growth and biomass of the plant. This study also revealed the mineral intake (potassium and sodium) of the tomato plant. The minerals intake was higher in the bacterized plant and the combined beneficial effect of microbial consortia helps the plants sustainability in saline soil. Results revealed that the treatment with SMC could alleviate the bad effects of salinity stress and helps in the better growth and development of tomato plants under the salt stress. Application of SMC appears to be the best and cost-effective method to alleviate salt stress and help in plant growth promotion under abiotic stresses (Kapadia et al., 2021). Zhuang et al. (2021) reported in a study, the rhizospheric microbiome (fungi and bacteria) of Jinxiang garlic was assessed under different growth period, different soil types, and agricultural practices (with and without microbial products), all different parameters were explored by using amplicon sequencing. High-efficiency top-down approaches based on high-throughput technology and synthetic community (SynCom) approaches were used to find efficient PGPB in the rhizosphere of garlic and improve crop yield. Results indicated that *Pseudomonas* species was a key plant growth-promoting rhizospheric bacteria for the garlic crop. Furthermore, artificial microbial consortia with six *Pseudomonas* strains isolated from the garlic rhizosphere were constructed, which showed that they have the ability to promote plant growth and development. A platform is established by microbial consortiums for the specialization in metabolic pathways. Stoichiometric management of the enzymes from each community is necessary for an effective metabolic pathway, and this necessitates control of the ratio between various populations. Different microbial species are combined to form SMC, which have a better probability of adapting to the soil environment than traditional natural consortia in a variety of situations (Pradhan et al., 2022). The advantage of SMC over natural consortia for plant development and growth primarily includes several factors such as host incompatibility, inability to adequately combat indigenous microorganisms.

12.3 Construction of synthetic microbial consortia

SMC is a promising way to improve plant growth and crop quality in sustainable agriculture but still its application in agricultural practices is very low in developing countries. SMCs have more advantages than single strains as an inoculant for the plant health and management. Several steps are involved in the formation of SMC such as isolation of microorganism from the plant, analysis of plant growth-promoting activities, identification by next generation sequencing, and network analysis of core microbial taxa. The improved crop growth, yield, and metabolite accumulation are the main aspects for constructing SMC (Fig. 12.1).

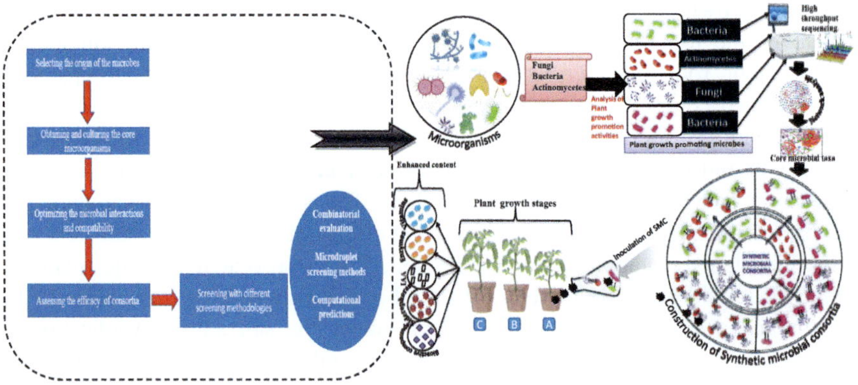

FIGURE 12.1

Construction of synthetic microbial consortia and their uses in plant growth promotion targeting to crop quality.

The method for generating SMC to encourage plant growth and biocides is still in its infancy. It is potential to construct SMC with a plethora of functions for promoting plant growth (Pradhan et al., 2022). In the past, a big part of constructing SMC was centered on mixing particular microbial genotypes with advantageous features. Currently, the typical SMC frequently contains PGPB and AMF, aiming to improve the nutrients (Zn, Ca, P, Mg, K, Cu, B, Mn, Na, N, and Fe) and metabolite contents (e.g., ascorbic acid, folic acid, volatile compound, anthocyanin, sugar, essential oil, vitamin, and zein), which represent higher neutraceutical values in crops (Kong et al., 2018). First, the construction of SMC depends on the origin of the microorganisms. The soil microbiome from high-quality crops is anticipated to be a perfect source for SMC, providing the same plants with higher growth and quality. Furthermore, because of its frequent interactions with plants, the rhizosphere is a hotspot for choosing SMC members. Furthermore, as they are more likely to survive in conditions, the endophytes that are advantageous to plants can also be employed to create SMC (Huang et al., 2018; Kong and Glick, 2017). There are primarily two methods for choosing SMC members: top-down and bottom-up, with the latter being more popular due to its logistical simplicity (Pradhan et al., 2022). By forming a SMC with four strains obtained from healthy tomato rhizospheres that showed protective activity against *Ralstonia solanacearum*, Lee et al. (2021) explored a top-down approach. The rapid development of synthetic biology enables the design and construction of artificial microbial consortia to get special functions. In China, the project named Design and Construction of Microbial Consortia was funded by the National Basic Research Program of China in January 2014. This project was proposed to address the major challenges in engineering natural microbial consortia and reconstructing artificial microbial consortia to meet the demand of industry (Ding et al., 2016) (Table 12.1).

Table 12.1 Some microbial consortia for plant growth promotion.

S. no.	Consortia name	Microbes	Crop	Effect on crop	References
Bacterial consortia					
1.	*Bacillus cereus*, *Bacillus subtilis*, and *Serratia* sp.	Bacteria + bacteria + bacteria	*Capsicum annuum* (red pepper)	Reduced the prevalence of *Phytophthora* blight	Zhang et al. (2019)
2.	*Variovorax paradoxus* + *Rhizobium leguminosarum* bv. *viciae* + AMF *Glomus*	Bacteria + bacteria + fungi	*Pisum sativum* (pea), *Brassica juncea* (brown mustard)	Phytoextraction and phytostabilization of toxic metals such as cadmium	Belimov et al. (2020)
3.	*Azospirillum lipoferum*, *R. leguminosarum*	Bacteria + bacteria	*Trifolium repens* (white clovers), *Cajanus cajan* (pigeon pea), *Cicer arietinum* (chickpea)	Enhanced nodulation process	Deanand et al. (2002)
4.	*Rhizobium*, *B. subtilis*, *B. megaterium*	Bacteria + bacteria + bacteria	*C. arietinum* (chickpea)	Increase root weight and seed yield	Elkoca et al. (2010)
5.	*Rizobium*, *B. subtilis*	Bacteria + bacteria	*P. sativum* (pea)	Improve root morphology, increase nodule formation	Schwartz et al. (2013)
6.	*Bacillus atrophoeus*, *Burkholderia cepacia*	Bacteria + bacteria	Gladiolus	Elicitations of defense enzyme for seasonal and unseasonal condition	Shanmugam et al. (2011)
7.	*Pseudomonas*, *Mesorhizobium*	Bacteria + bacteria	*C. arietinum* (chickpea)	Increased nodule formation	Malik and Sindhu (2011)
8.	*Burkholderia* sp., *Sinorhizobium meliloti*	Bacteria + bacteria	*C. cajan* (pigeon pea)	Seed yield	Pandey and Maheshwari (2007)

(*Continued*)

Table 12.1 Some microbial consortia for plant growth promotion. *Continued*

S. no.	Consortia name	Microbes	Crop	Effect on crop	References
9.	*Enterobacter* sp., *Bradyrhizobium* sp.	Bacteria + bacteria	*Vigna radiata* (green gram)	Nodulation	Gupta et al. (2015)
10.	*Pseudomonas, Azatobacter, Azospirillum*	Bacteria + bacteria + bacteria	*Solanum lycopersicum* (tomato)	Increase uptake of K, fruit lycopene, antioxidant property	Ordookhani et al. (2010)
11.	*Pseudomonas fluorescens, Rhizobacterium*	Bacteria + bacteria	*Beta vulgaris* (sugar beet)	Inhibit damping off disease	Dunne et al. (1998)
12.	*B. subtilis, Rhizobium tropici*	Bacteria + bacteria	*Phaseolus vulgaris* (kidney bean)	Increased yield	De et al. (2002)
13.	*Streptomyces lydicus, Rhizobium*	Bacteria + bacteria	*P. sativum* (pea)	Nodule colonization	Tokala et al. (2002)
14.	*Sinorhizobium fredii, P. fluorescens*	Bacteria + bacteria	*C. cajan* (pigeon pea)	Potential biocontrol efficacy against Fusarium wilt	Kumar et al. (2010)
Bacterial + fungal consortia					
15.	*Bacillus amyloliquefaciens* + *Pseudomonas chlororaphis* + *Pseudomonas azotoformans* + *Trichoderma harzianum* + AMF *Rhizophagus irregularis*	Bacteria + bacteria + bacteria + fungi + fungi	*S. lycopersicum* (tomato)	To control shootand root pathogens	Minchev et al. (2021)
16.	*Pseudomonas striata, Rhizobium, Glomus intraradices*	Bacteria + bacteria + fungi	*C. arietinum* (chickpea)	Increase plant growth, chlorophyll contain, and number of pods	Akhtar and Siddiqui (2008)

17.	*Pantoea dispersa* + AMF *Glomus monosporum*	Bacteria + fungi	*Ocimum tenuiflorum* (holy basil)	Reduced the recommended level of chemical fertilizers	Jyothi et al. (2018)
18.	*Paenibacillus lentimorbus*, *Piriformospora indica*	Bacteria + fungi	*C. arietinum* (chickpea)	Increasing nodulation to affect plant growth	Nautiyal et al. (2010)
19.	*Pseudomonas fluorescens*, *Rhizobium*, *Trichoderma*	Bacteria + bacteria + fungi	*C. arietinum* (chickpea)	Physiological defense against collar rot disease	Singh et al. (2008)
20.	*Penicillium* sp., *Pseudomonas* sp.	Bacteria + fungi	*Zea mays* (maize)	Increased root development	Guardiola et al. (2021)
21.	*B. subtilis*, *Trichoderma harzianum*	Bacteria + fungi	*P. vulgaris* (kidney bean)	Effective control of dry bean root rot	De et al. (2002)

AMF, *Arbuscular mycorrhizal fungus.*

12.4 Future prospective

To meet the world's growing food demand and improve agricultural sustainability, crop production must be increased as the population of the planet is growing at an alarming rate. The utilization of biofertilizers, biopesticides, and plant growth-promoting microorganisms is an efficient way of increasing crop production and for the eradication of crop diseases. Microbial consortia provide a platform to divide the labor of metabolism, and there is a need of a stoichiometric control over the enzymes for an efficient metabolic pathway that requires a definite ratio between the populations. There is a range of applications of SMC, designed by minimizing and utilizing the ecological microbial interactions. There are still some challenges and some points to be improved while developing SMC such as to understand natural microbial communities, to reduce the evolution of mutants in genetically modified consortia, to develop orthogonal cell signaling pathways, and to control the composition of microbial population. The future challenges are to control the populations tightly and to address this problem, researchers have to apply some innovative approaches such as the application of inducible signaling for controlling gene expression, the use of circuits to sense the changes in relative density of population between strains and separating microbial populations spatially to control relative ratio. With addition to these concerns, the focus on the transport of metabolites will also be needed for the appropriate development of consortia.

References

Ahmad, I., Khan, M.S.A., Aqil, F., Singh, M., Ahmad, F., Pichtel, J., 2011. Microbial applications in agriculture and the environment: a broad perspective. In: Microbes and Microbial Technology: Agricultural and Environmental Applications, Springer, New York, NY, pp. 1–27.

Akhtar, M.S., Siddiqui, Z.A., 2008. Biocontrol of root-rot disease complex of chickpea by *Glomus intraradices, Rhizobium*sp., *Pseudomonas straita*. Crop Prot. 27 (3), 410–417.

Bashan, Y., Prabhu, S.R., De, B.L.E., Kloepper, J.W., 2020. Disclosure of exact protocols of fermentation, identity of microorganisms within consortia, formation of advanced consortia with microbe-based products. Biol. Fertil. Soils 56, 443–445.

Belimov, A.A., Shaposhnikov, A.I., Azarova, T.S., Makarova, N.M., Safronova, V.I., Litvinskiy, V.A., et al., 2020. Microbial consortium of PGPR, rhizobia and arbuscular mycorrhizal fungus makes pea mutant SGECdt comparable with Indian Mustard in cadmium tolerance and accumulation. Plants 9 (8), 975.

Berg, G., Grube, M., Schloter, M., Smalla, K., 2014. Unraveling the plant microbiome: looking backand future perspectives. Front. Microbiol. 5, 148.

Bhardwaj, D., Ansari, M.W., Sahoo, R.K., Tuteja, N., 2014. Biofertilizers function as key player insustainable agriculture by improving soil fertility, plant tolerance and crop productivity. Microb. Cell Fact. 13, 66.

Carrion, V.J., Perez-Jaramillo, J., Cordovez, V., Tracanna, V., de Hollander, M., Ruiz-Buck, D., 2019. Pathogen-induced activation of disease-suppressive functions in the endophytic root microbiome. Science 366, 606–612.

Deanand, B., Patil, A., Kulkaarni, J., Algawadi, A., 2002. Effect of plant growth promoting rhizobacteria on growth and yield of pigeon pea (*Cajanus cajan* L.) by application of plant growth promoting rhizobacteria. Microbiol. Res. 159, 371–394.

De Souza, R., Ambrosini, A., Passaglia, L.M.P., 2015. Plant growth promoting bacteria as inoculants in agricultural soils. Genet. Mol. Biol. 38, 401–419.

De, J.C.E., Percich, J., Graham, P., 2002. Integrated management strategies of bean root rot with *Bacillus subtilis* and *Rhizobium* in Minnesota. Field Crop Res. 74, 107–115.

Ding, M.-Z., Song, H., Wang, E.-X., Liu, Y., Yuan, Y.-J., 2016. Design and construction of synthetic microbial consortia in China. Synth. Syst. Biotechnol. 1 (4), 230–235. Available from: https://doi.org/10.1016/j.synbio.2016.08.004.

Du Jardin, P., 2015. Plant biostimulants: definition, concept, main categories and regulation. Sci. Horticult. 196, 3–14.

Dunne, C., Moenne, L.Y., McCarthy, J., Higgins, P., Powell, J., Dowling, D.O.G.F., 1998. Combining proteolytic and phloroglucinol- producing bacteria for improved biocontrol of pythium-mediated damping-off of sugar beet. Plant Pathol. 47, 299–307.

Elkoca, E., Turan, M., Donmez, M.F., 2010. Effect of single, dual and triple inoculations with *Bacillus subtilis, Bacillus megaterium* and *Rhizobium leguminosarum*bv. Phaseoli on nodulation, nutrient uptake, yield and yield parameters of common bean (*Phaseolus vulgaris* l. ev, 'elkoca-05). J. Plant Nutr. 33, 2104–2119.

Finkel, O.M., 2020. A single bacterial genus maintains root growth in a complex microbiome. Nature 587, 103–108.

Gao, M., Sun, H., Shi, M., Wu, Q., Ji, D., Wang, B., et al., 2021. 2-Keto-L-gulonic acid improved the salt stress resistance of non-heading chinese cabbage by increasing L-ascorbic acid accumulation. Front. Plant Sci. 12, 697184.

Gianinazzi, S., Gollotte, A., Binet, M.N., van Tuinen, D., Redecker, D., Wipf, D., 2010. Mycorrhiza 20, 519–530.

Glick, B.R., 2012. Plant growth-promoting bacteria: mechanisms and applications. Scientifica 963401.

Großkopf, T., Soyer, O., 2014. Synthetic microbial communities. Curr. Opin. Microbiol. 18, 72–77.

Guardiola, M.C.E., Figueroa, M.M.L., Pacheco, M.A., Senes, G.C., 2021. Native microbial consortia improve maize shoot and root systems at early developmental stages in a seedbed assay. Sci. Fungorum 51, e1329.

Gupta, R., Bisaria, V.S., Sharma, 2015. Effect of agricultural amendments on *Cajanus cajan* (Pigeon pea) and its rizospheric microbial communities-a comparison between chemical fertilizers and bioinoculants. PLoS One 10, e0132770.

Hori, Y., Fujita, H., Hiruma, K., Narisawa, K., Toju, H., 2021. Synergistic and offset effects of fungal species combinations on plant performance. Front. Microbiol. 12, 713180.

Huang, L.H., Yuan, M.Q., Ao, X.J., Ren, A.Y., Zhang, H.B., Yang, M.Z., 2018. Endophytic fungi specifically introduce novel metabolites into grape flesh cells in vitro. PLoS ONE 13, e0196996.

Hu, J., Xue, Y., Guo, H., Gao, M., Li, J., Zhang, S., et al., 2017. Design and composition of synthetic fungal-bacterial microbial consortia that improve lignocellulolytic enzyme activity. Bioresour. Technol. 227, 247–255. Available from: https://doi.org/10.1016/j.biortech.2016.12.058.

Jyothi, E., Bagyaraj, D.J., Rao, E.V.S., 2018. Microbial consortia developed for *Ocimum tenuiflorum* reduces application of chemical fertilizers by 50% under field conditions. Med. Plants-Int. J. Phytomed. Relat. Indus. 10 (2), 138–144.

Kapadia, C., Sayyed, R.Z., El Enshasy, H.A., Vaidya, H., Sharma, D., Patel, N., et al., 2021. Halotolerant microbial consortia for sustainable mitigation of salinity stress, growth promotion, and mineral uptake in tomato plants and soil nutrient enrichment. Sustainability 13, 8369.

Kaur, S., Egidi, E., Qiu, Z., Macdonald, C.A., Verma, J.P., Trivedi, P., 2022. Synthetic community improves crop performance and alters rhizosphere microbial communities. J. Sustain. Agric. Environ. 1, 118–131.

Kong, Z., Glick, B.R., 2017. The role of plant growth-promoting bacteria in metal phytoremediation. Adv. Microb. Physiol. 71, 97.

Kong, Z., Hart, M., Liu, H., 2018. Paving the way from the lab to the field: using synthetic microbial consortia to produce high-quality crops. Front. Plant Sci. 9, 1467.

Kumar, H., Bajpai, V.K., Dubey, R.C., Maheshwari, D.K., Kang, S.C., 2010. Wilt disease management and enhancement of growth and yield of*Cajanus cajan* (L) var. Manak by bacterial combinations amended with chemical fertilizer. Crop Prot. 29, 591–598.

Lee, S.M., Kong, H.G., Song, G.C., Ryu, C.M., 2021. Disruption of *Firmicutes* and *Actinobacteria* abundance in tomato rhizosphere causes the incidence of bacterial wilt disease. ISME J. 15, 330–347.

Lugtenberg, B., 2015. Principles of Plant-Microbe Interactions: Microbes for Sustainable Agriculture. Springer International Publishing, Cham, p. 448.

Malik, D.K., Sindhu, S.S., 2011. Production of indole acetic acid by *Pseudomonas* sp.: effect of coinoculation with *Mesorhizobium* sp. *Cicer* on nodulation and plant growth of chickpea (*Cicer aerietinum*). Physiol. Mol. Biol. Plants 17, 25–32.

McCarty, N.S., Ledesma-Amaro, R., 2019. Synthetic biology tools to engineer microbial communities for biotechnology. Trends Biotechnol. 37 (2), 181–197.

Minchev, Z., Kostenko, O., Soler, R., Pozo, M.J., 2021. Microbial consortia for effective biocontrol of root and foliar diseases in tomato. Front. Plant Sci. 12.

Molina-Romero, D., Juarez-Sanchez, S., Venegas, B., Ortíz-Gonzalez, C.S., Baez, A., Morales-García, Y.E., et al., 2021. A bacterial consortium interacts with different varieties of maize, promotes the plant growth, and reduces the application of chemical fertilizer under field conditions. Front. Sustain. Food Syst. 4, 616757.

Nautiyal, C.S., Chauhan, P.S., Gupta, D.S.M., Seem, K., Varma, A., Stasson, W.J., 2010. Tripartite interactions among *Paenibacilluslentimorbus* NRRL B-30488, *Piriformospora indica* DSM11827, and *Cicer arietinum* L. World J. Microbiol. Biotechnol. 26, 1393–1399.

Ordookhani, K., Khavazi, K., Moezzi, A., Rejali, F., 2010. Influence of PGPR and AMF on antioxidant activity, lycopene and potassium contents in tomato. Afr. J. Agric. Res. 5, 1108–1116.

Pandey, P., Maheshwari, D.K., 2007. Bioformulation of *Burkholderia*sp: MSSP with multi-species consortium for growth promotion of *Cajanus cajan*. Can. J. Microbiol. 53 (2), 213–222.

Pradhan, S., Tyagi, R., Sharma, S., 2022. Combating biotic stresses in plants by synthetic microbial communities: principles, applications and challenges. J. Appl. Microbiol. 133 (5), 2742–2759.

Pringle, A., Bever, J.D., Gardes, M., Parent, J.L., Rillig, M.C., Klironomos, J.N., 2009. Mycorrhizal symbioses and plant invasions. Annu. Rev. Ecol. Syst. 40, 699–715.

Santoyo, G., Guzmán, G.P., Parra, C.F.I., Santos-V, S.D.L., Orozco-M, M.D.C., Glick, B.R., 2021. Plant growth stimulation by microbial consortia. Agronomy 11 (2), 219.

Schwartz, A., Ortiz, I., Maymon, M., Herbold, C., Fujishige, N., Vijanderan, J., et al., 2013. *Bacillus simplex*-a little known PGPB with anti-fungal activity-alters pea legume root architecture and nodule morphology when inoculated with *Rhizobium leguminosarum*bv. *Viciae*. Agronomy 3, 595.

Shanmugam, V., Kanoujia, N., Singh, M., Singh, S., Prasad, R., 2011. Biocontrol of vascular wilt and corm rot of gladiolus caused by *Fusarium oxysporum* f. sp. *Gladioli* using plant growth promoting rhizobacterial mixture. Crop Prot. 30, 807–813.

Singh, N., Pandey, P., Dubey, R., Maheshwari, D., 2008. Biological control of root rot fungus *Macrophominaphaseolina* and growth enhancement of *Pinus roxburghii*(Sarg.) by rhizosphere competent *Bacillus subtilis* BN1. World J. Microbiol. Biotechnol. 24, 1669–1679.

Singh, M., Singh, D., Gupta, A., et al., 2019. Plant growth promoting rhizobacteria: application in biofertilizers and biocontrol of phytopathogens. PGPR Amelioration in Sustainable Agriculture. pp. 41–66. Available from: https://doi.org/10.1016/b978-0-12-815879-1.00003-3.

Tabacchioni, S., Passato, S., Ambrosino, P., Huang, L., Caldara, M., Cantale, C., et al., 2021. Identification of beneficial microbial consortia and bioactive compounds with potential as plant biostimulants for a sustainable agriculture. Microorganisms 9, 426.

Tokala, R.K., Strap, J.L., Jung, C.M., Crawford, D.L., Salove, M.H., Deobald, L.A., et al., 2002. Novel plant-microbe rhizosphere interaction involving *Streptomyces lydicus* WYEC 108 and the pea plant (*Pisum sativum*). Appl. Environ. Microbiol. 68, 2161–2171.

Toushik, S.H., Lee, K.T., Lee, J.S., Kim, K.S., 2017. Functional applications of lignocellulolytic enzymes in the fruit and vegetable processing industries. J. Food Sci. 82 (3), 585–593.

Wang, Y., Li, H., Liu, Y., Zhou, M., Ding, M., Yuan, Y., 2022. Construction of synthetic microbial consortia for 2-keto-L-gulonic acid biosynthesis. Synth. Syst. Biotechnol. 7, 481–489.

Woo, S.L., Pepe, O., 2018. Microbial consortia: promising probiotics as plant biostimulants for sustainable agriculture. Front. Plant Sci. 9, 1801.

Yin, C., Hagerty, C.H., Paulitz, T.C., 2022. Synthetic microbial consortia derived from rhizosphere soil protect wheat against a soilborne fungal pathogen. Front. Microbiol. 3367.

Zhang, L.N., Wang, D.C., Hu, Q., Dai, X.Q., Xie, Y.S., Li, Q., et al., 2019. Consortium of plant growth-promoting rhizobacteria strains suppresses sweet pepper disease by altering the rhizosphere microbiota. Front. Microbiol. 10, 1668.

Zhuang, L., Li, Y., Wang, Z., Yu, Y., Zhang, N., Yang, C., et al., 2021. Synthetic community with six *Pseudomonas* strains screened from garlic rhizosphere microbiome promotes plant growth. Microb. Biotechnol. 14 (2), 488–502.

CHAPTER

Cyanobacterial mats and their application in sustainable agriculture

13

Jalaluddin and Kapil Deo Pandey

Laboratory of Algal Research, Centre of Advanced Study in Botany, Institute of Science, Banaras Hindu University, Varanasi, Uttar Pradesh, India

13.1 Introduction

Cyanobacteria are the most ancient oxygenic photoautotrophic on the Earth, dominating microbial communities in the world's most stressed environments, including hot springs, Antarctic water and ice shelves, and deserts (Eriksson and Clarke, 1996; Quesada and Vincent, 1997). They have probably evolved in the absence of the ozone shield during the pre-Cambrian period. Cyanobacteria are a valuable source of a variety of environmental, agricultural, pharmaceutical industrial, and natural compounds (Richa et al., 2011). Tightly integrated biogeochemical processes occur along a small upright chemical gradient in complex, semiclosed organo-sedimentary within the top 10 mm of microbial mats, which often makes them self-relevant ecosystem (Woebken et al., 2015).

The ability of microbial mats to trap small mineral particles and perform mineral cycling with ions in the surrounding environment have aided the preservation of cyanobacteria growing on siliciclastic sand. Fine grains coat the sheaths of filamentous cyanobacteria *Nodosilinea* sp. resides within the mat within 1 week. Cell growing on clay minerals rarely coat unsheathed cyanobacterium *Nostoc* sp. The sheathed filamentous cyanobacteria are perhaps better preserved in high-fluid-energy environments. These helped to develop a mechanistic model that explains the preservation of microbial fossils and nature in Ediacaran sand stones and siltstones, as well as in fine-granulated siliciclastic deposits with superbly preserved microbial mats (Newman et al., 2017).

The abundance of corals has declined a disappearing all over the world. The functional repercussions of modifications are a pressing concern that require in order to establish appropriate management measures for sustaining reef biological services (Webb et al., 2021a,b). The important reef functions in situ for five different benthic assemblages dominating shallow damaged Caribbean reef environments were investigated with reference to net community calcification (NCC), production (NCP), and nutrient recycling. In the experiment, a custom-made tent was placed over communities dominated by one of five functional groups: coral,

Sustainable Agricultural Practices. DOI: https://doi.org/10.1016/B978-0-443-19150-3.00012-6

grass, microalgae, bio-eroding sponges, and cyanobacterial mats. The chemical fluxes between these communities and water were measured during the day and night. By solving a system of differential equations characterizing the contribution of each process to the measured chemical fluxes, the measured fluxes were then translated into accountable biogeochemical processes. In comparison to those known for around the world, the estimated processes are minor. All communities showed negative or very low NCC rates. Net biomass production by photosynthesis is modest, while the remineralization of organic matter at night is slight, resulting in net heterotrophy by most groups. The amount and type of photosynthate, released by algal turfs and cyanobacterial mats on this reef, have enhanced heterotrophic activity and stimulated the proliferation of less diverse copiotrophic microbial populations. A multivariate pairwise analysis revealed the functional homogenization across substrate types (Webb et al., 2021a,b).

13.2 Cyanobacteria

The oxy-photosynthetic bacteria, cyanobacteria, have been probably originated around for at least 3.5 million years ago (Charpy et al., 2012). They have adapted to grow in any environment, from freshwater to marine ecosystems, as well as on land (Codd et al., 1999; Stal, 2000; Whitton and Potts, 2012). Cyanobacteria can withstand a wide range of salinity stress (Greeney et al., 2008). They have evolved several ecophysiological traits that enable them to survive in all environments subjected to natural and anthropologically caused changes throughout the course of their long existence (Hallock, 2005). They can survive a moderate higher temperature and increased sun radiation (Hallock, 2005). Cyanobacteria are economical and environmentally friendly alternative to chemical fertilizers being used as microbial inoculant in agro-ecosystem. In the rice cultivation, free-living and symbiotic cyanobacteria fix nitrogen enhances the pore sizes and make gummy compounds, excrete growth-promoting hormones, vitamins, and amino acids and their jelly boost water-holding capacity which improve soil fertility (Singh, 1961; Roger and Reynaud, 1982). Cyanobacteria can fix about 5–55 kg of nitrogen per hectare per crop, resulting in a 10%–16% improvement in grain output (Agarwal, 1979; Roger and Kulasooriya, 1980).

Tropical reefs are vulnerable to coral death and coral bleaching (Glynn, 1991; Lesser and Farrell, 2004). Degrading living, coral reefs provide nutrients to cyanobacterial, resulting in a phase shift from a coral reef to a cyanobacteria-dominated reef (Gorman-Smith et al., 2000; Gardner et al., 2003; Pandolfi et al., 2003). Though coral reefs are rich in filamentous cyanobacteria, little is known about their ecological roles. A survey of cyanobacteria-associated species with feeding preference was conducted to observe the rate of cyanobacteria as food and shelter for marine consumers. Forty-three invertebrate species herbivores were discovered in cyanobacterial mats. *Stylocheilus striatus* was numerous on cyanobacterial mats and solely on cyanobacteria. The macrograzers like urchins and fishes prefer algae over cyanobacteria while Sea hares were stimulated by extracts from the cyanobacterium *Lyngbya majuscula*. Certain small coral reef grazers feed on cyanobacteria extracellular release that are chemically protected against macrograzers. Cyanobacteria could have an indirect impact on distribution of cyanobacteria-dwelling organism's local biodiversity (Cruz-Rivera and Paul, 2002). Blooms are most common in limnetic zone during the summer and some are beginning to persist into the winter. Seven different benthic cyanobacterial mat (BCM) types were discovered, with the majority of them possessing a complex cyanobacterial consortium.

13.3 Benthic cyanobacteria

In addition, habitat characteristics in proximity to both deep and shallow BCMs, various environmental factors such as temperature, pH, salinity, chlorophyll, and radiation used in photosynthesis (PAR) were investigated (van Zanten et al., 2016). Stromatolite formation in the modern maritime is a dynamic balance between cyanobacterial mat and intermittent lithification (Cox and Reid, 2000). Cyanobacteria have played a crucial role in the carbon cycle. Biological processes linked to carbon and carbonate cycling revealed that some produce specialized carbonate structures while some remove carbonate substrates and others that combined, at the same time depending on their requirement (Golubic et al., 2019). The capacity to produce carbonate material is linked to cyanobacteria's photosynthetic activity and extracellular polymeric substances (Arp et al., 1999). Various cyanobacterial and bacterial symbionts can be found in marine sponges, for example, the sponge *Dysidea* contains the filamentous cyanobacterium *Oscillatoria spongeliae* in the Great Barrier Reef in Australia, as well as in three species of *Dysidea* discovered near Guam (Hinde et al., 1994). Predator and competitor, as well as resistance to microbial infections and cyanophages are all responsible for possible release of chemicals in the environment. Sponges can actively regulate their choice to cyanobacterial symbiont populations since symbiotic cyanobacteria can outgrow and kill their host sponge (Thacker and Starnes, 2003). *Phormidium* and *Plectonema* are common epiphytes on coral reef in the ocean. *Spirulina* frequently crawls inside of their sheaths while Coccoid epiphytic cyanobacteria that are tiny (<0.8 μm diameter) connect to broad sheaths of *L. majuscula* (>80 μm), having a huge variety of marine cyanobacterial cell size (Golubic and Seong-Joo, 1999).

13.4 Cyanobacterial mats

13.4.1 Benthic cyanobacterial mats

Microbial mats are often dominated by cyanobacteria which favor the occurrence of photosynthetic bacteria, sulfur bacteria, and other microorganisms. Near the shore, they usually form flat, extensive mats on sand or mud of several millimeters in thickness. Microbial mats are often found in soft, muddy floors of lagoons and they are constituted of various gliding filamentous cyanobacteria. Several metagenomics molecular approaches were used to explore the diversity of cyanobacterial mats seen in various natural habitats. Polyphasic approaches have been used on Tikehau atoll (French Polynesia) (Davoodi and Abed, 2003), in New Caledonia (Charpy et al., 2007), in the western Indian Ocean in Zanzibar (Tanzania) (Bauer et al., 2008), in La Reunion Island, and in Okinawa (Charpy et al., 2010) to explore the diversity and density of natural populations of benthic marine cyanobacteria. BCMs were dominated by filamentous nonheterocystous *Lyngbya, Microcoleus, Spirulina*, and *Oscillatoria*, as well as *Pseudanabaena* in Page Reef, Zanzibar. Unicellular species were also recovered, however, heterocystous species were infrequent (Bauer et al., 2008).

Cyanobacteria, often and significant component of coral reef ecosystems, are found in both benthic and planktonic forms. They contribute into primary production, nitrogen fixation, and coral reef formation significantly. Coral reef settlement and recruitment are inhibited by BCMs. This may impair coral's ability to recover from disturbances. BCMs are being more common component in coral reefs around the world (Brocke et al., 2015a).

On the island of Curaçao in the Caribbean, we performed a multiyear survey and found that BCM abundance was highest in sheltered reefs near urban areas. Macroalgae have a dense benthic cover and low coral cover were also characteristics of reefs with high BCM abundance. Nutrient contents in water molecules were typically low but increased dramatically just above BCM-covered substrata (both sandy and hard). This was true for both high and low BCM coverage areas, implying that BCM development is aided by a localized, substrate-linked release of nutrients via microbial organic matter breakdown. This idea was backed up by increased levels of organic material in reefs with high BCM sediments coverage, as well as an on-site investigation that revealed BCMs grew in days on organic-rich sediments (*Spirulina*). Probably the fertilizer runoff from cities promotes phototrophic blooms, as well as increased organic matter loading on the reef. Currents help the flow of organic stuff, and settling on the seafloor at low-hydrodynamic areas. The organic matter breakdown, increased nutrients are released, which drive the growth of BCMs. Improved and integrated fertilizer management in agro-ecosystem should minimize organic loading of sediments and gain of macroalgae on coral reefs, reducing BCM growth (Brocke et al., 2015a,b).

The cyanobacterial diversity of BCMs at four reef locations with total BCM abundance, the species composition of six common BCMs, and the nitrogen fixation rates over a diel cycle were determined. BCMs total nitrogen fixing potential was calculated (Brocke et al., 2018). Factors and mechanisms that drive cyanobacterial species in BMC spread are unknown.

Transcriptional patterns in response to experimental treatments, mat-forming cyanobacteria's circadian clocks have adapted to distinct life strategies (Hörnlein et al., 2020). Microbialites, endolithic cyanobacteria, symbiotic cyanobacteria, epiphytes, and microbial mats are all component of benthic cyanobacteria. Benthic cyanobacteria contribute to form microbial mats on the west coast of Bonaire BCMs (Rejmánkovά and Komárkovά, 2000; Charpy et al., 2012) through the dense structures dominated by cyanobacteria but also include photosynthetic bacteria and sulfur bacteria.

N_2-fixation gives this, by cyanobacteria mat communities, a distinct advantage while growing in N-limited situations, which are characteristics in marine habitats (Charpy et al., 2010). Biological N_2-fixation by BCMs appears to make a significant contribution to N supply in coral reef and N-cycle of the ecosystems. The oxygen-sensitive and energy-intensive, these bacteria segregate the carbon and nitrogen fixation processes either in space (heterocyst) or time (dark) (Berman-Frank et al., 2003). N_2-fixation that required for primary production is between 2% and 21% in Tikehau atoll lagoon (Charpy-Roubaud et al., 2001) and New Caledonia (Charpy et al., 2007). N_2-fixation rates of three different subenvironments (coral rubbles, microbial mats, and sandy bottoms) on La Reunion and Sesoko Islands were compared. The rate of N_2-fixation rates of microbial mats are one order of magnitude higher than the subenvironments and can contribute up to 95% of their primary production. As temperatures rise and nutrient levels increase, BCMs (Benthic Cyanobacterial Mats) can form dense proliferations in limnetic and benthic regions that were previously dominated by seagrass beds or hard-bodied corals (Paerl and Paul, 2012; Gurgel et al., 2020; Urrutia-Cordero et al., 2020; Benny et al., 2021). Although BCM proliferations are frequently observed, the extent of occurrence in time and space as well as the species involved remain unexplored.

BCMs, widespread from 45 to 90 m depth, were found during three deep dives expedition with a submersed off the west coast of Bonaire which supports the logic that these BCMs emerged only recently at these depths (Becking and Meesters, 2014). BCMs prevalence on shallow reefs (0–30 m) has been linked to nutrient loading and organic pollution in Bonairean reef (Heisler et al., 2008; Paerl et al., 2011; Brocke et al., 2015a,b). Macroalgae and BCMs on Curacao and Bonaire's shallow reefs in recent decades has led to an increase due to urbanization (Meltvedt and Jadot, 2014; Brocke et al., 2015a,b).

In a study was to map the distribution of deep BCMs along Klein Bonaire and Bonaire's west coast, bathymetry and bottom properties were collected throughout the coast. Deep BCMs were investigated to observe whether they are a common occurrence and whatever it is linked to metropolitan regions or other pollution sources along the coast. A high abundance of benthic primary producers, such as different ecotype cyanobacteria, is amazing at depths of down to 90 m. They require energy from light penetrating the water column for growth and proliferation, a high abundance of benthic cyanobacteria, is remarkable at depths of down to 90 m. In order to better understand grain size trends in stromatolites, trapping and binding capacities of filamentous cyanobacterial mats, as well as trapping in larger filamentous algal mats was investigated. Mats were cut into squares, and grains of various sizes (fine sand, coarse sand, and fine pebbles) were fed to their surface in saltwater tanks at angles ranging from 0 to 75 degrees (approximating the angle of lamina in normal stromatolites). The cyanobacterial mat's ability to trap grains was greatly influenced by filaments protruded from the sediment surface, grain size, and the mat's slope angle. Fine grains were trapped significantly more effectively by the cyanobacterial mats than bigger grains beyond the abiotic slide angle. Furthermore, over time, the cyanobacterial mats actively bonded grains of all sizes. The considerably larger algal mats, on the other hand, trapped medium and coarse grains from all sides. The findings suggested that the presence of detrital grains beyond the abiotic slide angle can be considered a bio signature in ancient stromatolites where biogenicity is questioned, and where coarse grains are present within stromatolite laminae at angles beyond the abiotic slide angle (e.g., most modern marine stromatolites). Typical cyanobacterial-type mats are likely not solely responsible for the construction, providing insight into stromatolite evolution (Frantz et al., 2015a,b) (Figs. 13.1 and 13.2).

13.4.2 Arctic and Antarctic cyanobacterial mats

Cyanobacterial mats are frequently observed in polar regions in freshwater ecosystem throughout. The majority of mats are several layers, three-dimensional formations with a gel-like matrix that contains mostly filamentous cyanobacteria. Analysis of the 18S r-RNA gene clone library to find eukaryotic communities of mats from the very high Arctic and Antarctic revealed 39 ribotypes with 99%

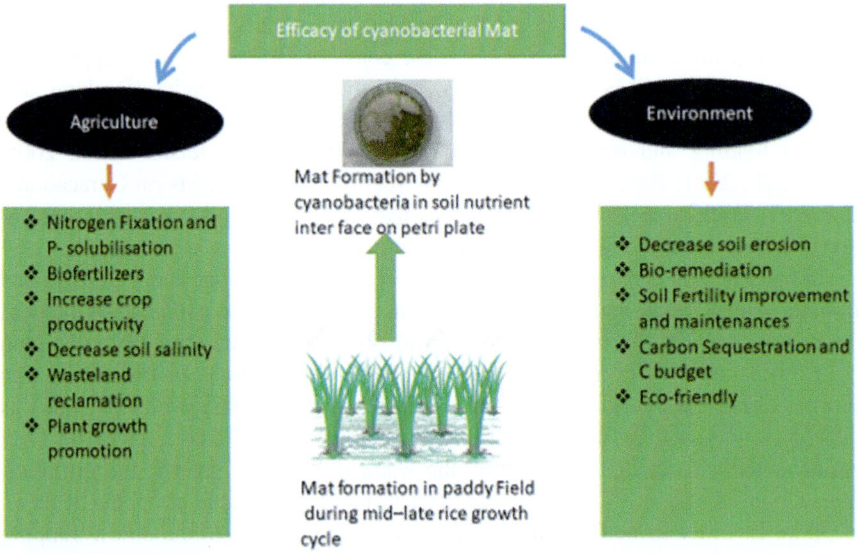

FIGURE 13.1

Efficacy of cyanobacterial mat.

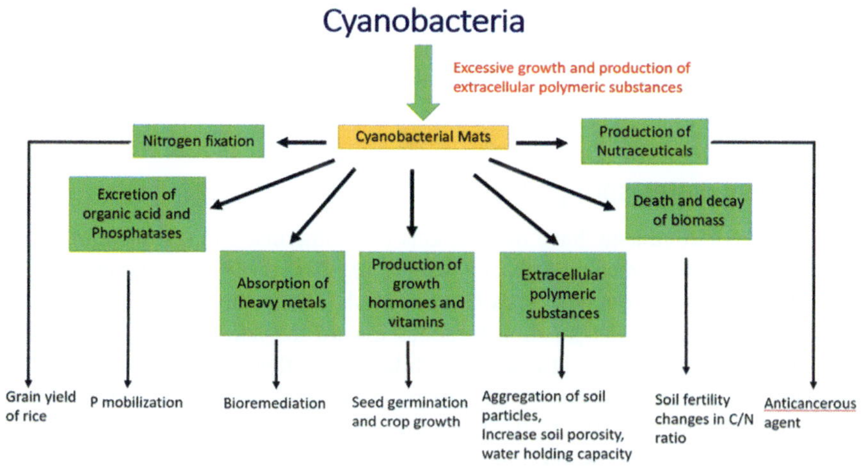

FIGURE 13.2

Broad spectrum on the role of cyanobacterial mat.

sequence similarity. These included taxa from the families *Prasinophyceae, Ulvophyceae, Trebouxiophyceae, Bacillariophyceae, Chrysophyceae, Ciliophora*, and *Cercozoa*, among other protist taxa. The eukaryotic taxa seemed to be habitat-specific, with limited overlap between communities found in lakes, ponds, and ice shelves. There were several ribotypes that were present in both Arctic and Antarctic mats, indicating that these species have spread globally and that the environmental filters affecting protist populations are comparable. Numerous of these eukaryotic species probably gain from sheltered, nutrient-rich microhabitats seen in cyanobacterial mat environments (Jungblut et al., 2012). In an effort to create biological wastewater cold-treatment methods for climatic locales, four filamentous cyanobacteria that form mats were obtained from Arctic and Antarctic habitats. These were tested for their capacity for growth and nutrient-stripping (Cassie, 2007).

Cyanobacteria are the major component of Arctic and Antarctic regions due to nutrient poor states of the soil and water. They grow in lakes, streams, weathered rock soils, and in ice-covered lakes. Most of the cyanobacterial diversity is psychrotolerant and only few are psychrophilic (Pandey et al., 2004). Cyanobacterial mats of *Nostoc commune* are evident on the seepage water near streams. Priyadarshani lakes (Schirmacher Oasis, Antarctica) bottom was covered with thick mats of *Oscillatoria, Synechococcus, Chroococcus, Nostoc* sp., and *Anabaena* sp. While growing in Priyadarshani lakes, they form benthic mat of benthic mat a few centimeters thick (Ingole and Parulekar, 1990). Cyanobacteria, while growing in harsh environment of Antarctic, have ability to maintain vital processes such as N_2-fixation which remains active at $-7°C$. Ellis-Evans and Bayliss (1993) investigated the micro gradient in cyanobacterial mats of lake and streams in Antarctica and metabolic potentials of these organisms. Standing biomass of cyanobacteria in Antarctica results of gradual acceleration over thousands of years. *Nostoc, Oscillatoria*, and *Phormidium* are dominant component of cyanobacterial diversity (Vincent, 2000).

13.4.3 Hot water springs mats

All the sites of continents have distributions of geothermal springs, most of which are related to recent volcanic activity (Ward et al., 2012). Hot water springs can be categorized into two common types (Renaut et al., 1986), the former often located on the fault line and fault junction locations toward the lake coast, with temperatures ranging from 33°C to 48°C, while other thermal springs, such the Mwanesis–Kwaipopei–Losara mat group in Kenya, may attain temperatures between 64°C and 97°C (McCall, 2010). Due to the distinctive thermophilic characteristics of the organisms surviving in niches, thermal springs have been of most significance. The presence of combined nitrogen, phosphorus, and free sulfide nutrients, along with temperature and pH combined influence the occurrence of thermophilic cyanobacteria. Above 75°C and below an approximate pH of 4, these organisms are not seen. Their diversity is typically restricted below pH 6 (Brock, 1973). Some of the springs are slightly alkaline in nature (pH 7.2) and temperature ranges between 35°C and 65°C and is always higher than the ambient air temperature. A large number of cyanobacteria species either forming a thick mat adhered to submerged rocks or floating forms were found along the thermal gradient (35°C–65°C) in stream. They were abundant and mostly colonized within the temperature range of 35°C–45°C, except *Synechococcus elongatus*, filamentous *Phormidium*, and *Mastigocladus laminosus*. There was a seasonal variation in the occurrence and population density of the other species. No new cyanobacterial species was found during intensive microscopic examination of the spring water and cyanobacterial mats during this investigation. Most of them were already reported in other thermal springs in India and in rest of the world (Mongra, 2012).

Miklavčič et al., (2010) investigated vertically stratified communities in a sand flat at Embryo data in Spain. The young mat contains mostly cyanobacteria belonging to genus *Spirulina*, whereas well-developed mat was dominated by coccoid cyanobacteria, *Microcoleus* and *Lyngbya*. In the lowermost layer, purple photosynthetic bacteria *Thiocapsa* and *Chromatium* were the major dominant microorganisms. The extracellular polymeric substances allowed them to attach sediment.

Molecular biology tools were applied to reveal the composition of cyanobacterial mat in the octopus spring of Yellow Stone National Park. Eleven different 16S r-RNA sequences of cyanobacteria were detected. *Synechococus lividus, Spirulina*, and *Microcoleus* were the dominant species (Harvey et al., 1999).

McGregor and Rasmussen (2008) reported cyanobacterial composition of a thermal spring microbial mat using polyphasic approach. Eight genera and ten species were identified based on morphological characters. *Pseudanabaena,*

Oscillatoria, Leptolyngbya were the dominant species of thermophilic cyanobacterial communities.

Hot water spring provides a unique environment where thermophilic and thermotolerant cyanobacteria form microbial mats. The upper layer supports the growth unicellular to filamentous and heterocystous cyanobacteria. The unabsorbed light intensities of higher wave lengths are utilized by an oxygenated photoautotrops. Photosynthetic bacteria utilize H2S generated by sulfur bacteria as a reductant. Thus, most of the elemental cycles, such as N, C, P, and S, are self-regulated in hot water springs by cyanobacterial mats.

13.4.4 Cyanobacterial mats in rice field

For the main objective to enhance grain yield and soil fertility, especially in rice farming, cyanobacteria are more appealing economically and environmentally than chemical fertilizers. In the rice ecosystem, free-living cyanobacteria dramatically increase nitrogen fixation, enhance the amount of nitrogen in the soil, and produce adhesive substances, excrete substances that promote growth, such as hormones, vitamins, and amino acids and boost water-holding capacity. All these improve soil fertility (Singh, 1961; Roger and Reynaud, 1982). Cyanobacteria have the capacity to boost soil phosphate availability by excreting organic acids, decrease soil salinity, raise soil microbial biomass when they die and decompose, and limit the growth of weeds (Wilson, 2006). The majority of the inherent texture of paddy soils support the population of cyanobacteria, which might potentially fix nitrogen for free (Mishra and Pabbi, 2004). Regarding cyanobacteria's needs for light, water, high temperature, and the availability of nutrients, the paddy field ecosystem offers a favorable habitat for their growth. This may be the reason why cyanobacterial growth is more prevalent in paddy soils than in upland soil (Singh, 1961).

Agricultural practices promote the growth of native strains that are already suited to the specific environmental conditions in the regions (Rogers, 1987). Application of cyanobacteria can fix roughly 25–55 kg of nitrogen per hectare per crop season increasing grain output by about 16% (Roger and Kulasooriya, 1980). Tropical rice field provide most favorable environment for the growth of cyanobacteria where N_2-fixing cyanobacteria form mats (*Aulosira, Scytonema, Tolypothrix*), balls (*Nostoc, Gleotrichia*), flakes (*Anabaena*), and crusts on alkaline usar soil (*N. commune, Microcoleus*). Mats are generally formed in the latter half of the rice crop cycle. Rigorous growth of cyanobacteria in the alkaline rice field soil reclaims the soil fertility (Singh, 1961; Roger and Kulasooriya, 1980). N_2-fixing mat forming cyanobacterium *Aulosira* is a potent nitrogen fixer of tropical and South Asian East countries rice field (Roger and Kulasooriya, 1980).

Aulosira fertilissima, a N_2-fixing, mat forming cyanobacterium exhibited diurnal pattern of diazotrophy (acetylene reduction). Rate of C_2H_2 reduction in the dark was about one-third of the day light fixation. Cultured cyanobacterium showed higher rate of C_2H_2 reduction as compared to natural mat (Banerjee et al., 2004). Organic carbon sources like glucose, sucrose, and fructose increased its heterocyst frequency and glutamine synthetase activity (Banerjee et al., 1989). Cyanobacterial mat was sensitive to UV and high light intensities evidenced by bleaching of phycocyanin pigment (Banerjee and Häder, 1996).

Pandey and Singh (1986) isolated partially derepressed mutants of mat-forming cyanobacterium *A. fertilissima* which could grow and reduce acetylene in the presence of nitrate and ammonia fertilizers. The mutant was partially blocked in nitrate reductase and glutamine synthetase activity. A new species of *Aulosira*, *A. boherensis*, was isolated from Czech Republic. 16S r-RNA revealed that new species was closest to *Nodularia* (Lukešová et al., 2009). *A. fertilissima* mat can scavenge nickel and chromium by immobilization (Banerjee et al., 2004).

13.5 Application of cyanobacterial mats in sustainable agriculture

There are many different habitats where microbial mats can be found. They are intricate, micro-scale ecosystems. The two major applications of the cyanobacterial mat include: it is used as an inoculant in biofertilizer technology in rice agriculture and its role in metal bioadsorption of heavy metals from the industrially polluted water bodies. In cyanobacterial inoculants, unicellular and filamentous N_2-fixing cyanobacteria either remain intact or are intermingled with the major inoculum of *Aulosira or Scytonema* sp. Among these two, the former is commonly practiced in agro-ecosystems, while the latter is largely neglected. Besides, cyanobacterial mats in a clumped condition like actinobacteria and fungi secrete certain metabolites, unidentified opens a new area in pharmaceuticals research and applications. Consortia of benthic cyanobacterial and microbial mat communities in Antarctic lakes indicate that these cyanobacteria are capable of nutrient recycling and scavenging, allowing the proliferation of large densities of organisms despite nutrient-poor surroundings (Andersen et al., 2011). In Lake Huron's submerged sinkholes, heterocystous and nonheterocystous filamentous cyanobacteria form a mat together with chemosynthetic bacteria in stratified layers (Biddanda et al., 2009; Nold et al., 2010). It is thought that Exopolysaccharide excretion serves multiple functions, including buffering against structural organization, nutrient storage (C, N, P, and some trace minerals), and environmental stressors (Flemming and Wingender, 2010). In cyanobacteria, EPS may have additional functions related to lithification (Dupraz et al., 2009), protection from drought (Tamaru et al., 2005), and UV radiation (Garcia-Pichel and Castenholz, 1993). On the other hand, a dense network of thin cyanobacterial filaments, especially *Leptolyngbya* sp., connects the cyanobacteria with the sediments and other filaments. The major factors that maintain and promote the cohesiveness of the microbial mats may be EPS and the thin filamentous cyanobacterial network. This cohesiveness is necessary to prevent the mats from being lifted by wind stress and working ice. Calcium carbonate was shown to be strongly linked to cyanobacterial cells, suggesting that the cells may contribute to the creation of this biomineral. In microbial mats, calcification is a frequent occurrence that appears to be regulated and controlled by the bacteria present in the mat (Al-Najjar et al., 2012; Stal, 2000). The inclusion of specific sediment particles by cyanobacterial mats, which would otherwise be blown away by current and wind, is known to improve sediment accumulation (Golubic et al., 2000). Mat-forming bacteria have been shown to create sticky excretions that can capture allochthonous mineral grains inside the mats (Gerdes and Krumbein, 1987). Around sediment particles, the interaction of physical and chemical forces can produce regions that favor the growth and spread of specific species (Stolz, 2000). The deeper particles may occasionally absorb enough solar radiation to melt a frozen mat locally and induce localized heating, which would encourage biological activity and the development of various microenvironments (Fritsen and Priscu, 1998). One method for sustaining life in one of the harshest conditions on earth appears to be the association of functionally varied microbes in organized habitats (Los Ríos et al., 2003). Mineral deposition and the buildup of empty cyanobacterial sheaths are common causes of microbial mat growth (Fenchel and Kühl, 2000; Paterson, 1994). Recent growth layers are represented by the mat's loose upper layers of upright filaments. The strata may have been compressed by silt accumulation and overgrowth of the younger filaments (Oppenheim and Paterson, 1990). Cyanobacteria are quite biologically diverse and can adapt to a wide range of conditions, even harsh ones, in marine ecosystems. Because of the peculiarities of their metabolism, cyanobacteria are one of the most important parts of food chains and can sometimes even directly affect the structure of a community. Cyanobacteria mat significantly contributes to the nitrogen and carbon cycles in marine ecosystems and produces physiologically active substances that allow them to interact with other creatures there. Cyanobacterial mats also serve as a grazing biomass by small invertebrates and herbivorous fishes.

13.6 Conclusion

Cyanobacteria are found in benthic mats in oceanic waters and on coral reefs. In the freshwater ecosystem, mat forming occurs in rice field and on the surface of soil near the seepages water of polar ecosystem. The mat adds to the nitrogen economy in the rice paddies. In hot water springs, cyanobacterial mat is self-reliant for their carbon, nitrogen, sulfur cycle. They are the feeder biomass in aquatic bodies by herbivores and in the soil, they improve the fertility of soil.

Acknowledgements

The authors thank the Head of the Department of Botany, the Institute of Science and Banaras Hindu University for providing basic facilities. The grants received from DBT (JRF/SRF) (CAS-Botany), DST-FIST, and DST-PURSA are also acknowledged.

References

Abed, R.M., Golubic, S., Garcia-Pichel, F., Camoin, G.F., Sprachta, S., 2003. Characterization of microbialite-forming cyanobacteria in a tropical lagoon: Tikehau Atoll, Tuamotu, French Polynesia. J. Phycol. 39 (5), 862–873.

Agarwal, A., 1979. Blue-green algae to fertilise Indian rice paddies. Nature 279 (5710), 181.

Al-Najjar, M.A., De Beer, D., Kühl, M., Polerecky, L., 2012. Light utilization efficiency in photosynthetic microbial mats. Environmental Microbiology 14 (4), 982–992.

Andersen, D.T., Sumner, D.Y., Hawes, I., Webster-Brown, J., McKay, C.P., 2011. Discovery of large conical stromatolites in Lake Untersee, Antarctica. Geobiology 9 (3), 280–293.

Arp, G., Reimer, A., Reitner, J., 1999. Calcification in cyanobacterial biofilms of alkaline salt lakes. Eur. J. Phycol. 34 (4), 393–403.

Banerjee, M., Häder, D.P., 1996. Effects of UV radiation on the rice field cyanobacterium, Aulosirafertilissima. Environmental and experimental botany 36 (3), 281–291.

Banerjee, M., Kumar, A., Kumar, H.D., 1989. Factors regulating nitrogenase activity and hydrogen evolution in Azolla-Anabaena symbiosis. International journal of hydrogen energy 14 (12), 871–879.

Banerjee, M., Mishra, S., Chatterjee, J., 2004. Scavenging of nickel and chromium toxicity in Aulosira fertilissima by immobilization: effect on nitrogen assimilating enzymes. Electron. J. Biotechnol. 7 (3), 13–14.

Bauer, K., Díez, B., Lugomela, C., Seppälä, S., Borg, A.J., Bergman, B., 2008. Variability in benthic diazotrophy and cyanobacterial diversity in a tropical intertidal lagoon. FEMS Microbiol. Ecol. 63 (2), 205–221.

Becking, L.E., Meesters, H.W.G., 2014. Bonaire deep reef expedition I (No. C006/14). IMARES.

Benny, N., Thomas, L.C., Padmakumar, K.B., 2021. Environmental influences on the cyanobacterial mat formation in the mangrove ecosystems along the southwest coast of India. Mar. Ecol. 42 (6), e12685.

Berman-Frank, I., Lundgren, P., Falkowski, P., 2003. Nitrogen fixation and photosynthetic oxygen evolution in cyanobacteria. Res. Microbiol. 154 (3), 157–164.

Biddanda, B.A., Nold, S.C., Ruberg, S.A., Kendall, S.T., Sanders, T.G., Gray, J.J., 2009. Great Lakes sinkholes: a microbiogeochemical frontier. Eos Trans. Am. Geophys. Union 90 (8), 61–62.

Brock, T.D., 1973. Evolutionary and ecological aspects of the cyanophytes. Bot. Monogr.

Brocke, H.J., Piltz, B., Herz, N., Abed, R.M., Palinska, K.A., John, U., et al., 2018. Nitrogen fixation and diversity of benthic cyanobacterial mats on coral reefs in Curaçao. Coral Reefs 37 (3), 861–874.

Brocke, H.J., Polerecky, L., De Beer, D., Weber, M., Claudet, J., Nugues, M.M., 2015a. Organic matter degradation drives benthic cyanobacterial mat abundance on Caribbean coral reefs. PLoS One 10 (5), e0125445.

Brocke, H.J., Wenzhoefer, F., De Beer, D., Mueller, B., Van Duyl, F.C., Nugues, M.M., 2015b. High dissolved organic carbon release by benthic cyanobacterial mats in a Caribbean reef ecosystem. Sci. Rep. 5 (1), 1–7.

Cassie, D.M., 2007. The population structure of Thamnolia subuliformis and Dicranum elongatum in northeastern coastal regions of Wapusk National Park, Manitoba (Master's thesis).

Charpy-Roubaud, C., Charpy, L., Larkum, A., 2001. Atmospheric dinitrogen fixation by benthic communities of Tikehau Lagoon (Tuamotu Archipelago, French Polynesia) and its contribution to benthic primary production. Mar. Biol. 139 (5), 991–998.

Charpy, L., Alliod, R., Rodier, M., Golubic, S., 2007. Benthic nitrogen fixation in the SW New Caledonia lagoon. Aquat. Microb. Ecol. 47 (1), 73–81.

Charpy, L., Casareto, B.E., Langlade, M.J., Suzuki, Y., 2012. Cyanobacteria in coral reef ecosystems: a review. J. Mar. Biol. 2012.

Charpy, L., Palinska, K.A., Casareto, B., Langlade, M.J., Suzuki, Y., Abed, R.M., et al., 2010. Dinitrogen-fixing cyanobacteria in microbial mats of two shallow coral reef ecosystems. Microb. Ecol. 59 (1), 174–186.

Codd, G., Bell, S., Kaya, K., Ward, C., Beattie, K., Metcalf, J., 1999. Cyanobacterial toxins, exposureroutes and human health. European Journal of Phycology 34 (4), 405–415.

Cox, D.R., Reid, N., 2000. The theory of the design of experiments. CRC Press.

Cruz-Rivera, E., Paul, V.J., 2002. Coral reef benthic cyanobacteria as food and refuge: diversity, chemistry and complex interactions. Proceedings of the Ninth International Coral Reef Symposium. Indonesian Institute of Sciences and State Ministry for Environment, Jakarta, Republic of Indonesia, 515–520.

Davoodi, M.H. R., & Abed, M.G. T. (2003). *Challenges of growth and globalization in the Middle East and North Africa*. International Monetary Fund.

Dupraz, C., Reid, R.P., Braissant, O., Decho, A.W., Norman, R.S., Visscher, P.T., 2009. Processes of carbonate precipitation in modern microbial mats. Earth-Sci. Rev. 96 (3), 141–162.

Ellis-Evans, J.C., Bayliss, P.R., 1993. Biologically active micro-gradients in cyanobacterial mats of Antarctic lakes and streams. Internationale Ver. für theoretische und Angew. Limnologie: Verhandlungen 25 (2), 948–952.

Eriksson, M.J., Clarke, A.K., 1996. The heat shock protein ClpB mediates the development of thermotolerance in the cyanobacterium *Synechococcus* sp. strain PCC 7942. J. Bacteriol. 178 (16), 4839–4846.

Fenchel, T., Kühl, M., 2000. Artificial cyanobacterial mats: growth, structure, and vertical zonation patterns. Microb. Ecol. 40 (2), 85–93.

Flemming, H.C., Wingender, J., 2010. The biofilm matrix. Nat. Rev. Microbiol. 8 (9), 623–633.

Frantz, C.M., Petryshyn, V.A., Corsetti, F.A., 2015a. Grain trapping by filamentous cyanobacterial and algal mats: implications for stromatolite microfabrics through time. Geobiology 13 (5), 409–423.

Frantz, L.A., Schraiber, J.G., Madsen, O., Megens, H.J., Cagan, A., Bosse, M., Groenen, M.A., et al., 2015b. Evidence of long-term gene flow and selection during domestication from analyses of Eurasian wild and domestic pig genomes. Nature genetics 47 (10), 1141–1148.

Fritsen, C.H., Priscu, J.C., 1998. Cyanobacterial assemblages in permanent ice covers on Antarctic lakes: distribution, growth rate, and temperature response of photosynthesis. J. Phycol. 34 (4), 587–597.

Garcia-Pichel, F., Castenholz, R.W., 1993. Occurrence of UV-absorbing, mycosporine-like compounds among cyanobacterial isolates and an estimate of their screening capacity. Appl. Environ. Microbiol. 59 (1), 163–169.

Gardner, T.A., Côté, I.M., Gill, J.A., Grant, A., Watkinson, A.R., 2003. Long-term regionwidedeclines in Caribbean corals. science 301 (5635), 958–960.

Gerdes, G., Krumbein, W.E. (Eds.), 1987. Biolaminated Deposits. Springer Berlin Heidelberg, Berlin, Heidelberg.
Glynn, S.M., 1991. Explaining science concepts: A teaching-with-analogies model. The psychology of learning science 219, 240.
Golubic, S., Le Campion-Alsumard, T., Campbell, S.E., 1999. Diversity of marine cyanobacteria. Bulletin-Institut Oceanographique Monaco-Numero, pp. 53–76.
Golubic, S., Schneider, J., Le Campion-Alsumard, T., Campbell, S.E., Hook, J.E., Radtke, G., 2019. Approaching microbial bioerosion. Facies 65, 1–17.
Golubic, S., Seong-Joo, L., 1999. Early cyanobacterial fossil record: preservation, palaeoenvironments and identification. European Journal of Phycology 34 (4), 339–348.
Golubic, S., Seong-Joo, L., Browne, K.M., 2000. Cyanobacteria: architects of sedimentary structures. Microbial Sediments. Springer, Berlin, Heidelberg, pp. 57–67.
Gorman-Smith, D., Tolan, P.H., Henry, D.B., 2000. A developmental-ecological model of the relation of family functioning to patterns of delinquency. Journal of quantitative criminology 16, 169–198.
Greeney, H.F., Dobbs, R.C., Martin, P.R., Gelis, R.A., 2008. The breeding biology of Grallaria and Grallaricula antpittas. Journal of Field Ornithology 79 (2), 113–129.
Gurgel, C.F.D., Camacho, O., Minne, A.J., Wernberg, T., Coleman, M.A., 2020. Marine heatwave drives cryptic loss of genetic diversity in underwater forests. Current Biology 30 (7), 1199–1206.
Hallock, P., 2005. Global change and modern coral reefs: new opportunities to understand shallow-water carbonate depositional processes. Sedimentary Geology 175 (1-4), 19–33.
Harvey, A.W., Mooi, R., Gosliner, T.M., 1999. Phylogenetic taxonomy and the status of Allonautilus Ward and Saunders, 1997. Journal of Paleontology 73 (6), 1214–1217.
Heisler, J., Glibert, P.M., Burkholder, J.M., Anderson, D.M., Cochlan, W., Dennison, W. C., et al., 2008. Eutrophication and harmful algal blooms: a scientific consensus. Harmful Algae 8 (1), 3–13.
Hinde, R., Pironet, F., Borowitzka, M.A., 1994. Isolation of Oscillatoria spongeliae, the filamentous cyanobacterial symbiont of the marine sponge Dysidea herbacea. Mar. Biol. 119 (1), 99–104.
Hörnlein, C., Confurius-Guns, V., Grego, M., Stal, L.J., Bolhuis, H., 2020. Circadian clock-controlled gene expression in co-cultured, mat-forming cyanobacteria. Sci. Rep. 10 (1), 1–17.
Ingole, B.S., Parulekar, A.H., 1990. Limnology of Priyadarshani Lake, Schirmacher Oasis, Antarctica. Polar Rec. 26 (156), 13–18.
Jungblut, A.D., Vincent, W.F., Lovejoy, C., 2012. Eukaryotes in Arctic and Antarctic cyanobacterial mats. FEMS Microbiol. Ecol. 82 (2), 416–428.
Lesser, M.P., Farrell, J.H., 2004. Exposure to solar radiation increases damage to both host tissues and algal symbionts of corals during thermal stress. Coral Reefs 23 (3), 367–377.
Los Ríos, A., Wierzchos, J., Sancho, L.G., Ascaso, C., 2003. Acid microenvironments in microbial biofilms of Antarctic endolithic microecosystems. Environ. Microbiol. 5 (4), 231–237.
Lukešová, A., Johansen, J.R., Martin, M.P., Casamatta, D.A., 2009. Aulosira bohemensis sp. nov.: further phylogenetic uncertainty at the base of the *Nostocales* (Cyanobacteria). Phycologia 48 (2), 118–129.
McCall, J., 2010. Lake Bogoria, Kenya: hot and warm springs, geysers and Holocene stromatolites. Earth-Sci. Rev. 103 (1–2), 71–79.
McGregor, G.B., Rasmussen, J.P., 2008. Cyanobacterial composition of microbial mats from an Australian thermal spring: a polyphasic evaluation. FEMS Microbiol. Ecol. 63 (1), 23–35.

Meltvedt, A., Jadot, C., 2014. Progression of the coral-algal phase shift in the Caribbean: A case study in Bonaire, Dutch Caribbean. Marine Technology Society Journal 48 (6), 33–41.

Miklavčič, D., Mir, L.M., Thomas Vernier, P., 2010. Electroporation-based technologies and treatments. The Journal of membrane biology 236, 1–2.

Mishra, U., Pabbi, S., 2004. Cyanobacteria: a potential biofertilizer for rice. Resonance 9 (6), 6–10.

Mongra, A.C., 2012. Distribution pattern of Cyanobacteria in hot water springs of Tattapani, Himachal Pradesh, India. JAIR 1 (7), 363–370.

Newman, S.A., Klepac-Ceraj, V., Mariotti, G., Pruss, S.B., Watson, N., Bosak, T., 2017. Experimental fossilization of mat-forming cyanobacteria in coarse-grained siliciclastic sediments. Geobiology 15 (4), 484–498.

Nold, S.C., Pangborn, J.B., Zajack, H.A., Kendall, S.T., Rediske, R.R., Biddanda, B.A., 2010. Benthic bacterial diversity in submerged sinkhole ecosystems. Appl. Environ. Microbiol. 76 (1), 347–351.

Oppenheim, D.R., Paterson, D.M., 1990. The fine structure of an algal mat from a freshwater maritime Antarctic lake. Can. J. Bot. 68 (1), 174–183.

Paerl, H.W., Hall, N.S., Calandrino, E.S., 2011. Controlling harmful cyanobacterial blooms in a world experiencing anthropogenic and climatic-induced change. Science of the total environment 409 (10), 1739–1745.

Paerl, H.W., Paul, V.J., 2012. Climate change: links to global expansion of harmful cyanobacteria. Water research 46 (5), 1349–1363.

Pandey, K.D., Shukla, S.P., Shukla, P.N., Giri, D.D., Singh, J.S., Singh, P., et al., 2004. Cyanobacteria in Antarctica: ecology, physiology and cold adaptation. Cell. Mol. Biol. 50 (5), 575–584.

Pandey, K.D., Singh, P.K., 1986. Isolation and characterization of partially derepressed mutants of Cyanobacterium *Aulosira* spec. Biochem. Physiol. der Pflanz. 181 (4), 219–225.

Pandolfi, J.M., Bradbury, R.H., Sala, E., Hughes, T.P., Bjorndal, K.A., Cooke, R.G., Jackson, J.B., et al., 2003. Global trajectories of the long-term decline of coral reef ecosystems. Science 301 (5635), 955–958.

Paterson, D.M., 1994. Microbiological mediation of sediment structure and behaviour. Microbial Mats. Springer, Berlin, Heidelberg, pp. 97–109.

Quesada, A., Vincent, W.F., 1997. Strategies of adaptation by Antarctic cyanobacteria to ultraviolet radiation. Eur. J. Phycol. 32 (4), 335–342.

Reid, R.P., Visscher, P.T., Decho, A.W., Stolz, J.F., Bebout, B.M., Dupraz, C., et al., 2000. The role of microbes in accretion, lamination and early lithification of modern marine stromatolites. Nature 406 (6799), 989–992.

Rejmánková, E., Komárková, J., 2000. A function of cyanobacterial mats in phosphorus-limited tropical wetlands. Hydrobiologia 431 (2–3), 135–153.

Renaut, R.W., Tiercelin, J.J., Owen, R.B., 1986. Mineral precipitation and diagenesis in the sediments of the Lake Bogoria basin, Kenya Rift Valley. Geological Society, London, Special Publications, 25 (1), 159–175.

Richa, R.R., Kumari, S., Singh, K.L., Kannaujiya, V.K., Singh, G., Kesheri, M., et al., 2011. Biotechnological potential of mycosporine-like amino acids and phycobiliproteins of cyanobacterial origin. Biotechnol. Bioinf. Bioeng. 1, 159–171.

Roger, P.A., Kulasooriya, S.A., 1980. Blue-Green Algae and Rice. International Rice Research Institute.

Roger, P.A., Reynaud, P.A., 1982. Free—living blue—green algae in tropical soils. Microbiology of Tropical Soils and Plant Productivity. Springer, Dordrecht, pp. 147–168.

Rogers, Jr. H., 1987. Theory of recursive functions and effective computability. MIT press.

Schneider, J., Le Campion-Alsumard, T., 1999. Construction and destruction of carbonates by marine and freshwater cyanobacteria. Eur. J. Phycol. 34 (4), 417–426.

Singh, R.N., 1961. Role of blue-green algae in nitrogen economy of Indian agriculture. *Role of blue-green algae in nitrogen economy of Indian agriculture.*

Stal, L.J., 2000. Cyanobacterial mats and stromatolites. The Ecology of Cyanobacteria: Their Diversity in Time and Space. Kluwer Academic Publishers.

Stolz, J.F., 2000. Structure of microbial mats and biofilms. Microbial Sediments. Springer, Berlin, Heidelberg, pp. 1–8.

Tamaru, Y., Takani, Y., Yoshida, T., Sakamoto, T., 2005. Crucial role of extracellular polysaccharides in desiccation and freezing tolerance in the terrestrial cyanobacterium Nostoc commune. Appl. Environ. Microbiol. 71 (11), 7327–7333.

Thacker, R.W., Starnes, S., 2003. Host specificity of the symbiotic cyanobacterium Oscillatoria spongeliae in marine sponges, *Dysidea* spp. Mar. Biol. 142 (4), 643–648.

Urrutia-Cordero, P., Zhang, H., Chaguaceda, F., Geng, H., Hansson, L.A., 2020. Climate warming and heat waves alter harmful cyanobacterial blooms along the benthic–pelagic interface. Ecology 101 (7), e03025.

Van Zanten, H.H., Mollenhorst, H., Klootwijk, C.W., Van Middelaar, C.E., De Boer, I.J., 2016. Global food supply: land use efficiency of livestock systems. The International Journal of Life Cycle Assessment 21, 747–758.

Vincent, W.F., 2000. Cyanobacterial dominance in the polar regions. The Ecology of Cyanobacteria. Springer, Dordrecht, pp. 321–340.

Ward, D.M., Castenholz, R.W., Miller, S.R., 2012. Cyanobacteria in geothermal habitats. Ecology of Cyanobacteria II. Springer, Dordrecht, pp. 39–63.

Webb, A.E., de Bakker, D.M., Soetaert, K., da Costa, T., van Heuven, S.M., van Duyl, F.C., et al., 2021b. Functional consequences of Caribbean coral reef habitat degradation. Biogeosci. Discuss. 2021, 1–23.

Webb, A.E., De Bakker, D.M., Soetaert, K., Da Costa, T., Van Heuven, S.M., Van Duyl, F.C., et al., 2021a. Quantifying functional consequences of habitat degradation on a Caribbean coral reef. Biogeosciences 18 (24), 6501–6516.

Whitton, B.A., Potts, M., 2012. Introduction to the cyanobacteria. Ecology of cyanobacteria II: their diversity in space and time. Springer Netherlands, Dordrecht, pp. 1–13.

Wilson, C., 2006. Learning phonology with substantive bias: An experimental and computational study of velar palatalization. Cognitive science 30 (5), 945–982.

Woebken, D., Burow, L.C., Behnam, F., Mayali, X., Schintlmeister, A., Fleming, E.D., et al., 2015. Revisiting N2 fixation in Guerrero Negro intertidal microbial mats with a functional single-cell approach. ISME J. 9 (2), 485–496.

CHAPTER

14 Beneficial soil microorganisms and their role in sustainable agriculture

Rostislav A. Streletskii[1], Angelika A. Astaykina[1], Andrey A. Belov[1], Vladimir S. Cheptsov[1] and Anna A. Vetrova[2]

[1]*Soil Science Faculty, Lomonosov Moscow State University, Moscow, Russia*

[2]*Federal Research Center "Pushchino Scientific Center for Biological Research of the Russian Academy of Sciences," G.K. Skryabin Institute of Biochemistry and Physiology of Microorganisms, Pushchino, Russia*

14.1 Introduction and definition of concepts

In recent decades, a significant amount of data has been accumulated on the role of dominant soil taxons of microorganisms in the formation and maintenance of soil fertility. In addition to carrying out fundamental processes that support the biological cycle, such as the nitrogen, carbon cycle, and soil formation, soil microorganisms also affect the well-being of phytocoenosis through biologically active substances, enzymes, and competition with pathogenic microflora. Thus technologies aimed at improving the quality of soil microbiome can be used to increase crop yields. This is most relevant for lands that have been in agricultural use for a long time, in which there has been a significant change in the overall structure of microbial communities due to tillage, fertilization, and pesticides. As a rule, such changes are characterized by a reduction in the number of microbial taxons and the replacement of dominants by simple heterotrophic genus. Similar changes are observed in the case of anthropogenic soil pollution. Reducing the abundance of beneficial microorganisms, such as, for example, free-living nitrogen fixers, ultimately leads to a decrease in soil fertility and the need to increase fertilizer doses. Due to the lack of competition from beneficial microorganisms, phytopathogens are becoming widespread, which has a critical impact on yields and requires an increase in the amount of pesticides used. Increasing the rate of application of pesticides and fertilizers leads to even greater degradation of the soil microbiota and a vicious circle is formed. The way out of this situation is the use of biological fertilizers and biopesticides. The purpose of using biological fertilizers based on beneficial strains of microorganisms is to increase crop yields by introducing a beneficial strain or artificial microbial consortiums into an existing microbial community.

Sustainable Agricultural Practices. DOI: https://doi.org/10.1016/B978-0-443-19150-3.00013-8

At the same time, the introduced strains become dominant for a certain period, and during this period, they perform useful functions. The introduction of biological fertilizers does not lead to a simplification of the structure of the soil microbial community, since during one growing season, the proportion of the applied strains will decrease and it will either disappear completely or become a normal component of the soil microbiota. In some cases, the displacement of the original strains is possible, which can be considered a positive phenomenon, since the components of biological fertilizers have been selected and have pronounced beneficial properties that wild strains cannot possess to the same extent.

Biological pesticides are created from microorganisms or other natural objects and are less toxic to nontarget organisms and decompose relatively quickly, disappearing without a trace from agrocenoses. In fact, their use does not carry the risks of soil and adjacent environment pollution, but at the same time allows you to control pests and plant diseases.

The use of biological fertilizers and pesticides is an essential component of a sustainable development strategy in agriculture, as it helps to conserve soils and leads to a reduction in xenobiotic pollution of the biosphere, as well as reduces greenhouse gas emissions. This reduces the risks associated with long-term negative changes in the quality of agricultural land and the environment in general. For example, 64% of agricultural land is currently at risk of being contaminated by at least one pesticide active ingredient. Thus biological technologies in the field of fertilizers and plant protection products in the future can help to avoid crop failures and starvation when existing agricultural methods have exhausted their capabilities and negative side effects outweigh the benefits.

However, the use of biological methods to improve soil fertility and control diseases and pests has its own characteristics. In particular, the effectiveness of biofertilizers will significantly depend on the composition and condition of native microflora, as well as on climatic and soil physicochemical conditions.

14.2 Change in soil microbial communities as a result of agricultural use

It has long been known that the type of tillage affects the abundance and taxonomic composition of microbial communities in arable soils. Compared to no-till, plowing consistently results in a reduction in microbial biomass and enzymatic activity (Zuber and Villamil, 2016). At the same time, in Ref. Nikitin et al. (2021), using the example of southern chernozems, it was shown that soils with traditional plowing with layer turnover have a higher diversity of micromycetes compared to no-till soils. Moreover, in no-till fields, the abundance of phytopathogenic and cellulolytic microorganisms are much higher, whereas in plowed areas, saprotrophic and oligotrophic microorganisms are much higher (Nikitin et al., 2021).

Application of mineral fertilizers leads to changes in soil microbial biomass, enzyme activity, community diversity, and composition. Many researchers have found that long-term application of mineral fertilizers can increase the biomass and diversity of the microbial communities of agrocenoses (Belay et al., 2002; Zhong et al., 2010; Geisseler and Scow, 2014). An extensive meta-analysis of 107 studies shows that microbial biomass carbon is higher by 15.1% in soils under long-term application of mineral fertilizers (Geisseler and Scow, 2014). However, the findings are in conflict with some previously published studies showing that nitrogen fertilization reduces the microbial biomass of soils of the unmanaged ecosystem, according to some data by an average of 15% (Liu and Greaver, 2010), and according to others by 5.8% (Lu et al., 2011). The revealed contradiction may be related, firstly, to the type of land use, for example, N applications to unmanaged ecosystem has effects on plant species diversity and composition, and secondly, the introduction of an additional source of nitrogen without lime additions, as a rule, reduces pH, which leads to a sharp decrease in the supply of nutrients in the cationic form to plants or even to the loss of nutrients, especially potassium, from plant roots (Sokolova et al., 1997). Thus the application of mineral fertilizers at low pH values of the soil solution, on uncultivated lands, and also in cases of short-term application (less than 10 years) leads to a decrease in microbial biomass. But the opposite effect is observed on agricultural lands with low natural fertility under long-term fertilization and when cultivating crops with a large amount of plant residues (Chernov and Semenov, 2021).

Numerous studies show that the application of mineral fertilizers leads to a change in the composition and taxonomic structure of the soil microbial communities. Tkhakakhova et al. (2016) studied the structure of microbial communities in agrochernozems under varying degrees of anthropogenic pressure. The authors came to the conclusion that the number of ammonifiers does not depend on land use systems and on the use of mineral fertilizers. However, according to the data of molecular genetic analysis, at the level of domains in arable soils, the representation of Archaea increases from 8.47% in the fallow to 20.63% in the variant with the application of mineral fertilizers at a dose of N60P60K60. At the phylum level, the percentage of Bacillota in agrogenic soils increases. At the family level, compared to the soil of the mowed steppe (fallow), the following changes occur in the taxonomic structure of the microbial community of arable soils:

1. The proportion of Conexibacteraceae, Patulibacteraceae, Hyphomicrobiaceae, Syntrophobacteraceae, [EB1017], and nonattributable families of the phyla Acidobacteriota, Actinomycetota, Pseudomonadota decreases;
2. The proportion of Geodermatophilaceae, Micrococcaceae, Rhodospirillaceae, Sphingomonadaceae, [Ellin5301], and nonattributed families of the phyla Acidobacteriota, Gemmatimonadota increases. Under the influence of mineral fertilizers in arable soils, the proportion of bacteria AKIW874, Solirubrobacteraceae, Enterobacteriaceae, Pseudomonadaceae, Gaiellaceae, and the unidentified family of the order Myxococcales decreases.

According to a metaanalysis from a dataset obtained from 102 peer-reviewed publications on the main biomes of the Earth (arid, continental, temperate, and tropical), compared to the soils of an elevated ecosystem, the microbial community has higher pH values and soils of agrocenoses. At the same time, the relative abundance of the pathogen of the phylum Verrucomicrobiota is higher in agricultural soils in continental, temperate, and tropical regions. Both groups are susceptible to oligotrophic microorganisms due to slow growth rate and distribution in soils with stable detection availability. Relative representation of Pseudomonadota is higher in natural soils in comparison with anthropogenically transformed soils. It is known that many representatives of the phylum are plant growth stimulants; therefore a decrease in the relative abundance of Pseudomonadota in soils of agroroses can contribute to crop productivity and soil fertility (Trivedi et al., 2016).

Differences in the modes of agricultural use affect the taxonomic structure of the microbial community of soils due to changes in the physicochemical conditions of the habitat of microorganisms. It has been found that, compared to the conventional agricultural management system (the use of mineral fertilizers, pesticides, and mechanical tillage), the alpha-diversity of the microbial community is higher with the organic type of land use, and the representation of bacteria of the genus *Streptomyces* and *Rhizobium* is increased (Zhou and Fong, 2021). The dominance of these taxa in the structure of the soil bacterial community contributes to the suppression of plant diseases and an increase in the rate of nutrient cycling (Kinkel et al., 2012; Caton et al., 2018).

A meta-analysis including a review of 1235 studies showed that the land use change leads to an increase of alpha-diversity and has the positive effect on beta diversity of soil microbial communities. But the application of N and N × P × K significantly decreases the richness and Shannon index. At the same time, according to the authors, anthropogenic factors such as warming (W), carbon-dioxide enrichment (eCO_2), altered precipitation, atmospheric nitrogen (N) deposition, nutrient fertilization, and land use change degrees influence the taxonomic structure of the prokaryotic and fungal communities rather than microbial diversity (Zhou et al., 2020).

Numerous studies confirm that the use of pesticides in agriculture, as well as the use of mineral fertilizers, leads to changes in soil microbial biomass, enzyme activity, community diversity, and composition. The impact of pesticides on the soil microbial community depends on the chemical groups of active substances. For example, according to Zhang et al. (2009), pyrethroids reduce the representation of bacteria of the phylum Bacillota and increase the representation of Gram-negative bacteria. Glyphosate, belonging to the organophosphate group, decreases the activity of phosphate enzyme up to 98% (Sannino and Gianfreda, 2001) and increases the relative abundance of (Lancaster et al., 2010). Neonicotinoids changes soil microbial community structure: for example, thiamethoxam and dinotefuran decrease the relative abundances of Gemmatimonadetes and [OD1] and increase the relative abundances of *Chloroflexi* and *Nitrospirae*, which indicated that these bacteria could be involved in the biodegradation of neonitocinoids

(Yu et al., 2020). Streletskii et al. (2022) provides data that under the influence of three pesticides belonging to different chemical classes, a change in dominants occurs in the soil microbial community, the representation of taxa changes, and the diversity of microorganisms, especially fungi, decreases.

Thus as a result of agricultural land use, the native soil microbial community is transformed. In most cases, changes in biomass, enzymatic activity, diversity, and structure of prokaryotic and fungal communities adversely affect soil health. Understanding the processes that underlie such changes contributes to the determination of indicators of the state of the soil microbial community. For example, an increase in the representation of certain taxa of microorganisms under the influence of pesticides indicates that bacteria/fungi can use the pesticide molecule as the only source of carbon; therefore these organisms can be potential degraders of these organic pollutants. Another way to restore the native state of soil microbiocenosis is the targeted introduction into the soil of specific strains of microorganisms capable of actively carrying out or stimulating certain processes, for example, plant nutrition and growth, nitrogen fixation, dissolution of phosphates, and reduction of greenhouse gas emissions. This approach is expressed in the development of biofertilizers-preparations consisting of one or more strains of microorganisms, including those obtained by selection and capable of carrying out the required processes.

14.3 Plant growth–promoting rhizobacteria

The ecosystem role of microorganisms in carbon and nitrogen fixation, decomposition of organic material, and solubilization of sparingly soluble compounds is well-known (Fitzsimons and Miller, 2010). At the same time, the role of soil microorganisms includes such processes that directly affects growth and germination of plants, like production of different biopolymers, that stimulating plant (or its parts) growth, or, opposite, suppressing introduction and/or growth of introduced microorganisms, including phytopathogenic representatives (Ramakrishna et al., 2019); those bioactive compounds are presented by different enzymes, polysaccharides, phytohormones, antibiotics, role of which we will discuss below. For today, due to intensive agriculture character and high probability of food crises in different parts of our planet, the studies of such microorganisms and their influence on growth and germination of higher plants appears to be an actual and prospective problem.

Among all bacteria positively affected on growth and germination of plants, it seems that most interesting group is plant growth–promoting rhizobacteria (non-taxonomic group; PGPR) due to its spatial proximity to the higher plants body and, hence, greater influence on the plant tissues and properties of soil surrounding the root system. The most studied PGPR are the representatives of Pseudomonadota and Bacillota phyla, among which beneficial effects on plants were shown for the representatives of *Bacillus*, *Pseudomonas*, *Acinetobacter*,

Serratia, *Burkholderia*, *Rhizobium*, and *Azotobacter* genera (Batista et al., 2018; Ramakrishna et al., 2019). PGPB belonging to Actinomycetota phyla are less studied. Representatives of genera *Arthrobacter*, *Agromyces*, *Corynebacterium*, *Mycobacterium*, *Micromonospora*, and *Streptomyces* positively affected growth and productivity of higher plants were found (Hamedi and Mohammadipanah, 2015; Sathya et al., 2017). Representatives of the specified phyla and genera are widespread in different soil and conjugated substrates and, hence, we could assume the detection of plant growth-promoting features among the other representatives of the other genera belongs to these phyla.

Research of the rhizobacteria, promoting plant growth, however, has a significant methodological limitation—the properties found in laboratory conditions often do not appear in natural habitats, which complicates development of efficient microbial systems for plant growth promotion. Detection of interactions between plants and root zone microorganisms as well as mechanisms of these interactions could be researched by modern molecular-biological techniques, such as genome annotations and detection of molecular responses of plant or bacteria on coexistence (Ramakrishna et al., 2019).

Significant limitation for "healthy" rhizosphere microbial community, promoting plant growth, formation is soil moisture: it was shown that rhizosphere colonization by bacteria, promoting the plant growth, highly depended on soil moisture level (Shrivastava and Kumar, 2015). Continuing this statement, it can be assumed that the functioning of root microbial communities and their positive effect on plant growth and development largely depend on soil moisture. In this regard, given the expanding areas of territories experiencing aridization and, consequently, moisture deficiency both for plants and soil microbial communities, it becomes obvious that drought-tolerant bacteria and rhizobacteria that promote plant growth and their possible impact on the drought tolerance of agricultural plants are becoming promising for further study.

There are quite a lot of materials regarding the study of PGPR effects on plants, in which decrease in the level of stress load on plants caused by water deficiency (both atmospheric and irrigated (Kaushal and Wani, 2016; Kasim et al., 2013; Delshadi et al., 2017; Mubeen et al., 2021) and due to high concentrations of water-soluble compounds, in particular salts (Numan et al., 2018; del Carmen Orozco-Mosqueda et al., 2020; Mokrani et al., 2020; Goswami and Suresh, 2020)) was observed.

However, for bacteria belonging to phyla, among which the overwhelming majority of PGPRs (Pseudomonadota and Bacillota) were found, high resistance to various physicochemical stresses, including drought, is not a characteristic property, although for individual representatives of these phyla isolated from extreme environments, resistance to various stress agents has been shown (Belov et al., 2019, 2020 and references therein). In particular, a number of species of the genus *Pseudomonas* (Pseudomonadota) with plant growth-promoting (PGP) properties have been shown to be highly resistant to a number of stress factors, including salinity and high pH (Vyas et al., 2009; Verma et al., 2015). Among stress-tolerant

PGP belonging to Bacillota phyla, bacteria of the genus *Bacillus* are considered as the most effective (Cherif-Silini et al., 2016; Radhakrishnan et al., 2017). At the same time, it should be noted, that representatives of this genus (and the phylum as a whole) are characterized by experiencing unfavorable conditions in the form of spores, that is, in a hypometabolic or metabolically inactive state (Nicholson et al., 2000; Filippidou et al., 2016; Nunes et al., 2018). Representatives of the Actinomycetota phyla are the most resistant and able to maintain metabolic activity under stress conditions, in particular, under drought, contrasting pH values, and exposure to high levels of insolation (Xu et al., 2018; Qin et al., 2016; Mohammadipanah and Wink, 2016). Since the same factors affect plants (Lakshmanan et al., 2014), it is promising to study actinobacteria that can improve the growth, development, and productivity of plants.

The known up-to-date mechanisms of the positive effect of PGPR on the growth and development of plants are the production of hydrolytic enzymes and polysaccharides that affect the decomposition and synthesis of organic compounds in the soil; antagonistic relationships between PGPR and pathogens of various diseases, due to both the limitation of available ecological niches and nutrients in the root zone and the production of toxins and antibiotics by "beneficial" microbial communities of the root zone (Pandey et al., 2019). The hydrolytic activity of exoenzymes of microorganisms of the root zone promotes the release of carbon contained in dead biomass and root exudates. The hydrolysis products of these compounds are involved in the metabolic processes of both plants and microorganisms. An increase in the productivity of root soil microbial communities that do not even directly interact with plants can lead to positive changes in the soil structure, the accumulation of organic matter in it (both humus and microbial biomass), which can be used by plants (Egamberdiyeva, 2007; Muniroh et al., 2019; Jadhav and Sayyed, 2016). Carbon and nitrogen bound into organic compounds of microbial biomass in the future can be used by plants, both during their oxidation and exudation by living microbial cells and during death and mineralization of microbial cells. Due to the extremely high diversity of metabolic processes implemented by prokaryotes (Kim and Gadd, 2019), eco-physiologically diverse soil prokaryotic communities of the root zone can have a beneficial effect on plant growth and development.

Microbial origin polysaccharides, in addition to their effect on soil structure, can contribute to water retention processes and, together with hydrolytic enzymes, prevent plant infection with various root infections (Jadhav et al., 2017; Saberi Riseh et al., 2021; Etesami and Maheshwari, 2018). In addition, the microbial communities of the root zone affect the solubility (increase or decrease) of various chemical compounds, in particular, phosphorus, potassium, and heavy metals (Saiyad et al., 2015; Sharma et al., 2016; Marciano Marra et al., 2012; Alaylar et al., 2020; Ahemad, 2019; Singh et al., 2018), and can also affect plants by synthesis of various phytohormonal compounds, such as auxins, gibberellins, and others (Kudoyarova et al., 2019; Nazli et al., 2020).

Distracting from agricultural plants and the obvious industrial significance of PGPR in ensuring the efficiency and ecologization of intensive farming, we note another important aspect of microbial communities in the root zone—the production of antagonist substances. Phytopathogenic microorganisms that obviously affect plant life are affected by various antimicrobial metabolites secreted by root microorganisms (Keohane et al., 2015; Chen et al., 2019). At the same time, an extremely interesting and important fact is the discovery of substances suppressing human pathogens among the metabolites of root-associated microbial communities (Mendes et al., 2013; Ramakrishna et al., 2019; and references therein). While it has been shown that the rhizosphere can serve as a habitat for human pathogens (Berg et al., 2005; Sharma et al., 2019; Ali et al., 2017), the discovery of natural metabolites aimed at combating them is of great importance for maintaining the health of the urban population during urban greening: due to the high population density of people in urban conditions, the ability of urban soils to self-purify, including from pathogenic and opportunistic pathogenic microorganisms, largely determines the quality of life of people in cities. The study of antimicrobial metabolites of rhizosphere communities and their ability to suppress the activity of bacteria pathogenic for humans, together with the study of both soil and microbial processes occurring in urbanized soils, is also seen as a promising area of research in this area.

Production of a wide range of different exoenzymes, antibiotics, and exopolysaccharides characteristic for Actinomycetota, combined with their wide abundance in various soils, resistance to oxidative stress (Feng and Wang, 2020), and wide biodiversity (Idris et al., 2017; Goodfellow et al., 2018), coupled with increasing rates of land degradation in recent decades (Li et al., 2021), reduction of arable lands (Farooqi et al., 2021), and climate aridity (Reynolds et al., 2007; Prăvălie, 2016; Zlobin, 2022), make it extremely promising to research and describe actinobacteria that promote plant growth and development. The most attractive ecotopes for isolating such organisms are the soils and sediments of hot arid deserts, where the combined stress, including the lack of available moisture, increased insolation and salinity, leads to the natural selection of resistant forms, which, in their existence, are associated with the vital activity of higher plants and, interacting with them, promote their growth under aggressive physicochemical conditions (ALKahtani et al., 2020; Bashan et al., 2012; Eida et al., 2018).

Since plants life is closely related to soils and the microbial communities inhabiting them, it is obvious that they exists in constant interaction. To describe microbial communities associated with plants, the term phytomicrobiome (Bhatt et al., 2020; Kumar et al., 2020; Lyu et al., 2021) was proposed, which includes various microorganisms (prokaryotes, as well as unicellular and multicellular eukaryotes and viruses) associated with plants.

In the modern biosphere, higher plants can be considered both consorts and edificators of most ecosystems: plants, due to their ability to photosynthesis, have the ability to synthesize carbohydrates and most other bioorganic compounds, which are further used both by higher organisms in food chains and by microorganisms. Moreover, the interaction of plants with microorganisms is remarkable in that microorganisms can consume both intravital plant secretions (in particular, root exudates) and dead (saprotrophic bacteria and fungi) or still alive (various phytopathogenic microorganisms) plant biomass (Perreault and Laforest-Lapointe, 2022.; Yadav, 2021; Middleton et al., 2021; Senthilkumar and Sankaranarayanan, 2021).

At the same time, interactions between plants and microbial communities of the host substrate are not limited exclusively to trophic interactions. Above, we mentioned the solubilization of sparingly soluble substances by root-associated communities and their influence on the soil structure and, as a result, its water-physical properties. Examples are widely known when the interactions of tissues of higher plants with microorganisms lead to the formation of symbiotic relationships between them. As an example, let us consider mycorrhizal fungi (MF) capable of forming arbuscular-vesicular mycorrhiza inside root tissues of woody plants (Willis et al., 2013) and symbiotic relationships between leguminous plants and Pseudomonadota of the genus *Rhizobium* (Hirsch et al., 2001; Prell and Poole, 2006; Gourion et al., 2015). In the first case, filamentous fungi improve the water nutrition of plants and contribute to mineral nutrition by improving the dissolution of various hard-to-reach forms of mineral compounds (in particular, phosphorus), and in the second case, bacterial cells localized in nodules carry out the process of nitrogen fixation and contribute to nitrogen nutrition of plants.

Discussing on the close functional relationship between plants and soil microbial communities, one cannot fail to mention the previously formulated plant holobiont theory, according to which mitochondria and chloroplasts of a eukaryotic plant cell are descendants of endosymbiotic proteobacteria and cyanobacteria, respectively (Lyu et al., 2021). Existing inside the host cell for a long time (about 1.5 million years (Gray, 2015)), endosymbiotic bacteria have lost part of the genome and, as a result, functions, while acquiring a narrow specialization—the implementation of reactions of oxygenic photosynthesis and the cycle of tricarboxylic acids and electron transport chain. Thus an eukaryotic cell acting as a single functional structure is the simplest example of a holobiont.

The accumulated data on the interaction of plants with microbial communities make it possible to expand the concept of a holobiont above the level of a eukaryotic cell and consider plants and their associated microbial communities as a functionally related structure. We present below the main most generalized provisions confirming this point of view:

1. The root zone of plants is the zone of the highest activity of microbial processes in soils, which suggests the role of the plant as a consort—the host organism around which the community is formed (Kuzyakov and Blagodatskaya, 2015; McNear, 2013; Solanki et al., 2020; Pronk et al., 2022).
2. Numerous endophytic microorganisms are known that are capable of penetrating and long-term existence in the tissues of higher plants. Such strategy can be considered as one of the intermediate stages of endosymbiosis, which once led to the emergence of a plant cell holobiont (Morales-Cedeño et al., 2021; Lodewyckx et al., 2002).

3. Both endophytic, rhizosphere and rhizoplane communities of microorganisms secrete various compounds to maintain their stability and protection against opportunistic, pathogenic, and other "foreign" microorganisms, which may indicate their coevolution (Barea et al., 2005; Hashem et al., 2016; Nanjundappa et al., 2019; Mamangkey et al., 2022).
4. The plant acts as a host organism, around which the corresponding community is formed, and it acts as the central element of the holobiont (Singh et al., 2020; Rossmann et al., 2020).
5. Root microbial communities and endophytic microorganisms increase the productivity and resistance of host plants, which indicates the optimization and improvement of its vital processes, that is, the synergy of the action of individual components in this system (Bhatt et al., 2020; Igiehon and Babalola, 2018; Nordstedt and Jones, 2020).

Thus plants in natural conditions are not so much individuals of modular organisms but functionally closely related with microbial communities that promote the growth and development of plants.

14.4 Nitrogen-fixing bacteria

The productivity of phytocenoses is determined by a combination of factors, of which the species and varietal properties of plants, the conditions of their mineral nutrition, and moisture supply should be noted as fundamental. In the context of an increasing population of the earth, degradation of agricultural areas and, as a result, an aggravation of the problem of food security (FAO, 2015; Krasilnikov, Taboada, 2022), the tasks of biological optimization of plant nutrition become extremely relevant.

Among the elements of mineral nutrition, which can rightfully be considered as the most essential, nitrogen should be noted. Being a part of amino acids and, consequently, protein molecules, as well as nucleotides, nucleosides, and many other organic compounds of a living cell, nitrogen plays an extremely important role in maintaining the viability of both a unicellular and a multicellular organism and its productivity (Kirby et al., 1987; Leghari et al., 2016). At the same time, the ability to biological nitrogen fixation (BNF) is a property that is characteristic only for prokaryotic organisms: bacteria (*Rhizobium*, *Frankia*, *Azotobacter*, *Azospirillum*, *Beijerinckia*, *Mycobacterium*, *Bacillus*, *Pseudomonas*, and others), cyanobacteria (*Anabaena*, *Nostoc*, and others) and archaea (Soumare et al., 2020).

Nitrogen fixation of prokaryotes is observed both in free-living soil bacteria and in bacteria spatially associated with plants in the root zone or directly interacting with them (legume-rhizobium symbiosis).

Free-living nitrogen-fixing bacteria are, in particular, representatives of the genera *Azotobacter*, *Azospirillum*, *Bacillus*, *Clostridium*, and *Pseudomonas*, capable of fixing in situ on average up to 60 kg/N/ha per year (Orr et al., 2011).

Associative diazotrophs, inhabit soil surround the root system, are mainly presented in the phylum Pseudomonadota (*Gluconacetobacter, Azospirillum, Burkholderia, Azotobacter, Serratia, Klebsiella, Pseudomonas, Pantoea*), less often representatives of the phylum Bacillota (*Bacillus, Paenibacillus*) (Van Dommelen and Vanderleyden, 2007; Abadi et al., 2021). Nitrogen-fixing representatives of the phyla Actinomycetota and Bacteroidota are, in particular, representatives of the genera *Brachybacterium, Chryseobacterium, Clavibacter, Curtobacterium, Kocuria, Microbacterium, Micrococcus, Paenarthrobacter, Pseudarthrobacter*, and *Rhodococcus* (Abadi et al., 2021). It is important to note that mentioned phyla (and genera) are often found in ecotopes characterized by a deficit of available moisture and show activity wide diapasons of temperature, salinity, and pH (Belov et al., 2019, 2020 and references therein), which indirectly indicates their functional role in plant communities of semiarid and arid regions. This confirms the data on the dominance of populations of the phylum Actinomycetota, Pseudomonadota, Bacteroidota, and Bacillota in the rhizosphere of plants studied in the Atacama, Namib, Sonoran, Sahara, and other deserts (Ayangbenro and Babalola, 2021). The role of cyanobacteria in nitrogen fixation is the most in primary soil formation in arid regions, in which there is a constant nitrogen deficiency: in these ecotopes, one of the main sources of biophilic elements are biological soil crusts, large ones consisting primarily of cyanobacteria (Su et al., 2011; Gallas and Pavao-Zuckerman, 2022; Wang et al., 2022). The intensity of associative nitrogen fixation in different soils is about 100 kg/N/ha per year (Roley et al., 2018 and references therein). Root and nodule symbioses with diazotrophic prokaryotes are the most effective and could provide fixation of 20–300 kg/N/ha per year (Lindström and Mousavi, 2020).

The process of molecular nitrogen fixation itself, catalyzed by nitrogenase enzymes, is extremely energy-consuming and anaerobic: it takes 16 ATP molecules to fix one N_2 molecule by molybdenum-dependent nitrogenase (Seefeldt et al., 2009), alternative nitrogenases are characterized by slightly different biochemistry but still require a large number of energy molecules for their work (Bellenger et al., 2020). This high energy dependence of the process of diazotrophy explains the tendency toward an increase in the productivity of nitrogen fixation in prokaryotes in a row free-living—associative—symbiotic. Obviously, the trophic interaction between plants and microbial communities greatly affects the efficiency of nitrogen fixation by root communities, but this process is also influenced by the exchange of various metabolites and the influence of both plants and communities on the soil conditions of the root zone.

No less obvious is the agricultural significance of the introduction of diazotrophs for increasing productivity, for example, of cereal plants. However, the creation of symbiotic communities on the roots of cereal or vegetable plants (analogous to nodules of legumes) faces a number of difficulties, both in the primary infection of the root with symbiotic bacteria and in the implementation of diazotrophy after infection and other problems (Ryu et al., 2020; Rosenblueth et al., 2018; Zaidi et al., 2017; Soumare et al., 2020; Igiehon and Babalola, 2018).

Nitrogen nutrition of agricultural plants is carried out from several soil pools: inorganic nitrogen (which accounts for a percentage) and organic nitrogen (soil organic matter, as well as living and dead biomass of various organisms, mainly prokaryotes) (Soumare et al., 2020). Even without being diazotrophs, saprotrophic prokaryotes of basal communities contribute to nitrogen labilization, both due to the hydrolysis of nitrogen-containing compounds and due to the mobilization of nitrogen-containing substances into their own biomass, from which they are then released after the death of the prokaryotic cell and become available for plants. Similarly, root communities can beneficially influence the supply of other minerals (potassium, phosphorus, iron, etc.) to plants. However, the intensity of these processes will depend on factors affecting the productivity and metabolic activity of microbial communities, in particular, the provision of available water.

14.5 Microorganisms from water-deficient soils

The study of changes occurring in soils and associated systems under water deficit is extremely relevant today (Degré et al., 2017; Carter, 2015). Arid ecotopes are the most common terrestrial ecosystems, accounting for about 30% of the land surface, of which 7% are hyperarid (Bull, 2011). It should also be taken into account that the availability of water for microbial cells is also limited by low temperatures, that is, cold ecosystems during annual and/or seasonal cycles are also characterized by a shortage of water available to microorganisms (Steven et al., 2006). The availability of water for microorganisms is also affected by the concentration of soluble compounds in soil solutions, in particular, when applying fertilizers and in the root zone of plants.

The lower limit of water activity (Aw), which calculates as the ratio water vapor pressure upon distilled water to water vapor pressure upon solution, at which the preservation of the reproductive activity of microbial cells is reliably shown, is currently about Aw 0.605: under these conditions, xerophilic microscopic fungi of the species *Xeromyces bisporus* are able to grow (Stevenson et al., 2015). At the same time, the Aw values at which microorganisms are able to survive in the inactive state are significantly lower. For example, it was shown that mycobacteria isolated from arid soils are able to develop up to a water activity of 0.80, and a change in the fatty acid composition of membranes is observed, which indicates active physiological processes of adaptation (Santos et al., 2015). On nutrient media with reduced water activity, the production of physiologically active substances (in particular, fumonisins) was shown by fungal strains that do not produce these compounds under conditions of normal water availability (Mogensen et al., 2009), which indicates a change in physiological processes as Aw decreases, which is certainly important for protecting plants from fungal infections during a drought period or when grown in an arid climate.

Representatives of novel bacterial taxa (mainly representatives of the phylum Actinomycetota) have been isolated from arid ecotopes, many of which have polyextremotolerant properties (Bull, 2011; Montero-Calasanz et al., 2012; Luo et al., 2012; Harwani, 2013; Santhanam et al., 2013). Despite a large number of studies of the microbial diversity of arid ecotopes and works to determine the limiting values of Aw for the growth of microorganisms cultivated from soils and rocks (Stevenson and Hallsworth, 2014; Lebre et al., 2017), studies of the effect of water activity on the taxonomic composition and metabolic activity of soil microbial communities in situ are practically absent.

Existing in conditions of water deficiency, bacteria forced to adapt to maintain metabolic activity expend more energy to provide the cell with the necessary amount of water, while bacteria that are not adapted to metabolism under these conditions go into a state of hydrobiosis, in which the cells of most bacteria are metabolically inactive but viable. For today, the Aw value of 0.88 is considered to be the point of hydrobiosis (Connon et al., 2007).

In the performed studies on microbial communities of arid natural environments, a significant biotechnological potential of the isolated strains were shown. The importance of further studies of xerotolerant microorganisms, especially prokaryotes, and their taxonomic diversity and metabolic features, both from a fundamental and applied point of view, were also noted (Margesin and Schinner, 2001; Mohammadipanah and Wink, 2016). Cell drying at low moisture availability leads to oxidative stress (Fredrickson et al., 2008; Fagliarone et al., 2017), which determines the relationship between the mechanisms of resistance of microorganisms to a wide range of extreme impacts: to low temperatures and low moisture availability (Gunde-Cimerman et al., 2003), ionizing radiation, UV radiation, and the presence of high concentrations of water-soluble salts and oxidizing compounds (Makarova et al., 2001; Shukla et al., 2007), which affects both microbial and plant cells.

Summing up the above, it should be noted that the soil microbial communities inhabiting the root zone of higher plants play an important role in their nutrition, protection from the penetration of pathogenic bacteria, and sustainable development. The main mechanisms of interaction between plants and bacteria include the two-way exchange of nutrients (carbohydrates and/or available forms of mineral compounds) as well as environment-forming effects due to the mutual influence of microbial communities and plants on soil conditions. Due to the observed climatic and social trends, namely the increase in average annual temperatures and the growth of the world's population, one of the solutions to overcome the food and environmental crises (including degradation and pollution of arable lands) arising in the existing system of intensive agriculture is the biologization of agriculture and the introduction of bacteria that contribute to plant growth in the agricultural system. Microorganisms of the extremobiosphere deserve special attention in this matter, since they have the ability to maintain metabolic activity under various stress conditions and stimulate plant growth by various mechanisms.

14.6 Biologically active substances released by soil microorganisms

Soil microbial communities are critical for plant growth and development and quickly adapt to various factors of environment (Abdul Rahman et al., 2021). The use of synthetic materials in agriculture leads to adverse environmental impacts on a global scale. The use of microorganisms is an integral part of plant nutrient management and pest and disease control practices. According to most researchers, soil microbial communities have a key influence on plants growth and development through biochemical, physical, and metabolic interactions with the host plant (Trivedi et al., 2021; Turner et al., 2013; Williams, 2013). There are thousands of clusters of biosynthetic genes in various bacteria and filamentous fungi. Many of them are not expressed in pure cultures but appear in natural ecosystems when studying microbial communities. Coding-specialized metabolites are often produced at very low concentrations, but nevertheless, they serve as communication signals for microorganisms in soil microbial communities. Many specialized metabolites have been identified that act as communication signals, including autoinductors, intergeneric, and interregal signals. According to Phelan et al. (2012), microbial genomes contain from 17% to 42% of predicted open reading frames (ORF) associated with the synthesis of such compounds. Thus the rhizosphere can be characterized as an ecosystem with a huge variety of different biologically active substances, for example, cyclic lipopeptides, siderophores, quorum-sensitive molecules, and antibiotics (Keohane et al., 2015). By mechanism of action, PGPRs can stimulate plant growth in the following ways: production of ACC deaminase to reduce ethylene levels in the roots of developing plants; production of plant growth regulators such as indoleacetic acid (Mishra et al., 2010), gibberellic acid, cytokinins (Castro et al., 2008), and ethylene (Saleem et al., 2007); asymbiotic nitrogen fixation (Ardakani et al., 2010); manifestation of antagonistic activity against phytopathogenic microorganisms by producing siderophores, b-1,3-glucanase, chitinases, antibiotics, fluorescent pigment, and cyanide (Pathma et al., 2011); solubilization of mineral phosphates and other nutrients (Hayat et al., 2010).

The following metabolites for signaling in the soil microbiome and in the plant-microorganism system can be listed:

1. Phytohormones: Organic compounds that are highly active at very low concentrations, including and regulating the physiological processes of plants. Phytohormones can be divided into auxins, gibberellins, cytokinins, and ethylene. They are synthesized in many organs and easily move not only between the cells and organs of a plant but also from one plant to another (e.g., gaseous ethylene). Phytohormones are very diverse in chemical nature—these are derivatives of amino acids (auxins) and nitrogenous bases of nucleotides (cytokinins), terpenoids (gibberellins, abscisic acid), alkenes (ethylene), and small proteins. They also include jasmonic and salicylic acids, brassinosteroids, and strigolactones (Neumann et al., 2009; Wani et al., 2016).

Auxins are indole derivatives, and about 80% of rhizospheric bacteria can synthesize and release auxins as a secondary metabolite (Patten and Glick, 1996). One of the most common is indolyl-3-acetic acid (IAA), (Spaepen et al., 2007). Other compounds that have auxin activity can be considered as precursors of IAA—indolylpyruvic acid, indolyl lactic acid, indolilacetonitrile, indolilacetaldehyde, indolilacetamide, tryptamine, tryptopol, or as products of its further transformation. IAA has a comprehensive effect on plant growth and development, including an increase in the root surface area and the number of root tips in many plants (Kloepper et al., 2007, Grobelak et al., 2015), as well as plant adaptation to salt stress (Iqbal et al., 2014; Fahad et al., 2015).

Cytokinins are a group of phytohormones, derivatives of the nitrogenous base of purine, that stimulate cell division (cytokinesis). Many phytopathogenic fungi and bacteria produce cytokinins. For example, a genetically modified strain of *Sinorhizobium meliloti*, with overexpression of cytokinin, increases drought tolerance in alfalfa (Xu et al., 2012).

Gibberellins are a large group of tetracyclic diterpenoid carboxylic acids (Hedden and Thomas, 2012). Among fungi, the ability to synthesize gibberellins is quite widespread. These substances contribute to the stimulation of plant growth (Dodd et al., 2010; Egamberdieva, 2013) and their resistance to environmental chemical influences (Calvo et al., 2014). They can also positively affect plants through the formation of certain microbial communities that improve plant metabolism and reduce the effects of abiotic factors.

Ethylene is a gaseous phytohormone; at optimal concentration during plant development, it regulates root and xylem formation, induces fruit ripening and flower wilt. Under conditions of abiotic stress, treatment of plants with strains of rhizospheric bacteria of the genera *Achromobacter*, *Azospirillum*, *Bacillus*, *Enterobacter*, *Pseudomonas*, and *Rhizobium* promotes the production of the 1-Aminocyclopropane-1-carboxylate C-deaminase gene (Duan et al., 2009; Ghosh et al., 2003; Govindasamy et al., 2008).

2. Volatile organic compounds are organic chemicals that have a high vapor pressure at room temperature. These include alkenes, alcohols, benzoids, aldehydes, ketones, or terpenes (Audrain et al., 2015). These are organic chemicals that have a high vapor pressure at room temperature. Such substances can be primary and secondary products of microbial metabolism (Venturi and Keel, 2016; Schmidt et al., 2015). These substances promote interaction both within microbial communities and in cellular interactions between microorganisms, plants, animals, or fungi (Audrain et al., 2015).
3. To obtain iron, microorganisms synthesize and produce siderophores. These substances are determined by microbial cell surface receptors and enter the cell (Faraldo-Gómez and Sansom, 2003). Therefore these molecules directly influence competitive and cooperative microbial interactions. In the work of Johnstone and Nolan, some siderophores have been shown to have other functional properties. These molecules can be used as signaling molecules, as antibiotics, as regulators of oxidative stress, and to act as sequestrants for

various metals and even heavy metal toxins (Johnstone and Nolan, 2015). Siderophores are divided into several types depending on the presence of functional groups: (1) siderophores based on peptides, (2) siderophores based on di- and triaminoalkanes, (3) siderophores based on citric acid, and (4) mixed siderophores (Cornelis and Andrews, 2010).

4. Antibiotics are a chemically heterogeneous group of organic substances synthesized by microorganisms and have a negative effect on the growth or metabolism of other microorganisms (Thomashow et al., 1997). These substances act as signaling molecules on plants and ectomycorrhizal fungi, which helps to stimulate plant growth (Brazelton et al., 2008; Jousset and Bonkowski, 2010; Riedlinger et al., 2006). The ability of rhizospheric bacteria to synthesize a wide variety of antibiotics determines their biocontrol potential in agriculture. Fluorescent pseudomonads are capable of synthesizing antibiotics, such as phenosine, pyoluteorin, 2,4-diacetylphloroglucinol, or pyrrolythrin (Vacheron et al., 2016; Haas and Keel, 2003). Bacillaene, prodiginines, actinorhodins, the calcium-dependent antibiotic, coelimycin, methylenomycin, etc. are antibiotics synthesized by microorganisms of the genera Streptomyces or Bacillus (Traxler and Kolter 2015; Solanki et al., 2015). The production of antibiotics depends on the interaction of microorganisms (Rosier et al., 2016).

Lipopeptides are a lipid molecule covalently linked to a short linear or cyclic peptide. (Raaijmakers et al., 2010). According to the chemical structure, cyclic lipopeptides are divided into lactones, in which the N-terminal amino acid residue of the peptide chain is linked by an amide bond to a β-hydroxy fatty acid, and the carboxyl group of the C-terminal amino acid covalently closes the ring by interacting with the β-hydroxyl group, and lactams, in which N-terminal amino acid residue of the peptide chain is connected by an amide bond with a β-amino fatty acid, and the C-terminus of the peptide chain is closed to the β-amino group of the fatty acid residue. Due to this chemical structure, these molecules exhibit amphipathic and biosurfactant characteristics. Strains of the genera *Bacillus*, *Streptomyces*, and *Pseudomonas* are capable of synthesizing lipopeptides that inhibit the growth and development of related phytopathogenic species or strains or have a broader spectrum of biocidal activity (bactericidal, fungicidal, insecticidal, and antiviral) (Raaijmakers et al., 2010). Some of them are synthesized on ribosomes with the participation of special plasmids (bacteriocinogenic factors), others are nonribosomal, with the participation of special multienzyme complexes called nonribosomal peptide synthetases. The wide structural diversity of lipopeptides is determined by differences in the composition and length of fatty acids and the structure of peptide rings (Berry et al., 2010; Raaijmakers et al., 2006). Of particular interest are the lipopeptides of bacilli and pseudomonades. Due to structural features, these compounds are amphiphilic and resistant to hydrolysis by peptidases and proteases, as well as insensitive to oxidation and relatively high temperatures. At the same time, their cysteine residues can be oxidized to sulfides and/or change the structure to characteristic intramolecular C−S bonds. (Ongena and Jacques, 2008). They play a role in motility, biofilm formation, and biocontrol activity (Debois et al., 2008; De Bruijn et al., 2007, 2008; Kuiper et al., 2004).

The family of phenazine molecules includes metabolites that share the phenazine core. Phenazine antibiotics are a group of low molecular weight heterocyclic nitrogen-containing compounds that differ in their physical and chemical properties depending on the location and type of functional groups (Mavrodi et al., 2006). A distinctive feature of these compounds is a wide spectrum of antifungal action. They prevent the development of not only phytopathogenic fungi such as *Monilinia fructigena, Rhizoctonia solani, Alternaria solani, Septoria tritici, Fusarium oxysporum, Pythium myriotylum, Candida albicans*, and others, but also a number of phytopathogenic bacteria—*Acidovorax avenae, Erwinia carotovora, Pseudomonas syringae*, and *Xanthomonas campestris* (Price-Whelan et al., 2006). The synthesis of phenazines was found in representatives of the genera *Pseudomonas, Burkholderia, Brevibacterium, Bacillus, Streptomyces, Methanosarcina*, and *Sorangium* (Zhang et al., 2015). Phenazines belong to a large class of p-diazine compounds. Synthesis of phenazine derivatives was found only in bacteria during aromatic pathway reactions. Most often, one type of bacteria produces several types of phenazine antibiotics. For example, *Pseudomonas chlororaphis* bacteria produce two types of these antibiotics, *Pseudomonas aeruginosa* produces six types, and *Pseudomonas fluorescens* bacteria produce only one type of phenazines.

Quorum sensing (QS) is one of the most common interaction systems in microbial communities found in gram-negative, gram-positive bacteria and archaea. QS is a special type of regulation of bacterial gene expression that depends on the density of their population. QS systems include low molecular weight signaling molecules (autoinductors) that easily diffuse from cells into the environment and back, and receptor regulatory proteins with which signaling molecules interact. Autoinductors include N-acyl-homoserine lactones, small peptides, furan derivatives, quinolones, butyrolactones, or unsaturated fatty acids. As the bacterial population increases and reaches a critical level, autoinducers accumulate to the required threshold and bind to the appropriate receptor proteins, which leads to a sharp activation (sometimes repression) of the transcription of certain sets of genes (LaSarre and Federle, 2013). With the help of signaling molecules of QS systems, intercellular communication of bacteria in populations occurs, providing a coordinated response of bacteria to changes in environmental conditions, such as luminescence, virulence, production of extracellular enzymes and antibiotics, or biofilm formation (Elias and Banin, 2012).

Thus microbial interactions are critical for the successful creation and maintenance of a microbial population, including aspects of interaction with plants. These interactions occur through environmental recognition, followed by the transfer of molecular and genetic information involving many mechanisms and classes of molecules.

14.7 Microbial growth regulators in sustainable agriculture and biological pesticides (biocontrol agent) based on microorganisms isolated from the soil

Soil microbial communities also play an important role in protecting plants from diseases and abiotic stresses (Frey-Klett et al., 2011) or increasing nutrient intake. Similar technologies are being used to create artificial communities for agricultural purposes. In addition to efforts to create a synthetic community for a particular habitat, there have been advances in designing microbes to participate in chemical signaling and thus to modulate microbial communities. For example, in early experiments, rhizosphere-associated *Rhizobium* species that exhibit plant growth-promoting properties were engineered to use the fungal siderophore to absorb iron, resulting in an increase in *Rhizobium* population density in the rhizosphere (Rajendran et al., 2007). Similarly, *Pseudomonas* strains have been engineered to produce the antifungal pyocyanin and then used to protect tomato plants from the fungal pathogen *F. oxysporum* (Chin-A-Woeng et al., 2001). Shoba et al. studied the effect of presowing treatment of spring wheat seeds with suspensions of various strains of actinomycetes. The authors demonstrated that actinomycetes can stimulate seed germination by consuming soil allelotoxins (Shoba et al., 2021). For example, Vurukonda et al. presented a review on plant growth–promoting rhizobacteria that may play an important role in mitigating the effects of drought in plants. Microorganisms are able to colonize the rhizosphere/endo-rhizosphere of plants and impart drought resistance by producing volatile compounds, phytohormones, siderophores, exopolysaccharides, 1-aminocyclopropane-1-carboxylate deaminase (ACC), causing the accumulation of osmolytes, antioxidants, increasing or suppressing gene regulation, responding to stress, and changes in root morphology upon acquiring drought resistance (Vurukonda et al., 2016).

Each of the organs of higher plants is a special ecological niche inhabited by microorganisms. To designate these niches, the terms phyllosphere (aerial parts of plants), phylloplana (surface of aboveground vegetative organs of plants), rhizosphere (a narrow zone of soil immediately surrounding plant roots), rhizoplane (surface of plant roots), and spermosphere (seeds) are accepted. Niches are caused by intravital secretions of plants that are used by microorganisms as nutrient substrates (e.g., stems and leaves secrete carbohydrates and organic acids.) At the same time, microorganisms settle on the surface of various plant organs, forming an epiphytic community, and penetrate into plant tissues, forming an endophytic community. Thus in the process of growth and development, plants act as centers for the formation of microbial communities. The active release of various organic compounds by plant roots into the environment provides soil microorganisms with nutrients, which creates favorable conditions for their existence in the zones of the rhizosphere and rhizoplane. Bacteria inhabiting the rhizosphere and rhizoplane form strong associations with the root system of plants and form specific rhizosphere bacterial communities. Such relationships are characterized by the terms "associative bacteria," "associative relationships," and "associative symbiosis." Associative plant-microbial relationships served as the starting point for the development of the concept of associative symbiosis, according to which the host plant (macrosymbiont) and its associated microorganisms (microsymbionts) form a multicomponent integral system with new properties determined by the interaction of partners.

According to the principle of action, the following groups of biological preparations for plant protection are subdivided:

1. Drugs with a cidal action that cause the death of an infectious agent—weed plants and pathogens of animal diseases.
2. Preparations of antagonist microorganisms that limit the spread of pests and diseases. For example, bacteria of the genus *Pseudomonas*, assimilating iron ions, convert them into siderophores that are not available to other microorganisms.
3. Preparations of antibiotics, toxicants, and antifeedants—metabolites of fungi and bacteria that inhibit the vital activity of other microorganisms. They have a neurotoxic or deterrent effect.

According to the current principle, microbial biological products are divided into viral, bacterial, fungal, actinomycete, as well as antibiotics, antifeedants, and toxicants:

1. Virus preparations for plant protection are produced mainly on the basis of baculoviruses, namely nuclear polyhedrosis viruses and granulosa viruses. In Western Europe, vaccination of plants with weakly pathogenic strains of viruses (preinoculation) has recently been widely used in order to develop induced (evoked) immunity.
2. Bacterial preparations are preparations containing live bacterial cells. Among the bacterial preparations for plant protection, the most common are preparations based on spore bacteria. Currently, two highly virulent bacterial cultures used for the preparation of effective biological products are allowed for use: bacteria from the genera *Pseudomonas* and *Bacillus*. Such drugs are widely used in the fight against fungal diseases, bacterioses and phytophages, insects, rodents.
3. Bacterial fungicides based on the use of certain types of fungal antagonist bacteria. For example, bacteria from the genus *Pseudomonas* are active in suppressing the development of root rot pathogens and plant wilt. The bacterium *P. fluorescens* produces the antibiotic pyrrolnitrin, which is active against the pathogen of cotton seedlings. Other antagonistic bacteria are effective against *Fusarium* rot of corn, rhizoctoniosis of wheat, oats and barley, root rot of carrots, and other diseases.

   Currently, about 70 types of biological preparations for plant protection are produced worldwide. Of these, almost 90% are based on the spore-forming bacterium *Bacillus thuringiensis*, which forms protein crystals with high insecticidal activity. Also, the bacterium strain *Bacillus subtilis* Ch-13 is a curative and immunizing fungicide, forming substances that stimulate plant growth and inhibit the development of phytopathogens.

4. Fungal preparations with insecticidal and fungicidal action based on microorganisms with the ability to parasitism and hyperparasitism. Currently, about a thousand species of fungi have been described that have antibiotic and pathogenic properties in relation to various pests and pathogens. The most common preparations is based on the fungus *Streptomyces avermitilis*. In such preparations, the active substance is not the fungus itself, but its metabolic products—avermectins (bioactive substances of the neurotoxin type). Also a widely used insect fungus is *Metarhizium anisopliae* P-72. This mushroom has the ability to control many types of insects from various orders. Avermectins, having a contact and systemic action, have strongly pronounced acaricidal properties, causing the death of many openly living sucking pests. At the same time, the mechanism of action of the drug on various pests is not the same, but the common effect is on the transmission of nerve impulses, which leads to paralysis and further death of the parasites.
5. Actinomycetes are used in preparations due to their antibiotic activity and ability to parasitize phytopathogens. For example, some actinomycetes of the genus *Actinomyces* are parasites of ticks, causing actinomycosis and, as a result, their rapid death. Representatives of the genus Streptomyces are powerful producers of antibiotics of the streptomycin series. These substances have a strong fungicidal effect and cause inhibition of growth and death of many phytopathogenic fungi.
6. Preparations of antibiotics, antifeedants, and toxicants have recently attracted the greatest attention of Russian manufacturers. The best known in this group are Agravertin and Fitoverm.

Endophytic microbial communities, composed of bacteria and fungi, are responsible for the ecological functioning of plants. Representatives of these communities are responsible for the synthesis of a wide variety of secondary metabolites responsible for plant growth and development. The interaction between members of endophytic communities is also primarily due to the relationship in nutrition, as well as self-regulation in plant-microorganism relationships (Hassani et al., 2018). The formation of secondary compounds with important functional ecological potential for plants is due to a set of closely related species that have complex interactions, such as a plant species and members of its microbiome (Netzker et al., 2015).

Vinale et al. (2017) and Wang et al. (2019) in their studies touched upon the issues of functional interaction within endophytic communities. These studies showed that the synthesis of secondary metabolites is significantly affected by the cultivation conditions of endophyte microorganisms of the Brazilian plant Lychnophora ericoides. Cocultivation of plant-beneficial fungi may represent an effective strategy to modulate the production of bioactive metabolites and possibly to identify new compounds. Chemical structures of secondary metabolites were determined by LC-Ms (liquid chromatography–mass spectrometry) and NMR (nuclear magnetic resonance) methods, and new substances such as harzianic and harziaphilic acids were identified (Fischer et al., 2019). At the same time, the ecological functional potential has not yet been determined for new compounds (Calcott et al., 2018).

Many microbial formulations containing individual strains or mixtures thereof have been developed and are currently in use. Such microbial consortia promote nitrogen fixation and phosphate dissolution and also synthesize siderophores, phytohormones, and exopolysaccharides. It should be noted that representatives of endophytic communities also take part in the processes of synthesis of lytic enzymes against pathogens and pests (Gupta et al., 2002; Sruthilaxmi and Babu, 2017).

Biologicals are used either as biofertilizers to stimulate plant growth or as bioagents for plant protection. Some microorganisms of the species *Azospirillum, Azotobacter, Rhizobium*, and *Acetobacter* are biofertilizers. Suman's work has shown that *Gluconacetobacter diazotrophicus* stimulates the growth of sugar cane, and strains of the genus *Pantoea* have been found to have multiactive properties in terms of increasing the yield and productivity of several plant crops (Suman et al., 2021, 2008). Some microorganisms inhabiting the plant endosphere make a positive contribution to agriculture (White et al., 2019).

The enhancing effect of the interaction of two or more members of microbial communities contributes to an increase in plant productivity, although biological preparations based on individual strains of microorganisms are most often used in agriculture (Bashan et al., 2014). Recently, complex bioproducts are often used to rejuvenate the rhizosphere, including combinations of microbes, plant extracts, and algae.

The strains of *Comamonas testosteroni, Pseudomonas putida, Enterobacter cloacae*, and *Citrobacter freundii*, which are members of the Mammoth P microbial association, increase the productivity of agricultural plants by a factor of 2 and solubilize phosphates. A more than twofold increase in the yield of field corn was revealed due to the use of the QuickRoots biopreparation based on the consortium of the filamentous fungus *Trichoderma virens* and *Bacillus amyloliquefaciens* (Monsanto BioAgAlliance, 2015).

For a number of microbial associations, stimulation of plant nutrition has been demonstrated (Shukla et al., 2008; Suman et al., 2008; Dal Cortivo et al., 2018). Interaction of strains *P. putida* KT2440, *Sphingomonas* sp. OF178, *Azospirillum brasilense* Sp7, and *Acinetobacter* sp. OF178 improves maize drought tolerance (Molina-Romero et al., 2017). Stimulation of the growth of tomatoes and a decrease in *Fusarium* were revealed when using biological products based on bacteria of the genus *Bacillus* (Tsolakidou et al., 2019; Menéndez and Paço, 2020) demonstrated that the use of microbial consortia has a more beneficial effect on plant growth and development compared to single-strain biologics. Microorganism strains *Trichoderma* and *Azotobacter* reduce stress conditions for plants, which improves their health (Woo and Pepe, 2018). Arbuscular MF promote the transfer of endophyte microorganisms in plants, the resistance of the latter to pathogens, and the preservation of moisture (Rouphael et al., 2015; Cameron et al., 2013). Through a genomic approach using multiplex amplicon sequencing of a community-based culture collection, Xu et al. (2016) identified the four most representative genera, *Bacillus*, *Chitinophaga*, *Rhizobium*, and *Burkholderia*, for bioinoculant development.

MF can also cooperate with bacteria on a reciprocal basis, especially bacteria that are able to dissolve phosphorus. The carbon released by the MF may allow phosphate solubilizing bacteria (PSB) to grow in their immediate vicinity, resulting in an increase in the availability of phosphorus to the MF as a result of an increase in the PSB population in the hyposphere. When the availability of phosphorus in the environment is not limited, PSB can significantly enhance the growth of MF hyphae and thus provide favorable reciprocity between microbial partners (Zhang and Wang, 2016).

Several microbial consortia have been field tested and some are commercially available. Most of these consortiums are mainly based on the addition of features beneficial to microbial plants. However, most field studies do not analyze interactions between plants and microbial communities and do not take into account plant genotypes and characteristics of soil agrosystems, which also have a significant impact on the interaction of individual microbial representatives within the community (Vacheron et al., 2013). An increase in plant growth can be due to both direct mechanisms (nitrogen fixation, phosphorus solubilization, synthesis of phytohormones, etc.) and indirect mechanisms of PGPRs influence (stimulation of mycorrhiza development, suppression of pathogens, or elimination of toxic compounds) (Bashan et al., 2008).

Therefore PGPRs that stimulate plant growth can be grouped according to their level of association with plant root cells, both extracellular (ePGPR) and intracellular rhizobacteria (iPGPR) (Martínez-Viveros et al., 2010). ePGPRs are present in the rhizoplane, rhizosphere, or localized between plant root cells; and iPGPRs usually occur within the specialized nodal structures of root cells. ePGPR includes microorganisms of the genera *Arthrobacter, Azotobacter, Agrobacterium, Azospirillum, Bacillus, Caulobacter, Chromobacterium, Erwinia, Burkholderia, Flavobacterium, Micrococcus, Pseudomonas*, and *Serratia* (Gray and Smith, 2005). iPGPR includes nitrogen-fixing bacteria and endophytes (Verma et al., 2010). Endophytes include bacteria that are able to stimulate plant growth by penetrating into root systems and forming nodules (*Allorhizobium, Azorhizobium, Bradyrhizobium, Mesorhizobium*, and *Rhizobium*) (Wang and Martinez-Romero, 2000). A number of micromycete strains (*Micromonospora, Streptomyces, Streptosporangium*, and *Thermobifida*) are used to colonize the plant rhizosphere by inhibiting the growth and development of pathogenic fungi (Franco-Correa et al., 2010).

One of the means of biological control of the development of *Fusarium* and armillary rot of pine *Pinus taeda* are representatives of the genus *Streptomyces* (de Vasconcellos and Cardoso, 2009). In the 2000s, a number of researchers demonstrated the high potential of a number of actinobacteria and *Pseudomonas solanacearum* in promoting the growth of tomatoes (Sabaratnam and Traquair, 2002), and *Colletotrichum musae* in bananas (Taechowisan et al., 2003). According to Terkina et al. (2006), actinomycetes are capable of synthesizing various antimicrobial substances. Vasconcellos et al. (2010) obtained active metabolites from the actinobacteria of the needles *Araucaria angustifolia*.

It is known that indoleacetic acid and chitinase are metabolites that contribute to the destruction of complex persistent soil organic substances. The work of Cao et al. demonstrated antagonistic activity of *Streptomyces griseorubiginosus* against *Fusarium oxysporum f. sp. cubense* (Cao et al., 2004). The role of PGPRs in relation to the resistance of tomatoes and peppers to water stress has been revealed (Mayak et al., 2004). It should be noted that bacteria of the genera *Pseudomonas* and *Bacillus* are able to colonize plant roots and synthesize growth-stimulating metabolites, which increases the yield of agricultural plants (Khalid et al., 2004). Research developments from China, Russia, and several other Western countries have now proven the potential use of PGPRs for plant disease control. The first commercial product of *B. subtilis* was developed in 1985 in the United States. A total of 60%–75% of US cotton, peanut, soybean, corn, vegetable, and small grain crops are now being treated with commercial *B. subtilis*, which is becoming effective against soilborne pathogens such as *Fusarium* and *Rhizoctonia* (Nakkeeran et al., 2005).

Thus rhizobacteria can act as phytostimulants, biofertilizers, destructors of toxicants, as well as antipathogenic and immunomodulatory factors for plants. The possibility of large-scale application of beneficial rhizobacteria through seed inoculation is an alternative to chemical fertilizers and pesticides, which often pollute the environment. The main areas of application of bacteria that stimulate plant growth include agriculture, horticulture, afforestation, and phytoremediation (technology for the restoration of contaminated objects using plants). Due to the high practical significance of this group of microorganisms, the range of known representatives and information about the mechanisms of positive influence is constantly expanding.

14.8 Funding

This chapter was funded by grants of the President of the Russian Federation to the research project «Synthesis of Auxin in Soils of Different Types» (MK-3790.2022.5) and "Stability and metabolic activity of soil prokaryotic communities under conditions of moisture deficit" (MK-664.2021.1.4).

References

Ahemad, M., 2019. Remediation of metalliferous soils through the heavy metal resistant plant growth promoting bacteria: paradigms and prospects. Arab. J. Chem. 12 (7), 1365–1377. Available from: https://doi.org/10.1016/j.arabjc.2014.11.020.

Ayangbenro, A.S., Babalola, O.O., 2021. Reclamation of arid and semi-arid soils: the role of plant growth-promoting archaea and bacteria. Curr. Plant. Biol. 25, 100173. Available from: https://doi.org/10.1016/j.cpb.2020.100173.

Abdul Rahman, N.S.N., Abdul Hamid, N.W., Nadarajah, K., 2021. Effects of abiotic stress on soil microbiome. Int. J. Mol. Sci. 22 (16), 9036. Available from: https://doi.org/10.3390/ijms22169036.

Ardakani, S.S., Heydari, A., Tayebi, L., Mohammadi, M., 2010. Promotion of cotton seedlings growth characteristics by development and use of new bioformulations. Int. J. Botany 6 (2), 95–100. Available from: https://doi.org/10.3923/ijb.2010.95.100.

Audrain, B., Farag, M.A., Ryu, C.M., Ghigo, J.M., 2015. Role of bacterial volatile compounds in bacterial biology. FEMS Microbiol. Rev. 39 (2), 222–233. Available from: https://doi.org/10.1093/femsre/fuu013.

Ali, M.A., Naveed, M., Mustafa, A., Abbas, A., 2017. The good, the bad, and the ugly of rhizosphere microbiome. In: Kumar, V., Kumar, M., Sharma, S., Prasad, R. (Eds.), Probiotics and Plant Health. Springer, Singapore. Available from: http://doi.org/10.1007/978-981-10-3473-2_11.

Alaylar, B., Egamberdieva, D., Gulluce, M., Karadayi, M., Arora, N.K., 2020. Integration of molecular tools in microbial phosphate solubilization research in agriculture perspective. World J. Microbiol. Biotechnol. 36 (7), 1–12. Available from: https://doi.org/10.1007/s11274-020-02870-x.

Abadi, V.A.J.M., Sepehri, M., Rahmani, H.A., Dolatabad, H.K., Shamshiripour, M., Khatabi, B., 2021. Diversity and abundance of culturable nitrogen-fixing bacteria in the phyllosphere of maize. J. Appl. Microbiol. 131 (2), 898–912. Available from: https://doi.org/10.1111/jam.14975.

ALKahtani, M.D.F., Fouda, A., Attia, K.A., Al-Otaibi, F., Eid, A.M., Ewais, E.E.-D., et al., 2020. Isolation and characterization of plant growth promoting endophytic bacteria from desert plants and their application as bioinoculants for sustainable agriculture. Agronomy 10, 1325. Available from: https://doi.org/10.3390/agronomy10091325.

Bull, A.T., 2011. Actinobacteria of the extremobiosphere. In: Horikoshi, K. (Ed.), Extremophiles Handbook. Springer, Tokyo. Available from: http://doi.org/10.1007/978-4-431-53898-1_58.

Berg, G., Eberl, L., Hartmann, A., 2005. The rhizosphere as a reservoir for opportunistic human pathogenic bacteria. Environ. Microbiol. 7 (11), 1673–1685. Available from: https://doi.org/10.1111/j.1462-2920.2005.00891.x.

Belay, A., Claassens, A., Wehner, F.C., 2002. Effect of direct nitrogen and potassium and residual phosphorus fertilizers on soil chemical properties, microbial components and maize yield under long-term crop rotation. Biol. Fertil. Soils 35 (6), 420–427. Available from: https://doi.org/10.1007/s00374-002-0489-x.

Barea, J.M., Pozo, M.J., Azcon, R., Azcon-Aguilar, C., 2005. Microbial co-operation in the rhizosphere. J. Exp. Botany 56 (417), 1761–1778. Available from: https://doi.org/10.1093/jxb/eri197.

Bashan, Y., de-Bashan, L.E., Prabhu, S.R., Hernandez, J.P., 2014. Advances in plant growth-promoting bacterial inoculant technology: formulations and practical perspectives (1998–2013. Plant Soil. 378 (1), 1–33. Available from: https://doi.org/10.1007/s11104-013-1956-x.

Bashan, Y., Puente, M.E., de-Bashan, L.E., Hernandez, J.P., 2008. Environmental uses of plant growth-promoting bacteria. Plant-Microbe Interact. 661 (2), 69–93.

Bashan, Y., Salazar, B.G., Moreno, M., Lopez, B.R., Linderman, R.G., 2012. Restoration of eroded soil in the Sonoran Desert with native leguminous trees using plant growth-promoting microorganisms and limited amounts of compost and water. J. Environ. Manag. 102, 26–36. Available from: https://doi.org/10.1016/j.jenvman.2011.12.032.

Batista, B.D., Lacava, P.T., Ferrari, A., Teixeira-Silva, N.S., Bonatelli, M.L., Tsui, S., et al., 2018. Screening of tropically derived, multi-trait plant growth-promoting rhizobacteria and evaluation of corn and soybean colonization ability. Microbiol. Res. 206, 33–42. Available from: https://doi.org/10.1016/j.micres.2017.09.007.

Bellenger, J.P., Darnajoux, R., Zhang, X., Kraepiel, A.M.L., 2020. Biological nitrogen fixation by alternative nitrogenases in terrestrial ecosystems: a review. Biogeochemistry 149 (1), 53–73. Available from: https://doi.org/10.1007/s10533-020-00666-7.

Belov, A.A., Cheptsov, V.S., Manucharova, N.A., Ezhelev, Z.S., 2020. Bacterial communities of Novaya Zemlya archipelago ice and permafrost. Geosciences 10 (2), 67. Available from: https://doi.org/10.3390/geosciences10020067.

Belov, A.A., Cheptsov, V.S., Vorobyova, E.A., Manucharova, N.A., Ezhelev, Z.S., 2019. Stress-tolerance and taxonomy of culturable bacterial communities isolated from a central Mojave Desert soil sample. Geosciences 9 (4), 166. Available from: https://doi.org/10.3390/geosciences9040166.

Berry, C., Fernando, W.D., Loewen, P.C., De Kievit, T.R., 2010. Lipopeptides are essential for Pseudomonas sp. DF41 biocontrol of *Sclerotinia sclerotiorum*. Biol. Control. 55 (3), 211–218. Available from: https://doi.org/10.1016/j.biocontrol.2010.09.011.

Bhatt, P., Verma, A., Verma, S., Anwar, M., Prasher, P., Mudila, H., et al., 2020. Understanding phytomicrobiome: a potential reservoir for better crop management. Sustainability 12 (13), 5446. Available from: https://doi.org/10.3390/su12135446.

Brazelton, J.N., Pfeufer, E.E., Sweat, T.A., Gardener, B.B.M., Coenen, C., 2008. 2, 4-Diacetylphloroglucinol alters plant root development. Mol. Plant-Microbe Interact. 21 (10), 1349–1358. Available from: https://doi.org/10.1094/MPMI-21-10-1349.

Carter, B.P., 2015. Investigations into practical applications for the critical water activity from dynamic dewpoint isotherms. PhD Dissertation .

Chernov, T.I., Semenov, M.V., 2021. Management of soil microbial communities: opportunities and prospects (a review). Eurasian Soil. Sci. 54 (12), 1888–1902. Available from: https://doi.org/10.1134/S1064229321120024.

Calvo, P., Nelson, L., Kloepper, J.W., 2014. Agricultural uses of plant biostimulants. Plant. soil. 383 (1), 3–41. Available from: https://doi.org/10.1007/s11104-014-2131-8.

Cameron, D.D., Neal, A.L., van Wees, S.C., Ton, J., 2013. Mycorrhiza-induced resistance: more than the sum of its parts? Trends Plant. Sci. 18 (10), 539–545. Available from: https://doi.org/10.1016/j.tplants.2013.06.004.

Cao, L., Qiu, Z., Dai, X., Tan, H., Lin, Y., Zhou, S., 2004. Isolation of endophytic actinomycetes from roots and leaves of banana (*Musa acuminata*) plants and their activities against *Fusarium oxysporum f. sp. cubense*. World J. Microbiol. Biotechnol. 20 (5), 501–504. Available from: https://doi.org/10.1023/B:WIBI.0000040406.30495.48.

Castro, R.O., Cantero, E.V., Bucio, J.L., 2008. Plant growth promotion by *Bacillus megaterium* involves cytokinin signalling. Plant. Signal. Behav. 3 (4), 263–265. Available from: https://doi.org/10.4161/psb.3.4.5204.

Caton, I.R., Caton, T.M., Schneegurt, M.A., 2018. Nitrogen-fixation activity and the abundance and taxonomy of nifH genes in agricultural, pristine, and urban prairie stream sediments chronically exposed to different levels of nitrogen loading. Arch. Microbiol. 200 (4), 623–633. Available from: https://doi.org/10.1007/s00203-018-1475-5.

Calcott, M.J., Ackerley, D.F., Knight, A., Keyzers, R.A., Owen, J.G., 2018. Secondary metabolism in the lichen symbiosis. Chem. Soc. Rev. 47 (5), 1730–1760. Available from: https://doi.org/10.1039/C7CS00431A.

Chen, Q.L., Cui, H.L., Su, J.Q., Penuelas, J., Zhu, Y.G., 2019. Antibiotic resistomes in plant microbiomes. Trends Plant. Sci. 24 (6), 530–541. Available from: https://doi.org/10.1016/j.tplants.2019.02.010.

Cherif-Silini, H., Silini, A., Yahiaoui, B., Ouzari, I., Boudabous, A., 2016. Phylogenetic and plant-growth-promoting characteristics of Bacillus isolated from the wheat rhizosphere. Ann. Microbiol. 66 (3), 1087–1097. Available from: https://doi.org/10.1007/s13213-016-1194-6.

Chin-A-Woeng, T.F., Thomas-Oates, J.E., Lugtenberg, B.J., Bloemberg, G.V., 2001. Introduction of the phzH gene of *Pseudomonas chlororaphis* PCL1391 extends the range of biocontrol ability of phenazine-1-carboxylic acid-producing Pseudomonas spp. strains. Mol. Plant-Microbe Interact. 14 (8), 1006–1015. Available from: https://doi.org/10.1094/MPMI.2001.14.8.1006.

Connon, S.A., Lester, E.D., Shafaat, H.S., Obenhuber, D.C., Ponce, A., 2007. Bacterial diversity in hyperarid Atacama Desert soils. J. Geophys. Res.: Biogeosci. 112 (G4). Available from: https://doi.org/10.1029/2006JG000311.

Cornelis, P., Andrews, S.C., 2010. Iron Uptake and Homeostasis in Microorganisms. Caister Academic Press, Norfolk, UK, pp. 1–85, ISBN 9781904455653.

Dal Cortivo, C., Barion, G., Ferrari, M., Visioli, G., Dramis, L., Panozzo, A., et al., 2018. Effects of field inoculation with VAM and bacteria consortia on root growth and nutrients uptake in common wheat. Sustainability 10 (9), 3286. Available from: https://doi.org/10.3390/su10093286.

De Bruijn, I., de Kock, M.J., de Waard, P., van Beek, T.A., Raaijmakers, J.M., 2008. Massetolide A biosynthesis in *Pseudomonas fluorescens*. J Bacteriol. Available from: https://doi.org/10.1128/JB.01563-07.

De Bruijn, I., De Kock, M.J., Yang, M., De Waard, P., Van Beek, T.A., Raaijmakers, J.M., 2007. Genome-based discovery, structure prediction and functional analysis of cyclic lipopeptide antibiotics in Pseudomonas species. Mol. Microbiol. 63 (2), 417–428. Available from: https://doi.org/10.1111/j.1365-2958.2006.05525.x.

de Vasconcellos, R.L.F., Cardoso, E.J.B.N., 2009. Rhizospheric streptomycetes as potential biocontrol agents of Fusarium and Armillaria pine rot and as PGPR for *Pinus taeda*. Biocontrol 54 (6), 807–816. Available from: https://doi.org/10.1007/s10526-009-9226-9.

Debois, D., Hamze, K., Guérineau, V., Le Caër, J.P., Holland, I.B., Lopes, P., et al., 2008. In situ localisation and quantification of surfactins in a *Bacillus subtilis* swarming community by imaging mass spectrometry. Proteomics 8 (18), 3682–3691. Available from: https://doi.org/10.1002/pmic.200701025.

Degré, A., van der Ploeg, M.J., Caldwell, T., Gooren, H.P., 2017. Comparison of soil water potential sensors: a drying experiment. Vadose Zone J. 16 (4), 1–8. Available from: https://doi.org/10.2136/vzj2016.08.0067.

del Carmen Orozco-Mosqueda, M., Glick, B.R., Santoyo, G., 2020. ACC deaminase in plant growth-promoting bacteria (PGPB): an efficient mechanism to counter salt stress in crops. Microbiol. Res. 235, 126439. Available from: https://doi.org/10.1016/j.micres.2020.126439.

Delshadi, S., Ebrahimi, M., Shirmohammadi, E., 2017. Influence of plant-growth-promoting bacteria on germination, growth and nutrients' uptake of *Onobrychis sativa* L. under drought stress. J. Plant. Interact. 12 (1), 200–208. Available from: https://doi.org/10.1080/17429145.2017.1316527.

Dodd, I.C., Zinovkina, N.Y., Safronova, V.I., Belimov, A.A., 2010. Rhizobacterial mediation of plant hormone status. Ann. Appl. Biol. 157 (3), 361–379. Available from: https://doi.org/10.1111/j.1744-7348.2010.00439.x.

Duan, J., Müller, K.M., Charles, T.C., Vesely, S., Glick, B.R., 2009. 1-aminocyclopropane-1-carboxylate (ACC) deaminase genes in rhizobia from southern Saskatchewan. Microb. Ecol. 57 (3), 423–436. Available from: https://doi.org/10.1007/s00248-008-9407-6.

Egamberdieva, D., 2013. The role of phytohormone producing bacteria in alleviating salt stress in crop plants. Biotechnological Techniques of Stress Tolerance in Plants. Stadium Press, Houston, TX, pp. 21–39.

Egamberdiyeva, D., 2007. The effect of plant growth promoting bacteria on growth and nutrient uptake of maize in two different soils. Appl. Soil. Ecol. 36 (2-3), 184–189. Available from: https://doi.org/10.1016/j.apsoil.2007.02.005.

Elias, S., Banin, E., 2012. Multi-species biofilms: living with friendly neighbors. FEMS Microbiol. Rev. 36 (5), 990–1004. Available from: https://doi.org/10.1111/j.1574-6976.2012.00325.x.

Etesami, H., Maheshwari, D.K., 2018. Use of plant growth promoting rhizobacteria (PGPRs) with multiple plant growth promoting traits in stress agriculture: action mechanisms and future prospects. Ecotoxicol. Environ. Saf. 156, 225–246. Available from: https://doi.org/10.1016/j.ecoenv.2018.03.013.

Eida, A.A., Ziegler, M., Lafi, F.F., Michell, C.T., Voolstra, C.R., Hirt, H., et al., 2018. Desert plant bacteria reveal host influence and beneficial plant growth properties. PLoS One 13 (12), e0208223. Available from: https://doi.org/10.1371/journal.pone.0208223.

FAO, 2015. World fertilizer trends and outlook to 2018. Food and Agriculture Organization of the United Nations—Rome 2015. Available from: http://www.fao.org/publication.

Feng, T., Wang, J., 2020. Oxidative stress tolerance and antioxidant capacity of lactic acid bacteria as probiotic: a systematic review. Gut Microbes 12 (1), 1801944. Available from: https://doi.org/10.1080/19490976.2020.1801944.

Faraldo-Gómez, J.D., Sansom, M.S., 2003. Acquisition of siderophores in gram-negative bacteria. Nat. Rev. Mol. Cell Biol. 4 (2), 105–116. Available from: https://doi.org/10.1038/nrm1015.

Fitzsimons, M.S., Miller, R.M., 2010. The importance of soil microorganisms for maintaining diverse plant communities in tallgrass prairie. Am. J. Botany 97 (12), 1937–1943. Available from: https://doi.org/10.3732/ajb.0900237.

Fagliarone, C., Mosca, C., Ubaldi, I., Verseux, C., Baqué, M., Wilmotte, A., et al., 2017. Avoidance of protein oxidation correlates with the desiccation and radiation resistance of hot and cold desert strains of the cyanobacterium Chroococcidiopsis. Extremophiles 21 (6), 981–991. Available from: https://doi.org/10.1007/s00792-017-0957-8.

Fahad, S., Hussain, S., Bano, A., Saud, S., Hassan, S., Shan, D., et al., 2015. Potential role of phytohormones and plant growth-promoting rhizobacteria in abiotic stresses: consequences for changing environment. Environ. Sci. Pollut. Res. 22 (7), 4907–4921. Available from: https://doi.org/10.1007/s11356-014-3754-2.

Farooqi, Z.U., Ahmad, Z., Ayub, M.A., Umar, W., Nadeem, M., Fatima, H., et al., 2021. Threats to arable land of the world: current and future perspectives of land use. In: Hasnat, G., Hossain, M. (Eds.), Examining International Land Use Policies, Changes, and Conflicts. IGI Global, pp. 186–209. Available from: http://doi.org/10.4018/978-1-7998-4372-6.ch010.

Filippidou, S., Wunderlin, T., Junier, T., Jeanneret, N., Dorador, C., Molina, V., et al., 2016. A combination of extreme environmental conditions favor the prevalence of endospore-forming firmicutes. Front. Microbiol. 7, 1707. Available from: https://doi.org/10.3389/fmicb.2016.01707.

Fischer, D., Gessner, G., Fill, T.P., Barnett, R., Tron, K., Dornblut, K., et al., 2019. Disruption of membrane integrity by the bacterium-derived antifungal jagaricin. Antimicrob. Agents Chemother. 63 (9), e00707–e00719. Available from: https://doi.org/10.1128/AAC.00707-19.

Franco-Correa, M., Quintana, A., Duque, C., Suarez, C., Rodríguez, M.X., Barea, J.M., 2010. Evaluation of actinomycete strains for key traits related with plant growth promotion and mycorrhiza helping activities. Appl. Soil. Ecol. 45 (3), 209–217. Available from: https://doi.org/10.1016/j.apsoil.2010.04.007.

Fredrickson, J.K., Li, S.M.W., Gaidamakova, E.K., Matrosova, V.Y., Zhai, M., Sulloway, H.M., et al., 2008. Protein oxidation: key to bacterial desiccation resistance? ISME J. 2 (4), 393–403. Available from: https://doi.org/10.1038/ismej.2007.116.

Frey-Klett, P., Burlinson, P., Deveau, A., Barret, M., Tarkka, M., Sarniguet, A., 2011. Bacterial-fungal interactions: hyphens between agricultural, clinical, environmental, and food microbiologists. Microbiol. Mol. Biol. Rev. 75 (4), 583–609. Available from: https://doi.org/10.1128/MMBR.00020-11.

Gray, M.W., 2015. Mosaic nature of the mitochondrial proteome: implications for the origin and evolution of mitochondria. Proc. Natl Acad. Sci. 112 (33), 10133–10138. Available from: https://doi.org/10.1073/pnas.1421379112.

Gallas, G., Pavao-Zuckerman, M., 2022. Spatial cover and carbon fluxes of urbanized Sonoran Desert biological soil crusts. Sci. Rep. 12 (1), 1–9. Available from: https://doi.org/10.1038/s41598-022-09769-7.

Geisseler, D., Scow, K.M., 2014. Long-term effects of mineral fertilizers on soil microorganisms – a review. Soil. Biol. Biochem. 75, 54–63. Available from: https://doi.org/10.1016/J.SOILBIO.2014.03.023.

Goswami, M., Suresh, D.E.K.A., 2020. Plant growth-promoting rhizobacteria—alleviators of abiotic stresses in soil: a review. Pedosphere 30 (1), 40–61. Available from: https://doi.org/10.1016/S1002-0160(19)60839-8.

Gray, E.J., Smith, D.L., 2005. Intracellular and extracellular PGPR: commonalities and distinctions in the plant–bacterium signaling processes. Soil. Biol. Biochem. 37 (3), 395–412. Available from: https://doi.org/10.1016/j.soilbio.2004.08.030.

Gupta, C., Dubey, R., Maheshwari, D., 2002. Plant growth enhancement and suppression of *Macrophomina phaseolina* causing charcoal rot of peanut by fluorescent Pseudomonas. Biol. Fertil. soils 35 (6), 399–405. Available from: https://doi.org/10.1007/s00374-002-0486-0.

Ghosh, S., Penterman, J.N., Little, R.D., Chavez, R., Glick, B.R., 2003. Three newly isolated plant growth-promoting bacilli facilitate the seedling growth of canola, *Brassica campestris*. Plant. Physiol. Biochem. 41 (3), 277–281. Available from: https://doi.org/10.1016/S0981-9428(03)00019-6.

Goodfellow, M., Nouioui, I., Sanderson, R., Xie, F., Bull, A.T., 2018. Rare taxa and dark microbial matter: novel bioactive actinobacteria abound in Atacama Desert soils. Antonie van. Leeuwenhoek 111 (8), 1315–1332. Available from: https://doi.org/10.1007/s10482-018-1088-7.

Gourion, B., Berrabah, F., Ratet, P., Stacey, G., 2015. Rhizobium–legume symbioses: the crucial role of plant immunity. Trends Plant. Sci. 20 (3), 186–194. Available from: https://doi.org/10.1016/j.tplants.2014.11.008.

Govindasamy, V., Senthilkumar, M., Gaikwad, K., Annapurna, K., 2008. Isolation and characterization of ACC deaminase gene from two plant growth-promoting rhizobacteria. Curr. Microbiol. 57 (4), 312–317. Available from: https://doi.org/10.1007/s00284-008-9195-8.

Grobelak, A., Napora, A., Kacprzak, M., 2015. Using plant growth-promoting rhizobacteria (PGPR) to improve plant growth. Ecol. Eng. 84, 22–28. Available from: https://doi.org/10.1016/j.ecoleng.2015.07.019.

Gunde-Cimerman, N., Sonjak, S., Zalar, P., Frisvad, J.C., Diderichsen, B., Plemenitaš, A., 2003. Extremophilic fungi in arctic ice: a relationship between adaptation to low temperature and water activity. Phys. Chem. Earth, Parts A/B/C 28 (28-32), 1273–1278. Available from: https://doi.org/10.1016/j.pce.2003.08.056.

Harwani, D., 2013. Biodiversity of rare thermophilic actinomycetes in the great Indian Thar desert: an overview. Indo Am. J. Pharm. Res. 3, 934–939. Available from: https://doi.org/10.13140/2.1.5102.4005.

Haas, D., Keel, C., 2003. Regulation of antibiotic production in root-colonizing Pseudomonas spp. and relevance for biological control of plant disease. Annu. Rev. Phytopathol. 41, 117. Available from: https://doi.org/10.1146/annurev.phyto.41.052002.095656.

Hedden, P., Thomas, S.G., 2012. Gibberellin biosynthesis and its regulation. Biochem. J. 444 (1), 11–25. Available from: https://doi.org/10.1042/BJ20120245.

Hamedi, J., Mohammadipanah, F., 2015. Biotechnological application and taxonomical distribution of plant growth promoting actinobacteria. J. Ind. Microbiol. Biotechnol. 42 (2), 157–171. Available from: https://doi.org/10.1007/s10295-014-1537-x.

Hassani, M., Durán, P., Hacquard, S., 2018. Microbial interactions within the plant holobiont. Microbiome 6 (1), 1–17. Available from: https://doi.org/10.1186/s40168-018-0445-0.

Hirsch, A.M., Lum, M.R., Downie, J.A., 2001. What makes the rhizobia-legume symbiosis so special? Plant. Physiol. 127 (4), 1484–1492. Available from: https://doi.org/10.1104/pp.010866.

Hayat, R., Ali, S., Amara, U., Khalid, R., Ahmed, I., 2010. Soil beneficial bacteria and their role in plant growth promotion: a review. Ann. Microbiol. 60 (4), 579–598. Available from: https://doi.org/10.1007/s13213-010-0117-1.

Hashem, A., Abd_Allah, E.F., Alqarawi, A.A., Al-Huqail, A.A., Wirth, S., Egamberdieva, D., 2016. The interaction between arbuscular mycorrhizal fungi and endophytic bacteria enhances plant growth of *Acacia gerrardii* under salt stress. Front. Microbiol. 7, 1089. Available from: https://doi.org/10.3389/fmicb.2016.01089.

Igiehon, N.O., Babalola, O.O., 2018. Rhizosphere microbiome modulators: contributions of nitrogen fixing bacteria towards sustainable agriculture. Int. J. Environ. Res. Public. Health 15 (4), 574. Available from: https://doi.org/10.3390/ijerph15040574.

Iqbal, N., Umar, S., Khan, N.A., Khan, M.I.R., 2014. A new perspective of phytohormones in salinity tolerance: regulation of proline metabolism. Environ. Exp. Botany 100, 34–42. Available from: https://doi.org/10.1016/j.envexpbot.2013.12.006.

Idris, H., Goodfellow, M., Sanderson, R., Asenjo, J.A., Bull, A.T., 2017. Actinobacterial rare biospheres and dark matter revealed in habitats of the Chilean Atacama Desert. Sci. Rep. 7 (1), 1–11. Available from: https://doi.org/10.1038/s41598-017-08937-4.

Jadhav, H.P., Sayyed, R.Z., 2016. Hydrolytic enzymes of rhizospheric microbes in crop protection. MOJ Cell Sci. Rep. 3 (5), 135–136. Available from: https://doi.org/10.15406/mojcsr.2016.03.00070.

Jadhav, H.P., Shaikh, S.S., Sayyed, R.Z., 2017. Role of hydrolytic enzymes of rhizoflora in biocontrol of fungal phytopathogens: an overview. In: Mehnaz, S. (Ed.), Rhizotrophs: Plant Growth Promotion to Bioremediation. Microorganisms for Sustainability, vol 2. Springer, Singapore. Available from: http://doi.org/10.1007/978-981-10-4862-3_9.

Johnstone, T.C., Nolan, E.M., 2015. Beyond iron: non-classical biological functions of bacterial siderophores. Dalton Trans. 44 (14), 6320–6339. Available from: https://doi.org/10.1039/C4DT03559C.

Jousset, A., Bonkowski, M., 2010. The model predator *Acanthamoeba castellanii* induces the production of 2, 4, DAPG by the biocontrol strain *Pseudomonas fluorescens* Q2-87. Soil. Biol. Biochem. 42 (9), 1647–1649. Available from: https://doi.org/10.1016/j.soilbio.2010.05.018.

Kaushal, M., Wani, S.P., 2016. Plant-growth-promoting rhizobacteria: drought stress alleviators to ameliorate crop production in drylands. Ann. Microbiol. 66 (1), 35–42. Available from: https://doi.org/10.1007/s13213-015-1112-3.

Kumar, M., Kumar, V., Prasad, R. (Eds.), 2020. Phyto-Microbiome in Stress Regulation. Springer. Available from: http://doi.org/10.1007/978-981-15-2576-6.

Kuzyakov, Y., Blagodatskaya, E., 2015. Microbial hotspots and hot moments in soil: concept and review. Soil. Biol. Biochem. 83, 184–199. Available from: https://doi.org/10.1016/j.soilbio.2015.01.025.

Keohane, C.E., Steele, A.D., Wuest, W.M., 2015. The rhizosphere microbiome: a playground for natural product chemists. Synlett 26 (20), 2739–2744. Available from: https://doi.org/10.1055/s-0035-1560711.

Khalid, A., Arshad, M., Zahir, Z.A., 2004. Screening plant growth-promoting rhizobacteria for improving growth and yield of wheat. J. Appl. Microbiol. 96 (3), 473–480. Available from: https://doi.org/10.1046/j.1365-2672.2003.02161.x.

Kim, B.H., Gadd, G.M., 2019. Prokaryotic metabolism and physiology. Camb. Univ. Press. Available from: https://doi.org/10.1017/9781316761625.

Kinkel, L.L., Schlatter, D.C., Bakker, M.G., Arenz, B.E., 2012. Streptomyces competition and co-evolution in relation to plant disease suppression. Res. Microbiol. 163 (8), 490–499. Available from: https://doi.org/10.1016/J.RESMIC.2012.07.005.

Kirby, E.G., Leustek, T., Lee, M.S., 1987. Nitrogen nutrition. Cell and Tissue Culture in Forestry. Springer, Dordrecht, pp. 67–88. Available from: http://doi.org/10.1007/978-94-017-0994-1_5.

Kloepper, J.W., Gutierrez-Estrada, A., McInroy, J.A., 2007. Photoperiod regulates elicitation of growth promotion but not induced resistance by plant growth-promoting rhizobacteria. Can. J. Microbiol. 53 (2), 159–167. Available from: https://doi.org/10.1139/w06-114.

Krasilnikov, P., Taboada, M.A., 2022. Fertilizer use, soil health and agricultural sustainability. Agriculture 12 (4), 462. Available from: https://doi.org/10.3390/agriculture12040462.

Kasim, W.A., Osman, M.E., Omar, M.N., El-Daim, A., Islam, A., Bejai, S., et al., 2013. Control of drought stress in wheat using plant-growth-promoting bacteria. J. Plant. Growth Regul. 32 (1), 122–130. Available from: https://doi.org/10.1007/s00344-012-9283-7.

Kudoyarova, G., Arkhipova, T., Korshunova, T., Bakaeva, M., Loginov, O., Dodd, I.C., 2019. Phytohormone mediation of interactions between plants and non-symbiotic growth promoting bacteria under edaphic stresses. Front. Plant. Sci. 10, 1368. Available from: https://doi.org/10.3389/fpls.2019.01368.

Kuiper, I., Lagendijk, E.L., Pickford, R., Derrick, J.P., Lamers, G.E., Thomas-Oates, J.E., et al., 2004. Characterization of two *Pseudomonas putida* lipopeptide biosurfactants, putisolvin I and II, which inhibit biofilm formation and break down existing biofilms. Mol. Microbiol. 51 (1), 97–113. Available from: https://doi.org/10.1046/j.1365-2958.2003.03751.x.

Liu, L., Greaver, T.L., 2010. A global perspective on belowground carbon dynamics under nitrogen enrichment. Ecol. Lett. 13 (7), 819–828. Available from: https://doi.org/10.1111/j.1461-0248.2010.01482.x.

LaSarre, B., Federle, M.J., 2013. Exploiting quorum sensing to confuse bacterial pathogens. Microbiol. Mol. Biol. Rev. 77 (1), 73–111. Available from: https://doi.org/10.1128/MMBR.00046-12.

Lakshmanan, V., Selvaraj, G., Bais, H.P., 2014. Functional soil microbiome: belowground solutions to an aboveground problem. Plant. Physiol. 166 (2), 689–700. Available from: https://doi.org/10.1104/pp.114.245811.

Lancaster, S.H., Hollister, E.B., Senseman, S.A., Gentry, T.J., 2010. Effects of repeated glyphosate applications on soil microbial community composition and the mineralization of glyphosate. Pest. Manag. Sci.: Former. Pesticide Sci. 66 (1), 59–64. Available from: https://doi.org/10.1002/ps.1831.

Lebre, P.H., De Maayer, P., Cowan, D.A., 2017. Xerotolerant bacteria: surviving through a dry spell. Nat. Rev. Microbiol. 15 (5), 285–296. Available from: https://doi.org/10.1038/nrmicro.2017.16.

Leghari, S.J., Wahocho, N.A., Laghari, G.M., Laghari, A.H., Bhabhan, G.M., Talpur, K.H., et al., 2016. Role of nitrogen for plant growth and development: a review. Adv. Environ. Biol. 10, 209–218.

Li, H., Yang, X., Zhang, K., 2021. Understanding global land degradation processes interacted with complex biophysics and socioeconomics from the perspective of the Normalized Difference Vegetation Index (1982–2015. Glob. Planet. Change 198, 103431. Available from: https://doi.org/10.1016/j.gloplacha.2021.103431.

Lindström, K., Mousavi, S.A., 2020. Effectiveness of nitrogen fixation in rhizobia. Microb. Biotechnol. 13 (5), 1314–1335. Available from: https://doi.org/10.1111/1751-7915.13517.

Lodewyckx, C., Vangronsveld, J., Porteous, F., Moore, E.R., Taghavi, S., Mezgeay, M., et al., 2002. Endophytic bacteria and their potential applications. Crit. Rev. plant. Sci. 21 (6), 583–606. Available from: https://doi.org/10.1080/0735-260291044377.

Lu, M., Yang, Y., Luo, Y., Fang, C., Zhou, X., Chen, J., et al., 2011. Responses of ecosystem nitrogen cycle to nitrogen addition: a meta-analysis. New Phytol. 189 (4), 1040. Available from: https://doi.org/10.1111/j.1469-8137.2010.03563.x.

Luo, X., Wang, J., Zeng, X.C., Wang, Y., Zhou, L., Nie, Y., et al., 2012. *Mycetocola manganoxydans* sp. nov., an actinobacterium isolated from the Taklamakan desert. Int. J. Syst. Evolut. Microbiol. 62 (Pt_12), 2967–2970. Available from: https://doi.org/10.1099/ijs.0.038877-0.

Lyu, D., Zajonc, J., Pagé, A., Tanney, C.,A.,S., Shah, A., Monjezi, N., et al., 2021. Plant holobiont theory: the phytomicrobiome plays a central role in evolution and success. Microorganisms 9 (4), 675. Available from: https://doi.org/10.3390/microorganisms9040675.

McNear Jr, D.H., 2013. The rhizosphere-roots, soil and everything in between. Nat. Educ. Knowl. 4 (3), 1.

Margesin, R., Schinner, F., 2001. Biodegradation and bioremediation of hydrocarbons in extreme environments. Appl. Microbiol. Biotechnol. 56 (5), 650–663. Available from: https://doi.org/10.1007/s002530100701.

Makarova, K.S., Aravind, L., Wolf, Y.I., Tatusov, R.L., Minton, K.W., Koonin, E.V., et al., 2001. Genome of the extremely radiation-resistant bacterium *Deinococcus radiodurans* viewed from the perspective of comparative genomics. Microbiol. Mol. Biol. Rev. 65 (1), 44–79. Available from: https://doi.org/10.1128/MMBR.65.1.44-79.2001.

Mamangkey, J., Mendes, L.W., Harahap, A., Briggs, D., Kayacilar, C., 2022. Endophytic bacteria and fungi from indonesian medicinal plants with antibacterial, pathogenic antifungal and extracellular enzymes activities: a review. Int. J. Sci. Technol. Manag. 3 (1), 245–255. Available from: https://doi.org/10.46729/ijstm.v3i1.428.

Marciano Marra, L., Fonsêca Sousa Soares, C.R., de Oliveira, S.M., Avelar Ferreira, P.A., Lima Soares, B., de Fráguas Carvalho, R., et al., 2012. Biological nitrogen fixation and phosphate solubilization by bacteria isolated from tropical soils. Plant. Soil. 357 (1), 289–307. Available from: https://doi.org/10.1007/s11104-012-1157-z.

Martínez-Viveros, O., Jorquera, M.A., Crowley, D.E., Gajardo, G.M.L.M., Mora, M.L., 2010. Mechanisms and practical considerations involved in plant growth promotion by rhizobacteria. J. Soil Sci. Plant Nutr. 10 (3), 293–319. Available from: https://doi.org/10.4067/S0718-95162010000100006.

Mavrodi, D.V., Blankenfeldt, W., Thomashow, L.S., 2006. Phenazine compounds in fluorescent Pseudomonas spp. biosynthesis and regulation. Annu. Rev. Phytopathol. 44, 417–445. Available from: https://doi.org/10.1146/annurev.phyto.44.013106.145710.

Mayak, S., Tirosh, T., Glick, B.R., 2004. Plant growth-promoting bacteria confer resistance in tomato plants to salt stress. Plant. Physiol. Biochem. 42 (6), 565–572. Available from: https://doi.org/10.1016/j.plaphy.2004.05.009.

Mendes, R., Garbeva, P., Raaijmakers, J.M., 2013. The rhizosphere microbiome: significance of plant beneficial, plant pathogenic, and human pathogenic microorganisms. FEMS Microbiol. Rev. 37 (5), 634–663. Available from: https://doi.org/10.1111/1574-6976.12028.

Menéndez, E., Paço, A., 2020. Is the application of plant probiotic bacterial consortia always beneficial for plants? Exploring synergies between rhizobial and non-rhizobial bacteria and their effects on agro-economically valuable crops. Life 10 (3), 24. Available from: https://doi.org/10.3390/life10030024.

Middleton, H., Yergeau, É., Monard, C., Combier, J.P., El Amrani, A., 2021. Rhizospheric plant–microbe interactions: miRNAs as a key mediator. Trends Plant. Sci. 26 (2), 132–141. Available from: https://doi.org/10.1016/j.tplants.2020.09.005.

Mishra, M., Kumar, U., Mishra, P.K., Prakash, V., 2010. Efficiency of plant growth promoting rhizobacteria for the enhancement of *Cicer arietinum* L. growth and germination under salinity. Adv. Biol. Res. 4 (2), 92–96.

Mogensen, J.M., Nielsen, K.F., Samson, R.A., Frisvad, J.C., Thrane, U., 2009. Effect of temperature and water activity on the production of fumonisins by *Aspergillus niger* and different Fusariumspecies. BMC Microbiol. 9 (1), 1–12. Available from: https://doi.org/10.1186/1471-2180-9-281.

Mohammadipanah, F., Wink, J., 2016. Actinobacteria from arid and desert habitats: diversity and biological activity. Front. Microbiol. 6, 1541. Available from: https://doi.org/10.3389/fmicb.2015.01541.

Mokrani, S., Nabti, E.H., Cruz, C., 2020. Current advances in plant growth promoting bacteria alleviating salt stress for sustainable agriculture. Appl. Sci. 10 (20), 7025. Available from: https://doi.org/10.3390/app10207025.

Molina-Romero, D., Baez, A., Quintero-Hernández, V., Castañeda-Lucio, M., Fuentes-Ramírez, L.E., Bustillos-Cristales, M.D.R., et al., 2017. Compatible bacterial mixture, tolerant to desiccation, improves maize plant growth. PLoS One 12 (11), e0187913. Available from: https://doi.org/10.1371/journal.pone.0187913.

Monsanto Bio Ag Alliance, 2015. QuickRoots: *Bacillus amyloliquefaciens* and *Trichoderma virens* Based Inoculant for Corn (Product Information). Monsanto bio ag, Belgium.

Montero-Calasanz, M., Göker, M., Pötter, G., Rohde, M., Spröer, C., Schumann, P., et al., 2012. *Geodermatophilus arenarius* sp. nov., a xerophilic actinomycete isolated from Saharan desert sand in Chad. Extremophiles 16 (6), 903–909. Available from: https://doi.org/10.1007/s00792-012-0486-4.

Morales-Cedeño, L.R., del Carmen Orozco-Mosqueda, M., Loeza-Lara, P.D., Parra-Cota, F.I., de Los Santos-Villalobos, S., Santoyo, G., 2021. Plant growth-promoting bacterial endophytes as biocontrol agents of pre-and post-harvest diseases: fundamentals, methods of application and future perspectives. Microbiol. Res. 242, 126612. Available from: https://doi.org/10.1016/j.micres.2020.126612.

Mubeen, M.U.H.A.M.M.A.D., Bano, A., Ali, B., Islam, Z.U., Ahmad, A., Hussain, S., et al., 2021. Effect of plant growth promoting bacteria and drought on spring maize (*Zea mays* L.). Pak. J. Bot. 53 (2), 731–739. Available from: https://doi.org/10.30848/PJB2021-2(38).

Muniroh, M.S., Nusaibah, S.A., Vadamalai, G., Siddique, Y., 2019. Proficiency of biocontrol agents as plant growth promoters and hydrolytic enzyme producers in *Ganoderma boninense* infected oil palm seedlings. Curr. Plant. Biol. 20, 100116. Available from: https://doi.org/10.1016/j.cpb.2019.100116.

Nakkeeran, S., Fernando, W.G., Siddiqui, Z.A., 2005. Plant growth promoting rhizobacteria formulations and its scope in commercialization for the management of pests and diseases. PGPR: Biocontrol and Biofertilization. Springer, Dordrecht, pp. 257–296. Available from: http://doi.org/10.1007/1-4020-4152-7_10.

Nanjundappa, A., Bagyaraj, D.J., Saxena, A.K., Kumar, M., Chakdar, H., 2019. Interaction between arbuscular mycorrhizal fungi and Bacillus spp. in soil enhancing growth of crop plants. Fungal Biol. Biotechnol. 6 (1), 1–10. Available from: https://doi.org/10.1186/s40694-019-0086-5.

Nazli, F., Mustafa, A., Ahmad, M., Hussain, A., Jamil, M., Wang, X., et al., 2020. A review on practical application and potentials of phytohormone-producing plant growth-promoting rhizobacteria for inducing heavy metal tolerance in crops. Sustainability 12 (21), 9056. Available from: https://doi.org/10.3390/su12219056.

Netzker, T., Fischer, J., Weber, J., Mattern, D.J., König, C.C., Valiante, V., et al., 2015. Microbial communication leading to the activation of silent fungal secondary metabolite gene clusters. Front. Microbiol. 6, 299. Available from: https://doi.org/10.3389/fmicb.2015.00299.

Neumann, K.H., Kumar, A., Imani, J., 2009. Plant cell and tissue culture – a tool in biotechnology. Principles and Practice. Springer. Available from: 10.1007/978-3-540-93883-5.

Nicholson, W.L., Munakata, N., Horneck, G., Melosh, H.J., Setlow, P., 2000. Resistance of Bacillus endospores to extreme terrestrial and extraterrestrial environments. Microbiol. Mol. Biol. Rev. 64 (3), 548–572. Available from: https://doi.org/10.1128/MMBR.64.3.548-572.2000.

Nikitin, D.A., Semenov, M.V., Zhelezova, A.D., Kutovaya, O.V., 2021. Influence of no-till technology on number and taxonomic composition of microscopic fungi in southern agrochernozemes. Mikologiya I Fitopatologiya 55, 189–202. Available from: https://doi.org/10.31857/S0026364821030077 (in Russian).

Nordstedt, N.P., Jones, M.L., 2020. Isolation of rhizosphere bacteria that improve quality and water stress tolerance in greenhouse ornamentals. Front. Plant. Sci. 11, 826. Available from: https://doi.org/10.3389/fpls.2020.00826.

Numan, M., Bashir, S., Khan, Y., Mumtaz, R., Shinwari, Z.K., Khan, A.L., et al., 2018. Plant growth promoting bacteria as an alternative strategy for salt tolerance in plants: a review. Microbiol. Res. 209, 21–32. Available from: https://doi.org/10.1016/j.micres.2018.02.003.

Nunes, I., Jurburg, S., Jacquiod, S., Brejnrod, A., Falcão Salles, J., Priemé, A., et al., 2018. Soil bacteria show different tolerance ranges to an unprecedented disturbance. Biol. Fertil. Soils 54 (2), 189–202. Available from: https://doi.org/10.1007/s00374-017-1255-4.

Ongena, M., Jacques, P., 2008. Bacillus lipopeptides: versatile weapons for plant disease biocontrol. Trends Microbiol. 16 (3), 115–125. Available from: https://doi.org/10.1016/j.tim.2007.12.009.

Orr, C.H., James, A., Leifert, C., Cooper, J.M., Cummings, S.P., 2011. Diversity and activity of free-living nitrogen-fixing bacteria and total bacteria in organic and conventionally managed soils. Appl. Environ. Microbiol. 77 (3), 911–919. Available from: https://doi.org/10.1128/AEM.01250-10.

Pandey, A., Tripathi, A., Srivastava, P., Choudhary, K.K., Dikshit, A., 2019. Plant growth-promoting microorganisms in sustainable agriculture. Role of Plant Growth Promoting Microorganisms in Sustainable Agriculture and Nanotechnology. Woodhead Publishing, pp. 1–19. Available from: http://doi.org/10.1016/B978-0-12-817004-5.00001-4.

Pathma, J., Kennedy, R., Sakthivel, N., 2011. Mechanisms of fluorescent pseudomonads that mediate biological control of phytopathogens and plant growth promotion of crop plants. Bacteria in Agrobiology: Plant Growth Responses. Springer, Berlin, Heidelberg, pp. 77–105. Available from: http://doi.org/10.1007/978-3-642-20332-9_4.

Patten, C.L., Glick, B.R., 1996. Bacterial biosynthesis of indole-3-acetic acid. Can. J. Microbiol. 42 (3), 207–220. Available from: https://doi.org/10.1139/m96-032.

Perreault, R., Laforest-Lapointe, I., 2022. Plant-microbe interactions in the phyllosphere: facing challenges of the anthropocene. ISME J. 16 (2), 339–345. Available from: https://doi.org/10.1038/s41396-021-01109-3.

Phelan, V.V., Liu, W.T., Pogliano, K., Dorrestein, P.C., 2012. Microbial metabolic exchange—the chemotype-to-phenotype link. Nat. Chem. Biol. 8 (1), 26–35. Available from: https://doi.org/10.1038/nchembio.739.

Prăvălie, R., 2016. Drylands extent and environmental issues. A global approach. Earth-Science Rev. 161, 259–278. Available from: https://doi.org/10.1016/j.earscirev.2016.08.003.

Prell, J., Poole, P., 2006. Metabolic changes of rhizobia in legume nodules. Trends Microbiol. 14 (4), 161–168. Available from: https://doi.org/10.1016/j.tim.2006.02.005.

Price-Whelan, A., Dietrich, L.E., Newman, D.K., 2006. Rethinking'secondary'metabolism: physiological roles for phenazine antibiotics. Nat. Chem. Biol. 2 (2), 71–78. Available from: https://doi.org/10.1038/nchembio764.

Pronk, L.J., Bakker, P.A., Keel, C., Maurhofer, M., Flury, P., 2022. The secret life of plant-beneficial rhizosphere bacteria: insects as alternative hosts. Environ. Microbiol. Available from: https://doi.org/10.1111/1462-2920.15968.

Qin, S., Li, W.J., Dastager, S.G., Hozzein, W.N., 2016. Actinobacteria in special and extreme habitats: diversity, function roles, and environmental adaptations. Front. Microbiol. 7, 1415. Available from: https://doi.org/10.3389/fmicb.2016.01415.

Raaijmakers, J.M., De Bruijn, I., De Kock, M.J., 2006. Cyclic lipopeptide production by plant-associated Pseudomonas spp.: diversity, activity, biosynthesis, and regulation. Mol. Plant-Microbe Interact. 19 (7), 699–710. Available from: https://doi.org/10.1094/MPMI-19-0699.

Raaijmakers, J.M., De Bruijn, I., Nybroe, O., Ongena, M., 2010. Natural functions of lipopeptides from Bacillus and Pseudomonas: more than surfactants and antibiotics. FEMS Microbiol. Rev. 34 (6), 1037–1062. Available from: https://doi.org/10.1111/j.1574-6976.2010.00221.x.

Radhakrishnan, R., Hashem, A., Abd_Allah, E.F., 2017. Bacillus: a biological tool for crop improvement through bio-molecular changes in adverse environments. Front. Physiol. 8, 667. Available from: https://doi.org/10.3389/fphys.2017.00667.

Rajendran, G., Mistry, S., Desai, A.J., Archana, G., 2007. Functional expression of *Escherichia coli* fhuA gene in Rhizobium spp. of *Cajanus cajan* provides growth advantage in presence of Fe3 + : ferrichrome as iron source. Arch. Microbiol. 187 (4), 257–264. Available from: https://doi.org/10.1007/s00203-006-0191-8.

Ramakrishna, W., Yadav, R., Li, K., 2019. Plant growth promoting bacteria in agriculture: two sides of a coin. Appl. Soil. Ecol. 138, 10–18. Available from: https://doi.org/10.1016/j.apsoil.2019.02.019.

Reynolds, J.F., Smith, D.M.S., Lambin, E.F., Turner, B.L., Mortimore, M., Batterbury, S.P., et al., 2007. Global desertification: building a science for dryland development. Science 316 (5826), 847–851. Available from: https://doi.org/10.1126/science.1131634.

Riedlinger, J., Schrey, S.D., Tarkka, M.T., Hampp, R., Kapur, M., Fiedler, H.P., 2006. Auxofuran, a novel metabolite that stimulates the growth of fly agaric, is produced by the mycorrhiza helper bacterium Streptomyces strain AcH 505. Appl. Environ. Microbiol. 72 (5), 3550–3557. Available from: https://doi.org/10.1128/AEM.72.5.3550-3557.2006.

Roley, S.S., Duncan, D.S., Liang, D., Garoutte, A., Jackson, R.D., Tiedje, J.M., et al., 2018. Associative nitrogen fixation (ANF) in switchgrass (*Panicum virgatum*) across a nitrogen input gradient. PLoS One 13 (6), e0197320. Available from: https://doi.org/10.1371/journal.pone.0197320.

Rosenblueth, M., Ormeño-Orrillo, E., López-López, A., Rogel, M.A., Reyes-Hernández, B. J., Martínez-Romero, J.C., et al., 2018. Nitrogen fixation in cereals. Front. Microbiol. 9, 1794. Available from: https://doi.org/10.3389/fmicb.2018.01794.

Rosier, A., Bishnoi, U., Lakshmanan, V., Sherrier, D.J., Bais, H.P., 2016. A perspective on inter-kingdom signaling in plant–beneficial microbe interactions. Plant. Mol. Biol. 90 (6), 537–548. Available from: https://doi.org/10.1007/s11103-016-0433-3.

Rossmann, M., Perez-Jaramillo, J.E., Kavamura, V.N., Chiaramonte, J.B., Dumack, K., Fiore-Donno, A.M., et al., 2020. Multitrophic interactions in the rhizosphere microbiome of wheat: from bacteria and fungi to protists. FEMS Microbiol. Ecol. 96 (4), fiaa032. Available from: https://doi.org/10.1093/femsec/fiaa032.

Rouphael, Y., Franken, P., Schneider, C., Schwarz, D., Giovannetti, M., Agnolucci, M., et al., 2015. Arbuscular mycorrhizal fungi act as biostimulants in horticultural crops. Sci. Horticulturae 196, 91–108. Available from: https://doi.org/10.1016/j.scienta.2015.09.002.

Ryu, M.H., Zhang, J., Toth, T., Khokhani, D., Geddes, B.A., Mus, F., et al., 2020. Control of nitrogen fixation in bacteria that associate with cereals. Nat. Microbiol. 5 (2), 314–330. Available from: https://doi.org/10.1038/s41564-019-0631-2.

Sabaratnam, S., Traquair, J.A., 2002. Formulation of a Streptomyces biocontrol agent for the suppression of Rhizoctonia damping-off in tomato transplants. Biol. Control. 23 (3), 245–253. Available from: https://doi.org/10.1006/bcon.2001.1014.

Saberi Riseh, R., Ebrahimi-Zarandi, M., Gholizadeh Vazvani, M., Skorik, Y.A., 2021. Reducing drought stress in plants by encapsulating plant growth-promoting bacteria with polysaccharides. Int. J. Mol. Sci. 22 (23), 12979. Available from: https://doi.org/10.3390/ijms222312979.

Saiyad, S.A., Jhala, Y.K., Vyas, R.V., 2015. Comparative efficiency of five potash and phosphate solubilizing bacteria and their key enzymes useful for enhancing and improvement of soil fertility. Int. J. Sci. Res. Publ. 5 (2), 1–6.

Saleem, M., Arshad, M., Hussain, S., Bhatti, A.S., 2007. Perspective of plant growth promoting rhizobacteria (PGPR) containing ACC deaminase in stress agriculture. J. Ind. Microbiol. Biotechnol. 34 (10), 635–648. Available from: https://doi.org/10.1007/s10295-007-0240-6.

Sannino, F., Gianfreda, L., 2001. Pesticide influence on soil enzymatic activities. Chemosphere 45 (4-5), 417–425. Available from: https://doi.org/10.1016/S0045-6535(01)00045-5.

Santhanam, R., Rong, X., Huang, Y., Andrews, B.A., Asenjo, J.A., Goodfellow, M., 2013. Streptomyces bullii sp. nov., isolated from a hyper-arid Atacama Desert soil. Antonie Van. Leeuwenhoek 103 (2), 367–373. Available from: https://doi.org/10.1007/s10482-012-9816-x.

Santos, R., de Carvalho, C.C., Stevenson, A., Grant, I.R., Hallsworth, J.E., 2015. Extraordinary solute-stress tolerance contributes to the environmental tenacity of mycobacteria. Environ. Microbiol. Rep. 7 (5), 746–764. Available from: https://doi.org/10.1111/1758-2229.12306.

Sathya, A., Vijayabharathi, R., Gopalakrishnan, S., 2017. Plant growth-promoting actinobacteria: a new strategy for enhancing sustainable production and protection of grain legumes. 3 Biotech. 7 (2), 1–10. Available from: https://doi.org/10.1007/s13205-017-0736-3.

Schmidt, R., Cordovez, V., De Boer, W., Raaijmakers, J., Garbeva, P., 2015. Volatile affairs in microbial interactions. ISME J. 9 (11), 2329–2335. Available from: https://doi.org/10.1038/ismej.2015.42.

Seefeldt, L.C., Hoffman, B.M., Dean, D.R., 2009. Mechanism of Mo-dependent nitrogenase. Annu. Rev. Biochem. 78, 701. Available from: https://doi.org/10.1146/annurev.biochem.78.070907.103812.

Senthilkumar, N.A.M., Sankaranarayanan, A., 2021. Senthilkumar. Plant-Microbe Interactions. Springer, US. Available from: https://doi.org/10.1007/978-1-0716-1080-0.

Sharma, A., Shankhdhar, D., Shankhdhar, S.C., 2016. Potassium-solubilizing microorganisms: mechanism and their role in potassium solubilization and uptake. Potassium Solubilizing Microorganisms for Sustainable Agriculture. Springer, New Delhi, pp. 203–219. Available from: https://doi.org/10.1007/978-81-322-2776-2_15.

Sharma, R., Bisaria, V.S., Sharma, S., 2019. Rhizosphere: a home for human pathogens. Plant Biotic Interactions. Springer, Cham, pp. 113–127. Available from: https://doi.org/10.1007/978-3-030-26657-8_8.

Shoba, S.A., Gracheva, T.A., Stepanov, A.L., Fedotov, G.N., Gorepekin, I.V., 2021. On the nature of the influence of some mycelial actinobacteria on the spring wheat seeds germination in soils. Doklady Biol. Sciences, Pleiades Publ. 498 (1), 85–88. Available from: https://doi.org/10.1134/S0012496621030030.

Shrivastava, P., Kumar, R., 2015. Soil salinity: a serious environmental issue and plant growth promoting bacteria as one of the tools for its alleviation. Saudi J. Biol. Sci. 22 (2), 123–131. Available from: https://doi.org/10.1016/j.sjbs.2014.12.001.

Shukla, M., Chaturvedi, R., Tamhane, D., Vyas, P., Archana, G., Apte, S., et al., 2007. Multiple-stress tolerance of ionizing radiation-resistant bacterial isolates obtained from various habitats: correlation between stresses. Curr. Microbiol. 54 (2), 142–148. Available from: https://doi.org/10.1007/s00284-006-0311-3.

Shukla, S.K., Yadav, R.L., Suman, A., Singh, P.N., 2008. Improving rhizospheric environment and sugarcane ratoon yield through bioagents amended farm yard manure in udic ustochrept soil. Soil. Tillage Res. 99 (2), 158–168. Available from: https://doi.org/10.1016/j.still.2008.02.007.

Singh, B.K., Liu, H., Trivedi, P., 2020. Eco-holobiont: a new concept to identify drivers of host-associated microorganisms. Environ. Microbiol. 22 (2), 564–567. Available from: https://doi.org/10.1111/1462-2920.14900.

Singh, V.K., Singh, A.K., Singh, P.P., Kumar, A., 2018. Interaction of plant growth promoting bacteria with tomato under abiotic stress: a review. Agric. Ecosyst. Environ. 267, 129–140. Available from: https://doi.org/10.1016/j.agee.2018.08.020.

Sokolova, T.A., Dronova, T.Y., Artyukhov, D.B., Korobova, N.L., 1997. Spatial and time variability of pH values in podzolic soils of the central forest biospheric reserve. Eurasian Soil. Sci. 30, 1199–1206.

Solanki, M.K., Singh, R.K., Srivastava, S., Kumar, S., Kashyap, P.L., Srivastava, A.K., 2015. Characterization of antagonistic-potential of two Bacillus strains and their biocontrol activity against *Rhizoctonia solani* in tomato. J. Basic. Microbiol. 55 (1), 82–90. Available from: https://doi.org/10.1002/jobm.201300528.

Solanki, S., Ameen, G., Sanyal, D., Jain, S., Elakhdar, A., Lall, S., et al., 2020. Friends and foes: phyto-microbial interactions in molecular perspective. Phyto-Microbiome Stress. Regul. 81–98. Available from: https://doi.org/10.1007/978-981-15-2576-6_5.

Soumare, A., Diedhiou, A.G., Thuita, M., Hafidi, M., Ouhdouch, Y., Gopalakrishnan, S., et al., 2020. Exploiting biological nitrogen fixation: a route towards a sustainable agriculture. Plants 9 (8), 1011.

Spaepen, S., Vanderleyden, J., Remans, R., 2007. Indole-3-acetic acid in microbial and microorganism-plant signaling. FEMS Microbiol. Rev. 31 (4), 425–448. Available from: https://doi.org/10.1111/j.1574-6976.2007.00072.x.

Sruthilaxmi, C.B., Babu, S., 2017. Microbial bio-inoculants in Indian agriculture: ecological perspectives for a more optimized use. Agric. Ecosyst. Environ. 242, 23–25. Available from: https://doi.org/10.1016/j.agee.2017.03.019.

Steven, B., Leveille, R., Pollard, W.H., Whyte, L.G., 2006. Microbial ecology and biodiversity in permafrost. Extremophiles 10 (4), 259–267. Available from: https://doi.org/10.1007/s00792-006-0506-3.

Stevenson, A., Burkhardt, J., Cockell, C.S., Cray, J.A., Dijksterhuis, J., Fox-Powell, M., et al., 2015. Multiplication of microbes below 0.690 water activity: implications for terrestrial and extraterrestrial life. Environ. Microbiol. 17 (2), 257–277. Available from: https://doi.org/10.1111/1462-2920.12598.

Stevenson, A., Hallsworth, J.E., 2014. Water and temperature relations of soil A ctinobacteria. Environ. Microbiol. Rep. 6 (6), 744–755. Available from: https://doi.org/10.1111/1758-2229.12199.

Streletskii, R., Astaykina, A., Krasnov, G., Gorbatov, V., 2022. Changes in bacterial and fungal community of soil under treatment of pesticides. Agronomy 12 (1), 124. Available from: https://doi.org/10.3390/agronomy12010124.

Su, Y.G., Zhao, X., Li, A.X., Li, X.R., Huang, G., 2011. Nitrogen fixation in biological soil crusts from the Tengger desert, northern China. Eur. J. Soil. Biol. 47 (3), 182–187. Available from: https://doi.org/10.1016/j.ejsobi.2011.04.001.

Suman, A., Govindasamy, V., Ramakrishnan, B., Aswini, K., SaiPrasad, J., Sharma, P., et al., 2021. Microbial community and function-based synthetic bioinoculants: a perspective for sustainable agriculture. Front. Microbiol. 12. Available from: https://doi.org/10.3389/fmicb.2021.805498.

Suman, A., Shrivastava, A.K., Gaur, A., Singh, P., Singh, J., Yadav, R.L., 2008. Nitrogen use efficiency of sugarcane in relation to its BNF potential and population of endophytic diazotrophs at different N levels. Plant. Growth Regul. 54 (1), 1–11. Available from: https://doi.org/10.1007/s10725-007-9219-6.

Taechowisan, T., Peberdy, J.F., Lumyong, S., 2003. Isolation of endophytic actinomycetes from selected plants and their antifungal activity. World J. Microbiol. Biotechnol. 19 (4), 381–385. Available from: https://doi.org/10.1023/A:1023901107182.

Terkina, I.A., Parfenova, V.V., Ahn, T.S., 2006. Antagonistic activity of actinomycetes of Lake Baikal. Appl. Biochem. Microbiol. 42 (2), 173–176. Available from: https://doi.org/10.1134/S0003683806020104.

Thomashow, L.S., Bonsall, R.F., Weller, D.M., 1997. Antibiotic production by soil and rhizosphere microbes in situ. Man. Environ. Microbiol 509, 1–24.

Tkhakakhova, A.K., Chernov, T.I., Ivanova, E.A., Kutovaya, O.V., Kogut, B.M., Zavalin, A.A., 2016. Changes in the metagenome of the prokaryotic community of chernozems under the influence of mineral fertilizers. Russian Agric. Sci. 42 (1), 62–65. Available from: https://doi.org/10.3103/s1068367416010201.

Traxler, M.F., Kolter, R., 2015. Natural products in soil microbe interactions and evolution. Nat. Product. Rep. 32 (7), 956–970. Available from: https://doi.org/10.1039/C5NP00013K.

Trivedi, P., Delgado-Baquerizo, M., Anderson, I.C., Singh, B.K., 2016. Response of soil properties and microbial communities to agriculture: implications for primary productivity and soil health indicators. Front. Plant. Sci. 7, 990. Available from: https://doi.org/10.3389/fpls.2016.00990.

Trivedi, P., Leach, J.E., Tringe, S.G., Sa, T., Singh, B.K., 2021. Plant–microbiome interactions: from community assembly to plant health. Nat. Rev. Microbiol. 19, 72. Available from: https://doi.org/10.1038/s41579-020-0412-1.

Tsolakidou, M.D., Stringlis, I.A., Fanega-Sleziak, N., Papageorgiou, S., Tsalakou, A., Pantelides, I.S., 2019. Rhizosphere-enriched microbes as a pool to design synthetic communities for reproducible beneficial outputs. FEMS Microbiol. Ecol. 95 (10), fiz138. Available from: https://doi.org/10.1093/femsec/fiz138.

Turner, T.R., James, E.K., Poole, P.S., 2013. The plant microbiome. Genome Biol. 14 (6), 1–10. Available from: https://doi.org/10.1186/gb-2013-14-6-209.

Vacheron, J., Desbrosses, G., Bouffaud, M.L., Touraine, B., Moënne-Loccoz, Y., Muller, D., et al., 2013. Plant growth-promoting rhizobacteria and root system functioning. Front. Plant. Sci. 4, 356. Available from: https://doi.org/10.3389/fpls.2013.00356.

Vacheron, J., Moënne-Loccoz, Y., Dubost, A., Gonçalves-Martins, M., Muller, D., Prigent-Combaret, C., 2016. Fluorescent Pseudomonas strains with only few plant-beneficial properties are favored in the maize rhizosphere. Front. Plant Sci. 7, 1212. Available from: https://doi.org/10.3389/fpls.2016.01212.

Van Dommelen, A., Vanderleyden, J., 2007. Associative nitrogen fixation. Biology of the Nitrogen Cycle. Elsevier, pp. 179–192. Available from: http://doi.org/10.1016/B978-044452857-5.50013-8.

Vasconcellos, R.L.F.D., Silva, M.C.P.D., Ribeiro, C.M., Cardoso, E.J.B.N., 2010. Isolation and screening for plant growth-promoting (PGP) actinobacteria from *Araucaria angustifolia* rhizosphere soil. Sci. Agricola 67, 743–746. Available from: https://doi.org/10.1590/S0103-90162010000600019.

Venturi, V., Keel, C., 2016. Signaling in the rhizosphere. Trends Plant. Sci. 21 (3), 187–198. Available from: https://doi.org/10.1016/j.tplants.2016.01.005.

Verma, J.P., Yadav, J., Tiwari, K.N., Lavakush, S., Singh, V., 2010. Impact of plant growth promoting rhizobacteria on crop production. Int. J. Agric. Res. 5 (11), 954–983. Available from: https://doi.org/10.3923/ijar.2010.954.983.

Verma, P., Yadav, A.N., Khannam, K.S., Panjiar, N., Kumar, S., Saxena, A.K., et al., 2015. Assessment of genetic diversity and plant growth promoting attributes of psychrotolerant bacteria allied with wheat (*Triticum aestivum*) from the northern hills zone of India. Ann. Microbiol. 65 (4), 1885–1899. Available from: https://doi.org/10.1007/s13213-014-1027-4.

Vinale, F., Nicoletti, R., Borrelli, F., Mangoni, A., Parisi, O.A., Marra, R., et al., 2017. Co-culture of plant beneficial microbes as source of bioactive metabolites. Sci. Rep. 7 (1), 1–12. Available from: https://doi.org/10.1038/s41598-017-14569-5.

Vurukonda, S.S.K.P., Vardharajula, S., Shrivastava, M., SkZ, A., 2016. Enhancement of drought stress tolerance in crops by plant growth promoting rhizobacteria. Microbiol. Res. 184, 13–24. Available from: https://doi.org/10.1016/j.micres.2015.12.003.

Vyas, P., Rahi, P., Gulati, A., 2009. Stress tolerance and genetic variability of phosphate-solubilizing fluorescent Pseudomonas from the cold deserts of the trans-Himalayas. Microb. Ecol. 58 (2), 425–434. Available from: https://doi.org/10.1007/s00248-009-9511-2.

Wang, B., Huang, Y., Li, N., Yao, H., Yang, E., Soromotin, A.V., et al., 2022. Initial soil formation by biocrusts: nitrogen demand and clay protection control microbial necromass accrual and recycling. Soil. Biol. Biochem. 167, 108607. Available from: https://doi.org/10.1016/j.soilbio.2022.108607.

Wang, E.T., Martinez-Romero, E., 2000. *Sesbania herbacea*–Rhizobium huautlense nodulation in flooded soils and comparative characterization of S. herbacea-nodulating Rhizobia in different environments. Microb. Ecol. 40 (1), 25–32. Available from: https://doi.org/10.1007/s002480000010.

Wang, Z.R., Li, G., Ji, L.X., Wang, H.H., Gao, H., Peng, X.P., et al., 2019. Induced production of steroids by co-cultivation of two endophytes from *Mahonia fortunei*. Steroids 145, 1–4. Available from: https://doi.org/10.1016/j.steroids.2019.02.005.

Wani, S.H., Kumar, V., Shriram, V., Sah, S.K., 2016. Phytohormones and their metabolic engineering for abiotic stress tolerance in crop plants. Crop J. 4 (3), 162–176. Available from: https://doi.org/10.1016/j.cj.2016.01.010.

White, J.F., Kingsley, K.L., Zhang, Q., Verma, R., Obi, N., Dvinskikh, S., et al., 2019. Endophytic microbes and their potential applications in crop management. Pest. Manag. Sci. 75 (10), 2558–2565. Available from: https://doi.org/10.1002/ps.5527.

Williams, S.C., 2013. The other microbiome. Proc. Natl Acad. Sci. 110 (8), 2682–2684. Available from: https://doi.org/10.1073/pnas.130092311.

Willis, A., Rodrigues, B.F., Harris, P.J., 2013. The ecology of arbuscular mycorrhizal fungi. Crit. Rev. Plant. Sci. 32 (1), 1–20. Available from: https://doi.org/10.1080/07352689.2012.683375.

Woo, S.L., Pepe, O., 2018. Microbial consortia: promising probiotics as plant biostimulants for sustainable agriculture. Front. plant. Sci. 9, 1801. Available from: https://doi.org/10.3389/fpls.2018.01801.

Xu, J., Li, X.L., Luo, L., 2012. Effects of engineered *Sinorhizobium meliloti* on cytokinin synthesis and tolerance of alfalfa to extreme drought stress. Appl. Environ. Microbiol. 78 (22), 8056–8061. Available from: https://doi.org/10.1128/AEM.01276-12.

Xu, L., Naylor, D., Dong, Z., Simmons, T., Pierroz, G., Hixson, K.K., et al., 2018. Drought delays development of the sorghum root microbiome and enriches for monoderm bacteria. Proc. Natl Acad. Sci. 115 (18), E4284–E4293. Available from: https://doi.org/10.1073/pnas.1717308115.

Xu, W., Liu, L., He, T., Cao, M., Sha, L., Hu, Y., et al., 2016. Soil properties drive a negative correlation between species diversity and genetic diversity in a tropical seasonal rainforest. Sci. Rep. 6 (1), 1–8. Available from: https://doi.org/10.1038/srep20652.

Yadav, A.N., 2021. Beneficial plant-microbe interactions for agricultural sustainability. J. Appl. Biol. Biotechnol. 9 (1), i–v. Available from: https://doi.org/10.7324/JABB.2021.91ed.

Yu, B., Chen, Z., Lu, X., Huang, Y., Zhou, Y., Zhang, Q., et al., 2020. Effects on soil microbial community after exposure to neonicotinoid insecticides thiamethoxam and dinotefuran. Sci. Total. Environ. 725, 138328. Available from: https://doi.org/10.1016/J.SCITOTENV.2020.138328.

Zaidi, A., Khan, M.S., Saif, S., Rizvi, A., Ahmed, B., Shahid, M., 2017. Role of nitrogen-fixing plant growth-promoting rhizobacteria in sustainable production of vegetables: current perspective. Microbial Strategies for Vegetable Production. Springer, Cham, pp. 49–79. Available from: http://doi.org/10.1007/978-3-319-54401-4_3.

Zhang, B., Bai, Z., Hoefel, D., Tang, L., Wang, X., Li, B., et al., 2009. The impacts of cypermethrin pesticide application on the non-target microbial community of the pepper plant phyllosphere. Sci. Total. Environ. 407 (6), 1915–1922. Available from: https://doi.org/10.1016/J.SCITOTENV.2008.11.049.

Zhang, C., Sheng, C., Wang, W., Hu, H., Peng, H., Zhang, X., 2015. Identification of the lomofungin biosynthesis gene cluster and associated flavin-dependent monooxygenase gene in *Streptomyces lomondensis* S015. PLoS One 10 (8), e0136228. Available from: https://doi.org/10.1371/journal.pone.0136228.

Zhang, H., Wang, X., 2016. Modular co-culture engineering, a new approach for metabolic engineering. Metab. Eng. 37, 114–121. Available from: https://doi.org/10.1016/j.ymben.2016.05.007.

Zhong, W., Gu, T., Wang, W., Zhang, B., Lin, X., Huang, Q., et al., 2010. The effects of mineral fertilizer and organic manure on soil microbial community and diversity. Plant. Soil. 326 (1), 511–522. Available from: https://doi.org/10.1007/s11104-009-9988-y.

Zhou, J., Fong, J.J., 2021. Strong agricultural management effects on soil microbial community in a non-experimental agroecosystem. Appl. Soil. Ecol. 165, 103970. Available from: https://doi.org/10.1016/j.apsoil.2021.103970.

Zhou, Z., Wang, C., Luo, Y., 2020. Meta-analysis of the impacts of global change factors on soil microbial diversity and functionality. Nat. Commun. 11 (1), 1–10. Available from: https://doi.org/10.1038/s41467-020-16881-7.

Zlobin, I.E., 2022. Linking the growth patterns of coniferous species with their performance under climate aridization. Sci. Total. Environ. 154971. Available from: https://doi.org/10.1016/j.scitotenv.2022.154971.

Zuber, S.M., Villamil, M.B., 2016. Meta-analysis approach to assess effect of tillage on microbial biomass and enzyme activities. Soil. Biol. Biochem. 97, 176–187. Available from: https://doi.org/10.1016/j.soilbio.2016.03.011.

CHAPTER

15 Evaluating chemical and physical properties of soil for maize production under agroforestry environments in South Africa

Thabo Nkuna[1,2], P. Maponya[2], C. Madakadze[1] and Z. Dube[3]

[1]*Department of Plant and Soil Sciences, Faculty of Natural and Agricultural Sciences, University of Pretoria, Hartfield, South Africa*

[2]*Department of Crop Science, Agricultural Research Council – Vegetable, Industrial and Medicinal Plants, Pretoria, South Africa*

[3]*Department of Agriculture, Faculty of Agriculture and Natural Science, University of Mpumalanga, Mbombela, South Africa*

15.1 Introduction

Agroforestry is documented as one of the strategies to achieve various benefits through interactive and intentional land use systems and technologies where trees are integrated with agricultural crops or animals. Beyond the beneficial elements, there are also harmful properties of the system on the agricultural crops. These include competition for light, water, and soil nutrients. Organic chemicals might also be discharged as leachates or leaf extract, root exudates, and/or products of decomposition which affect annual crops (Nyaga et al., 2017). Alebachew et al. (2015) reported that there was a significant reduction in crop biomass and yield when maize (*Zea mays*) were grown adjacent to Eucalyptus trees. They affirmed that the suppressive effect of Eucalyptus was higher where plant stands were close to Eucalyptus, and its effect decreased as distance was increased in all the three crops. Thus it expresses the allelopathic potential of Eucalyptus and other trees in agroforestry-based cropping system. Additionally, Iqbal et al. (2016) reported that the adverse effect of leaf extracts of Eucalyptus on root growth and nutrient uptake resulted in reduced shoot growth. They further reported that effective microorganism application reduced the adverse effect and enhanced shoot growth significantly through a better root system as a result of the decomposition of organic matter, which promotes the growth of crop plants.

Maize is a crop produced globally as major source of food to millions of people in Africa. The crop is reported to easily adapt to different environmental

Sustainable Agricultural Practices. DOI: https://doi.org/10.1016/B978-0-443-19150-3.00006-0

zones, hence its proposed use in agroforestry. However, the crop could be susceptible to allelochemicals released by trees, such as Eucalyptus and pine. Limited information on suitable soil is not yet investigated under agroforestry environment. The major threat to smallholder agricultural practices in South Africa is that of declining soil fertility which has led to poor agricultural productivity and decreased household food security. This problem has been aggravated by continued water and nutrient withdrawal by agroforestry trees (Piotrowska and Wilczewski, 2015). Most of the research conducted under agroforestry in South Africa has focused on tree integration with agricultural crops, therefore there is a need to understanding the dynamics allelochemicals in the soil released by both trees and weeds and have an effect on physical and chemical composition in the soil. These components play a major role of nutrients such as nitrogen (N), phosphorus (P), and metals (iron-Fe, Zn-zinc, copper-Cu), as well as organic matter, which substantially improves soil quality. However, they often contain high concentrations of heavy metals such as nickel (Ni), lead (Pb), and cadmium (Cd); its use in agriculture for consecutive years requires special caution because there is a risk of phytotoxicity symptoms occurrence due to metals accumulation, but also a possibility of metals' transfer to the food chain and a threat to human and animal health (Mcgrath et al., 1988).

Soil nutrient depletion as a result of continuous cultivation of soils without adequate addition of external inputs is a major challenge in the highlands of South Africa. Many interrelated factors, both natural and managerial, cause soil fertility decline. This decline may occur through leaching, soil erosion, and crop harvesting (Donovan and Casey, 1998). Unless the nutrients are replenished through the use of organic or mineral fertilizers, or partially returned through crop residues, or rebuilt more comprehensively through traditional fallow systems that allow restoration of nutrients and reconstruction of soil organic matter, soil nutrient levels decline continuously. The soils in the central highlands of South Africa are Humic Nitisols, with moderate to high inherent fertility (Jaetzold and Schmidt, 1983); however, their fertility has declined over time, with an annual net nutrient depletion exceeding 30 kg/N (Smaling, 1993) as a result of continuous cropping with insufficient external nutrient replenishment. For instance, a long-term trial in South Africa indicated that a fertile red soil lost about 1 Mg/ha of soil organic N and 100 kg/P/ha of soil organic P during 18 years of continuous maize-common bean rotation in the absence of nutrient inputs (Sanchez et al., 1997). During this period, maize yields without N and P fertilizer inputs decreased from 3 to 1 Mg/ha (Bekunda et al., 1997; Sanchez et al., 1997). In most smallholder farms, these deficiencies could be replenished through the use of mineral fertilizers and cattle manure. The situation is, however, further aggravated by the fact that even the farmers using mineral fertilizers hardly use the recommended rates (60 kg/N/ha) in the area, with most of them applying less than 20 kg/N/ha (Adiel, 2004). As a result, soil fertility has continued to decline as has the productivity of the land (Kapkiyai et al., 1998; Adiel, 2004).

Locally available organics could be used to curb this problem. For instance, Mugendi et al. (1999) reported that soil incorporation of calliandra and leucaena green biomass with or without fertilizer increased total soil nitrogen by 1%–8% over a period of 4 years. During the same period, total soil nitrogen declined by 2%–4% when biomass was not applied. Mutuo et al. (2000) reported that treatments that had received Tithonia biomass had a high residual effect of 50% yield increase above the control in western Kenya. Technologies that combine mineral fertilizers with organic nutrient sources can be considered as better options in increasing fertilizer use efficiency and providing a more balanced supply of nutrients (Donovan and Casey, 1998). Combination of organic and mineral fertilizer nutrient sources has been shown to result in synergistic effects and improved synchronization of nutrient release and uptake by crop (Pam and Novotny, 1997) leading to higher yields; especially when the levels of mineral fertilizers used are relatively low as is the case in most smallholder farms of central Kenya (Kapkiyai et al., 1998). Maize yields were increased with increasing rates of farmyard manure application; however, maize grain yields above 3.5 Mg/ha were only obtained when both farmyard manure and NP fertilizers were applied (Kihanda, 1996). Leucaena biomass combined with mineral fertilizer gave higher crop yields as compared to sole use of mineral fertilizer or sole leucaena biomass (Mugendi et al., 1999).

Historically, most smallholder farmers in South Africa were settled on marginal soils, characterized by heavy clay, high calcium content (calcareous), or high sand. The introduction of agricultural crops including maize in developmental projects in rural communities may be limited by these marginal soils. However, the responses of maize crops to marginal soils had given much attention. Thus there was a need to investigate the effects of marginal soils that are widely distributed among agroforestry areas in South Africa. The aim of the study was to match the production of maize crops to suitable soil type in smallholder farming systems under agroforestry environments of South Africa. Therefore the objective was to determine the compatibility of maize production under agroforestry environment.

15.2 Material and methods

15.2.1 Study location

The study was conducted in SAFCOL Tzaneen in Limpopo, South Africa Location (0°20′07″S; 37°37′14″E), Mopani District. According to Jaetzold and Schmidt (1983), the area is in upper midlands 2 and 3 (UM2–UM3) with an altitude of approximately 1500 m above sea level, annual mean temperature of about 30°C, and annual rainfall varying from 1200 to 1400 mm. The rainfall is bimodal, falling in two seasons, the long rains lasting from September through December and short rains from January through March. The soils are Humic Nitisols

(Jaetzold and Schmidt, 1983), which are deep and well weathered with moderate to high inherent fertility.

15.2.2 Experimental procedure

Soil samples were collected during summer season from tree plantations in Limpopo and kept in cold temperature until the time of analysis. The soil samples were randomly collected from the 20 cm topsoil part of each of the two-trial site, Safcol Block J4 and Safcol Block J25. Soil pH (H_2O) and (KCl) was measured for all samples: (H_2O) as follows, 10 g soil sample were shade-dried, sieved, and mixed with 25 mL deionized water in a beaker (Thomas, 1996). The mixtures were stirred for 5 seconds using a glass rod, incubated for 50 minutes and pH measured using a pH meter (Spectrum Technologies, Inc.). (KCl) as follows, 10 g sieved dry soil was mixed with 25 mL deionized water and mixed with KCl in –:– (v/v) ratio (Thomas, 1996). The mixture was stirred and measured.

The procedure was done according to the method of Mchlich (1974). A 100 mL of 5% dispersed solution of water was mixed with 880 mL of deionized water in a 1000 mL cylinder (blank mixture). In another reaction mixture, 50 g of soil and 100 mL of 5% dispersing solution were mixed. The suspension was quantitatively transferred to a 1000 mL cylinder and filled to the 1000 mL mark with deionized water equilibrated to room temperature. At the beginning of each set, temperature was read together with the hydrometer reading of the blank. Hydrometer was used to record the hydrometer reading at 40 seconds. This gave the amount of silt plus clay suspended. The sand would have settled to the bottom of the cylinder. The hydrometer reading was recorded again after 6 hours and 52 minutes. The reading will give the amount of clay in suspension. The reading gave the amount of clay in suspension. The silt would have settled to the bottom of the cylinder. Soil P, K, Ca, Mg, and Na were extracted using Mehlich III method and analyzed via inductively coupled plasma (ICP).

15.3 Results and discussion

Results appeared as follows: pH values for the soils were 5.77 and 5.1 (H_2O); however, 5.56 and 5.33 (KCl) for Blocks J4 and J25, respectively. Most changes appear in soil chemical properties were consistent with predictions based on nutrient budgets were P, K, Ca, Mg, and Na contributed to 4.5, 157, 883, 191, and 17.7 mg/kg at Block J4, whereas Block J25 resulted in 4.71, 49, 264, 107, and 8.1, respectively (Table 15.1). Soil texture results indicated that clay percentage at Block J4 was 50%, silt percentage was 30.2%, which gave the sandy texture a percentage of 19.8; and Block J25 results exhibited clay percentages of 46%, silt percentage 37.7%, and sandy texture 16.3% in the soils, respectively (Table 15.2). Under high soil pH, the nutrient elements phosphorus, potassium,

Table 15.1 Partitioning soil chemical properties and pH of Block J4 and Block J25.

	Method used						
	P-Bray 1 (mg/kg)	Ammonium acetate (mg/kg)				pH (KCl)	pH (H_2O)
Identity	Phosphorus	Calcium	Potassium	Magnesium	Sodium		
Safcol Block J4	4.5	883	157	191	17.7	5.56	5.77
Safcol Block J25	4.71	264	49	107	8.1	5.33	5.1

Table 15.2 Partitioning soil physical properties (soil texture) of Block J4 and Block J25.

Identity	Clay %	Sandy %	Silt %
Safcol Block J4	50	30.2	19.8
Safcol Block J25	46	16.3	17.7

sulfur molybdenum, and boron became available in toxic quantities, whereas nitrogen, calcium, and magnesium became unavailable to the plants. However, diverse group of microorganisms participate in solubilization of native reserve potassium in soils. However, the rhizosphere microorganisms extensively contribute in the mobilization of bound form of soil minerals in the soil (Supanjani et al., 2006; Sindhu et al., 2009). Different groups of soil microorganisms are able to solubilize mineral form of K to the form utilized by the crop plant (Zarjani et al., 2013; Gundala et al., 2013). The use of these microbial inoculants as potassium solubilizing biofertilizer are also economically viable and eco-friendly. The first evidence of solubilization of rock potassium by microbial involvement had been shown by Muentz (1980). Numerous fugal and bacterial species support plant growth by mobilization of insoluble forms of K and making it available for plant uptake.

Clay contributed a higher percentage in both the areas. Generally, clay soil has high organic matter and mineral elements, along with high capacity to retain water compared to other soils. Soil texture is an important factor where clays generally compromise the majority of cation exchange sites in the soils. This is because clay by virtue of their small particle size have the good surface area and, therefore, the most exchange sites. Consequently, clay soils have the greatest risk for excess sodium binding and dispersion (Leal et al., 2009). Clay mineralogy appears to be one of the most important indicators of soil quality, and proper knowledge of the clay mineralogy of a soil has significant practical implications on the use of fertilizers, on application and management, and bioavailability of heavy metals in soil. Furthermore, soil mineralogy-soil fertility relationship is a basic component in understanding and interpreting the results of fertility experiments and soil analytical data, especially in developing fertilizer recommendation systems. However, many soil fertility studies have hardly considered the role of soil mineralogy in understanding soil fertility and nutrient dynamics in agro ecosystems. The lack of integrated soil mineralogy-soil fertility studies is indicated by the lack of interest by soil fertility researchers in carrying out detailed results .

These results corroborates with Jat et al. (1993) and Mugendi et al. (1999) who reported a general reduction in pH after application of mineral fertilizer, and leucaena and calliandra biomass. There was also an increase in organic carbon, exchangeable Ca and K, after application of manure, and this is consistent with the work of Gao and Chang (1996). Changes in soil properties under organically and conventionally managed farming systems have been found to be more variable, perhaps due to differences in climate, crop rotation, soil type, or the length of time soil has been under a particular management (Werner et al., 1997). The lower pH in the combination of organics and mineral fertilizers compared to the sole application of organic inputs in the calliandra and leucaena treatments could be as a result of the H^+ ions, which are added on the cation exchange complex of soils from the mineral fertilizer (Tisdale et al., 1993). The pH increase with manure treatment corresponds with the findings by Eghball (2002) and could be attributed to the reduction of exchangeable aluminum in these acidic soils (pH 4.5–5.5) (Hue and Amien, 1989). This reduction is considered to occur through aluminum precipitation or chelation on organic colloids or by complexation of soluble aluminum by organic molecules, especially organic acids (Hue and Amien, 1989).

Attributes occurred on both soil sites allow maize cultivars to perform prodigious under agroforestry environment. Maize grown successfully in a variety of soils ranging from clay to clay loam that are high in soil nutrient accumulation, as indicated in both areas. However, soils with good organic matter content, having high water-holding capacity with acidic to neutral pH are considered good for high crop productivity (Leal et al., 2009). A mix of clay and organic matter, to hold water and give some depth and weight to the soil. Corn prefers a pH which is slightly acidic. One of the reasons ancient people used to grow corn, beans, and squash or pumpkins together was because they each use different nutrients, and the beans and squash would keep the soil covered, so the corn did not dry out.

15.4 Conclusion and recommendations

Findings in this study are of great importance in the use of maize cultivars as an alternative crop in marginal communities of South Africa. However, since the findings were accrued in soil analysis under agroforestry conditions, it is important that the study be empirically practiced under field conditions prior to rolling-out the findings to various communities. It is highly recommended that trend maize cultivars be planted in both areas under these environments while being monitored over an extended period in order to attain a more complete reflection on the impact of different treatments or crop practices might have on microbial diversity and activity as an indicators of soil fertility and health.

Acknowledgments

The authors wish to acknowledge the following: Agricultural Research Council, South African Forestry Company Limited (SAFCOL), Mountain to Ocean (MTO), Department of

Environment, Forestry and Fisheries (DEFF), Universities of Pretoria and Mpumalanga. A special thanks to Mr. Johan Habig (agritechnovation) and Dr. Mariette Marais (ARC-Biosystematics) for their excellent and professional work in analyzing the soil data.

References

Adiel, R.K., 2004. Assessment of On-Farm Adoption Potential of Nutrient Management Strategies in Chuka Division, Meru south, Kenya. MSc Thesis, Kenyatta University, Kenya.

Alebachew, M., Amare, T., Wendie, M., 2015. Investigation of the Effects of *Eucalyptus camaldulensis* on Performance of Neighbouring Crop Productivity in Western Amhara, Ethiopia. Amhara Agricultural Research Institute (ARARI), Adet Agricultural Research Center, Bahir Dar, Ethiopia.

Bekunda, A.M., Bationo, A., Ssali, H., 1997. Soil fertility management in Africa: a review of selected research trials. In: Buresh, R.J., Sanchez, P.A., Calhoun, F. (Eds.), Replenishing Soil Fertility in Africa. Soil Science Society of America, Madison, WI, Special publication No. 51.

Donovan, G., Casey, 1998. Soil Fertility Management in Sub-Saharan Africa. World Bank. Technical Paper, p. 408.

Eghball, B., 2002. Soil properties as influenced by phosphorus and nitrogen based manure and compost applications. Agron. J. 94, 128–135.

Gao, G., Chang, C., 1996. Changes in CEC and particle size distribution of soils associated with long-term annual application of cattle feed lot manure. Soil. Sci. 161, 115–120.

Gundala, P.B., Chinthala, P., Sreenivasulu, B., 2013. A new facultative alkaliphilic, potassium solubilizing, Bacillus spp. SVUNM9 isolated from mica cores of Nellore district, Andhra Pradesh, India. J. Microbiol. Biotechnol. 2 (1), 1–7.

Hue, Nv, Amien, I., 1989. Aluminium detoxification with green manures. Commun. Soil Sci. Plant. J. 20, 1499–1511.

Iqbal, M., Irshad, S., Nadeem, M., Fatima, T., Itrat, A.B., 2016. Salinity Effects on Wheat (Triticum aestivum L.) Characteristics: A Review Article. Department of Life Sciences, The Islamia University of Bahawalpur, Pakistan.

Jaetzold, R., Schmidt, H., 1983. Farm management handbook of Kenya, Natural Conditions and Farm Information, vol 11/C. East Kenya. Ministry of Agriculture, Kenya.

Jat, M., Jat, M.L., Jat, R.K., Singh, P., Jat, S.L., Sidhu, H.S., Jat, H.S., et al., 1993. Predicting yield and stability analysis of wheat under different crop management systems across agro-ecosystems in India. Agron. J. 85, 754–757. Available from: https://doi.org/10.2134/agronj1993.00021962008500030042x.

Kapkiyai, J.J., Karanja, N.K., Woomer, P., Qureshi, J.N., 1998. Soil organic carbon fractions in a long-term experiment and the potential for their use as a diagnostic assays in highland farming systems of central Kenya. Afr. Crop. Sci. J. 6, 19–28.

Kihanda, F.M., 1996. The Role of Farmyard Manure in Improving Maize Production in the Sub-Humid Central Highlands of Central Kenya. PhD Thesis, UK.

Leal, R.M., Herpin, U.W.E., Fonseca, D.A., Firme, A.F., Montes, C.R., Melfi, A.J., 2009. Sodality and salinity in a Brazilian oxisol cultivated with sugarcane irrigated with wastewater. Agric. Water Manag. 96 (2), 307–316.

Mcgrath, S.P., Brookes, P.C., Giller, K.E., 1988. Effects of potentially toxic metals in soil derived from past applications of sewage sludge on nitrogen fixation by *Trifolium repens* L. Soil. Biol. Biochem. 20 (4), 415–424.

Mchlich, A., 1974. Uniformity of soil test results as influenced by extractants and soil properties. In: Wehrmann, J. (Ed.), Proceedings Seventh International Colloquium Plant Analysis and Fertilizers Problem. Hanover, Germany, pp. 295–305.

Muentz, 1980. Surla décomposition desroches etla formationde la terrearable. CR Acad. Sci. 110, 1370–1372.

Mugendi, D.N., Nair, P.K.R., Mugwe, J.N., O'neill, M.K., Woomer, P.L., 1999. Calliandra and leucaena alley cropped with maize. Part 1. Soil fertility changes and maize production in the sub-humid highlands of Kenya. Agrof. Syst. 46, 39–50.

Mutuo, P.K., Mukalama, J.P., Agunda, J., 2000. On-farm testing of organic and inorganic phosphorous source on maize in Western Kenya. The Biology and Fertility of Tropical Soils. TSBF Report, p. 22.

Nyaga, J., Jefwa, J.M., Muthuri, C.W., Okoth, S.A., Matiru, V.N., Wachira, P., 2017. Influence of Soil Fertility Amendment Practices on Ex Situ Utilisation of Indigenous Arbuscular Mycorrhizal Fungi and Performance of Maize and Common Bean in Kenyan Highlands. Botany Department, Jomo Kenyatta University of Agriculture and Technology (JKUAT).

Piotrowska, A., Wilczewski, E., 2015. Influences of Catch Crop and Its Incorporation Time on Soil Carbon and Carbon-Related Enzymes. Available from: https://doi.org/10.1016/S1002-0160(15)30037-0 Corpus ID: 93061995.

Pam, A., Novotny, M., 1997. Macroporous polyacrylamide/poly(ethylene glycol) matrixes as stationary phases in capillary electrochromatography. Am. Chem. Soc. Available from: https://doi.org/10.1021/ac970626g. Submitted for publication.

Sanchez, P.A., Shepherd, K.D., Soule, M.J., Place, F.M., Buresh, R.J., Izac, A., et al., 1997. Soil fertility replenishment in Africa: an investment in natural resources capital. In: Buresh, R.J., Sanchez, P.A., Calhoun, F. (Eds.), Replenishing Soil Fertility in Africa. Soil Science Society of America, Madison, WI, Special publication No. 51.

Sindhu, S.S., Verma, M.K., Suman, M., 2009. Molecular genetics of phosphate solubilization in rhizosphere bacteria and its role in plant growth promotion. In: Khan, M.S., Zaidi, A. (Eds.), Phosphate Solubilising Microbes and Crop Productivity. Nova Science Publishers, New York, pp. 199–228.

Smaling, E., 1993. Soil nutrient depletion in sub-Saharan Africa. In: Van Reuler, H., Prins, W. (Eds.), The Role of Plant Nutrients for Sustainable Food Crop Production in Sub-Saharan Africa. VKP, Leidschendam, The Netherlands.

Supanjani, H.H.S., Jung, S.J., LEE, K.D., 2006. Rock phosphate potassium and rock solubilizing bacteria as alternative sustainable fertilizers, Agro. Sustain. Dev, 26. pp. 233–240.

Thomas, G.W., 1996. Soil pH and soil acidity. In: Sparks, D.L. (Ed.), Methods of Soil Analysis. Part3, Chemical Methods. SSSA Book Series, 5. Soil Science Society of America, Madison, WI, pp. 159–165.

Tisdale, S.L, Ghosal, P., Chakraborty, T., Banik, P., 1993. Phosphorus fixing capacity of the Oxic Rhodustalf—alfisol soil in the Chotanagpur plateau region of Eastern India. Agric. Sci. 2 (4). Available from: https://doi.org/10.4236/blr.2014.54024.

Werner, B., Parent, C., Paulin-Mohring, C., Saibi, A., 1997. LVDS I/O buffers with a controlled reference circuit. LVDS I/O buffers with a controlled reference circuit. Available from: https://doi.org/10.1109/ASIC.1997.617028.

Zarjani, J.K., Aliasgharzad, N., Oustan, S., Emadi, M., Ahmadi, A., 2013. Isolation and characterization of potassium solubilizing bacteria in some Iranian soils. Arch. Agron. Soil. Sci. 59 (12), 1713–1723.

CHAPTER

16 Cytological observations of intracellular microbes in plants, their roles in sustainable crop production, and effects of elevated carbon dioxide on rhizophagy in roots

April Micci[1], Kathryn Kingsley[1], Fernando Velazquez[1], Xiaoqian Chang[1], Ajay Kumar[2] and James F. White[1]

[1]*Department of Plant Biology, Rutgers University, New Brunswick, NJ, United States*

[2]*Amity Institute of Biotechnology, Amity University, Noida, Uttar Pradesh, India*

16.1 Introduction

Plants are inhabited by communities of nonpathogenic microbes (Arnold and Lutzoni 2007; Rosenblueth and Martinex-Romero 2006; Magnani et al., 2010; Johnston-Monje and Raizada 2011; Hardoim et al., 2015; Shehata and Dumigan et al., 2017; White et al., 2021). This microbiome consists of both bacterial and fungal components and may exist on plant surfaces and interiors. In a survey of seedling roots of 23 species of plants using the reactive oxygen stain diaminobenzidine tetrahydrochloride, White et al. (2014, 2018) found microscopic evidence for widespread occurrence of microbes within root cells (see also Table 16.1). However, we do not fully understand how this microbiome functions in plants. Research indicates that the microbiome inhabitants enhance host plant resistance to biotic and abiotic stresses (Kloepper, 1993; Redman et al., 2002; Waller et al., 2005; Weber et al., 2007; White and Torres, 2010; Torres et al., 2012; Verma et al., 2021; Rodriguez et al., 2009; Clay et al., 2005; Hamilton et al., 2012; Lata et al., 2018; Kuldau and Bacon, 2008). These studies suggest that the microbiome may possess defensive features for plant hosts. Plant microbiomes also possess nutritional properties (White et al., 2019; Reinhold-Hurek and Hurek, 2011; Puente and Bashan, 1994; Kumar et al., 2016; James, 2000; Hurek et al., 1994;

Sustainable Agricultural Practices. DOI: https://doi.org/10.1016/B978-0-443-19150-3.00016-3

Table 16.1 Survey of seedlings of 23 species in 16 families of seed plants for intracellular bacteria and evidence of bacterial oxidation/degradation.

Family	Species	Origin	Bacterial distribution in root cells	Cells showing bacterial degradation	Tissues where bacteria observed	Bacterial morphology
Agavaceae	*Agave chrysantha*	Sonoran desert, Arizona	Periplasmic space	Root hairs	Roots	Intercellular and intracellular: rod-like
Agavaceae	*Agave palmeri*	Sonoran desert, Arizona	Intercellular, periplasmic space, nuclei, cytoplasm	Root hairs, root cortex, root epidermis, root cap	Roots, shoots	Intercellular: rod-like. Intracellular coccoid
Agavaceae	*Agave schottii*	Sonoran desert, Arizona	Intercellular, periplasmic space, nuclei, cytoplasm	Root hairs, root cortex, root epidermis, root cap	Roots, shoots	Intercellular: rod-like. Intracellular: coccoid
Agavaceae	*Yucca schottii*	Sonoran desert, United States	Intercellular, periplasmic space, nuclei, cytoplasm	Root hairs, root cortex, root epidermis, root cap	Shoots, Roots	Intercellular: rod-like. Intracellular: coccoid
Anacardiaceae	*Rhus radicans*	New Jersey, United States	Intercellular, periplasmic space, cytoplasm	Root hairs, root epidermis	Roots	Intercellular: rod-like and coccoid. Intracellular: coccoid
Apiaceae	*Coriandrum sativum*	Commercial source; Mexico	Intercellular, root hairs, root epidermis	Root hairs, root epidermis	Roots	Intercellular and intracellular: coccoid
Araliaceae	*Hedera helix*	New Jersey, United States	Intercellular, periplasmic space, cytoplasm	Root hairs, root epidermis	Shoots, roots	Intercellular: rod-like. Intracellular: coccoid
Asparagaceae	*Ophiopogon japonicus*	New Jersey, United States	Intercellular, cytoplasm	Root cortex specialized oxidation parenchyma	Roots	Intercellular and intracellular: rod-like
Brassicaceae	*Brassica napus*	Commercial source: United States	Intercellular, root hairs	Root hairs	Roots	Intercellular: rod-like. Intracellular: coccoid and rod-like

Brassicaceae	*Lepidium virginicum*	New Jersey, United States	Intercellular, periplasmic space, cytoplasm	Root hairs, root epidermis	Shoots, roots	Intercellular: rod-like. Intracellular: coccoid
Cactaceae	*Cereus repandus*	Bonaire, Dutch Antilles	Periplasmic space, cytoplasm	Root hairs	Roots	Intracellular: coccoid
Cactaceae	*Ritterocereus griseus*	Bonaire, Dutch Antilles	Intercellular, periplasmic space, cytoplasm	Root hairs, root epidermis	Shoots, roots	Intercellular and intracellular: coccoid
Caprifoliaceae	*Lonicera japonica*	New Jersey, United States	Intercellular, periplasmic Space	Root hairs	Shoots, roots	Intercellular and intracellular: rod-like
Celastraceae	*Celastrus orbiculatus*	New Jersey	Intercellular, root hairs	Root hairs	Roots	Intercellular: rod-like. Intracellular: coccoid and rod-like
Cucurbitaceae	*Citrullus colocynthis*	Commercial source; Nigeria	Intercellular, periplasmic space, cytoplasm	Root hairs. root cortex, root epidermis, root cap	Roots	Intercellular: rod-like. Intracellular: coccoid and rod-like
Cucurbitaceae	*Cucurbita pepo*	Commercial source; United States	Intercellular, periplasmic space, cytoplasm	Root hairs. root cortex, root epidermis, root cap	Roots	Intercellular and intracellular: coccoid
Cyperaceae	*Fimbristylis cymosa*	Bonaire, Dutch Antilles	Intercellular, periplasmic space	Root hairs, root epidermis	Roots	Intercellular and intracellular: coccoid
Ericaceae	*Vaccinium oxycoccos*	New Jersey, United States	Intercellular, periplasmic space, cytoplasm	Root hairs. root cortex, root epidermis, root cap	Roots	Intercellular and intracellular: coccoid
Fabaceae	*Phaseolus acutifolius*	Commercial source, United States	Intercellular, periplasmic space, cytoplasm	Root hairs, root epidermis, root cap	Roots	Intercellular and intracellular: coccoid
Malvaceae	*Thespesia populnea*	St. John, US virgin Islands	Intercellular, periplasmic space	Root hairs	Roots	Intercellular and intracellular: rod-like

(*Continued*)

Table 16.1 Survey of seedlings of 23 species in 16 families of seed plants for intracellular bacteria and evidence of bacterial oxidation/degradation. *Continued*

Family	Species	Origin	Bacterial distribution in root cells	Cells showing bacterial degradation	Tissues where bacteria observed	Bacterial morphology
Moringaceae	*Moringa oleifera*	Commercial	Intercellular, periplasmic space, cytoplasm	Root hairs. root cortex, root epidermis, root cap	Shoots, roots	Intercellular: rod-like. Intracellular: coccoid and rod-like
Poaceae	*Festuca arundinacea*	Commercial, United States	Intercellular, periplasmic space, cytoplasm	Root Hairs, root epidermis	Shoots, roots	Intercellular: rod-like. Intracellular: coccoid
Poaceae	*Panicum virgatum*	New Jersey, United States	Intercellular, periplasmic space, cytoplasm	Root hairs, root epidermis	Roots	Intercellular: rod-like and coccoid. Intracellular coccoid

Modified from White, J.F., Torres, M.S., Verma, S.K., Elmore, M.T., Kowalski, K.P. and Kingsley, K.L., 2018. Evidence for widespread microbivory of endophytic bacteria in roots of vascular plants through oxidative degradation in root cell periplasmic spaces. In: A. Kumar, A. Singh and V. Singh, edsEds., Sustainable Agriculture: Food Security and Environmental Management. New York: Elsevier, https://doi.org/10.1016/B978-0-12-815879-1.00009-4, showing the bacterial morphology and distribution in plant tissues and cells. Note the variety of plant families that host intercellular and intracellular bacteria from commercial and wild seed.

Doebereiner, 1992; Döbereiner et al., 1994; Glick, 1995; James, 1998; James et al., 1994). "Associative nitrogen fixation," where endophytic bacteria fix nitrogen and stimulate plant growth, has been hypothesized to be responsible for efficient nitrogen use in stimulating plant growth and nutrition in many noncultivated plants without chemical nitrogen applications (James, 2000; Urquiaga et al., 1992; Taulé et al., 2012). Most research on plant microbiome nutritional effects on plants has focused on individual bacteria that are applied to plants. Limited information is available about how plants function to extract nutrients from microbes. The mechanisms of how plants manage and extract nutrients from microbes are a largely unanswered question. The work of Paungfoo-Lonhienne et al. (2010) demonstrated that tomato plants (*Solanum lycopersicum*) and *Arabidopsis thaliana* were capable of microbivory through "endocytosis" and degradation of microbes within root cells. Microbivory is a process of microbial consumption occurring in some heterotrophic protozoans and some small animals (e.g., nematodes) where the eukaryote engulfs and degrades prokaryotes or other microbes (Mikola, 1998). In plants, microbivory by root cells has been termed "rhizophagy" (etymology: root eating) (Paungfoo-Lonhienne et al., 2013). This work indicates that plants have a capability to obtain nutrients through lysis or other extraction of nutrients from microbes within its endosphere. Widespread "rhizophagy" or other mechanisms for nutrient extraction from microbes in plants could have important consequences for plant-based agriculture. Development of a thorough understanding of how plants employ microbes to obtain nutrients from them could lead to new strategies for plant management that use microbes as "living nutritional supplements" instead of chemical fertilizers (White et al., 2021). In this chapter, we will examine what is known regarding plant use of microbes as nutrient sources—and consequence of this method of nutrient acquisition in plants with regard to plant development, disease resistance, stress tolerance, and production of antioxidants and other chemical components of plants.

16.1.1 Roles of soil/plant bacteria in reducing disease caused by soil-borne fungal pathogens

Ascomycete and Basidiomycete fungi are common soil-borne phytopathogens. Agronomic practices impact the disease cycle, along with other environmental interactions which can change the soil microbe community. Selman Waksman started to research a group of filamentous bacteria, the actinomycetes, for the antibiotic properties in 1915. In the study of microbe interactions, we still use many of his original methods for testing inhibition, like dual culture. Recently, scientists at the Brazilian Biorenewables National Laboratory discovered that Volatile Organic Compounds (VOCs) of *Pseudomonas* could inhibit the growth of the ascomycete plant pathogen *Thielaviopsis ethacetica*, which was confirmed with transcriptome analysis and electron microscopy showing the altered growth of mycelia (Freitas et al., 2022). *Fusarium kuroshium*, an ascomycete phytopathogen of Avocado, was tested in classical antagonism assays against rhizosphere associated *Bacillus* species which reduced the growth of mycelia up to 48%, while VOC testing showed a nearly 70% reduction in mycelial growth (Guevara-Avendaño et al., 2020). *Rhizoctonia solani*, a basidiomycete pathogen of economic importance for several crops, including rice, has also been the subject of VOC testing with rhizosphere bacteria in double Petri dish assays with nine different strains of bacteria from eight different species, showing a range of growth inhibition from 8% to 100% (Wang et al., 2021).

Mechanisms of action against fungal diseases may be based on simple competition for resources, parasitism, or involve complex molecular interactions like siderophore production (Carmona-Hernandez et al., 2019). Antagonism can reduce the availability of polysaccharides to postharvest fungal pathogens and has been seen in strains of *Pseudomonas*, *Pantoea*, and *Bacillus* (Carmona-Hernandez et al., 2019). *Pseudomonas putida* species are known to colonize plant roots but lack genes to produce virulence factors such as enzymes for cellulose degradation in plant cell walls or the phospholipid bilayer of human cells. Other strains of *Pseudomonas*, such as *P. syringae* show the ability to impede fungal growth but are also known as plant pathogens. *P. syringae* has been extensively studied for its role in pathology with over 50 pathovars identified to cause disease in specific compatible host plants, but recent phylogenetic analysis has determined more members of this species live as commensals (Xin et al., 2018). Using *P. syringae* in food systems needs further work to determine safety, as immunocompromised consumers could be at a higher risk of bacterial infection through dietary exposure.

Bacterial endophytes may be passed through seed, with some studies showing that it is a consortium of bacterial endophytes that work together to provide protective or growth benefits. Quartana et al. (2022) shows that out of four *Pantoea* strains and one *Paenibacillus* strain isolated from durum wheat seedlings, combinations of two *Pantoea* strains were better than single isolates at improving plant vigor assessed by shoot height and root length while two other strains of *Pantoea* worked better at reducing fusarium foot rot disease caused by the ascomycete *Fusarium culmorum*. *F. culmorum* is pathogenic to many cereal crops but endophytic in the coastal dunegrass *Leymus mollis* (Rodriguez and Redman, 2008).

16.1.2 Roles of soil microbes in delivery of nutrients to plants via increased solubilization in soils

It has generally been believed that plants absorb most nutrients from soil liquid. It has been, in part, this belief that has resulted in the development and sustained use of fertilizers (e.g., ammonia, nitrate, phosphate, potash). The fact that chemical fertilization works to push crops to increase yields seems to provide support for this idea. It has been argued that microbes may play some role in increasing solubilization of immobilized nutrients. In this vein, microbes with significant phosphatase, or phosphorus solubilizing activity, are often sought as biofertilizers.

Similarly, microbes that show capability for nitrogen fixation are thought to fix nitrogen from the atmosphere, and these microbes may eventually be degraded in soils to release mineralized nitrogen. Minerals like iron have been shown to be absorbed directly into root tips. The root is thought to be the only plant organ that functions to acquire nutrients, and root hairs are the structures that extend into the soil to absorb additional nutrients in soil. This is the "dissolved nutrient hypothesis" (DNH) and is the main way that most people believe that plants obtain nutrients. It is all very reasonable and supported by decades of teaching and analysis—but may be only partially correct. Over the past several decades, it has become abundantly clear that plants contain communities of microbes (bacteria and fungi) as endophytes. Most of these endophytes colonize roots, while a smaller subset colonizes aerial parts of plants. The presence of nitrogen-fixing microbes within plant tissues has spurred some to hypothesize that plants may be absorbing nitrogen from these microbes, although mechanisms have not been clearly shown. Still some diazotrophs in plants have been shown to contribute nitrogen to plants (Roley et al., 2019); however, the DNH is still viewed by most as the main mechanism for nutrient acquisition in plants.

16.2 Microbial hormone production and modulation of development of plants

In 2012 White et al. reported that grass seeds (*Poa annua*, *Lolium perenne*, and *Festuca arundinacea*) that were rigorously surface disinfected (agitation in 3.5% NaOCl for 50 + minutes) showed altered development. Seedling roots, in particular, were impacted by removal of microbes. Without microbes on seeds, seedling roots often showed absence of the gravitropic response where roots failed to grow downward with gravity; although root hair initials were present on roots, they did not contain microbes internally and root hairs failed to elongate on roots. When microbes were reinoculated onto seeds prior to germination, seedlings recovered showing both gravitropism and elongation of root hairs from root hair initials that contained bacteria (Fig. 16.1).

Torres et al. (2012) further showed that presence of proteins in the medium around seedlings that lacked bacteria was enough to trigger gravitropism and root hair elongation. Verma et al. (2017) showed that bacteria in rice also modulate root development and without bacteria in root cells, root hairs did not elongate (Verma et al., 2018a,b). Since these observations, we have found that roots of dicots and monocots show loss of root hair elongation when all or most of the seed microbes are removed from plants (White et al., 2018). We have conducted dozens of microbial removal and replacement experiments and in every case, seedlings do not form root hairs when microbes are removed—and when microbes are added to axenic seedlings, root hairs form. This seems to be a basic phenomenon with regard to plant development—roots appear to require microbes

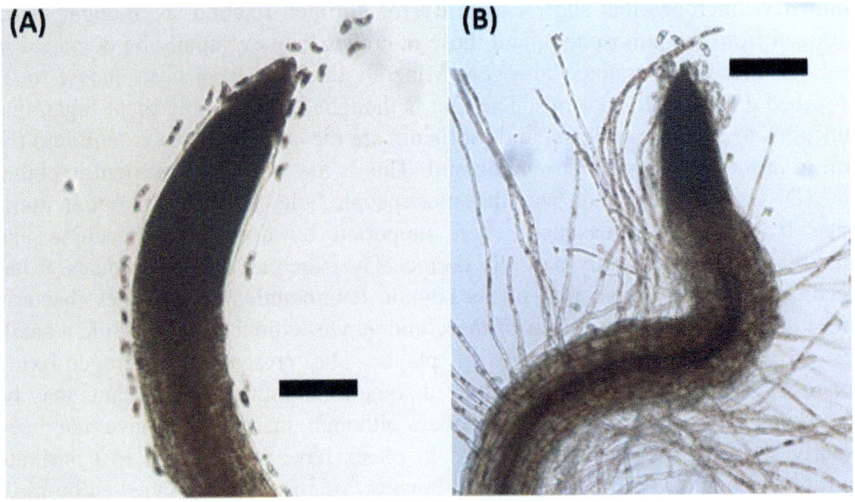

FIGURE 16.1

Bermuda grass (*Cynodon dactylon*) root tips stained for hydrogen peroxide using diaminobenzidine tetrahydrochloride (brown color). (A) Root where bacteria were removed by seed disinfection protocol showing absence of root hair formation (bar = 0.5 mm). (B) Root where bacteria were removed then Bacillus sp. were added to the seed prior to germination, showing abundant root hair formation (bar = 0.5 mm).

within root cells to trigger root hair formation. In a paper published in 2021, Chang, Kingsley, and White showed that bacteria within root hairs produce ethylene (Fig. 16.2) and nitric oxide, both of which are hormones that may trigger elongation of root hairs.

In the absence of bacteria in root cells, providing protein that is oxidized by plant-produced superoxide on/in root cells may result in formation of nitric oxide through partial oxidation—and this may explain how protein itself triggers root hair elongation. Whether these hormones are triggering gravitropism in roots, as well, is uncertain. As a result of these experiments, it is apparent that roots are plant organs that internalize soil microbes—and without microbes they do not develop properly. This linkage between root hair development and microbe internalization into root cells is possibly the result of several factors: (1) In early evolution of plants—soil microbes that surrounded and permeated plants became a reliable factor to the extent that their presence in plant tissues became important developmental modulators; (2) the nutritional function of root hairs involves the processing of microbes internally; and (3) root hairs may function to increase populations of compatible microbes, and growth of the hair is linked to ejection of microbes from hairs to increase levels of microbes in soils around plant roots. Studies have shown that microbes stimulate root branching and, in general, stimulate root growth (Verma et al., 2021). These effects are generally attributed to

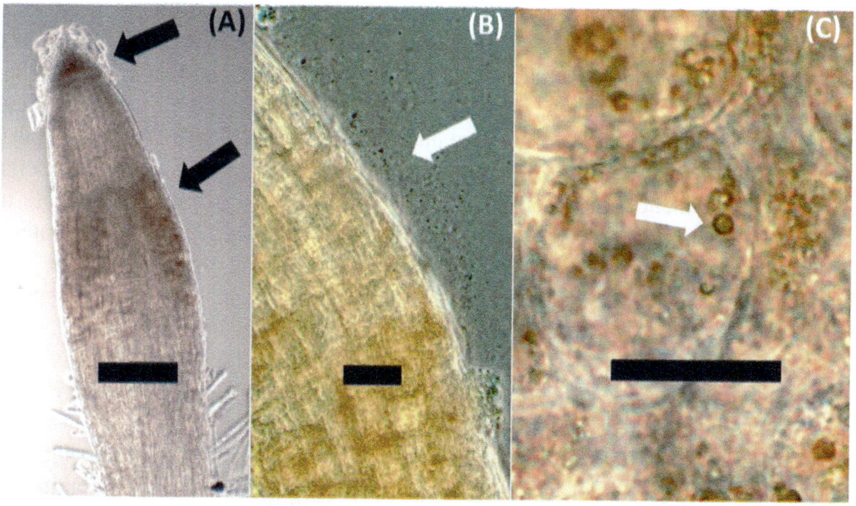

FIGURE 16.2

Tomato seedling root showing ethylene (purple-orange color) in tissues associated with bacteria after staining with ammonium molybdate. (A) Root showing root cap cells at tip with ethylene staining (upper arrow) and another zone below the root tip meristem where ethylene is produced (lower arrow) (bar = 1 mm). (B) Layer of bacteria (arrow) over the epidermis of the root in the area where bacteria are entering into epidermal cells of the root (bar = 25 μ). (C) Root cells showing ethylene (white arrow; purple color) around bacteria in root cell (bar = 10 μ).

hormones (e.g., auxin) production by microbes associated with plant roots; however, ethylene and nitric oxide could play roles in the gravitropic response of roots as well (Verma et al., 2021).

16.3 Mechanism of entry of bacteria into plant cells

Microbes are attracted to root tips by exudates (mostly carbohydrates, amino acids, and organic acids) that are secreted into the soil to attract and cultivate microbes at the root tip. It is unknown how plants internalize microbes into their tissues and cells. Paungfoo-Lonhienne et al. (2010) proposed that plant cells physically engulfed microbes in a process comparable to phagocytosis as seen in ameba, slime molds, or other animals. This possibility was also suggested by Paungfoo-Lonhienne et al. (2013). White et al. (2018) proposed that microbes enzymatically degraded plant cell walls to enter host cells by removing of organic acids (primarily butyric and propionic acids) at the root tip meristem and may trigger certain microbes to express increased virulence, resulting in use of cell wall degrading enzymes to penetrate cells. Certainly, many of the microbes that enter plant cells have cellulases and other enzymes that would be needed to degrade cell walls. However, over the past several years, we have made some observations that suggest another mechanism. It seems clear that microbes are attracted to, and cultivated on, growing root (and shoot) tips due to exudates that are released from the meristematic plant tissues (White et al., 2018). We also have observed that some bacteria degrade cell walls of plant cells to exit the cells. There is also evidence that charges on meristematic root tips could play a role in pulling into root tissues microbes that bear positive charges due to cation content. This idea may be referred to as the "attraction-of-opposites hypothesis." In support of this idea, root tip staining experiments using organic dyes that are positively or negatively charged suggest that root tips bear a negative charge. This may be due to absence of cations in meristematic cells since these cations have not yet been absorbed into plant tissues at the root tip meristem. Conversely, bacteria from soils would be expected to contain cations that have been absorbed from soils. Figure 16.3A shows the results of an expirement with the positively charged dye "crystal violet". The root tip meristem region was highly saturated,

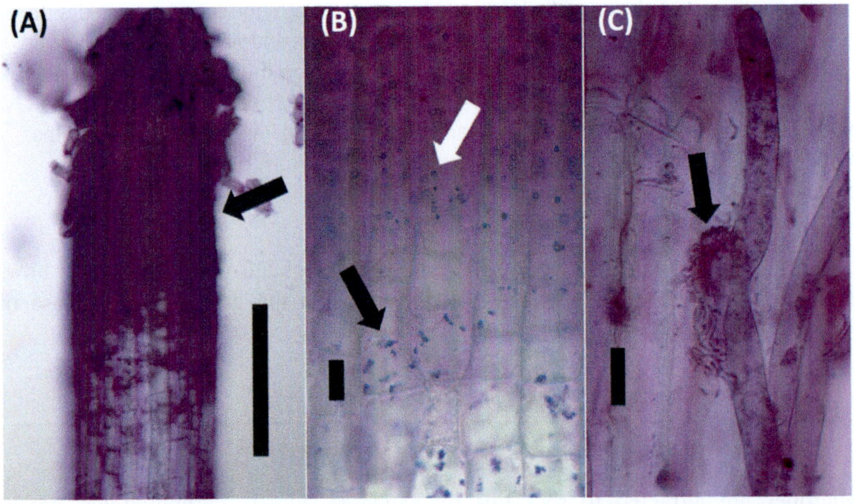

FIGURE 16.3

Annual bluegrass (*Poa annua*) root stained with positively charged dye crystal violet stain. (A) Root tip showing adherence of the dye to the region around the meristems (arrow; bar = 0.5 mm) showing positively charged stain around root tip where bacteria enter into root cells. (B) Close-up of root epidermal cells showing the transition from the tip charged area to older uncharged cells (bar = 20 μ). Bacteria within cells of the negatively charged cells do not stain with the dye (white arrow), while more staining is observed in bacteria (black arrow) in the noncharged cells. (C) Root hair showing bacteria (arrow) emerging from hair and causing curvature of the hair where bacteria emerged (bar = 20 μ). Bacteria are seen to stain with crystal violet as they emerge from the hair.

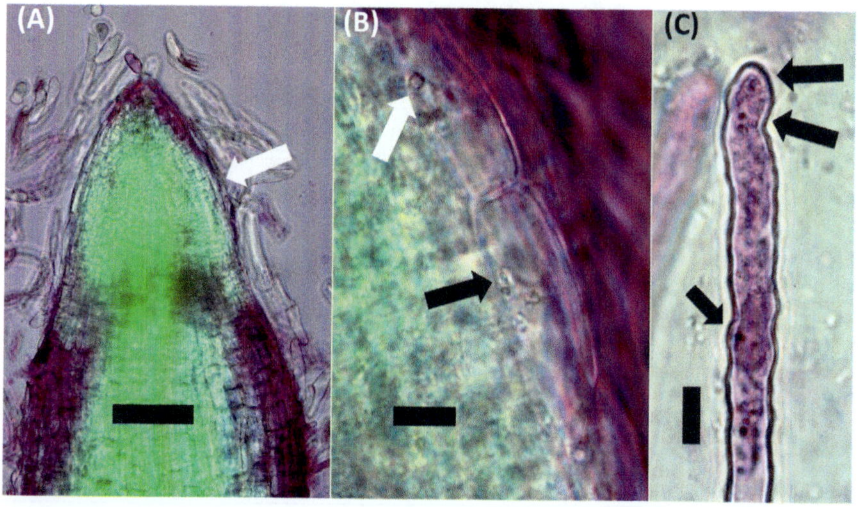

FIGURE 16.4

Clover (*Trifolium repens*) seedling root tissues stained with negatively charged Rose Bengal stain. (A) Root tip showing absence of staining in the meristem (arrow; bar = 100 μ). (B) Close-up of epidermal cells in the area where bacteria (arrows) are entering into epidermal cells (bar = 15 μ). (C) Root hair showing bacteria (arrows) within the root hair (bar = 10 μ).

while older parts of the root are not. Conversely, a similar experiment using the negatively charged dye "Rose Bengal" showed that staining did not occur at the root tip meristem (Fig. 16.4).

The negative charges on root tip cells could play a role in pulling bacterial cells from the soil that may bear a net positive charge due to their content of positively charged cations (e.g., Ca^{++}, Fe^{++}, Zn^{++}, Mn^{++}, Mg^{++}, etc.). Such positively charged cation-rich microbes may be pulled into the plant tissues preferentially over microbes with fewer cations. These positively charged microbes may fill the developing crevasses at the cell junctions of proliferating meristematic roots (and leaves) (Figs. 16.5 and 16.6). In this respect, it is notable that positively charged crystal violet did not stain the bacteria just after entry into the cells at the meristem tip (Fig. 16.5A; white arrow); while bacteria stain densely in older cells (Fig. 16.5B; black arrow) and after bacteria are ejected from root hair tips (Fig. 16.5C; black arrow). Such a pattern of staining may occur if bacteria charged with cations enter plant cells at the root tip, and as cations are extracted oxidatively from bacteria in the root tip, they become negatively charged and begin to stain with the positively charged crystal violet.

The actual internalization of microbes into plant cells may be in part a physical phenomenon, where bacteria that are pulled into, or otherwise enter, the space between epidermal cell junctions are pressed through the soft plant cell walls as the

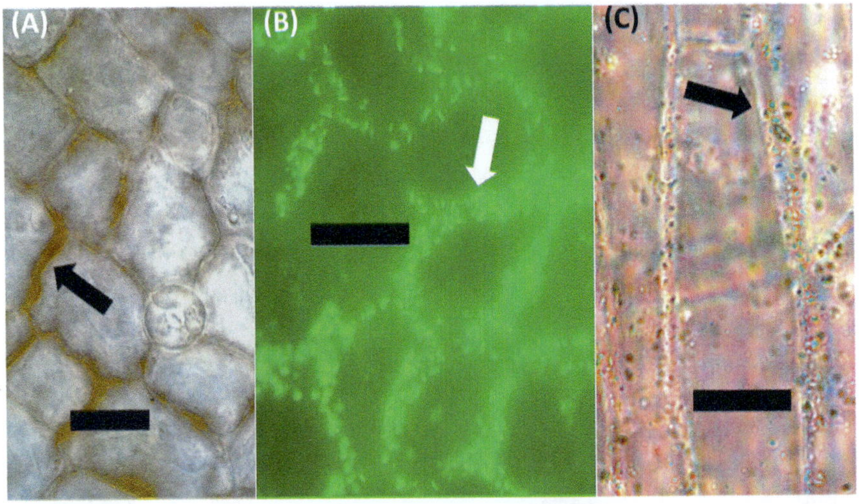

FIGURE 16.5

Young leaves of *Amaranthus caudatus* with and without bacterial biofilms of bacteria (*Bacillus* sp.) that situate over epidermal cell junctions (B and C). (A) Epidermis of leaf stained with diaminobenzidine tetrahydrochloride to show hydrogen peroxide (arrows) at cell junctions (bar = 40 μ). (B) Bacterial biofilm (arrow) formed over leaf epidermal cell junctions stained with nuclear stain Syto9 and visualized with fluorescence microscopy (bar = 40 μ). (C) Leaf epidermis showing bacteria (arrow) in the cell junctions (bar = 40 μ).

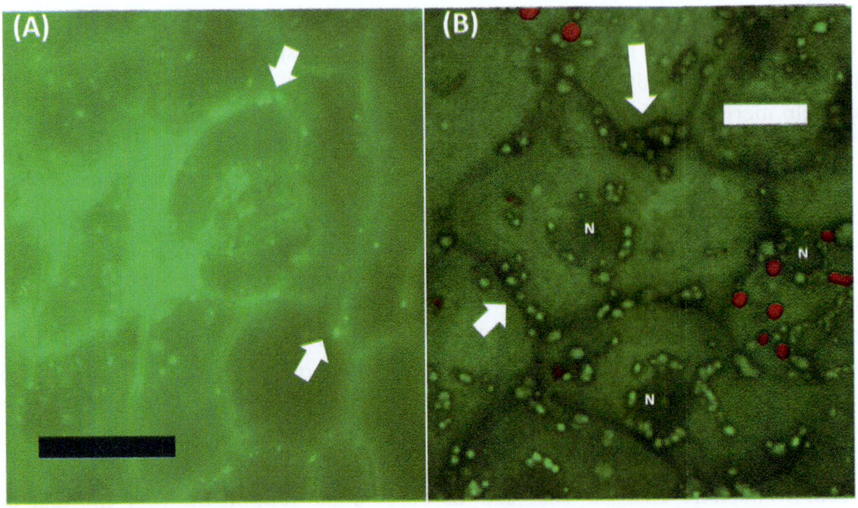

FIGURE 16.6

Amaranthus caudatus inoculated with *Bacillus* sp. from *Vanilla phaeantha* stained with nuclear stain Syto9 to show bacteria. (A) Leaf epidermis showing bacteria (arrows) aligning at cell junctions (bar = 40 μ). (B) Confocal image of the leaf epidermis showing bacteria (arrows) penetrating into leaf cells at the cell junctions (bar = 10 μ). Red structures in cells are chloroplasts in guard cells; nuclei (N) are seen to be surrounded by bacteria.

individual epidermal cells grow or expand. Fig. 16.6B (arrows) shows the penetration of bacteria (*Bacillus pumilus*) into root epidermal cells. The bacteria that enter at the root tip are later replicated in root hairs and are eventually ejected from tips of root hairs (White et al., 2021). This internalization mechanism may be more readily seen in developing leaves where bacteria are seen to grow above and at the epidermal cell junctions (Figs. 16.5A–C and 16.6). This pattern of entry into developing epidermal cells may explain why bacteria are often observed at margins of cells and may be seen to enter cells at junctions in growing tissues.

16.4 Plant cell cytoplasmic entry by bacteria

It is evident that some of these bacteria are internalized into the host cell cytoplasm. Ultrastructural studies of *Dahlia* seedling roots showed bacterial protoplasts (L-forms) and those containing what we interpret as endospores within the cytoplasm of root cells (Fig. 16.7). In these bacteria, thin cell walls were apparent but vesicular membranes were not observed (Fig. 16.7A; arrow).

What is more shocking is that at least some of these endophytic bacteria appear to colonize into plant cell nuclei (Figs. 16.8 and 16.9). Several studies (White et al., 2014; Micci et al., 2022) have shown that bacteria associate with

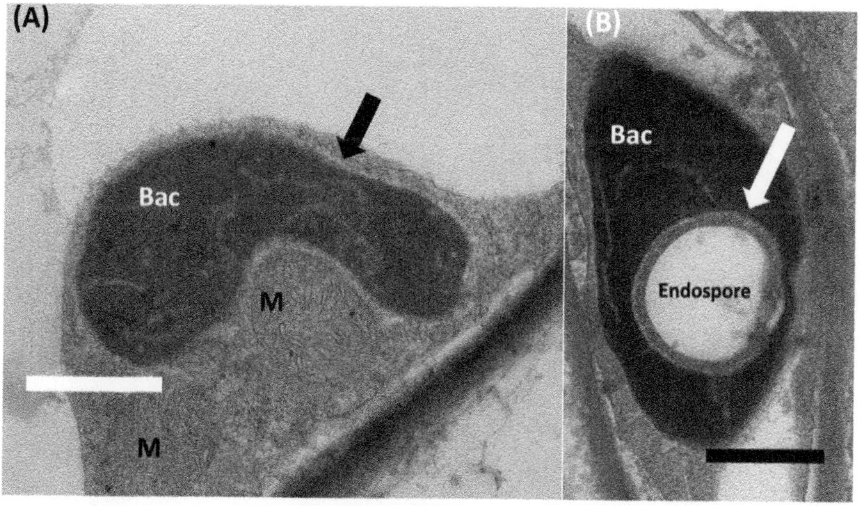

FIGURE 16.7

Ultrastructural image of Dahlia (*Dahlia variabilis*) seedling root cells showing what is interpreted to be intracellular bacteria (L-forms). (A) Dahlia seedling root cell showing what we interpret to be a bacterial protoplast (Bac) with a thin cell wall (arrow) near mitochondria (M) (bar = 0.5 μ). (B) Dahlia seedling root cell showing a bacterial protoplast (Bac) containing what we interpret to be an endospore (arrow) within the cytoplasm of the root cell (bar = 0.5 μ).

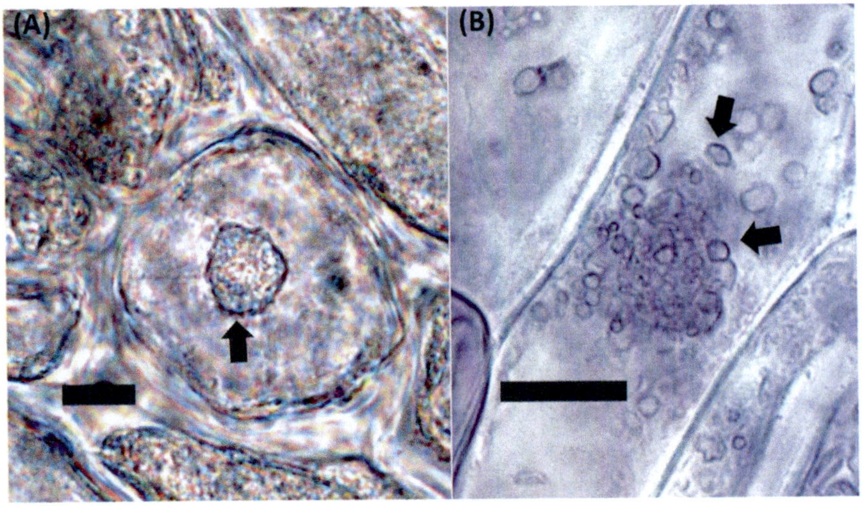

FIGURE 16.8

Agave spp. epidermal cells showing what we interpret to be bacteria on or within nuclei. (A) Epidermal cell of a leaf of an *Agave* sp. showing granular pattern of bacteria on, or within, the nucleus (arrow) (bar = 15 μ); stained with acidified diphenylamine. (B) Root cell of *Agave stricta* seedling showing oxidizing bacteria (arrows) exiting the nucleus (bar = 15 μ); stained with diaminobenzidine tetrahydrochloride.

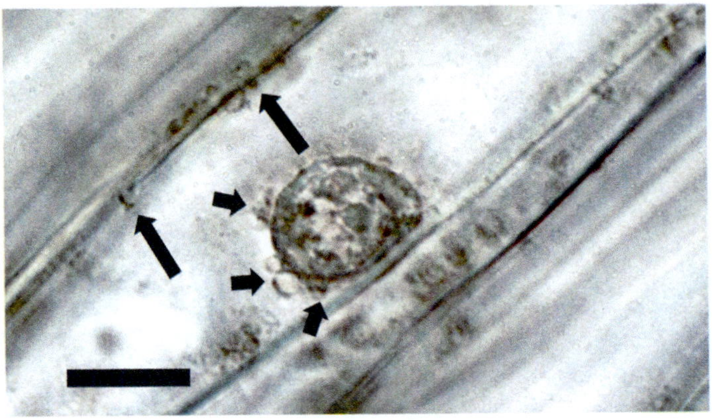

FIGURE 16.9

Agave stricta with bacteria (short arrows) nuclear colonization in root epidermal cell. Bacteria are also seen to be present in the periplasmic space (long arrows) (bar = 15 μ); stained with diaminobenzidine tetrahydrochloride (brown color).

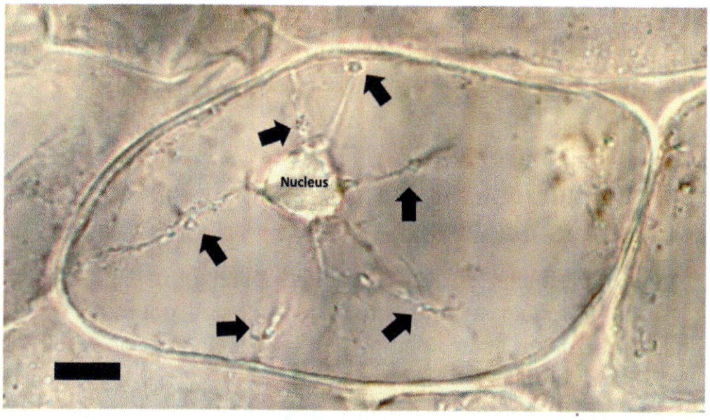

FIGURE 16.10

Bacterial tracks (arrows) from nucleus through cytoplasm of mint (bar = 15 μ).

nuclei (see Fig. 16.6B). Examination of *Agave* spp. and other monocots (e.g., grasses) suggest that nuclear colonization is common in monocots, especially in epidermal cells of roots and leaves (Micci et al., 2022).

Some of the nuclear-associated bacteria appear to be inside the nuclei; this appears more the rule than the exception. In *Agave stricta* root cells (Fig. 16.9), bacteria (arrows) can be seen to be located within the nucleus and on its surface. Bacteria appear to exit nuclei and move through the cytoplasm to the periplasmic space beneath the cell wall (Fig. 16.10).

In many instances where we have nuclear colonization by bacteria, we can observe what appears to be trails of bacteria (often as chains of rods) that extend from the nuclei to the periplasm of the cell. This can be seen in cells of the mint (Fig. 16.10; arrows) where several radiating trails of bacteria can be seen.

We interpret that the bacteria cultivated in the nucleus, move to the periplasm in these instances. This raises questions regarding the health or functioning of epidermal cells or other cells that have nuclear colonization. For example, some questions raised are: (1) How do the bacteria get into the nuclei? (2) Are there adaptions in the plant cell nuclei that enable proliferation of bacteria within nuclei? (3) Is there damage to plant chromosomes within the nuclei that contain bacteria? (4) How do the bacteria move to the periplasm? (5) Are bacteria in vesicles or some other membranous covering?

Bacteria may colonize nuclei due to low levels of oxygen there and low oxidative degradation (White et al., 2019, 2018; Mikola, 1998). Some bacteria have been reported to produce nucleomodulins that control gene expression in the host cell to reduce oxidative processes or modify activities of the cell to favor the nuclear bacteria (Bierne and Cossart, 2012). Nucleomodulins have been reported to be produced by *Agrobacterium tumefaciens* to control plant growth. Nuclear colonization also occurs in protozoans where Proteobacteria in family

Holosporaceae are phagocytized by the protozoan and then transported to the nucleus where bacteria multiply. It is unknown if any of this is occurring in plant cell nuclei where we observe bacteria.

16.4.1 Roles of microbes in the rhizophagy cycle to deliver soil nutrients to plants

In 2010 Paungfoo-Lonhiene et al. demonstrated that some plants absorbed soil microbes (yeasts and bacteria) into roots and degraded microbes as a source of nutrients; this process was later denominated "rhizophagy," or "root eating," but again a mechanism of nutrient extraction was not clearly articulated. In 2018 White, Kingsley et al. published a study that showed that numerous seedlings absorbed microbes into root cells and these intracellular microbes were exposed to plant-produced superoxide that appeared to strip cell walls from microbes and degrade some microbes, while other microbes survived the oxidation process within cells and were later ejected from tips of root hairs back into the soil (Figs. 16.11 and 16.12). Ejection of bacteria from root hairs is periodic, with ejections occurring at regular intervals when root hairs show growth spurts (Fig. 16.12). This cyclic process was referred to as the "rhizophagy cycle" because microbes alternated between a free-living phase in soils where nutrients

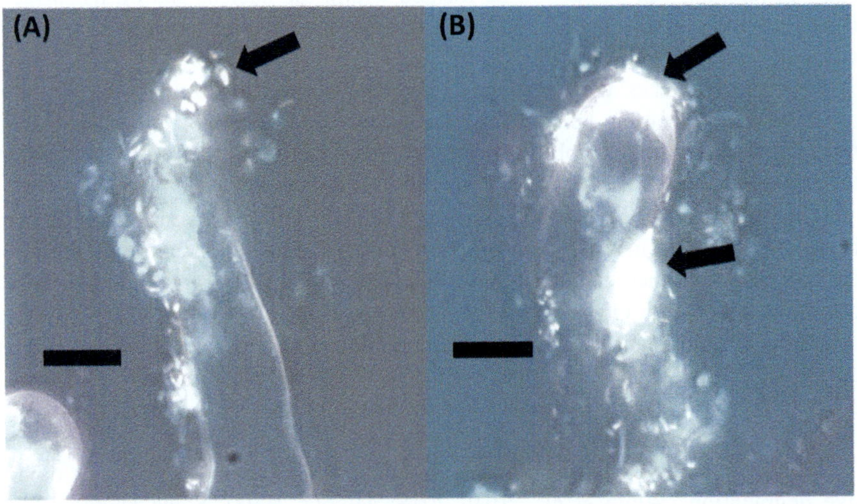

FIGURE 16.11

Yucca seedling root hair tips with bacteria (arrows). Stained with Syto9 a nucleic acid stain, visualized with fluorescence microscopy (bar = 10 μ). (A) shows bacterial cells with intact cell walls while (B) shows areas where cell walls have been stripped by oxidative processes and nucleic acids remain.

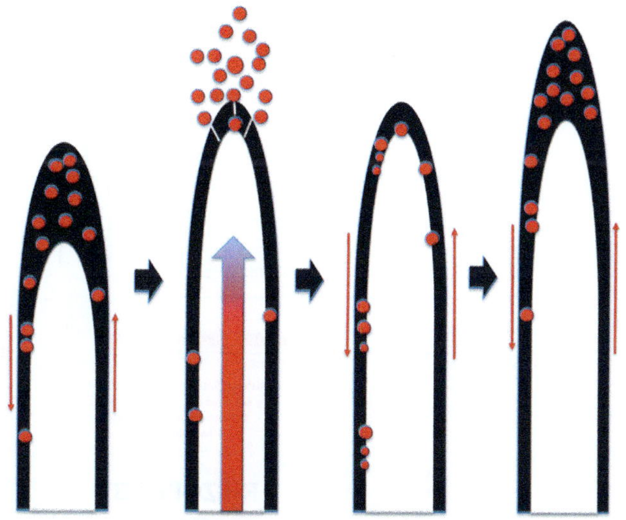

FIGURE 16.12

Diagrammatic representation of root hair microbe protoplast ejection (White et al., 2021). Microbes in tips are red to indicate superoxide exposure and nitrogen extraction; microbes cycling in hairs are blue to indicate replication and nitrogen fixation. (A) Cyclosis moves microbes to the periplasmic space at the top of the hair where they accumulate and secrete ethylene en mass. (B) Microbial ethylene triggers a growth spurt in the hair which manifests as an expansion wave that propagates from the base to the tip of the hair. The expansion wave (red arrow) causes growth in the hair and forces out the bacteria accumulated in the hair tip. (C) The remaining bacteria are taken back into the cyclosis stream and replicated. (D) As bacteria populations rebuild in root hairs, they again accumulate in the hair tips, subsequently triggering another ejection event.

are acquired and an endophytic phase in roots where nutrients may be extracted from bacteria (Fig. 16.13).

16.5 Nutrient absorption via rhizophagy cycle

Work by Hill et al. (2011) demonstrated transfer of nitrogen from microbes directly to plants. Beltran-Garcia et al. (2014) further showed that 15N-labeled bacteria transferred nitrogen contained within bacterial cells to plants, where it could be found in phaeophytin derived from chlorophylls. White et al. (2015) conducted experiments using grass seedlings to evaluate absorption of organic nitrogen in microbial proteins into roots. In these experiments, it was shown that seed-vectored bacteria on seedling grass roots were critical for nutrient absorption into roots.

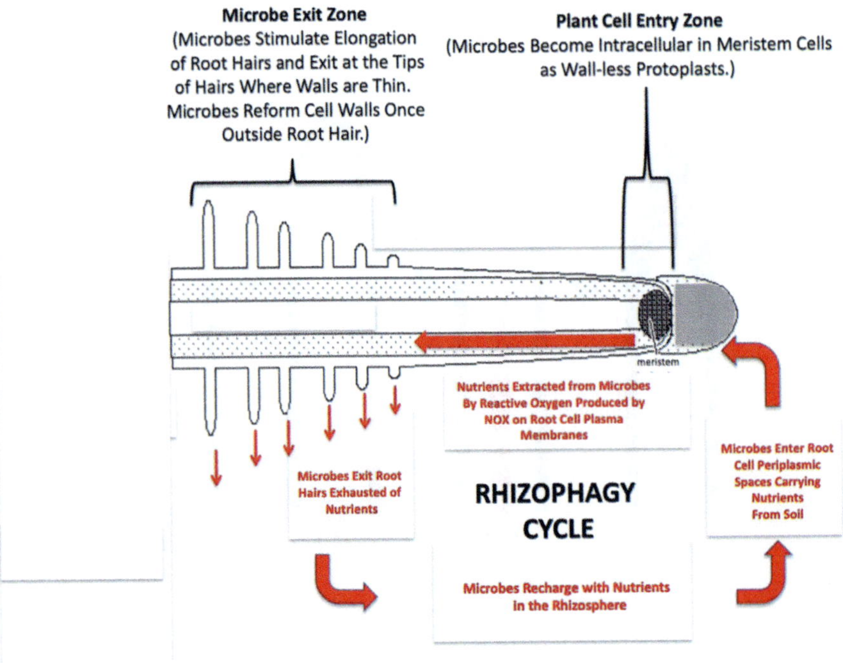

FIGURE 16.13

Diagrammatic representation of the rhizophagy cycle (White et al., 2018). Microbes alternate between a free-living phase in soils where bacteria acquire nutrients and an endosymbiotic phase in plant root cells where plants extract nutrients from microbes through exposure to superoxide produced on root cell plasma membranes.

Presence of bacteria on seedling roots resulted in an approximately 30% increase in absorption of nitrogen into plants (Fig. 16.14). Nutrient absorption experiments using plants with microbes reduced through seed surface sterilization continue to show that bacteria microbes in seeds are responsible for delivery of substantial amounts of nutrients to plants, including nitrogen, potassium, calcium, iron, phosphorus, and most macro and micro-nutrients (Figs. 16.15–16.18).

In the rhizophagy cycle process, bacteria or yeasts may be viewed as living packets of nutrients—where microbes enter the root cell at root tips where root cell walls are soft. Plant-produced superoxide strips all nutrients from cell walls of bacteria internalized into the root cell periplasm. The oxidized microbial cell wall remains are situated within the root cell periplasm beneath the plant cell wall—but just outside the root cell plasma membrane. On loss of cell walls, bacteria and yeasts become wall-less protoplasts, termed L-forms for bacteria (Errington, 2013) and "mycosomes" for fungi (Atsatt and Whiteside, 2014). These protoplast phases in plant cells are seen to form spherical phases that replicate rapidly, often in chains within the plant cell periplasm. The use of stains for

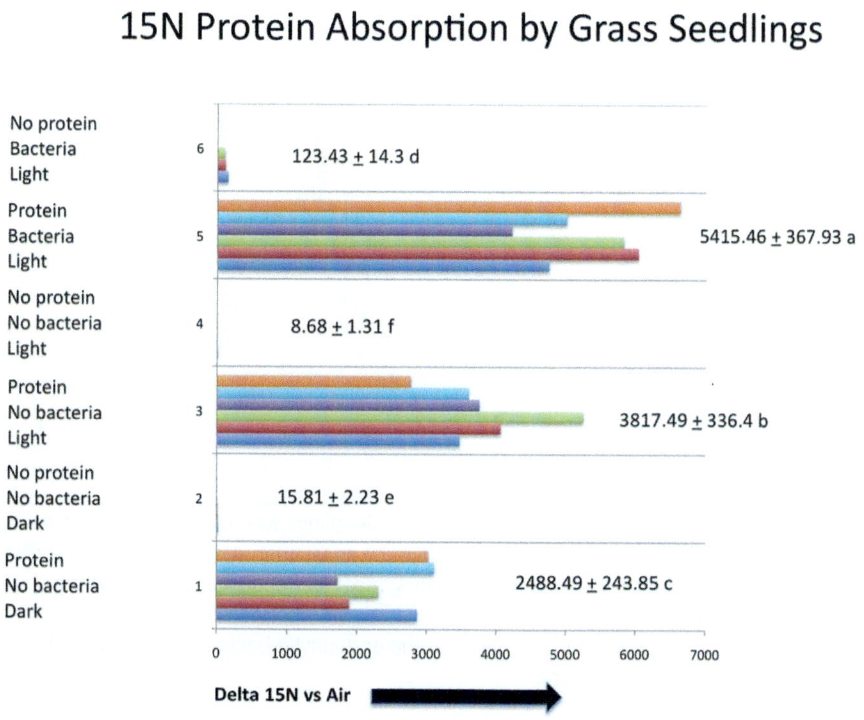

FIGURE 16.14

Results of 15N-labeled protein absorption experiment showing enhanced 15N absorption due to rhizobacteria and reactive oxygen effects (White et al., 2015). Averages with the same letter are not significantly different according to the Duncan's multiple range test ($P < .05$).

hydrogen peroxide, for example, diaminobenzidine tetrahydrochloride, or proteins, for example, aniline blue, that stain for hydrogen peroxide and proteins, respectively, are enough to visualize these bacteria in root cells (White et al., 2019, 2021). When using these stains, microbes are seen to be surrounded by reactive oxygen and to stain intensely blue in the cytoplasm with aniline blue. Bacterial cells within the periplasm are seen to frequently swell as they age and to lose staining of cytoplasm, eventually to disappear completely in the periplasm. We hypothesize that superoxide enters microbe cells by passing through the membrane envelope impacting on intracellular proteins that are oxidatively degraded. It is notable that prokaryote membranes are not oxidatively fortified with sterols, while plant membranes contain sterols that increase their resistance to the superoxide produced on them. As the internal proteins are denatured, cations and other nutrients are released from the proteins and diffuse from microbe cells into the periplasm where they may be absorbed by root cells, either via diffusion or nutrient transporters in plant cell membranes. Irizarry and White (2017) showed that

FIGURE 16.15

Winter wheat plants with and without bacteria. (A) Seedlings where bacteria were removed from seed of winter wheat by surface disinfection by agitation in 4% sodium hypochlorite and seedlings grown for 3 weeks in sterile potting mix. (B) Seedlings of plants where sterilized seeds were then treated with a bacteria (*Bacillus* sp.) isolated previously from winter wheat seeds; seedlings inoculated with the endophytic bacteria are substantially larger than plants without bacteria.

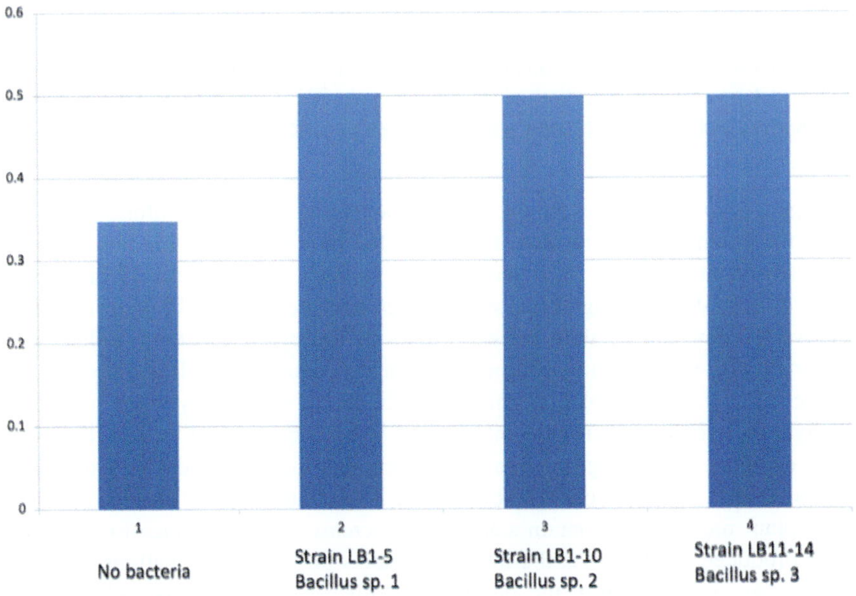

FIGURE 16.16

Nitrogen absorption into winter wheat seedlings in experiment from study in Fig. 16.15. Here strains LB1–5, LB1–10, and LB11–14 show enhanced content of nitrogen in leaves.

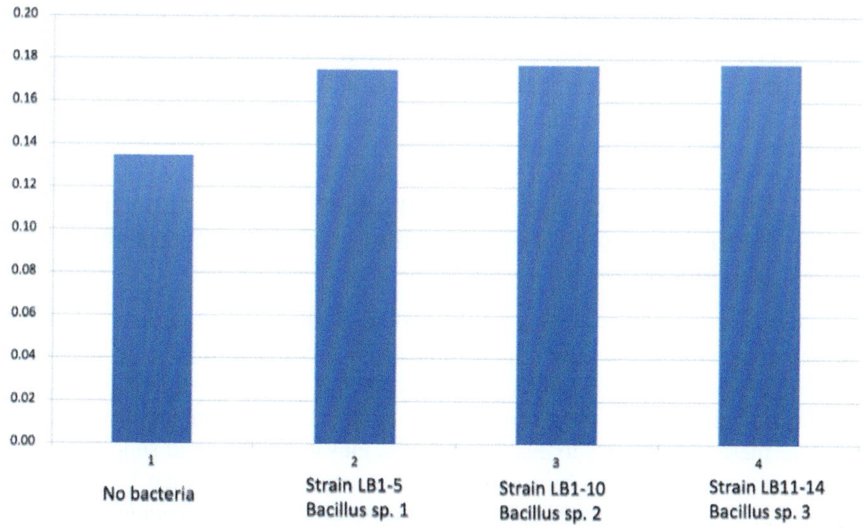

FIGURE 16.17

Phosphorus absorption into winter wheat seedlings in experiment from study in Fig. 16.15. Here strains LB1–5, LB1–10, and LB11–14 show enhanced content of phosphorus in leaves.

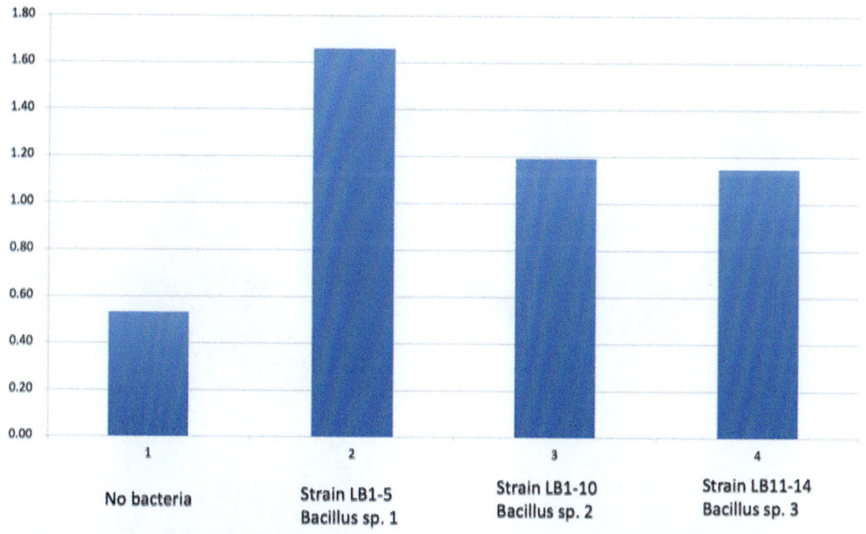

FIGURE 16.18

Potassium absorption into winter wheat seedlings in experiment from study in Fig. 16.15. Here strains LB1–5, LB1–10, and LB11–14 show enhanced content of nitrogen in leaves; with Bacillus stain LB1–5 showing absorption that is more than 3 × what is seen in sterile controls, while Bacillus strains LB1–10 and LB11–14 show only a 2 × increase in potassium absorption over controls.

after bacteria were provided to seedlings of cotton, nitrate transporters were upregulated in roots. On exposure to superoxide, the ultimate oxidized fate of organic nitrogen is expected to be nitrate (White et al., 2021).

16.5.1 Suppression of superoxide and nutrient extraction from bacteria in roots using elevated atmospheric carbon dioxide

Several experiments have been conducted to assess how plant nutrient extraction from bacteria is affected after suppression of superoxide formation in plants. In these experiments, elevated carbon dioxide content of air was seen to suppress production of superoxide by plant roots (White et al., 2019; Chiaranunt et al., 2020). In one of these experiments to evaluate whether elevated CO_2 could suppress bacterial replication in root cells by suppression of superoxide production by plant root cells, seeds of winter wheat (*Triticum aestivum*), tall fescue (*F. arundinacea*), and tomato (*Solanum tuberosum*) were lightly disinfected in 4% sodium hypochlorite for 5 minutes with continuous agitation before being placed on agarose in Petri dishes. Five Petri dishes for each plant species (with 5–10 seeds each) were placed in a sealed chamber containing dry ice to elevate carbon dioxide to approximately 180,000 ppm (~0.3 g dry ice/liter of air). The other Petri dishes were placed in a chamber with air (at approximately 410 ppm carbon dioxide). After 1 week, seedlings from both treatments were examined (Figs. 16.19–16.22). Staining using a

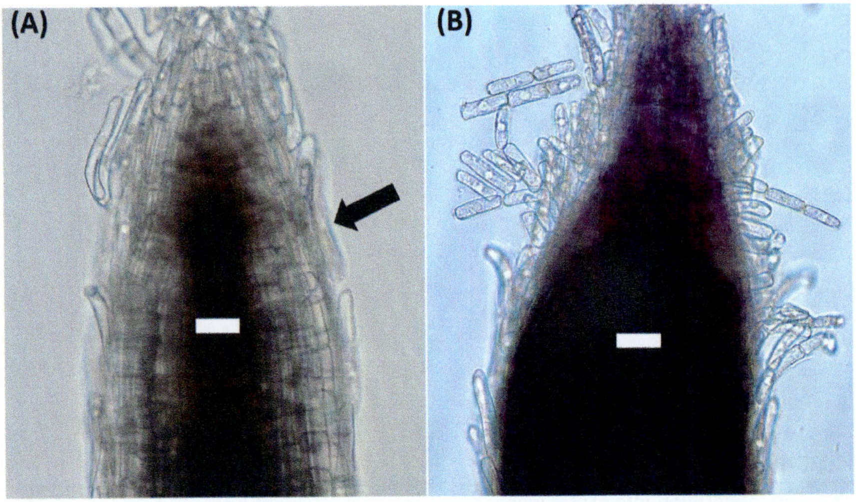

FIGURE 16.19

Wheat root tip from elevated carbon dioxide environment showing suppression of superoxide (purple color) in the epidermis and subepidermal layers of the root tip (A) compared to controls only in air (B) where superoxide is evident in the entire root tip (bars = 20 µ).

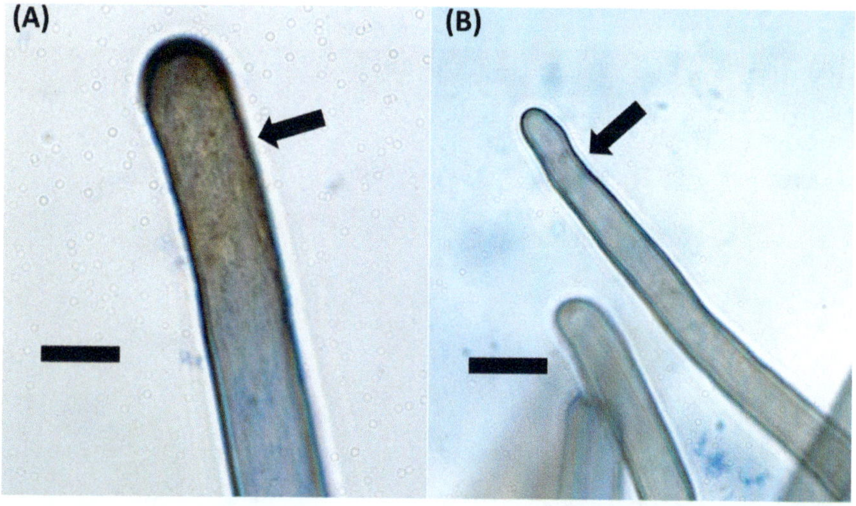

FIGURE 16.20

Tall fescue root hair tips in air (A) and in elevated carbon dioxide (B). (A) Root hair tip stained for hydrogen peroxide using diphenylamine benzidine tetrahydrochloride (DAB) showing brown-staining hydrogen peroxide around bacteria in the swollen root hair tip (arrow; bar = 10 μ). (B) Root hair showing absence of replication of bacteria within roots in the elevated carbon dioxide environment (bar = 20 μ). Bacteria were evident in root hairs as rods (not shown), but replication of bacteria within root hairs failed to occur and root hairs remain narrow and lack internal L-forms or other replication internally.

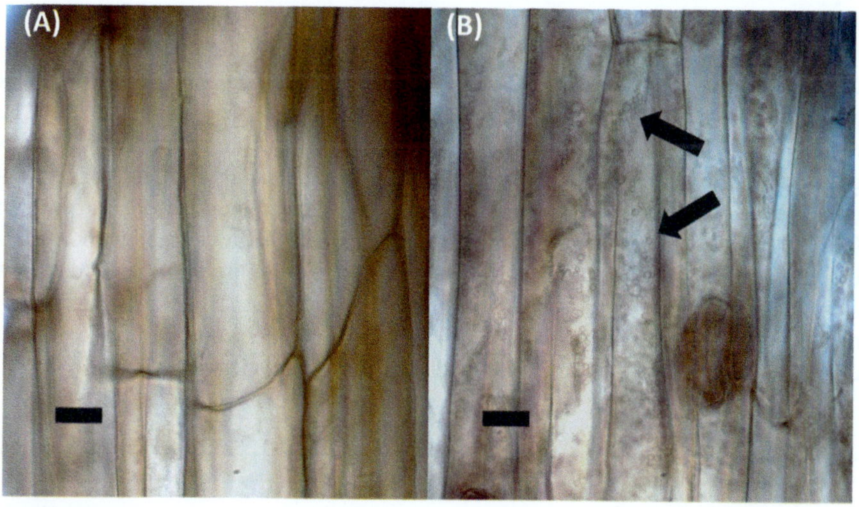

FIGURE 16.21

Tomato root epidermis with elevated carbon dioxide and air. (A) Root epidermis stained for hydrogen peroxide showing absence of bacteria (bar = 10 μ). (B) Root epidermis showing abundant presence of bacteria (arrows) within roots in the air only environment (bar = 10 μ).

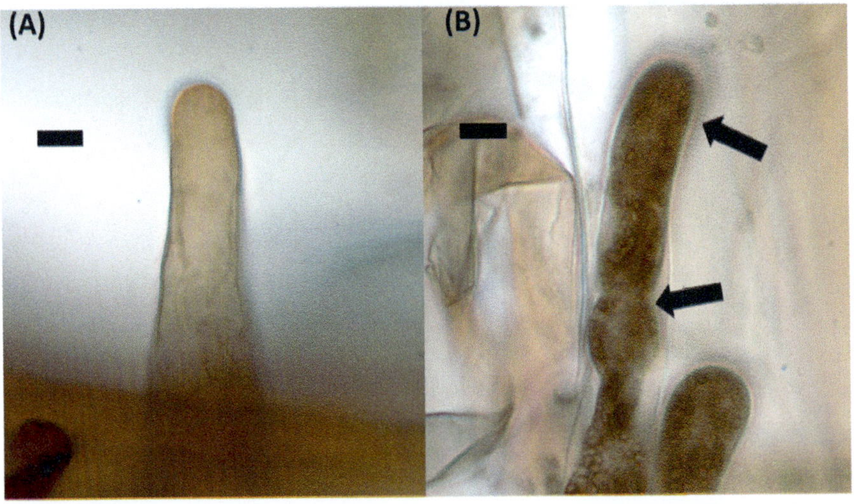

FIGURE 16.22

Tomato root hairs in elevated carbon dioxide (A) and air (B). (A) Root hair stained for hydrogen peroxide using DAB showing absence of bacteria (bar = 10 μ). (B) Root hair stained with DAB showing abundant presence of bacteria (arrows) within roots in the air only environment (bar = 10 μ).

superoxide stain nitro blue tetrazolium shows elevated carbon dioxide suppresses superoxide formation in outer layers of root tips where rhizophagy active microbes are present (Fig. 16.19). Root hairs that form in the elevated carbon dioxide treatments tend to show suppression in elongation of root hairs likely due to lack formation of wall-less L-forms in root hairs (Fig. 16.20). Because bacteria in root hairs modulate root hair elongation using ethylene and nitric oxide (White et al., 2021), repression of microbe replication in root hairs results in failure of root hairs to elongate. Even in tomato seedlings, elevated carbon dioxide results in failure of replication of bacteria in root cells (Figs. 16.21 and 16.22) and repressed root hair growth (Table 16.2). Growth of tomato plants for several weeks in an elevated carbon dioxide environment as above results in plants that are diminished in size (data not shown) with drastic decreases in potassium absorption (Fig. 16.23). These experiments collectively suggest that many nutrients may be impacted by oxidative extraction of nutrients from microbes in the rhizophagy cycle within roots.

16.6 The crisis of elevated carbon dioxide in earth's atmosphere

Recent research has shown that elevated atmospheric carbon dioxide may be reducing nutrient absorption into plants (Lenka et al., 2019). A recent study by

Table 16.2 Tomato seedling root hair lengths after seven days in air and air + elevated carbon dioxide.

Seedling number	Air (nonelevated CO_2)	Elevated CO_2 + air (0.3 g dry ice/L air)
1	1666.85 ± 225.23 μm (N = 16)	258.46 ± 71.04 μm (N = 14)
2	673.03 ± 152.23 μm (N = 16)	148.43 ± 77.86 μm (N = 13)
3	733.52 ± 148.26 μm (N = 16)	443.45 ± 165.27 μm (N = 16)
4	785.71 ± 141.60 μm (N = 16)	318.75 ± 90.14 μm (N = 16)
5	650.54 ± 182.32 μm (N = 16)	335.85 ± 94.29 μm (N = 16)
6	650.14 ± 155.97 μm (N = 16)	142.54 ± 63.37 μm (N = 13)
7	714.06 ± 159.99 μm (N = 16)	183.21 ± 83.93 μm (N = 16)
8	657.33 ± 185.26 μm (N = 16)	228.43 ± 66.10 μm (N = l6)
Means	691.40 ± 48.80 μm	257.39 ± 104.16 μm

N, Number of observations.

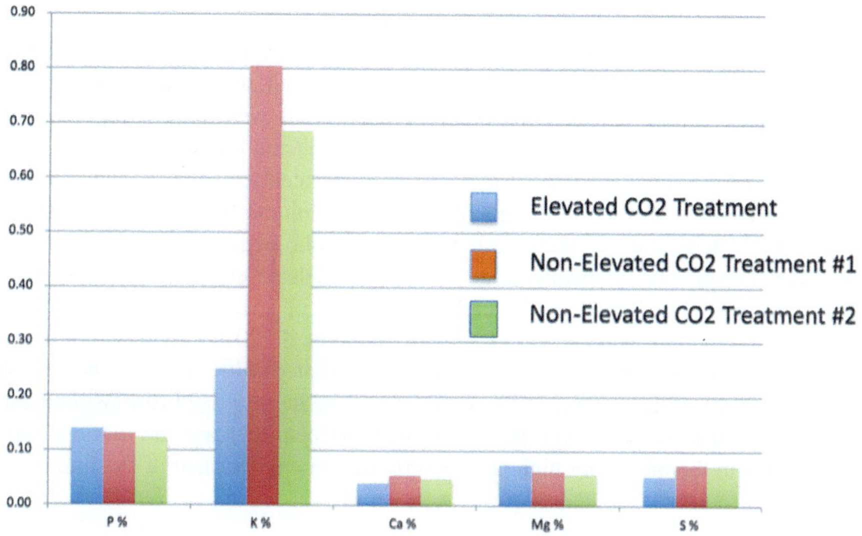

FIGURE 16.23

Effect of elevated carbon dioxide on absorption of nutrients into tomato seedlings. This experiment suggests that suppression of rhizophagy cycle nutrient extraction from bacteria in roots suppresses absorption of nutrients into plants.

Boufeldja et al. (2022) focused on nutritional content of tomatoes grown with elevated carbon dioxide levels, finding proteins, minerals, and carotenoids levels decreased compared to tomatoes grown in ambient conditions. Reduced absorption into plants could be the result of suppression of superoxide and reduced extraction of nutrients from microbes in the rhizophagy cycle. Even modest increases in carbon dioxide in the atmosphere may result in much higher levels of carbon dioxide in soils where respiration by bacteria and fungi may produce carbon dioxide. One study examining effects of elevated carbon dioxide around fruits showed that a 10% carbon dioxide level around pears resulted in observable depressions of reactive oxygen species in the fruits (Wang et al., 2021). The levels of carbon dioxide that we have used in our experiments are very likely much higher than what is expected in the next few decades; however, carbon dioxide levels in soils are significantly higher than that in the air around plants. Even modest increases in atmospheric levels of carbon dioxide could be magnified in soils around roots, resulting in reduced oxidative extraction of nutrients from bacteria in the rhizophagy cycle in roots. More research is needed to evaluate the importance of elevated greenhouse gasses on the oxidative extraction of nutrients from bacteria.

16.7 Conclusions

As we look to the future of sustainable agricultural practices in a world where atmospheric carbon dioxide peeks over 450 ppm, we must explore in more detail the relationships between plant and soil microbes associated with disease protection, nutrient solubilization, and phytohormone production, as well as how the entry of endophytes into plants is affected. We must also understand the interactions of multiple stress interactions on these systems, such as elevated temperature and changes in weather patterns leading to drought stress and intermittent flooding. Studies show increased soil temperatures increase both the metabolic activities of soil microbes and evapotranspiration of plants (Liu et al., 2009; Selsted et al., 2012) while elevated carbon dioxide increases the cellulose content of soils (Weber et al., 2011) and decreases the nitrogen content of leaves (Olsrud et al., 2010). Evidence of nitrogen fixation in the phyllosphere, particularly associated with floral trichomes of *Humulus lupulus* (Micci et al., 2022), may lead to the development of biostimulants to alleviate foliar effects of nitrogen deficiency and help increase the nutrient content of fruits in the inevitably high atmospheric carbon environment of the future.

References

Arnold, A.E., Lutzoni, F., 2007. Diversity and host range of foliar fungal endophytes: are tropical leaves biodiversity hotspots? Ecology 88 (3), 541–549.

Atsatt, P.R., Whiteside, M.D., 2014. Novel symbiotic protoplasts formed by endophytic fungi explain their hidden existence, lifestyle switching, and diversity within the plant kingdom. PLoS One 9 (4), e95266.

Beltran-Garcia, M.J., White, J., James, F., Prado, F.M., Prieto, K.R., Yamaguchi, L.F., et al., 2014. Nitrogen acquisition in *Agave tequilana* from degradation of endophytic bacteria. Sci. Rep. 4 (1), 6938.

Bierne, H., Cossart, P., 2012. When bacteria target the nucleus: the emerging family of nucleomodulins. Cell. Microbiol. 14 (5), 622–633.

Boufeldja, L., Brandt, D., Guzman, C., Vitou, M., Boudard, F., Morel, S., et al., 2022. Effect of elevated carbon dioxide exposure on nutrition-health properties of micro-tom tomatoes. Molecules 27 (11), 3592.

Carmona-Hernandez, S., Reyes-Pérez, J.J., Chiquito-Contreras, R.G., Rincon-Enriquez, G., Cerdan-Cabrera, C.R., Hernandez-Montiel, L.G., 2019. Biocontrol of postharvest fruit fungal diseases by bacterial antagonists: a review. Agronomy 9 (3), 121.

Chiaranunt, P., Kingsley, K., White, J.F., 2020. Hungry plants: rhizophagy as a model for P and Zn uptake in rice. *Ecological Society of America Annual Meeting*, Aug 3–6 July 2020.

Clay, K., Holah, J., Rudgers, J.A., Moran, N.A., 2005. Herbivores cause a rapid increase in hereditary symbiosis and alter plant community composition. Proc. Natl Acad. Sci. - PNAS 102 (35), 12465–12470.

Doebereiner, J., 1992. History and New Perspectives of Diazotrophs in Association With Non-Leguminous Plants. Balaban, pp. 1–13.

Döbereiner, J., Baldani, V., Olivares, F.L., Reis, V.M., 1994. Endophytic diazotrophs: the key to BNF in gramineous plants. Nitrogen Fixation With Non-Legumes. *The American University in Cairo Press, Egypt*, pp. 395–408.

Errington, J., 2013. L-form bacteria, cell walls and the origins of life. Open. Biol. 3 (1), 120143.

Freitas, C.S.A., Maciel, L.F., Corrêa, D.S., Renato, A., Costa, O.M.M.M., et al., 2022. Bacterial volatile organic compounds induce adverse ultrastructural changes and DNA damage to the sugarcane pathogenic fungus *Thielaviopsis ethacetica.* Environ. Microbiol. 24 (3), 1430–1453.

Glick, B.R., 1995. The enhancement of plant growth by free-living bacteria. Can. J. Microbiol. 41 (2), 109–117.

Guevara-Avendaño, E., Bravo-Castillo, K., Monribot-Villanueva, J., Kiel-Martínez, A.,L., Ramírez-Vázquez, M., Guerrero-Analco, J., et al., 2020. Diffusible and volatile organic compounds produced by avocado rhizobacteria exhibit antifungal effects against *Fusarium kuroshium.* Braz. J. Microbiol. 51 (3), 861–873.

Hamilton, C.E., Gundel, P.E., Helander, M., Saikkonen, K., 2012. Endophytic mediation of reactive oxygen species and antioxidant activity in plants: a review. Fungal Diversity 54 (1), 1–10.

Hardoim, P.R., Overbeek, V., Leonard, S., Berg, G., Pirttilä, A.M., Compant, S., et al., 2015. The hidden world within plants: ecological and evolutionary considerations for defining functioning of microbial endophytes. Microbiol. Mol. Biol. Rev. 79 (3), 293–320.

Hill, P.W., Quilliam, R.S., Deluca, T.H., Farrar, J., Farrell, M., Roberts, P., et al., 2011. Acquisition and assimilation of nitrogen as peptide-bound and D-enantiomers of amino acids by wheat. PLoS One 6 (4), e19220.

Hurek, T., Reinhold-Hurek, B., Montagu, M.V., Kellenberger, E., 1994. Root colonization and systemic spreading of Azoarcus sp. strain BH72 in grasses. J. Bacteriol. 176 (7), 1913–1923.

Irizarry, I., White, J.F., 2017. Application of bacteria from non-cultivated plants to promote growth, alter root architecture and alleviate salt stress of cotton. J. Appl. Microbiol. 122 (4), 1110–1120.

James, E., 1998. Infection and colonization of sugar cane and other graminaceous plants by endophytic diazotrophs. Crit. Rev. Plant. Sci. 17 (1), 77–119.

James, E.K., 2000. Nitrogen fixation in endophytic and associative symbiosis. Field Crop. Res. 65 (2), 197–209.

James, E.K., Reis, V.M., Olivares, F.L., Baldani, J.I., Döbereiner, J., 1994. Infection of sugar cane by the nitrogen-fixing bacterium *Acetobacter diazotrophicus*. J. Exp. Botany 45 (6), 757–766.

Johnston-Monje, D., Raizada, M.N., 2011. Conservation and diversity of seed associated endophytes in Zea across boundaries of evolution, ethnography and ecology. PLoS One 6 (6), e20396.

Kloepper, J.W., 1993. Plant growth-promoting rhizobacteria as biological control agents. In: Metting, F.B. (Ed.), Soil Microbial Ecology: Applications in Agricultural and Environmental Management. M. Dekker, New York, pp. 255–274.

Kuldau, G., Bacon, C., 2008. Clavicipitaceous endophytes: their ability to enhance resistance of grasses to multiple stresses. Biol. Control. 46 (1), 57–71.

Kumar, A., Vandana, Singh, M., Singh, P.P., Singh, S.K., Singh, P.K., et al., 2016. Isolation of plant growth promoting rhizobacteria and their impact on growth and curcumin content in *Curcuma longa* L. Biocatal. Agric. Biotechnol. 8, 1–7.

Lata, R., Chowdhury, S., Gond, S.K., White, J.F., 2018. Induction of abiotic stress tolerance in plants by endophytic microbes. Lett. Appl. Microbiol. 66 (4), 268–276.

Lenka, N.K., Lenka, S., Singh, K.K., Kumar, A., Aher, S.B., Yashona, D.S., et al., 2019. Effect of elevated carbon dioxide on growth, nutrient partitioning, and uptake of major nutrients by soybean under varied nitrogen application levels. J. Plant. Nutr. Soil. Sci. 182 (4), 509–514.

Liu, W., Zhang, Z., Wan, S., 2009. Predominant role of water in regulating soil and microbial respiration and their responses to climate change in a semiarid grassland. Glob. Change Biol. 15 (1), 184–195.

Magnani, G.S., Didonet, C.M., Cruz, L.M., Picheth, C.F., Pedrosa, F.O., Souza, E.M., 2010. Diversity of endophytic bacteria in Brazilian sugarcane. Genet. Mol. Res. 9 (1), 250–258.

Micci, A., Zhang, Q., Chang, X., Kingsley, K., Park, L., Chiaranunt, P., et al., 2022. Histochemical evidence for nitrogen-transfer endosymbiosis in non-photosynthetic cells of leaves and inflorescence bracts of angiosperms. Biology 2022 (11), 876. Available from: https://doi.org/10.3390/biology11060876.

Mikola, J., 1998. Effects of microbivore species composition and basal resource enrichment on trophic-level biomasses in an experimental microbial-based soil food web. Oecologia 117 (3), 396–403.

Olsrud, M., Carlsson, B.Å., Svensson, B.M., Michelsen, A., Melillo, J.M., 2010. Responses of fungal root colonization, plant cover and leaf nutrients to long-term exposure to elevated atmospheric CO_2 and warming in a subarctic birch forest understory. Global Change Biol. 16 (6), 1820–1829.

Paungfoo-Lonhienne, C., Rentsch, D., Robatzek, S., Webb, R.I., Sagulenko, E., Näsholm, T., et al., 2010. Turning the table: plants consume microbes as a source of nutrients. PLoS One 5 (7), e11915.

Paungfoo-Lonhienne, C., Schmidt, S., Webb, R.I., Lonhienne, T.G.A., 2013. Rhizophagy—a new dimension of plant–microbe interactions. Molecular Microbial Ecology of the Rhizosphere. John Wiley & Sons, Inc, Hoboken, NJ, USA, pp. 1199–1207.

Puente, M., Bashan, Y., 1994. The desert epiphyte *Tillandsia recurvata* harbours the nitrogen-fixing bacterium *Pseudomonas stutzeri*. Can. J. Botany 72 (3), 406–408.

Quartana, C., Faddetta, T., Anello, L., Di Bernardo, M., Petralia, R., Campanella, V., 2022. Activity of bacterial seed endophytes of landrace durum wheat for control of Fusarium foot rot. Phytopathologia Mediterranea 61 (1), 95–106.

Redman, R.S., Sheehan, K.B., Stout, R.G., Rodriguez, R.J., Henson, J.M., 2002. Thermotolerance generated by plant/fungal symbiosis. Science 298 (5598), 1581.

Reinhold-Hurek, B., Hurek, T., 2011. Living inside plants: bacterial endophytes. Curr. Opin. Plant. Biol. 14 (4), 435–443.

Rodriguez, R., Redman, R., 2008. More than 400 million years of evolution and some plants still can't make it on their own: plant stress tolerance via fungal symbiosis. J. Exp. Botany 59 (5), 1109–1114.

Rodriguez, R.J., Woodward, C., Kim, Y.O., Redman, R.S., 2009. Habitat-adapted symbiosis as a defense against abiotic and biotic stresses. Defensive Mutualism in Microbial Symbiosis. CRC Press, pp. 353–364.

Roley, S.S., Xue, C., Hamilton, S.K., Tiedje, J.M., Robertson, G.P., 2019. Isotopic evidence for episodic nitrogen fixation in switchgrass (*Panicum virgatum* L.). Soil. Biol. Biochem. 129, 90–98.

Rosenblueth, M., Martinex-Romero, E., 2006. Bacterial endophytes and their interactions with hosts. Mol. Plant-Microbe Interact. 19 (8), 827–837.

Selsted, M.B., Van Der Linden, L., Ibrom, A., Michelsen, A., Larsen, K.S., Pedersen, J.K., et al., 2012. Soil respiration is stimulated by elevated CO_2 and reduced by summer drought: three years of measurements in a multifactor ecosystem manipulation experiment in a temperate heathland (CLIMAITE). Glob. Change Biol. 18 (4), 1216–1230.

Shehata, H.R., Dumigan, C., Watts, S., Raizada, M.N., 2017. An endophytic microbe from an unusual volcanic swamp corn seeks and inhabits root hair cells to extract rock phosphate. Sci. Rep. 7 (1), 13479.

Taulé, C., Mareque, C., Barlocco, C., Hackembruch, F., Reis, V.M., Sicardi, M., et al., 2012. The contribution of nitrogen fixation to sugarcane (*Saccharum officinarum* L.), and the identification and characterization of part of the associated diazotrophic bacterial community. Plant. Soil. 356 (1–2), 35–49.

Torres, M.S., White, J.F., Zhang, X., Hinton, D.M., Bacon, C.W., 2012. Endophyte-mediated adjustments in host morphology and physiology and effects on host fitness traits in grasses. Fungal Ecol. 5 (3), 322–330.

Urquiaga, S., Cruz, K.H.S., Boddey, R.M., 1992. Contribution of nitrogen fixation to sugar cane: nitrogen-15 and nitrogen-balance estimates. Soil. Sci. Soc. Am. J. 56 (1), 105–114.

Verma, S.K., Kingsley, K., Bergen, M., English, C., Elmore, M., Kharwar, R.N., et al., 2018b. Bacterial endophytes from rice cut grass (*Leersia oryzoides* L.) increase growth, promote root gravitropic response, stimulate root hair formation, and protect rice seedlings from disease. Plant. Soil. 422 (1), 223–238.

Verma, S.K., Kingsley, K.L., Bergen, M.S., Kowalski, K.P., White, J.F., 2018a. Fungal disease prevention in seedlings of rice (*Oryza sativa*) and other grasses by growth-promoting seed-associated endophytic bacteria from invasive *Phragmites australis*. Microorganisms 6 (1), 21.

Verma, S.K., Kingsley, K., Irizarry, I., Bergen, M., Kharwar, R.N., White Jr, J.F., 2017. Seed-vectored endophytic bacteria modulate development of rice seedlings. J. Appl. Microbiol. 122 (6), 1680–1691.

Verma, H., Kumar, D., Kumar, V., Kumari, M., Singh, S.K., Sharma, V.K., et al., 2021. The potential application of endophytes in management of stress from drought and salinity in crop plants. Microorganisms 9 (8), 1729.

Waller, F., Achatz, B., Baltruschat, H., Fodor, J., Becker, K., Fischer, M., et al., 2005. The endophytic fungus *Piriformospora indica* reprograms barley to salt-stress tolerance, disease resistance, and higher yield. Proc. Natl Acad. Sci. - PNAS 102 (38), 13386–13391.

Wang, E., Liu, X., Si, Z., Xu, L., Bi, J., Dong, W., et al., 2021. Volatile organic compounds from rice rhizosphere bacteria inhibit growth of the pathogen *Rhizoctonia solani*. Agriculture 11 (4), 368.

Wang, D., Li, W., Li, D., Li, L., Luo, Z., 2021. Effect of high carbon dioxide treatment on reactive oxygen species accumulation and antioxidant capacity in fresh-cut pear fruit during storage. Sci. Hortic. 281, 109925.

Weber, O.B., Muniz, C.R., Vitor, A.O., Freire, F.C.O., Oliveira, V.M., 2007. Interaction of endophytic diazotrophic bacteria and *Fusarium oxysporum f. sp. cubense* on plantlets of banana 'Maça'. Plant. Soil. 298 (1/2), 47–56.

Weber, C.F., Zak, D.R., Hungate, B.A., Jackson, R.B., Vilgalys, R., Evans, R.D., et al., 2011. Responses of soil cellulolytic fungal communities to elevated atmospheric CO_2 are complex and variable across five ecosystems. Environ. Microbiol. 13 (10), 2778–2793.

White, J.F., Chang, X., Kingsley, K.L., Zhang, Q., Chiaranunt, P., Micci, A., et al., 2021. Endophytic bacteria in grass crop growth promotion and biostimulation. Grass Res. 1 (1), 1–9. Available from: https://doi.org/10.48130/GR-2021-0005.

White, J.F., Chen, Q., Torres, M.S., Mattera, R., Irizarry, I., Tadych, M., et al., 2015. Collaboration between grass seedlings and rhizobacteria to scavenge organic nitrogen in soils. AoB Plants 7, 2015. Available from: https://doi.org/10.1093/aobpla/plu093. plu093.

White, J.F., Kingsley, K.L., Verma, S.K., Kowalski, K.P., 2018. Rhizophagy cycle: an oxidative process in plants for nutrient extraction from symbiotic microbes. Microorganisms 2018 (6), 95. Available from: https://doi.org/10.3390/microorganisms6030095.

White, J.F., Kingsley, K.L., Zhang, Q., Verma, R., Obi, N., Dvinskikh, S., et al., 2019. Review: endophytic microbes and their potential applications in crop management. Pest. Manag. Sci. 75 (10), 2558–2565. Available from: https://doi.org/10.1002/ps.5527.

White Jr, J.F., Torres, M.S., 2010. Is plant endophyte-mediated defensive mutualism the result of oxidative stress protection? Physiol. Plant. 138 (4), 440–446.

White Jr, J.F., Torres, M.S., Somu, M.P., Johnson, H., Irizarry, I., Chen, Q., et al., 2014. Hydrogen peroxide staining to visualize intracellular bacterial infections of seedling root cells. Microsc. Res. Tech. 77 (8), 566–573.

White, J.F., Torres, M.S., Verma, S.K., Elmore, M.T., Kowalski, K.P., Kingsley, K.L., 2018. Evidence for widespread microbivory of endophytic bacteria in roots of vascular plants through oxidative degradation in root cell periplasmic spaces. In: Kumar, A., Singh, A., Singh, V. (Eds.), Sustainable Agriculture: Food Security and Environmental Management. Elsevier, New York. Available from: https://doi.org/10.1016/B978-0-12-815879-1.00009-4.

Xin, X.F., Kvitko, B., He, S.Y., 2018. *Pseudomonas syringae*: what it takes to be a pathogen. Nat. Rev. Microbiol. 16 (5), 316–328.

CHAPTER

Endophytic fungi: diversity and their relevance in sustainable agriculture

17

Sunil Kumar[1,2], Yashoda Nandan Tripathi[1], Vaishali Shukla[1], Rahul Prasad Singh[1], Ajay Kumar[3] and Ram Sanmukh Upadhyay[1]

[1]*Department of Botany, Centre of Advanced Study, Institute of Science, Banaras Hindu University, Varanasi, Uttar Pradesh, India*

[2]*Central Ayurveda Research Institute—Central Council for Research in Ayurvedic Sciences, Bhubaneswar, Odisha, India*

[3]*Amity Institute of Biotechnology, Amity University, Noida, Uttar Pradesh, India*

17.1 Introduction

A family of fungi is one of the most abundant taxonomic groups of microbes on earth that are eukaryotic and heterotrophic, including mildews, molds, mushrooms, yeast, and puffballs. As well as being beneficial to plant growth and crop production, they also provide crop protection. It is reported that fungi can be found in soil, water, dead and decaying organic matter in animals and on their skin that have a high capacity and flexibility to adapt to diverse undesirable or unfavorable conditions (Karun et al., 2018; Rana et al., 2019). Therefore the agriculture sector must input excess use of chemical fertilizers, insecticides, and herbicides to fulfill global food security challenges. These agrochemical inputs in agriculture extend significant environmental damage and cause health problems in humans and animals. In the last decades, cognition of organic farming has been raised considerably due to the hazardous consequences of chemical application on agricultural land. Organic inputs in agriculture provide food safety from chemicals, the high nutritional value of foods, healthy soil and ecological harmony of natural microflora. Many plant growth-promoting microbes are used as biofertilizers for accomplishing long-lasting sustainability in the agriculture sector. In addition to improving plant growth and health, fungi can also be used for nutrient uptake, water uptake, stress tolerance, and biocontrol (Yadav et al., 2020a,b). Plant endophytes exist in the presence of their plant hosts, whereby the plant provides adequate nutrients and a suitable habitat for the microbes to survive. They live within the tissues of the plant without causing disease symptoms. Several plant growth-promoting (PGP) attributes are produced by the beneficial fungi. For example, micronutrients are soluble (phosphorus, potassium, and zinc) and plant growth regulators are generated (auxin, gibberellins, cytokinin, and ethylene) or indirectly via the synthesis of siderophores, antagonistic

Sustainable Agricultural Practices. DOI: https://doi.org/10.1016/B978-0-443-19150-3.00008-4

substances, antibiotics, and enzymes that lyse cell walls (cellulases, glucanase, and glycosidase) (Abo Nouh, 2019). A number of processes for sustainable development involve fungi in agriculture, medicine, and industry. The use of fungi and their products in the medical field is used to control diseases in both humans and animals. Currently, fungi are being used as a biological pest control method in many regions of the world. The beneficial fungi play a critical role in numerous physiological processes, including mineral and water uptake, stomatal movement, photosynthesis, and biosynthesis of compounds as well as in mitigating environmental stresses like drought, heat, salinity, cold and heavy metal exposure (Begum et al., 2019). Mycofungicides and fungal biofertilizers aided agricultural practices in managing plant diseases while protecting the environment and improving crop yield (Kour et al., 2020). Various micro fungicides have been developed over the last few years, for examples, Ketomium, formulated from *Chaetomium globosum* and *Ch. cupreum*, Promote, constructed from *Trichoderma harzianum*, *Trichoderma viride*, SoilGard (constructed from *Gliocladium virens*), Trichodex, derived from *T. harzianum* and Ketomium, derived from *C. globosum* (Kaewchai, 2009). Several fungal biofertilizers possessing diverse PGP characteristics have been reported such as *Trichoderma*, mycorrhizal fungi (arbuscular mycorrhiza, e.g., *Glomus intraradices*), ectomycorrhiza (*Pisolithus tinctorius*), and those that form mutualistic associations with plants, while *Glomus* species and *Trichoderma* species are commonly used as biocontrol and biostimulants for horticultural/agricultural crops (Frąc et al., 2018; Rana et al., 2019).

This chapter deals with the diversity of beneficial endophytic fungi from different habitats and their applications in plant growth promotion and soil health. Fungal endophytes are a significant component of plant micro-ecosystems and have been found in large number of plant species examined. They have been reported to solubilize insoluble phosphates and produce PGP hormones including auxins, cytokinins, and gibberellins. Furthermore, they provide protection to plants against pathogens by producing antagonistic compounds, inducing host defense mechanisms, or providing competition for nutrients and colonization sites.

17.2 Biodiversity and ecology of fungal endophytes

The diversity and distribution of microorganisms or microbes are critical topics in microbial ecology and the trending topic in this area is fungi. Fungi, decomposers, mutualisms, and pathogens are highly diverse. Various fungal species have been recognized in diverse air, soil, plant-associated, and extreme environments like deep oceans, hot springs, glaciers, deserts, mines, and coastal regions. The discovery of fungal endophytes on land plants has improved over the past few years. Fungal endophytes have been isolated from boreal forests, tropical climates, diverse xeric environments, extreme arctic environments, ferns, gymnosperms, and angiosperms (Selim et al., 2017). In addition to protecting their host from attacking

Table 17.1 Fungal endophytes with different agricultural use plant species.

Fungal endophytes	Plant host	Category	References
Penicillium, Fusarium, Colletotrichum, Alternaria, Xylaria, and *Cladosporium*	*Capsicum annuum*	Systemic and nonsystemic	Paul et al. (2012)
Phyllosticta sp.	*Coffea arabica*	Systemic	Hata and Sone (2008)
Beauveria bassiana, Lecanicillium lecanii, Paecilomyces spp.	*Gossypium hirsutum*	Systemic	Sword et al. (2012)
Trichoderma harzianum	*Amaranthus* spp.	Systemic	Uppala et al. (2010)
Acremonium zeae, Alternaria alternata, Aspergillus favus, Aspergillus niger, Colletotrichum graminicola, Fusarium verticillioides, Saccharomyces cerevisiae, and *Trichoderma koningii*	*Zea mays*	Systemic and nonsystemic	Orole and Adejumo (2011)
Aspergillus sp.	*Moringa olifera*	Systemic	Mahdi et al. (2014)
Chaetomium globosum, Penicillium chrysogenum, Fusarium oxysporum, Cladosporium cladosporioides	*Oryza sativa*	Systemic and nonsystemic	Naik et al. (2009)
Colletotrichum sp. *C. gloeosporioides*	*Triticum aestivum*	Systemic and nonsystemic	Larran et al. (2002)
Penicillium sp.	*Lycopersicum esculentum*	Systemic	Larran et al. (2001)

phytopathogens, endophytic fungi play an important role in releasing macronutrients such as phosphorus, potassium, zinc, and oxidizing atmospheric nitrogen; as well as producing hydrolytic enzymes, ammonia, siderophore, and hydrogen cyanide (HCN) (Rana et al., 2016; Verma et al., 2016). A rhizosphere is divided into ecto- and endorhizospheres. The endorhizosphere refers to the multilayered microenvironment composed of mucus on the surface of the roots, epidermis around the roots including root hairs, and cortical cells around the roots (Bolton et al., 1993). Endophytes are microbial organisms that have developed a mutualistic relationship with plants, whereby the plants provide the endophytes with sufficient nutrients and habitat to survive (Table 17.1). In the agricultural field, fungal endophytes contain compounds that can be used in the production of fertilizer and pesticide (Zhao et al., 2012).

17.3 Biological classification of fungal endophytes

There are so many species of fungal endophyte that isolation, characterization, and identification are difficult operations and only a few have been described to date because of the vast taxonomic diversity of the community (Hawksworth, 2004). Endophytic fungi are classified and differentiated based on their functional characteristics, including their mode of transmission, host type they could occupy, mode of reproduction, nutrition source and expression of symptoms in host plants, general morphology and colonized parts of plant helping clarify their unique symbiotic and ecological functions (Bamisile et al., 2018). Several factors contribute to endophytic association in plants, including the host type, host species, inoculum type and density, and the stage of development of the host at the time of inoculation (Surjit and Rupa, 2014). There are two classes of endophytes: clavicipitaceous and nonclavicipitaceous (NC). Clavicipitaceous endophytes colonize grasses while NC endophytes inhabit vascular and nonvascular plants (Rodriguez et al., 2009). A total of three subgroups of NC endophytes have been identified: class 2, class 3, and class 4. The class 2 endophytes offer both nonhabitat adaptation (NHA) fitness and habitat-adapted (HA) traits to plants that are primarily driven by environmental factors like temperature, pH, salinity, and humidity (Rodriguez et al., 2008) and characterized by a wider range of hosts, extensive in planta colonization, a lesser diversity of nature and both mode of transmission to host generations. However, type 3 and type 4 NC endophytes can infect a wide range of hosts, transmit horizontally, and have fitness effects similar to those of NHA. Other characteristics of class 1 endophytes include narrow host ranges, extensive colonization of their hosts, less diversity, vertical and horizontal transmission modes, and preferential use of benefits from NHA. There are a number of fungal genera in this group that belong to the class *Ascomycota*, as well as a few genera of *Basidiomycota* (Johnson et al., 2014). The most important characteristic of these endophytic fungi is their ability to colonize roots, stems, and leaves, and to cause extensive infection of plants. Fungal endophytes can be vertically transmitted or horizontally transmitted based on their mode of transmittance to their host. The vertical transmission occurs through seeds and can be directly transmitted from the host to the progeny. For example, *Epichloe* endophytes are seed-transmitted endophytes (Schardl et al., 2013). The horizontal mode of transmission is different from the vertical mode, because this mode of transmission transmits fungal endophytes to multiple plants within a population, particularly when spores are involved (Wiewióra et al., 2015).

17.4 Endophytic fungi in sustainable agriculture

Since plant diseases and pests are responsible for reducing crop yields by 30%–50% worldwide each year (Mousa and Raizada, 2013), the use of endophytes as a means of enhancing yield and quality of cultivated plants represents an important avenue in combating plant disease and pest extermination. There are many endophytes that provide beneficial biological properties to the hosts, such as deterring pathogenic microbes, insects, and other herbivores while increasing plant growth and development (Sudha et al., 2016). Plant endophytes play a crucial role in their growth, as they produce auxins, cytokinins, gibberellins, siderophores, ammonia, HCN, and diverse hydrolytic enzymes that protect host plants from stress and provide them with nutrients (Fig. 17.1). Endophytic fungi are associated with almost all studied plants. They perform an important ecological role and possess a variety of interactions with host plants. There are several mechanisms by which endophytes promote and protect plant growth and forest health from pests and pathogens (Busby et al., 2016; Fesel and Zuccaro, 2016). Endophytes produce secondary metabolites or biochemicals that suppress or reduce the impact of plant pathogens, such as volatile compounds that prevent pathogen development (Kottb et al., 2015). Endophytes can also induce plant defense mechanisms to protect their hosts by acquiring resistance or generating resistance in their host plant (Sun et al., 2020). Some endophytes have biocontrol potential by secreting antibacterial and antifungal compounds, from which pathogens are inhibited, or they may exhibit mycoparasitic activity, from which one fungus is parasitized by another fungus (De Silva et al., 2019). An application of endophytes from *Theobroma cacao* leaves protected leaves from leaf necrosis and mortality when under stress from *Phytophthora* sp. (Arnold et al., 2003). In addition to safeguarding host plants from pathogens directly, some endophytes have PGP properties that can

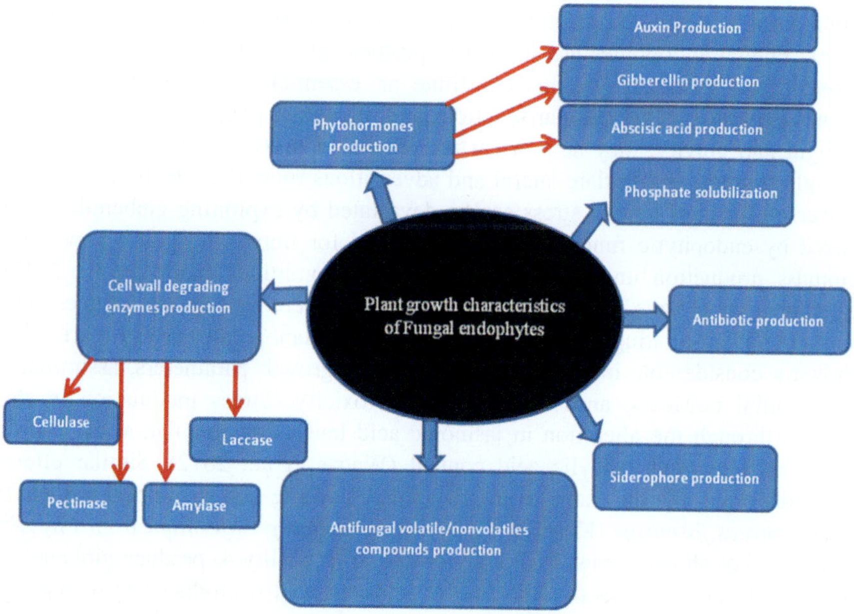

FIGURE 17.1

The diagrammatic representation of role of fungal endophytes in sustainable agriculture.

strengthen the host plant by providing nutrients and nutrient cycling. For instance, plants grow more rapidly when nourished by PGP endophytes (Saleem et al., 2007). Endophytes can also influence plant hormones such as auxin, cytokinin, ethylene, and gibberellin, producing other bioactive compounds (Joseph and Priya, 2011; Parthasarathi et al., 2012). A plant may harbor many endophytes, which can remain localized and act as a tissue-specific barrier against disease (Wu et al., 2013) or can spread systemically to other parts of the plant (Chutulo and Chalannavar, 2018; Hodgson et al., 2014). These symbiotic and perhaps mutualistic interactions occur in wild and grown plants (Arnold et al., 2000). In 300,000 different plant species, there are estimated to be more than 1 million endophytic species, but only a small fraction of them have been isolated and their roles explored (Fouda et al., 2015). In *Glycine max* growing under salinity stress, the endophytic fungus *Penicillium funiculosum* has diverse growth-promoting properties (Khan et al., 2011b). *Phoma* sp. isolated from *Tinospora cordifolia* and *Calotropis procera* has been found to promote maize growth with a variety of growth-promoting traits in plants growing under salinity stress (Kedar et al., 2014).

17.5 Role of phytohormones in agriculture

Phytohormones regulate plant enlargement and plant responses toward the biotic stress. Fungal endophytes are studied to promote plant growth by producing different plant hormones. Auxins constitute an essential group among phytohormones exploiting other cellular functions and regulating plant growth. In response to light and gravity, they help orient the growth of roots and shoot, discriminate vascular tissues, and initiate lateral and adventitious roots (Cecchetti et al., 2008). Adverse effects of abiotic stress can be devastated by exploiting gibberellins produced by endophytic fungi, which are essential for improving plant growth and biomass production under environmental stress conditions (Khan et al., 2015). Cucumber roots have been found to synthesize gibberellic acid and indole acetic acid (IAA). Inoculating these strains in cucumber plants under drought stress has shown a considerable increase in plant biomass, growth parameters, assimilation of essential nutrients, and reduced sodium toxicity. Stress modulation is also assured through the alteration in jasmonic acid level, downregulation of abscisic acid, and increased salicylic acid content (Waqas et al., 2012). Similar effects were observed by the same research group working on cucumber endophyte *Paecilomyces formosus* (Khan et al., 2012). *Fusarium proliferatum* from *Physalis alkekengi* has shown robust PGP activity due to its ability to produce gibberellins (Rim et al., 2005). Bioactive gibberellic acids were also produced by species of *Aspergillus*, *Cladosporium*, and *Talaromyces* from soybean (Hamayun et al., 2009a; Khan et al., 2011a,b). *Galactomyces geotrichum* from *Trapa japonica* has been reviewed to produce enormous quantity of jasmonic acid, thereby causing systemic resistance in soybean plants (Waqas et al., 2014). *Penicillium citrinum* and *Aspergillus terreus* inoculated in sunflower plants revealed alteration in salicylic acid and jasmonic acid levels of the plants (Waqas et al., 2015). Numerous fungal endophytes, such as *Neurospora crassa* (Rademacher, 1994), *P. citrinum* (Khan et al., 2008), *Scolecobasidium tshawytschae* (Hamayun et al., 2009a), *Arthrinium phaeospermum* (Khan et al., 2008), *Chrysosporium pseudomerdarium* (Hamayun et al., 2009b), *Cladosporiums phaerospermum* (Hamayun et al., 2009c), *Cladosporium* sp. (Hamayun et al., 2009c), *Fusarium fujikuroi*, *Sphaceloma manihoticola* (Shweta et al., 2010), *Phaeosphaeria* sp., *Penicillium* sp. (Hamayun et al., 2010), *Aspergillus fumigatus* (Khan et al., 2011a), *Exophiala* sp. (Khan et al., 2011b), and *P. funiculosum*, have been examined and illustrated as immaculate gibberellins producers. IAA influences plant cell division, elongation, and differentiation to produce seed and tuber germination, improves the rate of xylem and root growth, and begins lateral and adventitious root construction, among many other functions (Kumar et al., 2016). On the other hand, gibberellins (GAs) are a class of tetracyclic diterpenoid hormones and maintain an essential role in plant growth (Martínez et al., 2016). Plant growth promotion was obeyed in *V. radiata* by *Geotrichum candidum* isolated from *Bruguiera cylindrica* (George et al., 2021) and *C. tropicalis* from rice plants by the production of IAA (Amprayn et al., 2012).

This IAA exhibit capability recreates a vital role in PGP in crops. In cereals, this ability has been associated with improved growth in wheat plants, thanks to fungi *T. harzianum*, *T. aureoviride* (Ripa et al., 2019), *Penicillium roqueforti* (Ikram et al., 2018), and *Alternaria alternata* isolated from *Elymus dahuricus* seeds and which also improves the photosynthetic ability and accumulation of nutrients in plant tissues (Qiang et al., 2019). *A. niger* (Lubna et al., 2018) or *Bipolaris* sp. isolated from *C. sativa* having the ability to produce GAs (Asaf et al., 2019). 1-Aminocyclopropane-1-carboxylic acid (ACC) is direct precursor of ethylene. ACC deaminase assists plants in diminishing many of the manifestations of abiotic stress and encouraging growth (generally improving the length of roots and aerial structures) and fostering adaptation and survival by degradation of ethylene (del Carmen Orozco-Mosqueda et al., 2020).

17.6 Phosphate solubilization

After nitrogen, phosphorus is the second most important macronutrient for plants (Khan et al., 2014). This macronutrient plays a critical role in chemical structures and makes up 0.2% of the plant's dry weight. A surprising amount of phosphorus is stored in soil (400–1200 mg/kg) in an insoluble form complexed with iron, calcium, and aluminum is not available to plants (Behie and Bidochka, 2014). There is a low concentration of soluble phosphorus in soil, which is not sufficient for plants metabolism processes and its deficiency may result in slower growth and decreased leaf biomass (Avdalović et al., 2015). A variety of chemical fertilizers had been used to meet the requirement; however, these products are not environmentally friendly. Therefore fungi are seen to be a more sustainable option because they can provide phosphorus in a soluble form without harming the environment. The fungi in the soil assimilate the phosphorus present in the soil by producing various organic acids such as tartaric acid, succinic acid, oxalic acid, malic acid, 2-ketogluconic acid, glyoxylic acid, gluconic acid, fumaric acid, citric acid, and alpha-ketobutyric acid. Many fungi such as *Penicillium sclerotiorum* from tea roots (Nath et al., 2015), *Byssochlamys nivea* from *Pistacia vera* roots (Dolatabad et al., 2017), and many other fungi from wheat roots were isolated and had high potential as biofertilizers in different soils for managing sustainable agroecosystems (Spagnoletti et al., 2017). Besides *Epicoccum nigrum* and *Paecilomyces variotii*, two fungi isolated from succulent plant *Caralluma acutangula* could solubilize phosphate in vitro (Ali et al., 2019); similarly, *Mucoromycotina guilliermondii* was isolated from pteridophyte *Drynaria quercifolia* (Aban et al., 2017) and from *Frankincense* tree (*Boswellia sacra*) *Preussia* sp. (Khan et al., 2016), to cite a few examples of fungi with potential use in agriculture, after being tested on crops. Once it became known that endophytic fungi could solubilize phosphates in vitro, different agricultural crops inoculated with the fungi have reported considerable increases in the nutrients they provide to plants (Fig. 17.2). It has been demonstrated that the filamentous fungus

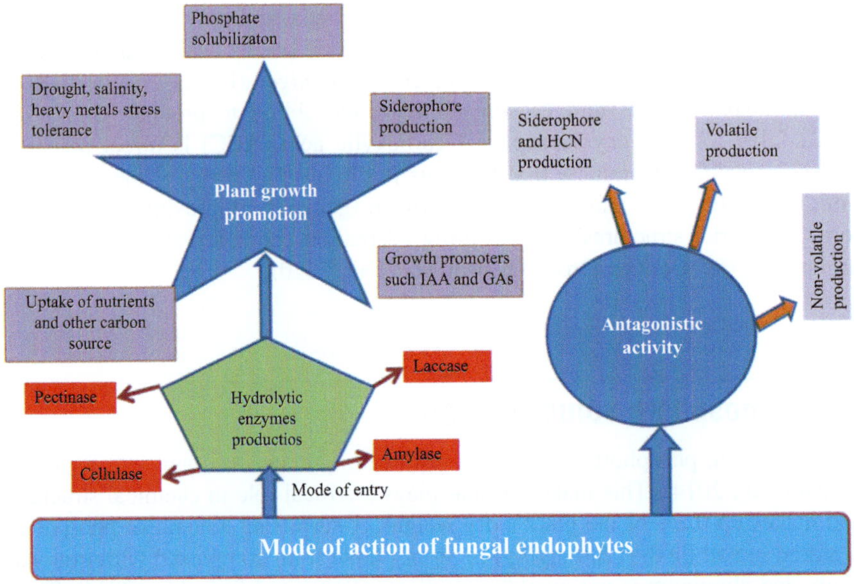

FIGURE 17.2

Action mechanisms of fungal endophytes against the biotic/abiotic stresses in agricultural crop.

Penicillium crustosum can increase maize growth via its phosphorus-solubilization capability (Hassan, 2017); *Byssochlamys brongniartii* causes an increase in the diameter of its fruits in *Capsicum chinense* (Toscano-Verduzco et al., 2020), while *Penicillium bilaiae* causes an increase in wheat and lentil shoot growth (Wakelin et al., 2007); and *Trichoderma asperellum* significantly enhances cucumber seedling growth (Qi and Zhao, 2013).

17.7 Siderophore production

The iron deficiency leads to chlorosis in plants and reduces metabolic activity and biomass in microbial species. Plants and microorganisms have developed chelation strategies to improve metal availability in stressful environments (Satapute et al., 2019; Hider and Kong, 2010). Endophytic fungi in plants synthesize a wide range of hydroxamate and carboxylate siderophores, including coprogens, fusaronines, and ferrichromes (Fig. 17.2) (Haas, 2003; Khan et al., 2018). These siderophores are critical for the transport of iron to bacteria and actinomycetes (Sheldon and Heinrichs, 2015; Jogaiah et al., 2016). Several studies have shown that the majority of siderophores have an endophytic nature, being linked to mycorrhizae as well as orchidaceous fungi (Jogaiah et al., 2016; Tripathi et al., 2017). *Penicillium chrysogenum* that produces bioactive siderophores might be an effective biotherapeutic candidate for the development of resistance against other phytopathogens in crop plants due to its ability to inhibit virulent plant pathogens like *Ralstonia solanacearum*, which causes bacterial wilt in groundnuts, and *Xanthomonas oryzae*, which causes bacterial blight in rice (Chowdappa et al., 2020).

17.8 Production of extracellular enzymes

Generally, the fungal endophytes enter inside the host tissues by natural opening or by degradation of host cell wall containing compounds by secretion of extracellular enzymes such as cellulose, pectinase, amylase, laccase protease, lipase, phenol oxidase, and lignin catabolic enzymes to make their establishment easy inside the host cell (Khan et al., 2017; Saxena et al., 2015). Cellulase degrades the β 1–4 linkage of cellulose of plants and pectinase degrades pectin. Various macromolecules, including proteins, carbohydrate-based polymers, lignin have been targeted by extracellular enzymes, which break them down into simpler molecules (Fig. 17.2). Extracellular enzymes production has been measured quantitatively and qualitatively from agar plate-based methods to advanced spectrophotometric analyses (Yadav et al., 2017).

17.9 Protective efforts of endophytic fungi against plant diseases

17.9.1 Biocontrol properties of endophytes

Many plant pathogens affecting several crops are a significant and chronic threat to food production and ecosystem stability worldwide (Compant et al., 2005). Field and postharvest plant diseases are estimated to reduce food production by 10%–16% around the world (Lo Presti et al., 2015; Strange and Scott, 2005). A critical component of the world's biological diversity is the diversity of microbes. Recent technologies have shown that there are still numerous microbes that have not been discovered and that it is still unclear how these microbes are involved in society (Kumawat et al., 2019). The three main types of pesticides: fungicides, herbicides, and insecticides are used to combat the pathogen, weeds, and insects, but these all are causes a negative impact on environmental health, including humans and animals (Baron et al., 2019). Biofertilizers enhancing growth and productivity, keeping soil healthy and fertile, as well as biocontrol agents that prevent pathogens from attacking plants has been known to be important in agriculture. A biocontrol strategy aims to lessen the negative impact of pathogenic organisms on plants by using beneficial organisms, their genes, or products as a means of counteracting those (Fig. 17.2) (Vinale et al., 2008). There are many fungi that are important biocontrol agents, such as *Aspergillus, Fusarium, Gliocladium, Petriella,* and *Trichoderma* (De Silva et al., 2019). Hajji-Hedfi et al. (2018) have demonstrated *Verticillium leptobactrum* biocontrol ability against wilt disease caused by *Fusarium oxysporum* F. *lycopersici*. Additionally, there are many fungal biocontrols that can be available as commercial products, such as *Verticillium lecanii, T. polysporum, T. gamsii, T. asperellum, T. viride, T. erinaceum, T. asperellum and T. harzianum Purpureocilium lilacinum, Phlebiopsis gigantean, Paecilomyces lilacinus, Metarhizium anisopliae, Gliocladium catenulatum, Coniothyrium minitans, Candida oleophila, Beauveria bassiana, Aureobasidium pullulans*, and *Ampelomyces quisqualis* (Kumar et al., 2021; Larran et al., 2016; Shukla et al., 2022; Tranier et al., 2014; Yadav et al., 2023). As a safe and environmentally friendly strategy for sustainable agriculture, the use of fungi is beneficial. In addition, fungi could be explored for their possible uses in enhancing agricultural productivity, nano-agriculture, and metabolite production (Singh et al., 2019).

17.9.2 Competing with pathogens for space and nutrients

As beneficial fungi can invade shared habitats such as plant tissues, the rhizosphere, or the phyllosphere, they can deprive pathogens of space and nutrients. Theoretically, all fungal agents can accomplish this in some form, depending on their properties and adaptation to their hosts. If beneficial fungi are to significantly outcompete invading pathogens for niche and nutrients, they must have effective colonization strategies and remain at high populations within overlapping habitats. The *Trichoderma* sp. are ubiquitous in almost all ecosystems (crop fields, forests, marshes, and deserts), all climatic zones (tropical, temperate, and subarctic) and also in unusual places such as the atmosphere, lakes, marine bivalves, shellfish, and termites (Mukherjee et al., 2013; Waghunde et al., 2016). *Trichoderma* sp. ubiquitous occurrence is due to the fact that they possess a wide variety of strategies for colonizing such miscellaneous niches, such as the ability to intensively colonize the root system of plants mediated by their high activity hydrophobin genes (such as *TasHyd1* and *Tvhydii1*) during root colonization, allowing them to attach filamentous hyphae to hydrophobic root surfaces (Guzmán-Guzmán et al., 2017). In addition, *Trichoderma* strains are resistant to antimicrobial compounds made by plants (such as phytoalexins, phenols, and flavonoids), which makes them better suited for inhabiting the rhizosphere and root systems (Shoresh et al., 2005; Błaszczyk et al., 2014). Researchers found that the most common cause of death in some pathogenic fungi was "starvation," as observed by both in vitro and in vivo studies on the interactions between *Trichoderma* species (*T. harzianum, T. viride*, and *T. erinaceum*) and some root pathogens (*Fusarium acuminatum, A. alternata, A. infectoria, S. sclerotiorum*) (Mokhtar and Aid, 2013; Kumar et al., 2021). There is evidence that *Trichoderma* species alter the transcriptome and architecture of host cells, which restricts the growth of fungi (Hermosa et al., 2014; Kumar et al., 2021).

17.9.3 Mycoparasitism

As the name suggests, mycoparasitism refers to the parasitic nature of the relationship between a fungus and another fungus (mycoparasite). The mycoparasite receives at least some of its nutrients from another fungus in this manner. To apply mycoparasitism, it is necessary to penetrate the host hyphae by means of host-specific organs, such as the haustorium and secrete enzymes or secondary metabolites that degrade fungal structures followed by nutrient or metabolite uptake from the host fungus (Daguerre et al., 2014). *Trichoderma* is a typical mycoparasitic fungus. *Trichoderma* secretes enzymes that degrade the cell wall of fungal cells (endochitinases, β-1,3-glucanases, and proteases) to assimilate it (Lopes et al., 2012; Vos et al., 2015). The hyphae of *Trichoderma* initially bind to the hyphae of the pathogen and penetrate into the host cell (forming hooks and appressorium-like bodies), destroying chitin by the action of the enzymes chitinase and glucanase. Consequently, the growth of the mycoparasitic fungus induces the release of antibiotic compounds that penetrate into the affected hyphae and prevent resynthesis of the cell wall (Toghueo et al., 2016), due to the fact that antibiosis is accompanied by mycoparasitism. A variety of *Trichoderma* species has been demonstrated to protect plants from pathogenic fungi, including *Rhizoctonia solani, S. sclerotiorum* (Zhang and Zhuang, 2020; Kumar et al., 2021).

17.10 Alleviation of abiotic stress on agricultural crops via fungal endophytes

Plant growth and development are limited by a variety of extreme factors, including environmental stresses and stresses caused by human communities. Plants are able to cope with abiotic stress in two ways: (1) they can counter the negative effects of stress by activating their response systems as soon as stress is present (Meena et al., 2017) and (2) endophytes synthesize biochemical compounds in response to stress and act as antistress agents (Schulz et al., 2002). There are two major classes of fungi associated with plants: (1) endophytic fungi, which live entirely inside of the tissues of plants and can be associated with roots, stems, or leaves; and (2) mycorrhizal fungi, which live only in roots, but extend out into the rhizosphere (Rana et al., 2019). After inoculation with *Phoma glomerata* and *Penicillium* sp., cucumber plants were significantly more able to absorb nutrients such as magnesium, potassium, and calcium, plant biomass, growth parameters and reduced sodium toxicity, compared with uninoculated plants (Waqas et al., 2012). The endophyte microbiome is crucial to plant adaptation against biotic and abiotic stress (Fig. 17.3). In addition to these factors, environmental stress, temperature, and moisture content are also important for endophytic microbial diversity (Kumar et al., 2020). The morphology of the root system is altered during

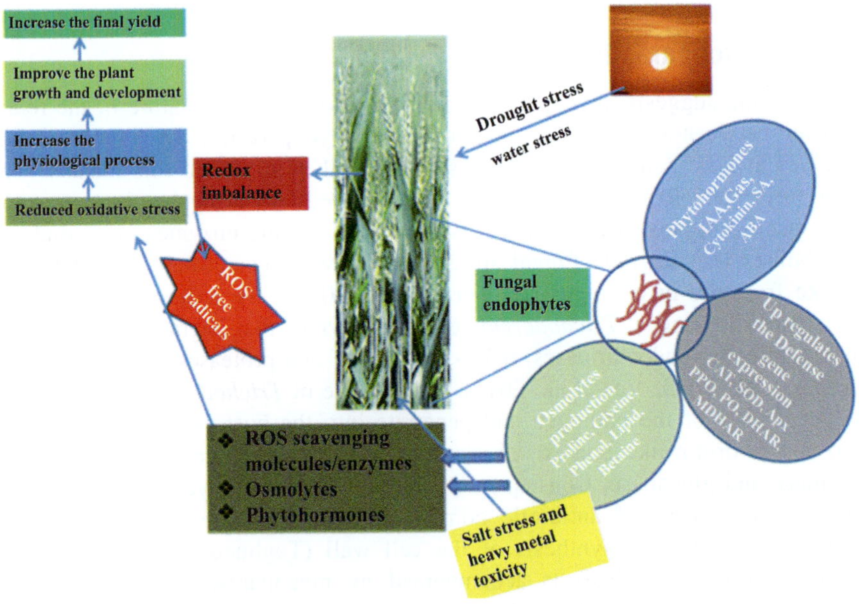

FIGURE 17.3

Role of fungal endophytes in alleviation of abiotic stress on agricultural crops.

Table 17.2 Role of fungal endophytes in sustainable agriculture under abiotic stress (Verma et al., 2021).

Fungal endophytes	Type of stress	Mechanism of stress amelioration	Benefited plant
Phoma species	Drought	Increased proline peroxidase (POD), catalase (CAT), superoxide dismutase (SOD)	*Pinus tabuliformis*
Glomus mosseae	Drought	Increment of peroxidase activity	*Poncirus trifoliata*
Ampelomyces sp.	Drought	Enhanced plant growth, stress tolerance, recovery, and fruit yield	Tomato plants
Epichloë bromicola	Salinity	Content, antioxidant capacity, and glycine betaine content	*Hordeum brevisubulatum* seedling
Curvularia sp.	Salinity	Elevates antioxidant enzymes (SOD and APX)	Poplar plant
Piriformospora indica	Salinity	Modulation of the expression levels of the major Na^+ and K^+ ion channels and balanced ion homeostasis of Na^+/K^+	*Arabidopsis thaliana*
Trichoderma harzianum	Salinity	Reduces lipid peroxidation	*Lycopersicum esculentum* seed
P. indica		Enhanced plant growth and attenuated the NaCl-induced lipid peroxidation, metabolic heat efflux, and fatty acid desaturation in leaves. In addition, significantly elevated the amount of ascorbic acid and increased the activities of antioxidant enzymes catalase, ascorbate peroxidase, dehydroascorbate reductase, monodehydroascorbate reductase, and glutathione reductase	*Hordeum vulgare* seedling

Modified from Verma, H., Kumar, D., Kumar, V., Kumari, M., Singh, S.K., Sharma, V.K., et al., 2021. The potential application of endophytes in management of stress from drought and salinity in crop plants. Microorganisms 9 (8), 1729.

drought conditions, leading to the secretion of root exudates, changes that affect the chemical compounds in exudates, thus affecting the diversity and abundance of microbes (Table 17.2). Similarly, seasonal conditions affect the composition of microbes by affecting the concentrations of amino acids, proteins, sugar, and organic acids (Kuffner et al., 2012).

17.10.1 Drought stress

Abiotic stress such as drought can suppress plant growth, development, and productivity. Plants are under drought conditions when either there is a limited supply of water to the roots or when the transpiration rate is very high (Anjum et al., 2011). During droughts, germination rates are decreased, membrane integrity is compromised, photosynthesis is inhibited, and reactive oxygen species are increased (Greenberg et al., 2008). Compared with nonsymbiotic plants, plants harboring endophytes (rice, tomato, dune grass, and panic grass) consumed significantly less water and produced significantly more biomass (Table 17.2). The endophytic strain *Phoma* spp. improves the seedling growth of *Pinus tabuliformis* under drought stress condition (Zhou et al., 2021). By inoculating microbial endophytes or exogenously providing phytohormones, plants are significantly better able to adapt to drought conditions such as photosynthesis, chlorophyll contents, root growth, water status, antioxidant enzymes, and nutrient uptake (Fig. 17.3) (Khan et al., 2016; Naveed et al., 2014a,b).

17.10.2 Salinity stress

A salinification of soil occurs when a buildup of water-soluble salts threatens the environment, agricultural production, and economics. In the first stages, salinization affects soil organisms and decreases soil productivity, but in the later stages, it destroys vegetation and other soil organisms, making fertile soil barren and desertified (Jones et al., 2019). The amount of irrigated land affected by salinity is estimated at about 20% (45 million hectares), which produces about one-third of all global food (Shrivastava and Kumar, 2015). This results in a significant loss of agricultural land in the European Union, the Mediterranean countries in particular, which is the cause of desertification. In Spain, about 3% of irrigated land (3.5 million hectares) is severely affected, reducing its agricultural potential, while another 15% are at high risk (Stolte et al., 2015). Soil salinity affects a plant's ability to survive by affecting its chemical, morphological, and physiological processes (Otlewska et al., 2020). In these circumstances, microbial endophytes would appear to represent an appropriate alternative for managing salinity stress (Fig. 17.3). Numerous studies report that fungal endophytes confer some of the benefits that plants need to adapt to environmental stresses, like drought, heat, and salinity, to the plants they inhabit (Lata et al., 2018; Sadeghi et al., 2020). A significant increase in proline content and antioxidant enzyme activities has been observed in model plants (cucumber and tomato) associated with *P. macrospinosa* under all salinity and drought stresses in comparison to the endophyte-free plants, while *Neocamarosporium* species have indicated significant increasing proline content only under very high levels of salinity and drought stress (Table 17.2) (Moghaddam et al., 2021).

17.10.3 Heavy metal toxicity

Abiotic stress such as heavy metal (HM) toxicity is a major cause of crop loss that affects about 25%–80% of various crops. Plant roots are very toxic to heavy metals and have a poor growth rate when exposed to them (Singh et al., 2011).

Acidic soils have been impoverished due to heavy metal toxicity, which has restricted crop productivity and interfered with many biological processes, such as nutrient uptake, protein and nitrogen metabolism, photosynthesis, and respiration (Zhang et al., 2009). Some endophytic species of fungi are extraordinarily resistant to toxic metals (Domka et al., 2019). Microorganisms may resolve HM toxicity through biosorption of the metal ions, immobilization of the metal ions outside the cell, or deposition of metal ions on the cell wall, complexation, crystallization, and changing the valence of metal ions (Fig. 17.3). There are various mechanisms by which HMs can be immobilized, including pH changes, the production of siderophores and polysaccharides, as well as dealkylation and methylation (Ayangbenro and Babalola, 2017; Aryal, 2021). As a result of this adaptation, fungi that colonize metalliferous habitats have an edge over nonadapted fungi (Gadd, 2016). The dark septet's endophyte, *Exophiala pisciphila*, associated with *Zea mays* root showed a significant increase in antioxidant enzyme activity when grown in Cd-stressed soil (Wang et al., 2016). A number of studies have been conducted on the ability of fungi to remove Cd from various polluted sites. Many typical endophytic fungi have been demonstrated to be capable of actively removing Cd from various polluted sites. *Aspergillus* (Manguilimotan and Bitacura, 2018; Alothman et al., 2020), *Alternaria, Microdochium, Bipolaris, Alternaria, Pleosporales, Fusarium, Paecilomyces, Clonostachys, Epicoccum* (Shadmani et al., 2020, 2021), Microsporum, *Terichoderma* (Mohammadian Fazli et al., 2015), *Fomitopsis, Trichoderma, Rhizopus, Beauveria, Paraphaeos, Pyrenochaeta, Rhizopycnis* (An et al., 2015), *Penicillium* (Manguilimotan and Bitacura, 2018; Alothman et al., 2020), *Candida*, Cryptococcus, *Exophiala* (Zhao et al., 2015), *Rhodotorula* (Li and Yuan, 2008), *Zygosaccharomyces* and *Saccharomyces* (Li et al., 2014). *Arabidopsis arenosa* is negatively affected by the endophytic *Mucor* sp., decreasing the plant's Zn and Fe uptake, and simultaneously increasing root-to-shoot Zn, Fe, and Cd translocation.

17.11 Conclusion

Through the extensive use of chemicals to increase agriculture productivity, the delicate ecological balance has been disturbed, resulting in pathogen resistance and health risks for other living beings including humans. A growing interest has been shown in finding eco-friendly and safe ways to increase agriculture productivity. There are several key functions of fungal endophytes in sustainable agriculture, including the production of phytohormones, the solubilization of phosphates, the production of siderophores, and the protection of the plants from biotic (plant pathogens), abiotic (salinity stress, drought stress, heavy metal toxicity) stress factors and promote the plant growth. Due to the beneficial activities carried out by fungal endophytes, studies on the plant–fungus association have increased sharply in recent years. Researchers have used genetically modified endophytes to improve plant productivity and defensive properties in recent years.

Acknowledgments

Sunil Kumar is thankful to Central Ayurveda Research Institute, Bhubaneswar, Odisha, CCRAS New Delhi for providing necessary facilities and grammarly tools during this study.

References

Aban, J.L., Barcelo, R.C., Oda, E.E., Reyes, G.A., Balangcod, T.D., Gutierrez, R.M., et al., 2017. Auxin production, phosphate solubilisation and ACC deaminase activity of root symbiotic fungi (RSF) from *Drynaria quercifolia* L. Bull. Environ. Pharmacol. Life Sci. 6 (5), 18–23.

Abo Nouh, F.A., 2019. Endophytic fungi for sustainable agriculture. Microb. Biosyst. 4, 31–44.

Ali, S., Khan, S.A., Hamayun, M., Iqbal, A., Khan, A.L., Hussain, A., et al., 2019. Endophytic fungi from *Caralluma acutangula* can secrete plant growth promoting enzymes. Fresenius Environ. Bull. 28, 2688–2696.

Alothman, Z.A., Bahkali, A.H., Khiyami, M.A., Alfadul, S.M., Wabaidur, S.M., Alam, M., et al., 2020. Low cost biosorbents from fungi for heavy metals removal from wastewater. Sep. Sci. Technol. 55 (10), 1766–1775.

Amprayn, K.O., Rose, M.T., Kecskés, M., Pereg, L., Nguyen, H.T., Kennedy, I.R., 2012. Plant growth promoting characteristics of soil yeast (*Candida tropicalis* HY) and its effectiveness for promoting rice growth. Appl. Soil Ecol. 61, 295–299.

Anjum, S.A., Wang, L.C., Farooq, M., Hussain, M., Xue, L.L., Zou, C.M., 2011. Brassinolide application improves the drought tolerance in maize through modulation of enzymatic antioxidants and leaf gas exchange. J. Agron. Crop Sci. 197, 177–185.

An, H., Liu, Y., Zhao, X., Huang, Q., Yuan, S., Yang, X., et al., 2015. Characterization of cadmium-resistant endophytic fungi from *Salix variegata* Franch. in three Gorges Reservoir Region, China. Microbiol. Res. 176, 29–37.

Arnold, A.E., Maynard, Z., Gilbert, G.S., Coley, P.D., Kursar, T.A., 2000. Are tropical fungal endophytes hyperdiverse? Ecol. Lett. 3 (4), 267–274.

Arnold, A.E., Mejía, L.C., Kyllo, D., Rojas, E.I., Maynard, Z., Robbins, N., et al., 2003. Fungal endophytes limit pathogen damage in a tropical tree. Proc. Natl. Acad. Sci. 100 (26), 15649–15654.

Aryal, M., 2021. A comprehensive study on the bacterial biosorption of heavy metals: materials, performances, mechanisms, and mathematical modellings. Rev. Chem. Eng. 37 (6), 715–754.

Asaf, S., Khan, A.L., Waqas, M., Kang, S.M., Hamayun, M., Lee, I.J., et al., 2019. Growth-promoting bioactivities of *Bipolaris* sp. CSL-1 isolated from *Cannabis sativa* suggest a distinctive role in modifying host plant phenotypic plasticity and functions. Acta Physiol. Plant 41 (5), 1–16.

Avdalović, J., Beškoski, V., Gojgić-Cvijović, G., Mattinen, M.L., Stojanović, M., Zildžović, S., et al., 2015. Microbial solubilization of phosphorus from phosphate rock by iron-oxidizing *Acidithiobacillus* sp. B2. Miner. Eng. 72 (17–22).

Ayangbenro, A.S., Babalola, O.O., 2017. A new strategy for heavy metal polluted environments: a review of microbial biosorbents. Int. J. Environ. Res. 14 (1), 94.

Bamisile, B.S., Dash, C.K., Akutse, K.S., Keppanan, R., Wang, L., 2018. Fungal endophytes: beyond herbivore management. Front. Microbiol. 9, 544.

Baron, N.C., Rigobelo, E.C., Zied, D.C., 2019. Filamentous fungi in biological control: current status and future perspectives. Chil. J. Agric. Res. 79, 307–315.

Begum, N., Qin, C., Ahanger, M.A., Raza, S., Khan, M.I., Ashraf, M., et al., 2019. Role of arbuscular mycorrhizal fungi in plant growth regulation: implications in abiotic stress tolerance. Front. Plant Sci. 10, 1068.

Behie, S.W., Bidochka, M.J., 2014. Nutrient transfer in plant–fungal symbioses. Trends Plant Sci. 19 (11), 734–740.

Bolton, M.D., Gui, M.W., Phillips, R., 1993. Review of miniature soil probes for model tests. In: Proceedings of the 11th Southeast Asian Geotechnical Conference, pp. 85–90.

Busby, P.E., Ridout, M., Newcombe, G., 2016. Fungal endophytes: modifiers of plant disease. Plant Mol. Biol. 90 (6), 645–655.

Błaszczyk, L., Siwulski, M., Sobieralski, K., Lisiecka, J., Jedryczka, M., 2014. *Trichoderma* spp, application and prospects for use in organic farming and industry. J. Plant Protec. Res. 54 (4), 309–317.

De Silva, N.I., Brooks, S., Lumyong, S., Hyde, K.D., 2019. Use of endophytes as biocontrol agents. Fungal Biol. Rev. 33 (2), 133–148.

del Carmen Orozco-Mosqueda, M., Glick, B.R., Santoyo, G., 2020. ACC deaminase in plant growth-promoting bacteria (PGPB): an efficient mechanism to counter salt stress in crops. Microbiol. Res. 235, 126439.

Cecchetti, V., Altamura, M.M., Falasca, G., Costantino, P., Cardarelli, M., 2008. Auxin regulates *Arabidopsis* anther dehiscence, pollen maturation, and filament elongation. Plant Cell 20 (7), 1760–1774.

Chowdappa, S., Jagannath, S., Konappa, N., Udayashankar, A.C., Jogaiah, S., 2020. Detection and characterization of antibacterial siderophores secreted by endophytic fungi from *Cymbidium aloifolium*. Biomolecules 10 (10), 1412.

Chutulo, E.C., Chalannavar, R.K., 2018. Endophytic mycoflora and their bioactive compounds from *Azadirachta indica*: a comprehensive review. J. Fungus 4 (2), 42.

Compant, S., Duffy, B., Nowak, J., Clément, C., Barka, E.A., 2005. Use of plant growth-promoting bacteria for biocontrol of plant diseases: principles, mechanisms of action, and future prospects. Appl. Environ. Microbiol. 71, 4951–4959.

Daguerre, Y., Siegel, K., Edel-Hermann, V., Steinberg, C., 2014. Fungal proteins and genes associated with biocontrol mechanisms of soil-borne pathogens: a review. Fungal Biol. Rev. 28, 97–125.

Dolatabad, H.K., Javan-Nikkhah, M., Shier, W.T., 2017. Evaluation of antifungal, phosphate solubilisation, and siderophore and chitinase release activities of endophytic fungi from *Pistacia vera*. Mycol. Prog. 16 (8), 777–790.

Domka, A.M., Rozpaądek, P., Turnau, K., 2019. Are fungal endophytes merely mycorrhizal copycats? The role of fungal endophytes in the adaptation of plants to metal toxicity. Front. Microbiol. 10, 371.

Fesel, P.H., Zuccaro, A., 2016. Dissecting endophytic lifestyle along the parasitism/mutualism continuum in *Arabidopsis*. Curr. Opin. Microbiol. 32, 103–112.

Fouda, A.H., Hassan, S.E.D., Eid, A.M., Ewais, E.E.D., 2015. Biotechnological applications of fungal endophytes associated with medicinal plant *Asclepias sinaica* (Bioss.). Ann. Agric. Sci. 60 (1), 95–104.

Frąc, M., Hannula, S.E., Bełka, M., Jędryczka, M., 2018. Fungal biodiversity and their role in soil health. Front. Microbiol. 9, 707.

Gadd, G.M., 2016. Fungi and industrial pollutants. Environmental and Microbial Relationships. Springer, Cham, pp. 99–125.

George, T.K., SubaidaBeevi, S., Asok, A.K., Shaikmoideen, J.M., 2021. Lant growth promoting endophytic yeast *Geotrichum candidum* (jx 477426) from roots of *Bruguiera cylindrica*. J. Microbiol. Biotechnol. Food Sci. 2021, 267–272.

Greenberg, B., Huang, X.D., Yu, X.M., Chang, P.C., Wu, S.S., Gerhardt, K., et al., 2008. Phytoremediation of salt-impacted soils: greenhouse and field trials using plant growth promoting rhizobacteria (PGPR) to improve plant growth and salt phytoaccumulation. In: 31st AMOP Technical Seminar on Environmental Contamination and Response.

Guzmán-Guzmán, P., Alemán-Duarte, M.I., Delaye, L., Herrera-Estrella, A., Olmedo-Monfi, V., 2017. Identification of effector-like proteins in *Trichoderma* spp. and role of a hydrophobin in the plant-fungus interaction and mycoparasitism. BMC Genet. 18, 16.

Haas, H., 2003. Molecular genetics of fungal siderophore biosynthesis and uptake: the role of siderophores in iron uptake and storage. Appl. Microbiol. Biotechnol. 62 (4), 316–330.

Hajji-Hedfi, L., Regaieg, H., Larayedh, A., Chihani, N., Horrigue-Raouani, N., 2018. Biological control of wilt disease complex on tomato crop caused by *Meloidogyne javanica* and *Fusarium oxysporum* f. sp. *lycopersici* by *Verticillium leptobactrum*. Environ. Sci. Pollut. Res. 25, 18297–18302.

Hamayun, M., Afzal Khan, S., Ahmad, N., Tang, D.S., Kang, S.M., Na, C.I., et al., 2009a. *Cladosporium sphaerospermum* as a new plant growth-promoting endophyte from the roots of *Glycine max* (L.) Merr. World J. Microbiol. Biotechnol. 25 (4), 627–632.

Hamayun, M., Khan, S.A., Khan, A.L., Rehman, G., Sohn, E.Y., Shah, A.A., et al., 2009b. *Phoma herbarum* as a new gibberellin-producing and plant growth-promoting fungus. J. Microbiol. Biotechnol. 19 (10), 1244–1249.

Hamayun, M., Khan, S.A., Khan, M.A., Khan, A.L., Kang, S.M., Kim, S.K., et al., 2009c. Gibberellin production by pure cultures of a new strain of *Aspergillus fumigatus*. World J. Microbiol. Biotechnol. 25 (10), 1785–1792.

Hamayun, M., Khan, S.A., Khan, A.L., Rehman, G., Kim, Y.H., Iqbal, I., et al., 2010. Gibberellin production and plant growth promotion from pure cultures of *Cladosporium* sp. MH-6 isolated from cucumber (*Cucumis sativus* L.). Mycologia 102 (5), 989–995.

Hassan, S.E.D., 2017. Plant growth-promoting activities for bacterial and fungal endophytes isolated from medicinal plant of *Teucrium polium* L. J. Adv. Res. 8 (6), 687–695.

Hata, K., Sone, K., 2008. Isolation of endophytes from leaves of *Neolitsea sericea* in broadleaf and conifer stands. Mycoscience 49 (4), 229–232.

Hawksworth, D.L., 2004. 'Misidentifications' in fungal DNA sequence databanks. N. Phytol. 161, 13–15.

Hermosa, R., Cardoza, R.E., Rubio, M.B., Gutiérrez, S., Monte, E., 2014. Secondary metabolism and antimicrobial metabolites of *Trichoderma*. Biotechnology and Biology of *Trichoderma*. Elsevier, pp. 125–137.

Hider, R.C., Kong, X., 2010. Chemistry and biology of siderophores. Nat. Prod. Rep. 27 (5), 637–657.

Hodgson, S., de Cates, C., Hodgson, J., Morley, N.J., Sutton, B.C., Gange, A.C., 2014. Vertical transmission of fungal endophytes is widespread in forbs. Ecol. Evol. 4 (8), 1199–1208.

Ikram, M., Ali, N., Jan, G., Jan, F.G., Rahman, I.U., Iqbal, A., et al., 2018. IAA producing fungal endophyte *Penicillium roqueforti* Thom., enhances stress tolerance and nutrients uptake in wheat plants grown on heavy metal contaminated soils. PLoS One 13 (11), e0208150.

Jogaiah, S., Kurjogi, M., Govind, S.R., Huntrike, S.S., Basappa, V.A., Tran, L.S.P., 2016. Isolation and evaluation of proteolytic actinomycete isolates as novel inducers of pearl millet downy mildew disease protection. Sci. Rep. 6 (1), 1–13.

Johnson, J.M., Alex, T., Oelm€uller, R., 2014. *Piriformospora indica*: the versatile and multifunctional root endophytic fungus for enhanced yield and tolerance to biotic and abiotic stress in crop plants. J. Trop. Agric. 52 (2), 103–122.

Jones, P., Garcia, B.J., Furches, A., Tuskan, G.A., Jacobson, D., 2019. Plant host-associated mechanisms for microbial selection. Front. Plant Sci. 862.

Joseph, B., Priya, R.M., 2011. Bioactive compounds from endophytes and their potential. Am. J. Biochem. Mol. Biol. 1 (3), 291–309.

Kaewchai, S., 2009. Mycofungicides and fungal biofertilizers. Fungal Divers 38, 25–50.

Karun, N.C., Bhagya, B.S., Sridhar, K.R., 2018. Biodiversity of macrofungi in *Yenepoya campus*. Southwest India Microb. Biosyst. 3, 1–11.

Kedar, A., Rathod, D., Yadav, A., Agarkar, G., Rai, M., 2014. Endophytic *Phoma* sp. isolated from medicinal plants promote the growth of *Zea mays*. Nus. Biosci. 6 (2).

Khan, A.L., Al-Harrasi, A., Al-Rawahi, A., Al-Farsi, Z., Al-Mamari, A., Waqas, M., et al., 2016. Endophytic fungi from Frankincense tree improves host growth and produces extracellular enzymes and indole acetic acid. PLoS One 1, e0158207.

Khan, S.A., Hamayun, M., Khan, A.L., Lee, I.J., Shinwari, Z.K., Kim, J.G., 2012. Isolation of plant growth promoting endophytic fungi from dicots inhabiting coastal sand dunes of Korea. Pak. J. Bot. 44 (4), 1453–1460.

Khan, A.L., Hamayun, M., Kim, Y.H., Kang, S.M., Lee, I.J., 2011a. Ameliorative symbiosis of endophyte (*Penicillium funiculosum* LHL06) under salt stress elevated plant growth of *Glycine max* L. Plant Physiol. Biochem. 49 (8), 852–861.

Khan, A.L., Hamayun, M., Kim, Y.H., Kang, S.M., Lee, J.H., Lee, I.J., 2011b. Gibberellins producing endophytic *Aspergillus fumigatus* sp. LH02 influenced endogenous phytohormonal levels, isoflavonoids production and plant growth in salinity stress. Process. Biochem. 46 (2), 440–447.

Khan, S.A., Hamayun, M., Yoon, H., Kim, H.Y., Suh, S.J., Hwang, S.K., et al., 2008. Plant growth promotion and *Penicillium citrinum*. BMC Microbiol. 8 (1), 1–10.

Khan, M.I.R., Iqbal, N., Masood, A., Mobin, M., Anjum, N.A., Khan, N.A., 2016. Modulation and significance of nitrogen and sulfur metabolism in cadmium challenged plants. Plant. Growth Regul. 78 (1), 1–11.

Khan, M.I.R., Nazir, F., Asgher, M., Per, T.S., Khan, N.A., 2015. Selenium and sulfur influence ethylene formation and alleviate cadmium-induced oxidative stress by improving proline and glutathione production in wheat. J. Plant Physiol. 173, 9–18.

Khan, A.L., Shahzad, R., Al-Harrasi, A., Lee, I.J., 2017. Endophytic microbes: a resource for producing extracellular enzymes. Endophytes: Crop Productivity and Protection. Springer, Cham, pp. 95–110.

Khan, A., Singh, P., Srivastava, A., 2018. Synthesis, nature and utility of universal iron chelator–siderophore: a review. Microbiol. Res. 212, 103–111.

Khan, M.D., Zaidi, A., Ahmad, E., 2014. Mechanism of phosphate solubilization and physiological functions of phosphate-solubilizing microorganisms. Phosphate Solubilizing Microorganisms. Springer, Cham, pp. 31–62.

Kottb, M., Gigolashvili, T., Großkinsky, D.K., Piechulla, B., 2015. *Trichoderma volatiles* effecting *Arabidopsis*: from inhibition to protection against phytopathogenic fungi. Front. Microbiol. 6, 995.

Kour, D., Rana, K.L., Yadav, A.N., Yadav, N., Kumar, M., Kumar, V., et al., 2020. Microbial biofertilizers: bioresources and eco-friendly technologies for agricultural and environmental sustainability. Biocatal. Agric. Biotechnol. 23, 101487.

Kuffner, M., Hai, B., Rattei, T., Melodelima, C., Schloter, M., Zechmeister-Boltenstern, S., et al., 2012. Effects of season and experimental warming on the bacterial community in a temperate mountain forest soil assessed by 16S rRNA gene pyrosequencing. FEMS Microbiol. Ecol. 82 (3), 551–562.

Kumar, A., Droby, S., Singh, V.K., Singh, S.K., White, J.F., 2020. Entry, colonization, and distribution of endophytic microorganisms in plants. Microbial Endophytes. Woodhead Publishing, pp. 1–33.

Kumar, S., Shukla, V., Dubey, M.K., Upadhyay, R.S., 2021. Activation of defense response in common bean against stem rot disease triggered by *Trichoderma erinaceum* and *Trichodermaviride*. J. Basic Microbiol. 61, 910–922.

Kumar, V., Yadav, A.N., Saxena, A., Sangwan, P., Dhaliwal, H.S., 2016. Unravelling rhizospheric diversity and potential of phytase producing microbes. SM J. Biol. 2 (1), 1009.

Kumawat, N., Kumar, R., Khandkar, U.R., Yadav, R.K., Saurabh, K., Mishra, J.S., et al., 2019. Silicon (Si)-and Zinc (Zn)-solubilizing microorganisms: role in sustainable agriculture. Biofertilizers for Sustainable Agriculture and Environment. Springer, Cham, pp. 109–135.

Larran, S., Monaco, C., Alippi, H.E., 2001. Endophytic fungi in leaves of *Lycopersicon esculentum* mill. World J. Microbiol. Biotechnol. 17 (2), 181–184.

Larran, S., Perello, A., Simon, M.R., Moreno, V., 2002. Isolation and analysis of endophytic microorganisms in wheat (*Triticum aestivum* L.) leaves. World J. Microbiol. Biotechnol. 18 (7), 683–686.

Larran, S., Simon, M.R., Moreno, M.V., Siurana, M.S., Perelló, A., 2016. Endophytes from wheat as biocontrol agents against tan spot disease. Biol. Control. 92, 17–23.

Lata, R., Chowdhury, S., Gond, S.K., White Jr, J.F., 2018. Induction of abiotic stress tolerance in plants by endophytic microbes. Lett. Appl. Microbiol. 66 (4), 268–276.

Li, C., Jiang, W., Ma, N., Zhu, Y., Dong, X., Wang, D., et al., 2014. Bioaccumulation of cadmium by growing *Zygosaccharomyces rouxii* and *Saccharomyces cerevisiae*. Bioresour. Technol. 155, 116–121.

Li, Z., Yuan, H., 2008. Responses of *Rhodotorula* sp. Y11 to cadmium. Biometals 21 (6), 613–621.

Lopes, F.A., Steindorff, A.S., Geraldine, A.M., Brandao, R.S., Monteiro, V.N., Lobo Jr., M., et al., 2012. Biochemical and metabolic profiles of *Trichoderma* isolates isolated from common bean crops in the Brazilian Cerrado, and potential antagonism against *Sclerotinia sclerotiorum*. Fung. Biol. 116, 815–824.

Lubna, Asaf, Hamayun, S., Gul, M.H., Lee, I.J., Hussain, A., 2018. *Aspergillus niger* CSR3 regulates plant endogenous hormones and secondary metabolites by producing gibberellins and indoleacetic acid. J. Plant Interact. 13 (1), 100–111.

Mahdi, T., Mohamed, I., Yagi, S., 2014. Endophytic fungal communities associated with ethno-medicinal plants from Sudan and their antimicrobial and antioxidant prospective. Ecosystems 4 (5).

Manguilimotan, L.C., Bitacura, J.G., 2018. Biosorption of cadmium by filamentous fungi isolated from coastal water and sediments. J. Toxicol. 2018.

Martínez, C., Espinosa-Ruiz, A., Prat, S., 2016. Gibberellins and plant vegetative growth. Annu. Rev. Plant Biol. 49, 285–322.

Meena, M., Swapnil, P., Zehra, A., Aamir, M., Dubey, M.K., Goutam, J., 2017. Beneficial microbes for disease suppression and plant growth promotion. Plant-Microbe Interactions in Agro-Ecological Perspectives. Springer, Singapore, pp. 395–432.

Moghaddam, M.S.H., Safaie, N., Soltani, J., Hagh-Doust, N., 2021. Desert-adapted fungal endophytes induce salinity and drought stress resistance in model crops. Plant Physiol. Biochem. 160, 225–238.

Mohammadian Fazli, M., Soleimani, N., Mehrasbi, M., Darabian, S., Mohammadi, J., Ramazani, A., 2015. Highly cadmium tolerant fungi: their tolerance and removal potential. J. Environ. Health Sci. Eng. 13 (1), 1–9.

Mokhtar, H., Aid, D., 2013. Contribution in isolation and identification of some pathogenic fungi from wheat seeds, and evaluation of antagonistic capability of *Trichoderma harzianum* against those isolated fungi in vitro. Agric. Biol. J. N. Am. 4 (2), 145–154.

Mousa, W.K., Raizada, M.N., 2013. The diversity of anti-microbial secondary metabolites produced by fungal endophytes: an interdisciplinary perspective. Front. Microbiol. 4, 65.

Mukherjee, P.K., Horwitz, B.A., Singh, U.S., Mukherjee, M., Schmoll, M., 2013. Trichoderma in agriculture, industry and medicine: an overview. In: Mukherjee, P.K., Horwitz, B.A., Singh, U.S., Mukherjee, M., Schmoll, M. (Eds.), *Trichoderma*: Biology and Applications. CABI, Nosworthy, Way, Wallingford, Oxon, UK, pp. 1–9.

Naik, B.S., Shashikala, J., Krishnamurthy, Y.L., 2009. Study on the diversity of endophytic communities from rice (*Oryza sativa* L.) and their antagonistic activities in vitro. Microbiol. Res. 164 (3), 290–296.

Nath, R., Sharma, G.D., Barooah, M., 2015. Plant growth promoting endophytic fungi isolated from tea (*Camellia sinensis*) shrubs of Assam, India. Appl. Ecol. Environ. Res. 13, 877–891.

Naveed, M., Hussain, M.B., Zahir, Z.A., Mitter, B., Sessitsch, A., 2014a. Drought stress amelioration in wheat through inoculation with *Burkholderia phytofirmans* strain PsJN. Plant. Growth Regul. 73, 121–131.

Naveed, M., Mitter, B., Reichenauer, T.G., Wieczorek, K., Sessitsch, A., 2014b. Increased drought stress resilience of maize through endophytic colonization by *Burkholderia phytofirmans* PsJN and *Enterobacter* sp. FD17. Environ. Exp. Bot. 97, 30–39.

Orole, O.O., Adejumo, T.O., 2011. Bacterial and fungal endophytes associated with grains and roots of maize. J. Ecol. Nat. Environ. 3 (9), 298–303.

Otlewska, A., Migliore, M., Dybka-Stępień, K., Manfredini, A., Struszczyk-Świta, K., Napoli, R., et al., 2020. When salt meddles between plant, soil, and microorganisms. Front. Plant Sci. 1429.

Parthasarathi, S., Sathya, S., Bupesh, G., Samy, R.D., Mohan, M.R., Kumar, G.S., et al., 2012. Isolation and characterization of antimicrobial compound from marine *Streptomyces hygroscopicus* BDUS 49. World J. Fish. Mar. Sci. 4 (3), 268–277.

Paul, N.C., Deng, J.X., Sang, H.K., Choi, Y.P., Yu, S.H., 2012. Distribution and antifungal activity of endophytic fungi in different growth stages of chili pepper (*Capsicum annuum* L.) in Korea. J. Plant Pathol. 28 (1), 10–19.

Presti, L.L., Lanver, D., Schweizer, G., Tanaka, S., Liang, L., Tollot, M., et al., 2015. Fungal effectors and plant susceptibility. Annu. Rev. Plant Biol. 66, 513–545.

Qiang, X., Ding, J., Lin, W., Li, Q., Xu, C., Zheng, Q., et al., 2019. Alleviation of the detrimental effect of water deficit on wheat (*Triticum aestivum* L.) growth by an indole acetic acid-producing endophytic fungus. Plant Soil 439 (1), 373–391.

Qi, W., Zhao, L., 2013. Study of the siderophore-producing *Trichoderma asperellum* Q1 on cucumber growth promotion under salt stress. J. Basic Microbiol. 53 (4), 355–364.

Rademacher, W., 1994. Gibberellin formation in microorganisms. Plant. Growth Regul. 15 (3), 303–314.

Rana, K.L., Kour, D., Yadav, A.N., Kumar, V., Dhaliwal, H.S., 2016. Biotechnological applications of endophytic microbes associated with barley (*Hordeum vulgare* L.) growing in Indian Himalayan regions. In: Proceeding of 86th Annual Session of NASI and Symposium on "Science, Technology and Entrepreneurship for Human Welfare in the Himalayan Region," vol. 80.

Rana, K.L., Kour, D., Yadav, A.N., 2019. Endophytic microbiomes: biodiversity, ecological significance and biotechnological applications. Res. J. Biotechnol. 14, 142–162.

Rim, S.O., Lee, J.H., Choi, W.Y., Hwang, S.K., Seok, J.S., Lee, I.J., et al., 2005. *Fusarium proliferatum* KGL0401 as a new gibberellin-producing fungus. Microbiol. Biotechnol. 15 (4), 809–814.

Ripa, F.A., Cao, W.D., Tong, S., Sun, J.G., 2019. Assessment of plant growth promoting and abiotic stress tolerance properties of wheat endophytic fungi. Biomed. Res. Int. 2019.

Rodriguez, R.J., Henson, J., VanVolkenburgh, E., Hoy, M., Wright, L., Beckwith, F., et al., 2008. Stress tolerance in plants via habitat-adapted symbiosis. ISME J. 2 (4), 404–416.

Rodriguez, R.J., White Jr, J.F., Arnold, A.E., Redman, A.R.A., 2009. Fungal endophytes: diversity and functional roles. New Phytol. 182 (2), 314–330.

Sadeghi, F., Samsampour, D., Seyahooei, M.A., Bagheri, A., Soltani, J., 2020. Fungal endophytes alleviate drought-induced oxidative stress in mandarin (*Citrus reticulata* L.): toward regulating the ascorbate–glutathione cycle. Sci. Hortic. 261, 108991.

Saleem, M., Arshad, M., Hussain, S., Bhatti, A.S., 2007. Perspective of plant growth promoting rhizobacteria (PGPR) containing ACC deaminase in stress agriculture. J. Ind. Microbiol. Biotechnol. 34 (10), 635–648.

Satapute, P., Kamble, M.V., Adhikari, S.S., Jogaiah, S., 2019. Influence of triazole pesticides on tillage soil microbial populations and metabolic changes. Sci. Total Environ. 651, 2334–2344.

Saxena, A., Raghuwanshi, R., Singh, H.B., 2015. *Trichoderma* species mediated differential tolerance against biotic stress of phytopathogens in *Cicer arietinum* L. J. Basic Microbiol. 55 (2), 195–206.

Schardl, C.L., Florea, S., Pan, J., Nagabhyru, P., Bec, S., Calie, P.J., 2013. The epichloae: alkaloiddiversity and roles in symbiosis with grasses. Curr. Opin. Plant Biol. 16 (4), 480–488.

Schulz, B., Boyle, C., Draeger, S., Römmert, A.K., Krohn, K., 2002. Endophytic fungi: a source of novel biologically active secondary metabolites. Mycol. Res. 106 (9), 996–1004.

Selim, K.A., Nagia, M.M., Ghwas, D.E.E., 2017. Endophytic fungi are multifunctional biosynthesizers: ecological role and chemical diversity. Endophytic Fungi: Diversity, Characterization and Biocontrol. Nova Publishers, NewYork, pp. 39–92.

Shadmani, L., Jamali, S., Fatemi, A., 2021. Isolation, identification, and characterization of cadmium-tolerant endophytic fungi isolated from barley (*Hordeum vulgare* L.) roots and their role in enhancing phytoremediation. Braz. J. Microbiol. 52 (3), 1097–1106.

Sheldon, J.R., Heinrichs, D.E., 2015. Recent developments in understanding the iron acquisition strategies of gram positive pathogens. FEMS Microbiol. Rev. 39 (4), 592–630.

Shoresh, M., Yedidia, I., Chet, I., 2005. Involvement of jasmonic acid/ethylene signaling pathway in the systemic resistance induced in cucumber by *Trichoderma asperellum* T203. Phytopathology 95, 76–84.

Shrivastava, P., Kumar, R., 2015. Soil salinity: a serious environmental issue and plant growth promoting bacteria as one of the tools for its alleviation. Saudi J. Biol. Sci. 22 (2), 123–131.

Shukla, V., Kumar, S., Tripathi, Y., Upadhyay, R., 2022. *Bacillus subtilis*-and *Pseudomonas fluorescens*-mediated systemic resistance in tomato against sclerotium rolfsii and study of physio-chemical alterations. Front. Fungal Biol. 3, 851002. Available from: https://www.frontiersin.org/articles/10.3389/ffunb.2022.851002.

Shweta, S., Zuehlke, S., Ramesha, B.T., Priti, V., Kumar, P.M., Ravikanth, G., et al., 2010. Endophytic fungal strains of *Fusarium solani*, from *Apodytes dimidiata* E. Mey. ex Arn (Icacinaceae) produce camptothecin, 10-hydroxycamptothecin and 9-methoxycamptothecin. Phytochemistry 71 (1), 117–122.

Singh, L.P., Gill, S.S., Tuteja, N., 2011. Unraveling the role of fungal symbionts in plant abiotic stress tolerance. Plant Signal. Behav. 6 (2), 175–191.

Singh, V.K., Singh, M., Singh, S.K., Kumar, C., Kumar, A., 2019. Sustainable agricultural practices using beneficial fungi under changing climate scenario. Climate Change and Agricultural Ecosystems. Woodhead Publishing, pp. 25–42.

Spagnoletti, F.N., Tobar, N.E., Di Pardo, A.F., Chiocchio, V.M., Lavado, R.S., 2017. Dark septate endophytes present different potential to solubilize calcium, iron and aluminum phosphates. Agric. Ecosyst. Environ. 111, 25–32.

Stolte, J., Tesfai, M., Øygarden, L., Kværnø, S., Keizer, J., Verheijen, F., et al., 2015. Soil threats in Europe: status, methods, drivers and effects on ecosystem services. A review report, deliverable 2.1 of the RECARE Project, vol EUR 27607. Office for Official Publications of the European Community, Luxembourg, pp. 69–78.

Strange, R.N., Scott, P.R., 2005. Plant disease: a threat to global food security. Annu. Rev. Phytopathol. 43.

Sudha, V., Govindaraj, R., Baskar, K., Al-Dhabi, N.A., Duraipandiyan, V., 2016. Biological properties of endophytic fungi. Braz. Arch. Biol. Technol. 59.

Sun, X., Wang, G., Xiao, H., Jiang, J., Xiao, D., Xing, B., et al., 2020. Strepimidazoles A–G from the plant endophytic *Streptomyces* sp. PKU-EA00015 with inhibitory activities against a plant pathogenic fungus. J. Nat. Prod. 83 (7), 2246–2254.

Surjit, S.D., Rupa, G., 2014. Beneficial properties, colonization, establishment and molecular diversity of endophytic bacteria in legumes and non legumes. Afr. J. Microbiol. Res. 8 (15), 1562–1572.

Sword, G., Ek-Ramos, M.J., Lopez, D.C., Kalns, L., Zhou, W., Valencia, C., 2012. Fungal endophytes and their potential for biocontrol in cotton. In: Entomological Society of America Annual Meeting.

Toghueo, R.M.K., Eke, P., Zabalgogeazcoa, Í., de Aldana, B.R.V., Nana, L.W., Boyom, F. F., 2016. Biocontrol and growth enhancement potential of two endophytic *Trichoderma* spp. from *Terminalia catappa* against the causative agent of common bean root rot (*Fusarium solani*). Biol. Control. 96, 8–20.

Toscano-Verduzco, F.A., Cedeño-Valdivia, P.A., Chan-Cupul, W., Hernández-Ortega, H.A., Ruiz-Sánchez, E., Galindo-Velasco, E., et al., 2020. Phosphates solubilization, indol-3-acetic acid and siderophores production by *Beauveria brongniartii* and its effect on growth and fruit quality of *Capsicum chinense*. J. Hortic. Sci. Biotech. 95, 235–246.

Tranier, M.S., Pognant-Gros, J., Quiroz, R.D.L.C., González, C.N.A., Mateille, T., Roussos, S., 2014. Commercial biological control agents targeted against plant-parasitic root-knot nematodes. Braz. Arch. Biol. Technol. 57, 831–841.

Tripathi, S., Mishra, S.K., Varma, A., 2017. Mycorrhizal fungi as control agents against plant pathogens. Mycorrhiza-Nutrient Uptake, Biocontrol, Ecorestoration. Springer, Cham, pp. 161–178.

Uppala, S., Beena, S., Chapala, M., Bowen, K.L., 2010. Bioefficacy of endophytes in the management of leaf blight disease of amaranth. Plant Growth Promotion by Rhizobacteria for Sustainable Agriculture. Scientific Publishers, Jodhpur, pp. 524–530.

Verma, H., Kumar, D., Kumar, V., Kumari, M., Singh, S.K., Sharma, V.K., et al., 2021. The potential application of endophytes in management of stress from drought and salinity in crop plants. Microorganisms 9 (8), 1729.

Verma, P., Yadav, A.N., Khannam, K.S., Kumar, S., Saxena, A.K., Suman, A., 2016. Molecular diversity and multifarious plant growth promoting attributes of Bacilli associated with wheat (*Triticum aestivum* L.) rhizosphere from six diverse agro-ecological zones of India. J. Basic Microbiol. 56 (1), 44–58.

Vinale, F., Sivasithamparam, K., Ghisalberti, E.L., Marra, R., Woo, S.L., Lorito, M., 2008. *Trichoderma*–plant–pathogen interactions. Soil Biol. Biochem. 40, 1–10.

Vos, C.M., De Cremer, K., Cammue, B.P.A., De Coninck, B., 2015. The toolbox of *Trichoderma* spp. in the biocontrol of *Botrytis cinerea* disease. Mol. Plant Pathol. 16, 400–412.

Waghunde, R.R., Shelake, M.R., Sabalpara, N.A., 2016. *Trichoderma*: a significant fungus for agriculture and environment. Afric. J. Agric. Res. 11, 1952–1965.

Wakelin, S.A., Gupta, V.V., Harvey, P.R., Ryder, M.H., 2007. The effect of *Penicillium* fungi on plant growth and phosphorus mobilization in neutral to alkaline soils from southern Australia. Can. J. Microbiol. 53 (1), 106–115.

Wang, J.L., Li, T., Liu, G.Y., Smith, J.M., Zhao, Z.W., 2016. Unraveling the role of dark septate endophyte (DSE) colonizing maize (*Zea mays*) under cadmium stress: physiological, cytological and genic aspects. Sci. Rep. 6 (1), 1–12.

Waqas, M., Khan, A.L., Hamayun, M., Shahzad, R., Kang, S.M., Kim, J.G., et al., 2015. Endophytic fungi promote plant growth and mitigate the adverse effects of stem rot: an example of *Penicillium citrinum* and *Aspergillus terreus*. J. Plant Interact. 10 (1), 280–287.

Waqas, M., Khan, A.L., Kamran, M., Hamayun, M., Kang, S.M., Kim, Y.H., et al., 2012. Endophytic fungi produce gibberellins and indoleacetic acid and promotes host-plant growth during stress. Molecules 17 (9), 10754–10773.

Waqas, M., Khan, A.L., Kang, S.M., Kim, Y.H., Lee, I.J., 2014. Phytohormone-producing fungal endophytes and hardwood-derived biochar interact to ameliorate heavy metal stress in soybeans. Biol. Fertil. Soils 50 (7), 1155–1167.

Wiewióra, B., Żurek, G., Pańka, D., 2015. Is the vertical transmission of Neotyphodium lolii in perennial ryegrass the only possible way to the spread of endophytes? PLoS One 10 (2), e0117231.

Wu, L., Han, T., Li, W., Jia, M., Xue, L., Rahman, K., et al., 2013. Geographic and tissue influences on endophytic fungal communities of *Taxus chinensis* var. mairei in China. Curr. Microbiol. 66 (1), 40–48.

Yadav, M., Divyanshu, K., Dubey, M.K., Rai, A., Kumar, S., Tripathi, Y.N., et al., 2023. Plant growth promotion and differential expression of defense genes in chilli pepper against *Colletotrichum truncatum* induced by *Trichoderma asperellum* and *T. harzianum*. BMC Microbiol. 23 (1), 54.

Yadav, A.N., Kour, D., Kaur, T., Devi, R., Yadav, N., 2020a. Agriculturally important fungi for crop productivity: current research and future challenges. In: Yadav, A.N., Mishra, S., Kour, D., Yadav, N., Kumar, A. (Eds.), Agriculturally Important Fungi for Sustainable Agriculture: Volume 1: Perspective for Diversity and Crop Productivity. Springer, Cham.

Yadav, A.N., Kour, D., Kaur, T., Devi, R., Yadav, N., 2020b. Functional annotation of agriculturally important fungi for crop protection: current research and future challenges. In: Yadav, A.N., Mishra, S., Kour, D., Yadav, N., Kumar, A. (Eds.), Agriculturally Important Fungi for Sustainable Agriculture: Volume 2: Functional Annotation for Crop Protection. Springer, Cham, pp. 1−15.

Yadav, A.N., Kumar, R., Kumar, S., Kumar, V., Sugitha, T.C.K., Singh, B., et al., 2017. Beneficial microbiomes: biodiversity and potential biotechnological applications for sustainable agriculture and human health. J. Appl. Biol. 5 (6), 4−7.

Zhang, W., Feng, H., Chang, J., Qu, J., Xie, H., Yu, L., 2009. Heavy metal contamination in surface sediments of Yangtze River intertidal zone: an assessment from different indexes. Environ. Pollut. 157 (5), 1533−1543.

Zhang, Y., Zhuang, W.Y., 2020. *Trichoderma brevicrassum* strain TC967 with capacities of diminishing cucumber disease caused by *Rhizoctonia solani* and promoting plant growth. Biol. Control. 142, 104151.

Zhao, J., Fu, Y., Luo, M., Zu, Y., Wang, W., Zhao, C., et al., 2012. Endophytic fungi from pigeon pea (*Cajanus cajan* (L.) Millsp.) produce antioxidant cajaninstilbene acid. J. Agric. Food Chem. 60, 4314−4319.

Zhao, D., Li, T., Wang, J., Zhao, Z., 2015. Diverse strategies conferring extreme cadmium (Cd) tolerance in the dark septate endophyte (DSE), *Exophiala pisciphila*: evidence from RNA-seq data. Microbiol. Res. 170, 27−35.

Zhou, X.R., Dai, L., Xu, G.F., Wang, H.S., 2021. A strain of *Phoma* species improves drought tolerance of *Pinus tabulaeformis*. Sci. Rep. 11 (1), 1−11.

Index

Note: Page numbers followed by "*f*" and "*t*" refer to figures and tables, respectively.

A

C

D

E

F

G

N

O

Q

R

S

CPI Antony Rowe
Eastbourne, UK
May 23, 2024